Metal–Organic Frameworks for Environmental Sensing

ACS SYMPOSIUM SERIES **1394**

Metal–Organic Frameworks for Environmental Sensing

Smita S. Kumar, Editor
J.C. Bose University of Science and Technology, YMCA
Faridabad, Haryana, India

Pooja Ghosh, Editor
Indian Institute of Technology Delhi
New Delhi, India

Lakhveer Singh, Editor
SRM University-AP, Amaravati
Andhra Pradesh, India

American Chemical Society, Washington, DC

Library of Congress Cataloging-in-Publication Data

Names: Kumar, Smita S., editor. | Ghosh, Pooja, editor. | Singh, Lakhveer, editor.
Title: Metal-organic frameworks for environmental sensing / Smita S. Kumar, Pooja Ghosh, Lakhveer Singh, editor.
Description: Washington, DC : American Chemical Society, [2021] | Series: ACS symposium series ; 1394 | Includes bibliographical references and index.
Identifiers: LCCN 2021049643 (print) | LCCN 2021049644 (ebook) | ISBN 9780841298101 (hardcover) | ISBN 9780841298095 (ebook other)
Subjects: LCSH: Environmental monitoring. | Metal-organic frameworks. | Biosensors. | Chemical detectors.
Classification: LCC QH541.15.M64 M47 2021 (print) | LCC QH541.15.M64 (ebook) | DDC 363.7/063--dc23/eng/20211208
LC record available at https://lccn.loc.gov/2021049643
LC ebook record available at https://lccn.loc.gov/2021049644

The paper used in this publication meets the minimum requirements of American National Standard for Information Sciences—Permanence of Paper for Printed Library Materials, ANSI Z39.48n1984.

Foreword

The ACS Symposium Series is an established program that publishes high-quality volumes of thematic manuscripts. For over 40 years, the ACS Symposium Series has been delivering essential research from world leading scientists, including 36 Chemistry Nobel Laureates, to audiences spanning disciplines and applications.

Books are developed from successful symposia sponsored by the ACS or other organizations. Topics span the entirety of chemistry, including applications, basic research, and interdisciplinary reviews.

Before agreeing to publish a book, prospective editors submit a proposal, including a table of contents. The proposal is reviewed for originality, coverage, and interest to the audience. Some manuscripts may be excluded to better focus the book; others may be added to aid comprehensiveness. All chapters are peer reviewed prior to final acceptance or rejection.

As a rule, only original research papers and original review papers are included in the volumes. Verbatim reproductions of previous published papers are not accepted.

ACS Books

Contents

Indexes

Preface

Metal-organic frameworks (MOFs) are an evolving class of porous coordination polymers composed of organic linkers with metal nodes, having a competitive advantage over the conventional porous materials like activated carbon and zeolites. Currently, MOFs are receiving increased research attention owing to their unique features such as high surface area, adjustable porosity, excellent chemical stability, and modifiable compositions—making them a potential candidate in the field of heterogeneous catalysis and application in frontier domains of energy and environment.

Over the past decade, tremendous advances have been made in the development of novel MOFs and their composites, having desirable characteristics for a wide range of applications in biomass conversion, gas capture and storage, catalysis, chemical sensing, wastewater remediation, air purification, and others. Unfortunately, a lack of recent compilation in this emerging area of research makes it difficult for those interested in this field to get holistic information in one place. Therefore, the purpose of this book is to bring together the scientists working in this area to share the recent research trends in engineering exceptionally stable MOF-based catalysts for applications in biomass conversion, carbon capture, energy applications, environmental sensing, and remediation.

A total of 40 chapters (in three volumes) are included in this book. Volume 1 deals with the applications of MOFs for biomass conversion, carbon capture, and energy applications. Volumes 2 and 3 are focused on application of MOFs in environmental sensing and environmental remediation, respectively.

In this second volume, 12 chapters are included that aim at facilitating the understanding of the emerging applications of MOFs as environmental sensors. The chapters included in this volume highlight its application as electrochemical, fluorescent, colorimetric, luminescent, and radioluminescent sensors for detection of environmental contaminants. The advantages of MOFs over the conventional sensors, along with the recent advances and challenges, have been discussed comprehensively.

Hopefully, the three volumes of the book will be of great interest to graduate and undergraduate students, researchers, and engineers, either involved or interested in MOFs and their applications.

We thank all the authors for their contributions and patience during the peer review process. We are also grateful to the reviewers for their time and effort in reviewing the chapters and providing useful feedbacks. We are also thankful to the ACS Books staff (Amanda Koenig, Sara Tenney, Tracey Glazener, Nikki Lazenby, Harry Weisbecker, Lake Lloyd, and Grace Taylor) for their expert handling of the manuscripts and their editorial efforts.

Smita S. Kumar

Department of Environmental Sciences and Centre for Energy Studies

J.C. Bose University of Science and Technology, YMCA

Mathura Road, Sector 6, Faridabad, Haryana 121006, India

Email: smita@jcboseust.ac.in

Email: smitajcboseust@gmail.com

Pooja Ghosh
Centre for Rural Development and Technology
Indian Institute of Technology Delhi
Hauz Khas, New Delhi 110016, India
Email: poojaghosh@iitd.ac.in
Email: pooja.ghosh9@gmail.com

Lakhveer Singh
Department of Environmental Sciences
SRM University-AP, Amaravati
Andhra Pradesh 522502, India
Email: lucki.chem09@gmail.com

Chapter 1

Metal–Organic Frameworks as Sensors of Biomolecules

Zhenbao Liu[*,1,3] **and Yanfei Liu**[*,2]

[1]**Department of Pharmaceutics, Xiangya School of Pharmaceutical Sciences, Central South University, Changsha 410013, Hunan Province, P. R. China**

[2]**Department of Pharmaceutical Engineering, College of Chemistry and Chemical Engineering, Central South University, Changsha 410083, Hunan Province, P. R. China**

[3]**Molecular Imaging Research Center of Central South University, Changsha 410008, Hunan Province, P. R. China**

[*]**Email: zhenbaoliu@csu.edu.cn**

[*]**Email: Liuyf@csu.edu.cn**

Metal-organic framework (MOF), a novel hybrid porous material which is bridged by the metal center and organic linkers, has become a promising candidate in biosensing applications due to their superior properties. Many types of MOFs were applied for sensing biomacromolecules (nucleic acid, enzyme/protein) and small biomolecules (dopamine, amino acid, glucose, and H_2O_2). Herein, the unique properties of MOF, such as rational aperture design, diversity, easy to functionalize, high Brunauer-Emmett-Teller (BET)/Langmuir surface area, etc., have been highlighted and the relationships between properties and detection performances have been discussed. Meanwhile, an update of MOF for biosensing has been provided.

Introduction

MOF, a branch of coordination polymer, is organic-inorganic hybrid material consists of organic linker molecules and inorganic metal center. Since Yaghi et al. (*1*) formally proposed the concept of "MOF" for the first time and greatly improved the apparent surface area, pore volume, and stability of MOF-5 (2), the biological performances of MOF had been greatly optimized, and the applications had been greatly developed. Compared with traditional materials, MOF has many advantages, such as large specific surface area, high porosity and loading efficiency, easy functionalization, tunable pore sizes, better biocompatible and biodegradability, flexible networks and topological diversity *etc.* (*2–4*) Benefiting from the advantages mentioned, MOF has gained considerable attention in energy storage (*5–10*), drug delivery (*11–13*), separation (*14–16*), catalytic (*17–23*), purification (*24*), sensing (*25–28*), storage of gas (*29*), imaging (*30*) and *etc.* Among which, MOFs are very promising biomedical candidates for bisensing due to their unique properties.

MOFs have been widely used as novel probes in the field of biosensing. In 2002, Wang et al. (*31*) designed the first sensor involving MOFs, and Asefa *et al.* (*32*) were the first persons who prepared biosensing of MOF in 2008. Since then, MOFs have attracted great interest as biosensing materials. For example, as the most widely used optical sensor, fluorescence can be turned by the metal, ligand unit (*33*) or the guest molecule (*34*). Meanwhile, MOF possesses adjustable size, low toxicity, and biodegradability.

Especially, benefiting from quenching and intrinsic fluorescence properties of MOF, it has been developed as a fluorescence quencher of dye-labeled DNA or other recognition elements, which enable selectively detection of various biomolecules including biomacromolecule (nucleic acid, enzyme-activity/proteins) and small biomolecules (dopamine, amino acid, glucose and H_2O_2). Additionally, through reasonable designing, MOFs can be integrated with fluorescence properties, when the MOF doped with Au, Mn^{2+}, Gd^{3+}, SPIO, Fe_3O_4, Fe^{3+}/Fe^{2+}, fluorescent dyes, upconversion nanomaterials (NaYF4: Yb/Er, Gd^{3+} and Tm^{3+}), quantum dots as well as other materials (graphene, dots, nanotubes), MOF can provide signal for optical detection.

In recent years, the emerging functional MOFs have shown great advantages in the field of biosensing. Although previous applications and challenges have been summarized and presented, with the rapid development of MOF, the applications are also rapidly updated (*35*, *36*). Therefore, the unique properties of MOF, such as rational aperture design, diversity, easy to functionalize, high BET/Langmuir surface area *etc.*, have been summarized. Meanwhile, we also updates and highlights recent research developments in the biosensing applications related to MOF. Furthermore, the future development directions and challenges in this field are further discussed. We hope these knowledges will arouse wide interest and inspire readers to realize more biological applications based on the unique properties of MOF.

The Properties of MOF

Compared with traditional materials, MOF has the following unique advantages.

Rational Aperture Design

MOF is able to be rationally designed with predictable porosities and structures/topologies. For example, MOF compounds of various pore sizes can be obtained by modulating ligands. Yaghi et al. (37) used MOF-5 as a structural prototype to expand the length of terephthalic acid, thus successfully constructed a series of isoreticular MOF materials, IRMOF-n, with the uniform periodic pores (from 3.8 Å to 28.8 Å) to realize the transition of MOF material from micropores to mesopores. Meaningfully, the pore size and function were changed without altering the original cubic topology. Also, the open space of the material reached 91.1% of the crystal volume. Similarly, Yaghi et al. (*38*) reported an isoreticular series of MOF-74 structures, IRMOF-74-I to XI, through systematically expanding MOF-74 structure from its original linker of one phenylene ring (I) to eleven (XI) with apertures ranging from 14 to 98 Å. The pore apertures of IRMOF-74-IX and IRMOF-74-VII are large enough even to hold natural proteins. The pore size of the MOF is adjustable by selecting ligands of different lengths. This means that it is entirely possible to achieve "tailoring" for the guest molecule which is irreplaceable with other materials. This means that it is possibly to design a suitable hole for the dye molecules or other luminescent particles by adjusting the length of the ligand to wrap it, which will be very beneficial for sensing applications. Simultaneously, large holes can accommodate more signal molecules for signal enhancement.

Diversity

Figure 1 shows some common MOF structures. Since Yaghi et al.(*1*) put forward the concept of "MOF", thousands of MOFs have been synthesized. Typically, Yaghi's group designed and synthesized the MOF (2), ZIF (*39*), COF (*40*), IRMOF (*37*), and MTV-MOF (*41*) series of materials, and extended these basic science to the application field. Not only that, there was also a MIL-n family represented by Ferey, and the famous Materials of Institut Lavoisier (MIL) materials include MIL-100 (*42*) and MIL-101 (with hierarchical pore system, accessibility metal active sites and giant pores) (*43*), MIL-53 (highly flexible) (*44*), MIL-125 (Ti) (highly porous, the first photoactive Ti-based MOF) (*45*). Besides, University of Oslo (UiO)-n series were discovered by Lillerud et al. (*46*), Dresden University of Technology (DUT)-n series were presented by the Stefan Kaskel's group (*47*), Hong Kong University of Science and Technology (HKUST)-n were prepared, *etc.* (*48*) The magnitude of the MOF family is largely due to the diversity of organic ligands and metal center species. This diversity is beyond the reach of other materials. More importantly, the diversity of MOF offers the possibility of selecting materials for different biological applications.

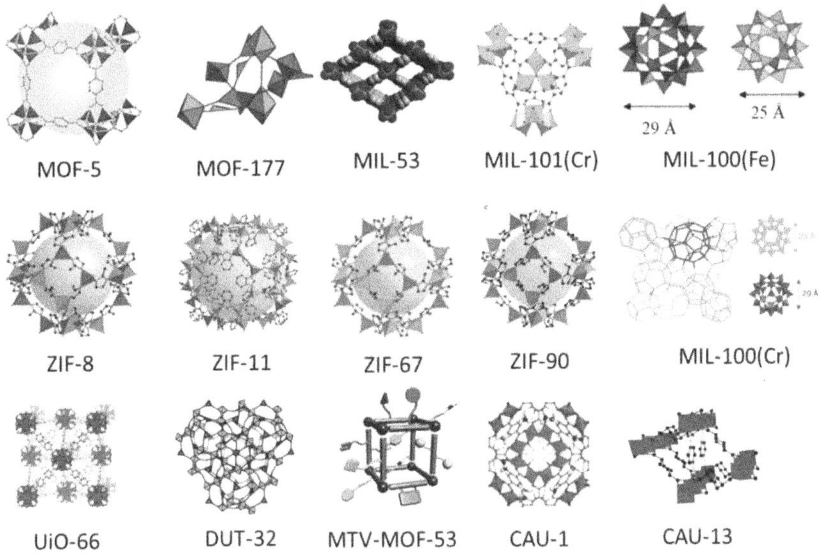

Figure 1. Summary of some common MOF structure diagrams.

Easy to Functionalize

A large number of functional sites which are mainly derived from the employ of the organic linker with specific sites exist on the surface of MOF. For instance, amine and pyridyl sites can be introduced purposefully into the pore surfaces of the MOFs through functional linker for recognizing small molecules and diverse applications (*49*). Typically, Sergio et al. synthesized a series of functionalized UiO-66 MOFs, such as UiO-66–R (R=NH_2, Br, or NO_2), through reticular chemistry. Particularly, UiO-66-NH_2 was suitable for post-synthetic modification with various anhydrides to generate functionalized and novel frameworks (*50*) (Figure 2). Alternatively, multiple functional sites can also be introduced on the surface of the pore by a multivariate method (*41*). Importantly, rich functional sites lay a solid foundation for further functionalization and broader application. Also, this is an irreplaceable part of MOF. For example, numerous functional sites increase the possibility of molecular connections and facilitate covalent linkages to enhance the stability of the linkage, and further modifications play a significant role in the functionalization of MOF.

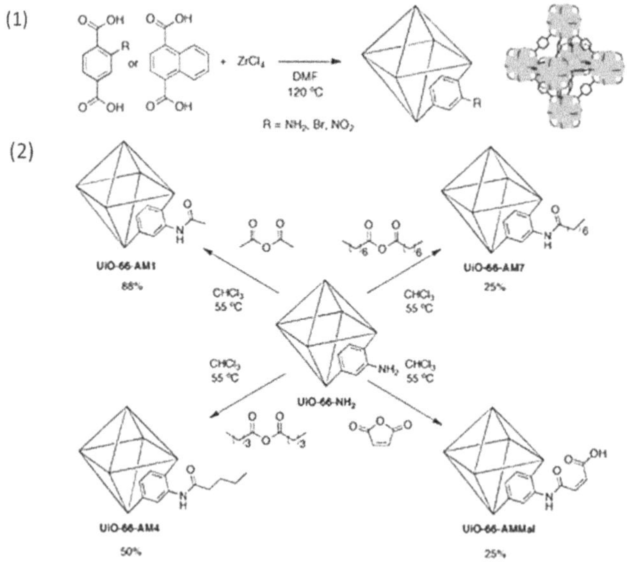

Figure 2. (1) Synthesis of isoreticular UiO-66 functionalized analogs UiO-66-R (R=NH_2, Br, or NO_2); (2) Post-synthetic modification of UiO-66-NH_2 with various anhydrides. (Reprinted with permission from ref. (50). Copyright 2010, The Royal Society of Chemistry) (50).

High BET/Langmuir Surface Area

Since MOF was first proposed, it is known for its high specific surface area. For example, MOF-5, due to its regular branched structure, ensures a large surface area (Langmuir specific surface area was up to 2900 m^2/g) while forming a 3D structure compared to most porous zeolite materials with higher porosity (55% to 61%) (2). In 2004 and 2005, Ferey et al. reported two molecular sieve-type metal-organic framework materials, MIL-100 (Cr) (*42*) and MIL-101(Cr) (*43*), with large pore reach 2.5~2.9 and 2.9~3.4 nm, respectively. And the volume of individual cages was 380 nm^3 and 702 nm^3, respectively, especially the Langmuir specific surface area of MIL-101 was 5900±300 m^2/g. In 2010, Yaghi's group synthesized MOF-210 by ligand extension and mixing of different ligands, with a BET surface area of 6240 m^2/g and Langmuir surface area of 10400 m^2/g, as well as an average porosity of 90% (*51*). In the same work, MOF-200 possessed Langmuir surface areas and BET surface areas of 10400 and 4530 m^2/g, respectively. The ultrahigh surface areas presented by MOF-200 and MOF-210 reached near the ultimate limit for solid materials. More than that, Omar K et al. (52) designed MOF NU-109E and NU-110E with BET being 7010 and 7140 m^2/g, respectively. (Some MOFs with large specific surface area are listed in Table 1).

With the development of MOF, various MOFs with high surface area were synthesized, which will provide a large platform for the attachment of macromolecules such as proteins, small molecules, and nucleic acids, creating basic conditions for sensing.

Table 1. Parameters of metal-organic frameworks

MOFs	*Topology*	*Linkers*	S_{BET} *surface area* (m^2g^{-1})	S_{Lang} *surface area* (m^2g^{-1})	V_{pore} (cm^3g^{-1})	*Refs.*
UiO-66	FCU	BDC	-	1187	-	(46)
ZIF-90(Zn)	SOD	CAIM	1270	1,320	-	(53)
MIL-100(Cr)	MTN	BTC	3100	-	1.61	(54)
MOF-5	PCU	BDC	3800	4400	1.55	(2)
MIL-101(Cr)	MTN-E	BDC	4230	5900	2.15	(43)
MOF-205	ITH-D	BTB, NDC	4460	6170	2.16	(51)
MOF-177	QOM	BTB	4500	5340	1.89	(55)
MOF-200	QOM	BBC	4530	10400	3.59	(51)
UMCM-2	UMT	BTB, T2DC	5200	6060	2.32	(56)
MOF-210	TOZ	BTE, BPDC	6240	10400	3.60	(51)
NU-109E	RHT	L^6	7010	-	3.75	(52)
NU-110E	RHT	L^6	7140		4.40	(52)

Quenching Ability

Similar to some traditional carriers/platforms, such as graphene (*57–59*), graphene oxide (GO) (*60*), gold nanoparticles (GNPs) (*61*), single-walled carbon nanotubes (*62, 63*), carbon nanoparticles (*64*), etc., MOF also own the property of fluorescence quenching ability. The mechanisms mainly based on fluorescence resonance energy transfer (FRET), photoinduced electron transfer (PET), and electronic energy transfer, *etc.* (*65*) The fluorescence of the fluorescent dye-labeled probe is quenched by the MOF in the absence of target molecules, however, when the target present, the fluorescence recovered as the probe formed a complex with the target that weakens the bonding force of the probe to the MOF platform and released from the platform. Furthermore, the guest molecules loaded on MOF induced the change of luminescence or fluorescence intensity.

Fluorescent Property

On the one hand, MOF is a fluorescence quenching platform, on the other hand, fluorescence can also be derived from the MOF, such as organic linker containing conjugated π moieties or aromatic, metal ions including inorganic clusters, lanthanides and so on (*66*). In detail, the production of fluorescence mainly has the following forms: (1) Direct excitation from organic ligands (*67*). It is caused by the transition from the lowest excited singlet state to the ground state, that is, the transition of $\pi \rightarrow \pi^*$ or $n \rightarrow \pi^*$. (The organic ligand is usually a rigid organic molecule having a π-conjugated system, and the rigid skeleton is bonded to a carboxyl group or a heterocyclic group capable of coordinating with the metal); (2) Emission from center metal. Mainly refers to the fluorescence emission of lanthanides, such as Eu^{3+} (red), Tb^{3+} (green), Sm^{3+} (yellow), Tm^{3+} (blue light), Yb^{3+} (near-infrared fluorescence), Nd^{3+} (near-infrared fluorescence) and Er^{3+} (near-infrared fluorescence) (fluorescence is enhanced by antenna effects in lanthanide MOFs); (3) Fluorescence from charge transfer. Fluorescence caused by the allowable transition from the charge transfer excited state to the ground state, including ligand-metal charge transfer (LMCT) (*68*) and metal-ligand charge transfer (MLCT) (*69*); (4) Fluorescence emission from guest molecules. Due to the highly regular pore structure and adjustable pore size, MOFs can be used as rigid or flexible frameworks to encapsulate guest fluorescent species such as lanthanide ions (*70*) and fluorescent dyes (*71*).

Sensing of Biomolecules

Since Asefa and coworker (*32*) first proposed biosensing using MOF, MOFs have attracted great interest as biosensing materials. According to the target of the interest, it can be divided into the following categories: (1) Nucleic acids sensing; (2) Enzyme/proteins-activity sensing; (3) Small-biomolecule sensing; (4) Others. MOF-based biosensing applications are summarized in Figure 3.

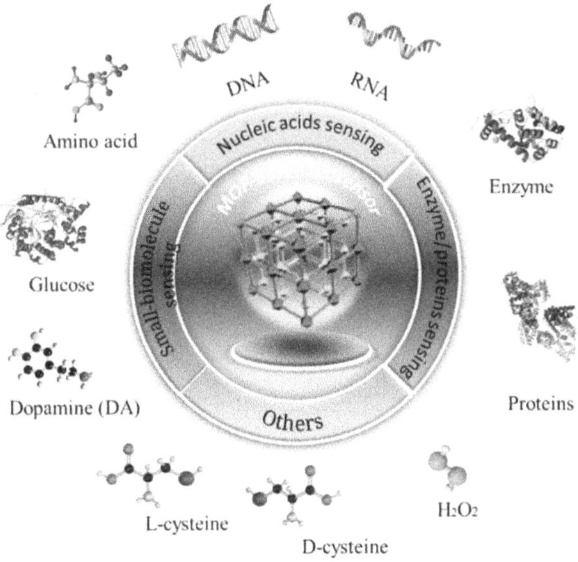

Figure 3. MOF-based biosensing applications.

Nucleic Acids (DNA/RNA) Sensing

Nucleic acid-MOF hybrids were first created by Mirkin et al. (72), in which UiO-66 MOF functionalized with oligonucleotides. DNA/RNA has been incorporated on MOF surfaces for DNA/RNA detection and particle stabilization. For the fluorescence quenching platforms, there is a common mechanism: after the addition of the probe complementary to the target, the fluorescence intensity was enhanced ("turn-on") or decreased ("turn off") to determine whether there was target molecules (73–75), we systematically summarized MOF-based sensing platforms for virus detection, and the dominant mechanisms were based on fluorescent signal recovery (Figure 4) (76). Generally, "turn off" will generate a false response signal due to various environmental factors, making its sensitivity worse than "turn-on". Therefore, for biosensing, "turn-on" strategy is more practical.

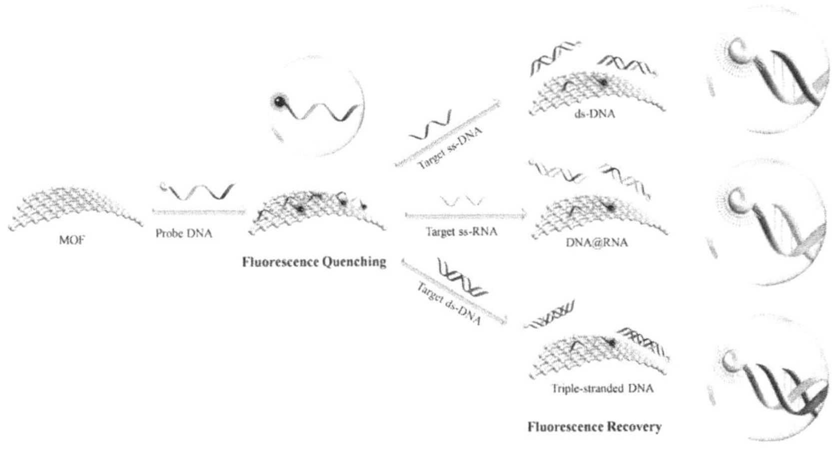

Figure 4. Schematic diagram of the nucleic acid detection mechanisms based on MOF. (Reprinted with permission from ref. (76). Copyright 2020, Elsevier).

DNA Sensing (ss-DNA/ds-DNA Sensing)

In 2013, Chen et al. (*77*) proposed and proved for the first time that MOF had a good fluorescence quenching ability. In the experiment, they synthesized MOF ($[Cu(H_2DTOA)]_n$) with H_4DTOA as the ligand and Cu^{2+} as the coordination center, MOF was used as the fluorescence quenching platform to effectively detect HIV-1 DNA sequences and thrombin. Later, MIL-101 was introduced in Fang's laboratory as a quenching platform to decrease the high background fluorescence of SYBR Green I (SG)/probe DNA complex (*78*). Thereby, the fluorescence signal was greatly improved. Since then, various types of MOF had been used for DNA sensing.

Peptide nucleic acids (PNA) has a higher affinity with DNA and RNA compared with natural oligonucleotides (*79*). Mejia-Ariza et al. (*80*) detected DNA with PNA-modified MOF (MIL-88A) by flow cytometry. Both non-covalent and covalent interactions had been used, leading to biotin-modified and alkyne-modified MIL-88A, respectively. Moreover, DNA sensing platform could be used for discriminating between fully complementary, randomized DNA targets and single-base mismatched DNA.

Qiu's group (*81*) synthesized 2D ultrathin (<10 nm) nanosheets of MOF (MOF-NSs), Cu-TCPP, by a one-step method with surfactant-assisted for synchronous fluorometric detection of the DNAs of *Listeria monocytogenes, Salmonella enterica, and Vibrio parahemolyticus*. And the detection limits was 35 pM, 28 pM and 15 pM, respectively. Note here that, the system used "turn-on" mechanism and presented three different colors when the corresponding target presented.

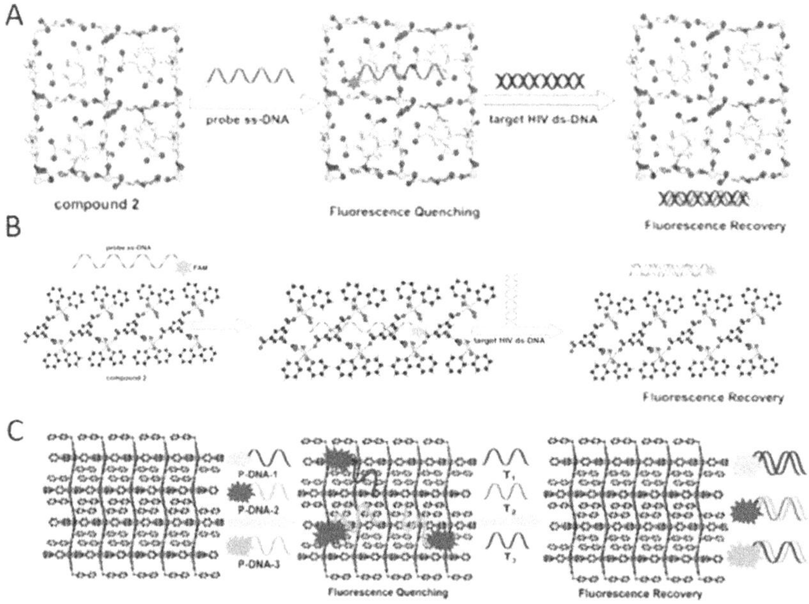

Figure 5. (A) Proposed mechanism for the detection of target HIV ds-DNA sequences based on a fluorescent biosensor formed from compound 2 and fluorophore-labeled probe ss-DNA; (Reprinted with permission from ref. (82). Copyright, Elsevier). (B) Mechanism for the target HIV ds-DNA based on fluorescent biosensor; (Reprinted with permission from ref. (83). Copyright 2019, Springer). (C) Proposed mechanism for the synchronous fluorescence detection of T1, T2, and T3 based on a fluorescent biosensor formed from MOF 1 and fluorophore-labeled P-DNAs. (Reprinted with permission from ref. (84). Copyright, Royal Society of Chemistry).

Unlike the ss-DNA sensing, ds-DNA sensing mechanism can be interpreted as the formation of a rigid triple-strand structure via Hoogsteen hydrogen bonding when the target ds-DNA interacts with the TFO probe or the ss-DNA probe, causing the dye-labeled probe to release from the MOF surface resulting in fluorescence recovery.

Human immunodeficiency virus 1 (HIV-1) is a virus that causes defects in the human immune system, which serious threaten human health. Zhao *et al.* (*82*) synthesized zinc(II)-based 2D MOF, $\{[Zn(HCbdcp)_2]\cdot H_2O\}_n$, for selective and sensitive sensing of HIV-1 ds-DNA sequences (Figure 5A). The detection limit reached as low as 10 pM (S/N=3). The MOF could quench the fluorescence (turn off) of FAM via a PET process. When target HIV ds-DNA is presented, the hybridization of ds-DNA adsorbed P-DNA would form triplex DNA, leading to the release of the dye-labeled DNA to the solution, thus resulting in the recovery of fluorescence (turn-on). Sun et al. (*83*) synthesized four water-stable zwitterionic zinc carboxylate compounds for the detection of HIV-1 ds-DNA with linear concentration range in the 0-60 nM and detection limit of 7.4 nM (S/N=3) (Figure 5B).

RNA Sensing

Various types of MOFs have been used for RNA sensing. Xie's (*85*) group synthesized a water-stable and soluble 3D MOF ([Cu(Dcbcp)(bpe)]n, bpe = 1,2-bis (4-pyridyl) ethylene, Dcbcp = N-(3,5-dicarboxylbenzyl)-(3-carboxyl) pyridinium). First, MOF has a high affinity toward carboxyfluorescein (FAM) or 5(6)-carboxyrhodamine, triethylammonium salt (ROX)-tagged single-stranded probe DNA (P-DNA) through electrostatic interactions, stacking, etc., different affinities of MOF with P-DNA and the ds-DNA/RNA were formed, which can be applied for simultaneous detection of Zika virus (ZIKV) and Dengue virus (DENV) RNA sequences. Compared with single detection method, simultaneous detection provides lower detection limits: 121 and 184 pM, respectively, while it is 332 and 192 pM respectively under single detection. Meanwhile, the principle of fluorescence quenching and recovery was the same as ss-DNA, which followed the "turn-on" mechanism.

Zwitterionic copper (II) MOF has been applied to synchronous detection of ebolavirus-encoded miRNA-like fragment and ebolavirus conserved RNA sequences (*86*). A 3D dysprosium (Dy) MOF as double-probe for Ebolavirus RNA sequences. MOF-based sensor was developed from dual-probes to tripel-probes for RNA detection. Xie et al. (*84*) reported a synchronous DNAs@MOF hybridization platform to simultaneously detect three conserved RNA sequences of ZIKV with high selectivity and sensitivity, moreover, low detection limits of 0.56 ± 0.02, 0.16 ± 0.04, and 0.19 ± 0.05 nM were also obtained. (Figure 5C), Chen et al. reported an effective fluorescence sensing platform formed by water-stable 3D dysprosium-based MOF and probe ss-DNA, which can be used for the detection of Ebolavirus RNA sequences with the detection limit down to 160 pM (*87*).

Moreover, Li et al. (*88*) synthesized a new type of multifunctional iron-based MOF (PdNPs@Fe-MOFs) by assembly palladium nanoparticles on the surface of Fe-MIL-88-NH2 MOFs microcrystals for sensitive detection of microRNA-122. In this work, two hairpin probes: signal probes and capture probes, were designed. When target miR-122 triggered the hybridization between capture probes and signal probes, the next reaction process was further initiated, resulting in numerous tracer indicators anchored onto the sensing interfaces. So that the detection signal was enhanced (turn-on). Specifically, a low detection limit of 0.003 fM (S/N =3) was obtained with a wide detection range from 0.01 fM to 10 pM.

Unlike the general turn-on mechanism, Chang and co-workers fabricated a novel homogeneous electrochemical biosensor, MB@UIO and TMB@UIO, for simultaneously and accurately detecting multiple tumor biomarkers, let-7a and miRNA-21, with detection limits down to 3.6 and 8.2 fM, respectively (*89*). The generation of RNA-DNA complexes was attributed to the hybridization and recognition of PMB with target miRNAs, simultaneously, the release of electroactive dyes from MOF (UIO-66-NH2) leading to the generation of the electrochemical signal. Compared with the example just mentioned, this produces an electrical signal rather than a fluorescent signal (Figure 6).

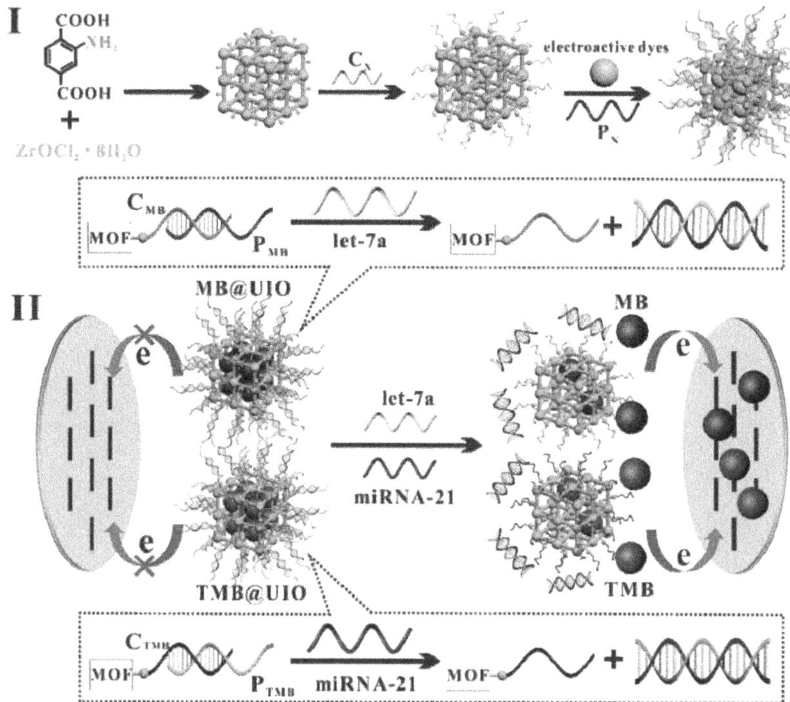

Figure 6. Schematic illustration of nucleic acid-functionalized MOFs and the principle of the MOF-Based homogeneous electrochemical biosensor for multiple detection of miRNAs. (Reprinted with permission from ref. (89). Copyright, American Chemical Society).

ds-DNA and microRNA Sensing

DNA and RNA hybrid and multifunctional sensing were gradually emerged. For example, in 2015, Yang et al. (*90*) reported a water-stable 3D Cu-based MOF platform for human immunodeficiency virus **1** ds-DNA and Sudan virus RNA sequences, with detection limits of 196 and 73 pM, respectively. And it's a "turn-on" mechanism in fluorescence recover of FAM by PET process.

Apart from this, Song et al. (*91*) reported a novel 2D fluorescent MOF sensing platform made from Mn ions and flexible ligand 2,2-dimethylsuccinate (MnDMS) for intracellular microRNA and DNA sensing (Figure 7). The study indicated that MnDMS could cause high-efficiency fluorescence quenching of fluorophore-labeled microRNA (miRNA) and ss-DNA. A detection limit of 0.2 pM and good linearity from 1 pM to 200 pM of target ss-DNA was obtained. Due to hybridization chain reaction (HCR) amplification technology, HCR/MnDMS system can be used to detect miRNA in living cells. The experimental results strongly proved that Mn-based MOF could also be used as a good quenching platform. Meanwhile, the HCR technology also provided a new idea for the amplification of fluorescence intensity.

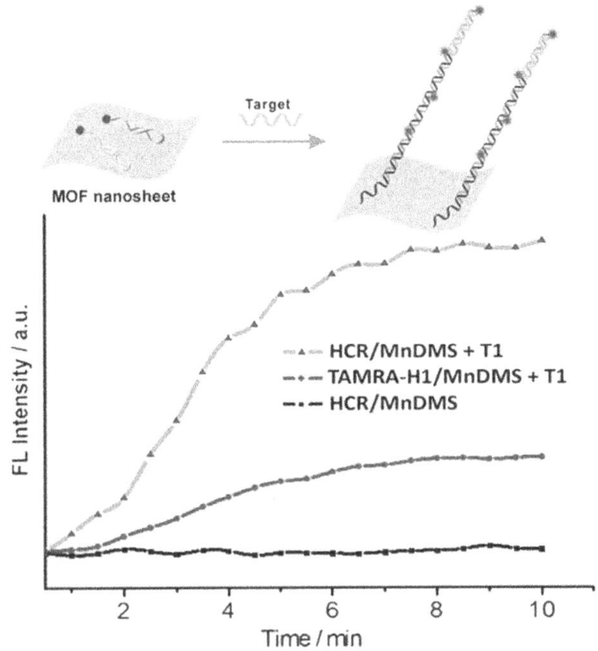

Figure 7. Schematic illustration of the HCR/MnDMS nanosheet composite for DNA detection. (Reprinted with permission from ref. (91) Copyright, ELSEVIER).

Enzyme-Activity/Proteins Sensing

As typical biomacromolecules, enzyme/protein immobilization improves biocatalysis, sensing, biofuel, and biomedicine by enhancing enzyme stability and cost-efficiency (*92*, *93*). However, how to immobilize and encapsulate enzymes/proteins for MOF remains a challenge (*94*). MOFs have shown great promise for protein/enzyme sensing because of its chemically compatible and exhibited excellent catalytic performance (used to enhance electrochemical signal). Compared with the fluorescent mechanism of common nucleic acid sensing, most of the sensing of protein/enzyme macromolecules tends to be photoelectrochemical.

Proteins play important roles in cell growth, metabolic modulation, signal transduction, and other biological functions (*95*). The abnormal expression of the protein may lead to many diseases, such as cancer, obesity, *etc.* Therefore, the detection of proteins is of great significance.

Zhang et al. (*96*) developed a label-free PEC sensing of a phosphoprotein using a novel Zr-based porphyrinic MOFs (PCN-222) for label-free phosphoprotein detection. Note here that, the detected electrical signal changes by the inhibition of the photoelectric conversion reaction by phosphoprotein. What's more, this biosensor showed high sensitivity for detecting α-casein, detection limit was obtained to be 0.13 $\mu g{\cdot}mL^{-1}$ (S/N =3) with a better linear range between the concentration of the α-casein and photoelectric from 0 to 2 $\mu g{\cdot}mL^{-1}$. Unlike other mechanisms, this was a photoelectrochemical strategy based on charge recombination inhibition (Figure 8).

Zhou et al. (*97*) obtained 2D Zn-MOF-on-Zr-MOF by MOF-on-MOF method and utilized it as an aptasensor platform to detect the cancer biomarker, protein tyrosine kinase-7 (PTK7), with an ultralow detection limit of 0.84 $pg{\cdot}mL^{-1}$. The mechanism is that the specific binding of PTK7 to the Sgc8 aptamer triggers structural changes in the aptamer, which in turn leads to changes in the electrochemical signal. The highlight is that the integration of the two MOFs can enhance the binding interaction between the formed g-quadruplex structure and the MOF matrix, the detection showed good reproducibility, stability and high sensitivity (Figure 9).

Also, abnormal aberrant enzyme activities are coupled with various diseases, including cancers, neurodegenerative diseases and diabetes (*98*, *99*). Therefore, accurate sensing of kinase activity and their potential inhibitors evaluation are not only significant for the drug discoveries, therapies, and kinase-targeted diagnosis but also helpful for elaborating the molecular signal transduction mechanism.

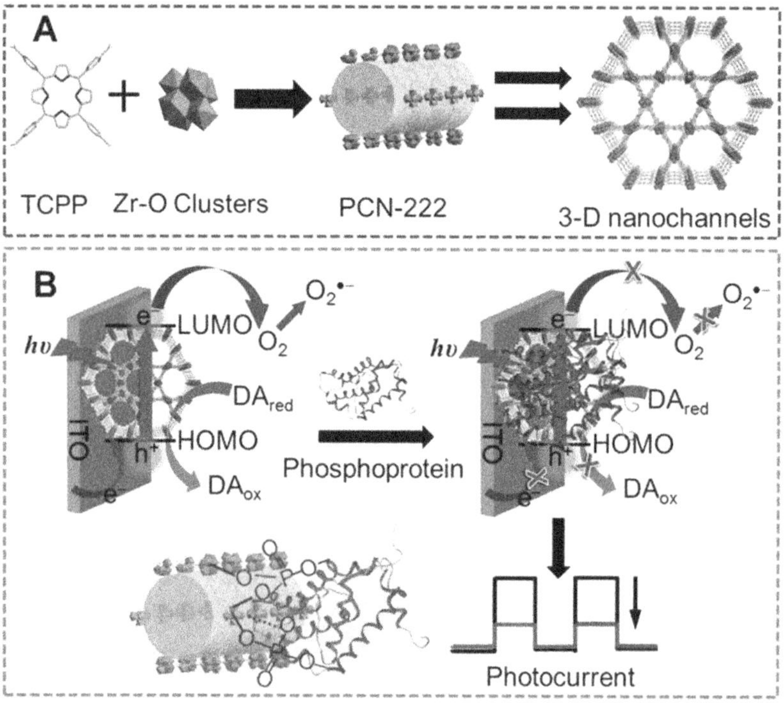

Figure 8. Schematic illustration for the construction of PCN-222 (A) and the mechanism of charge recombination suppression-based photoelectrochemical strategy for the detection of phosphoprotein (B). (Reprinted with permission from ref. (96) Copyright, American Chemical Society).

Based on the surface defect recognition and multiple signal amplification, Wang et al. (*100*) reported a highly sensitive photoelectrochemical biosensor on the basis of MOF (UiO-66) for detection and inhibition activity of a kinase, indeed, which was a type of enzyme that catalyzes the transfer of phosphate groups from phosphate-donating molecules with high-energy to specific substrates. Notably, this detection was based on phosphorylation reaction in the presence of protein kinase A (PKA) appears. What's more, the detection limit was 0.0049 $U{\cdot}mL^{-1}$ (S/N=3) and the visible light PEC biosensor provides a simple and cost-effective detection manner for PKA activity assays. In 2017, Wang et al. (*101*) designed a fluorescence sensing assay through using pyrophosphate ion (PPi)-triggered competitive replacement of ss-DNA from (Ce^{3+}/Ce^{4+})-based MOF (MVCM) for alkaline phosphatase (ALP) detection. The detection limit toward ALP was calculated to be 0.18 $mU{\cdot}mL^{-1}$ and the reaction was completed in an extremely brief time (2 min) which was different from the conventional aptamer-targeted conformational change.

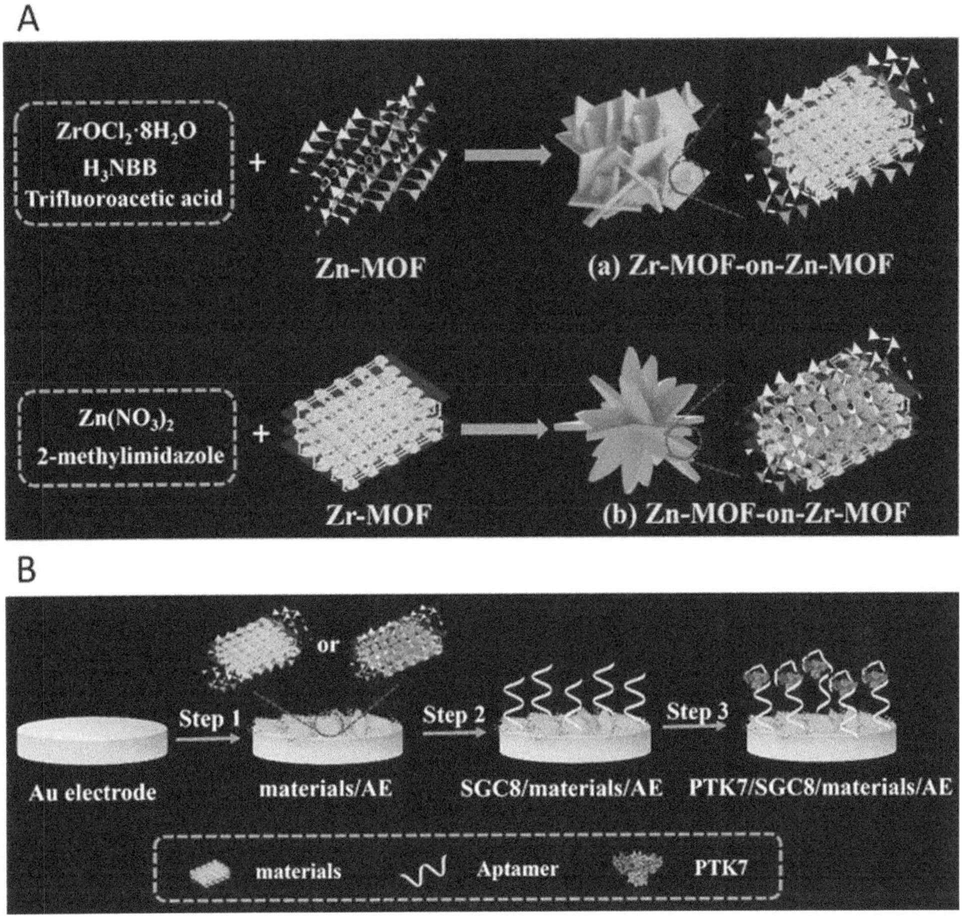

Figure 9. (A) The schematic diagram of the synthesis procedures for two kinds of MOFs, including (a) Zn-MOF-on-Zr-MOF and (b) Zr-MOF-on-Zn-MOF hybrids, and (B) the detection procedure toward PTK7, including the Au electrode (AE) modified with Zn-MOF-on-Zr-MOF or Zr-MOF-on-Zn-MOF hybrid (Step 1), the SGC8 aptamer immobilization on the modified AE (Step 2), and the PTK7 detection (Step 3). (Reprinted with permission from ref. (97) Copyright, ELSEVIER).

Small Biomolecules Sensing

It is necessary to achieve ultrasensitive and selective detection of small biomolecules, include glucose, H_2O_2, dopamine (DA), amino acids, cysteine (CySH), formaldehyde (FA), etc.

Glucose/H_2O_2 Sensing

Glucose, as one of the most important indicators for the detection of diabetes, has received increasing attention (*102*). Liu and co-workers (*19*) reported a facile approach to establish mimic multienzyme systems with hierarchically porous (HP) biomimetic MOF and enzymes for biocatalyze and small-biomolecule detection. The MOF (HP-PCN-224 (Fe)) not only acted as a matrix for enzyme immobilization or an effective enzyme mimic for catalyzing but also as colorimetric sensors for the detection of glucose and uric acid (UA). Here, GOx@HP-PCN-224 (Fe) was used for the detection of glucose and uricase@HP-PCN-224 (Fe) for UA. Different from the colorimetric sensing, Xu et al. (*103*) achieved "All-in-One" multifunction by combining *in situ* enzyme immobilization with O_2-sensitive luminescent MAF-2 (Cu (I) triazolate framework). More importantly, the composite, GOx@MAF-2, rapidly response toward dissolved O_2, which realized ratiometric and direct sensing of glucose with a low detection limit of 1.4 μM. Also, the accurate detection of glucose was realized. Similarly, Zhu et al. (*104*) designed a system, Co-TCPP (Fe)@luminol@GOD, which could catalyze glucose to release H_2O_2 in the presence of glucose oxidase (GOD) on Co-TCPP (Fe) and the decomposition of H_2O_2, as well as oxidize luminol to generate a strong chemiluminescence signal. Most importantly, the sensors exhibited many advance performances, such as good sensitivity, the limit of detection (LOD) was 10.667 μg/L, satisfactory accuracy (recoveried 89-121.2%), low cost, and simple operation, etc.

MOF was demonstrated to show great potential in electrochemical sensors (*105*, *106*). However, the poor conductivity of MOF seriously suppresses their electrocatalytic performance. Currently, the most universal strategy to strengthen the poor conductivity of MOFs is to couple conductive materials (such as graphene, Au) with MOF. Generally speaking, electrochemical detection of glucose can be achieved through non enzymatic and enzymatic electrochemical biosensors. Compared with enzymatic electrochemical biosensors, non-enzymatic electrochemical biosensors are favored for its low price and eliminating the immobilization process of enzymes.

For example, Zhang et al. (*107*) designed an efficient electrocatalyst by orientationally growing Ni_2P nanocomposite *in situ* on graphene surface (Ni_2P/G) as a glucose sensor in alkaline media. The detection range was from 5 μM to 1.4 mM, with a detection limit of 0.44 μM. Here, Ni-MOF-74 was selected as the precursor to form uniform Ni_2P nanoparticles, while graphene was selected as the support for enhancing their conductivity and immobilizing the particles (Figure 10A).

Chen's group used graphene as a conductive medium (*108*), sandwich-like graphene@ZIF-67 heterostructure hybrids were obtained with *in situ* synthesis method for non-enzymatic highly sensitive glucose sensing in human serums. Benefiting from the synergistic effect and enhanced electron transfer ability between two materials, the system showed a detection limit of 0.36 μM (S/N=3) and a linear range of 1-805.5 μM (Figure 10B).

In addition to graphene, the method of using Au as a substrate has gradually attracted attention. Simultaneously, combined with the advantages of multi-metal MOF and nonenzymatic, electrochemical sensing is more attractive. For instance, Li et al. (*109*) successfully prepared a novel vertical 2D NiCo bimetal-MOF (NiCo-MOF) parallelogrammic nanosheets array on the surface of nanoporous gold with a bottom-up strategy as non-enzymatic glucose biosensing. Most importantly, the strong electron interaction between Ni, Co, and N, enhanced electrochemical catalytic oxidation of glucose. The system showed a wide linear range from 1 μM to 8 mM, and a low detection limit of 0.29 μM (Figure 10C).

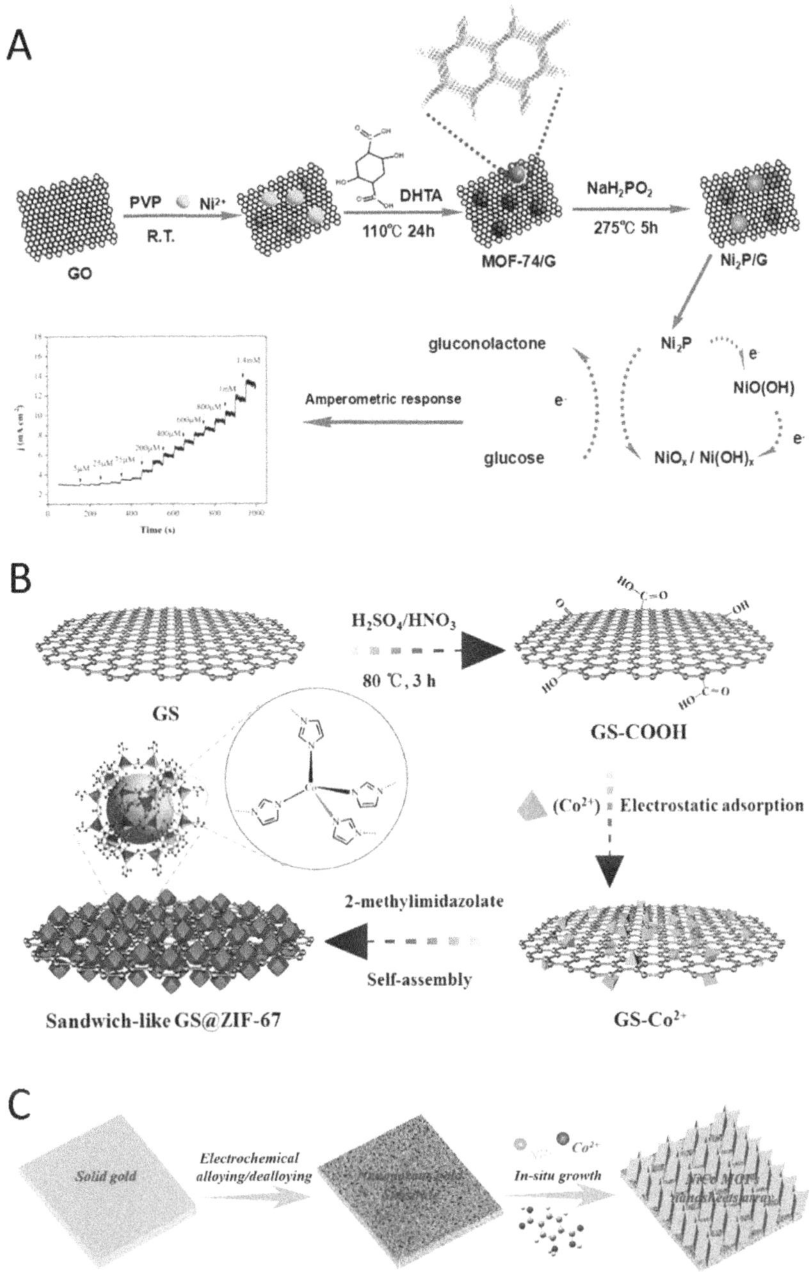

Figure 10. (A) Schematic diagram of the synthesized process of Ni_2P/graphene; (Reprinted with permission from ref. (107) Copyright, American Chemical Society) (B) Illustration of the fabrication of sandwich-like GS@ZIF-67 hybrids; (Reprinted with permission from ref. (108) Copyright, American Chemical Society) (C) Schematic illustration for the synthesis of vertical NiCo-MOF nanosheets array. (Reprinted with permission from ref. (109) Copyright, ELSEVIER).

H_2O_2, an indispensable member in reactive oxygen species (ROS) family, contribute to oxidative stress in living organisms (*110*). The overexpression of H_2O_2 will cause many diseases, for instance, cancer (*111*), diabetes (*112*), Alzheimer's disease (*113*). Therefore, the detection of H_2O_2 is important. Yu and co-workers had designed a highly chemiluminescent (CL) MOF, MIL-101 (Cr), by loading luminol for detection of pyrophosphate ions and hydrogen peroxide (*114*). The unsaturated metal sites of MIL-101 (Cr) and the amino groups of luminol were immobilized by Lewis acid-base interactions. Particularly, the MOF loaded with luminescent materials showed strong CL in the alkaline solution of hydrogen peroxide (H_2O_2). In turn, pyrophosphate (PPi) acted as a quenching agent for this CL. Furthermore, Mostakim SK's group developed a boronic acid-functionalized Zr (IV) MOF, Zr-UiO-66-B$(OH)_2$, as sensitive and selective sensing for the detection of H_2O_2 in N′-a-hydroxythylpiperazine-N′-ethanesulfanic acid (HEPES) buffer (pH 7.4), even in presence of interfering relevant analytes and ROS (*115*). The LOD was 0.015 μM, which was lower than other H_2O_2 sensors based on MOF. The "turn-on" of fluorescence could be attributed to oxidation of $-B(OH)_2$ into –OH in the present of H_2O_2.

Designing a platform with multiple functions of sensing has always been the goal of researchers. Luo et al. (*116*) obtained a composite, Hemin@HKUST-1, with 1,3,5-benzene tricarboxylate as a ligand, Cu^{2+} as the center and the Hemin as a catalyst, for detecting glucose and H_2O_2 simultaneously. The mechanism was as follows: (1) Glucose oxidizes in the presence of glucose oxidase and hydrogen peroxide; (2) Hemin catalyzes the chemiluminescence reaction of H_2O_2 -luminol system; (3) When the concentration of glucose or hydrogen peroxide changes, the fluorescence intensity also changes. Similarly, Yi et al. (*117*) also adopted H_2O_2-luminol system to design MIL-53 (Fe) to catalyze chemiluminescence for detection of glucose with CL intensity 20 times higher than that in the luminol–H_2O_2 system. Mao et al. (*118*) also used H_2O_2-luminol system to prepare β-CD functionalized MOF-235, for the detection of glucose and H_2O_2. Compared with the previous system, the synergistic effect of MOF-235 and β-CD exhibited higher catalytic activity with a good linear relationship (r=0.9987) at the glucose concentration of 0.01 to 3 μM, and LOD for glucose was 10 nM. This system was 30 times higher than that in the luminol-H_2O_2 system.

There are various sensing methods, including colorimetric, ratiometric and direct, electrochemical, electrocatalytic, *etc.* MOF can not only serve as a carrier but also assist in catalysis, etc. to produce synergy in the sensing process. It can be seen that the electrochemical detection method based on MOF plays an important role in small molecule sensing.

Cysteine Sensing

Cysteine (CySH), an essential nonessential amino acid, which plays an important role in cellular activities including detoxification metabolism and protein synthesis (*119*).

In 2015, Yang *et al.* (*120*) demonstrated that chemiluminescence (CL) of luminol could be catalyzed by the Co-MOF and designed a Co-MOF-luminol CL system to selective detect l-cysteine. Presumably, the mechanism was that cobalt-peroxide complexes were formed between exposed reactive metal sites and oxygen-associated free radicals. Notably, the result suggested that CySH could be selectively detected in the range of 0.1–10 μM, meanwhile, it had a lower detection limit of 18 nM. Owing to the inherent system structure of Co-BTC which could overcome interferences from other complex matrixes, promotes a better selectivity of CySH.

Subsequently, He et al. in 2019 employed PCN-222, a porphyrini MOF, for sensitive and fast determination of cysteine (Cys) based on the "turn-on" mechanism (*121*). In short, after the metallization of Hg^{2+}, the fluorescence (FL) of PCN-222 was quick quenched ("turn off "). However, in the presence of Cys, the strong combination of Cys and Hg^{2+} made the Hg leave the MOF, so that the fluorescence was restored ("turn-on") (Figure 9). Based on fluorescence measurement, the detection limit was as low as 60 nm. Specifically, when Cys was added as low as 100 nM, the color change of mercury-metalized PCN-222 could be easily observed by the naked eye, which means, the sensing could also be done by visual means. Furthermore, compared with other similar sensors, this work had many advantages, such as fast response (1 min) to Cys, cyclic usage and facile preparation, high sensitivity for visual sensing and FL, and no extra additives (Figure 11).

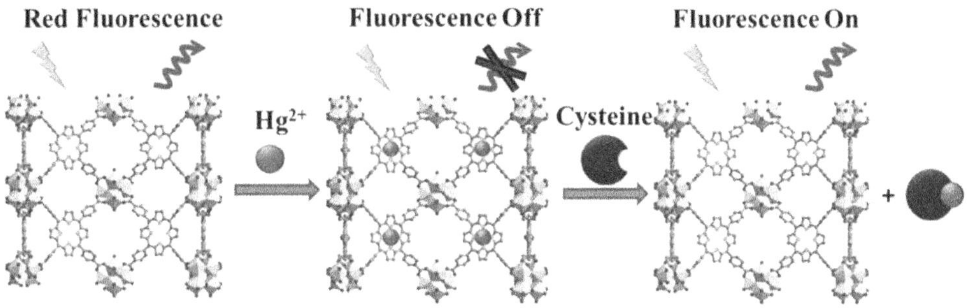

Figure 11. The proposed mechanism for the FL sensing of Cys by using the probe of PCN-222/Hg^{2+}. (Reprinted with permission from ref. (121) Copyright, ELSEVIER).

Dopamine Sensing

Dopamine (DA), a catecholamine neurotransmitter, plays an important role in the central nervous system and immune system. Lesions of DA cause various diseases such as schizophrenia (*122*).

Du and his workers prepared a boric acid-modified lanthanide MOF, Eu-MOF via a one-pot method as a ratiometric fluorescence probe for the detection of DA in the serum of human samples (*123*). In this work, the probe was characterized by high selectivity and sensitivity due to its combination of specific fluorescence technology and the nature of boric acid. Under a single excitation of 275 nm, Eu-MOF probe showed dual emission response signals of Eu^{3+} ion and 3,5-dicarboxybenzeneboronic acid (DBA) which was caused by antenna effect. When DA is added, boric acid covalently connects with the *cis*-glycol compound of DA. As a result, the emission at 363 nm gradually increased, while at 615 nm, the emission gradually decreased, and the color change of the mixture could be observed by the naked eye. What's more, under optimum measurement conditions, the concentration detection range of DA was 0-500 μM and detection limit was 0.015 μM (Figure 12).

Besides DA, this sensor could also detect other *cis*-diol substances, for example, gallic acid and catechol. Furthermore, this sensor provided a novel method for detecting DA and other similar molecules.

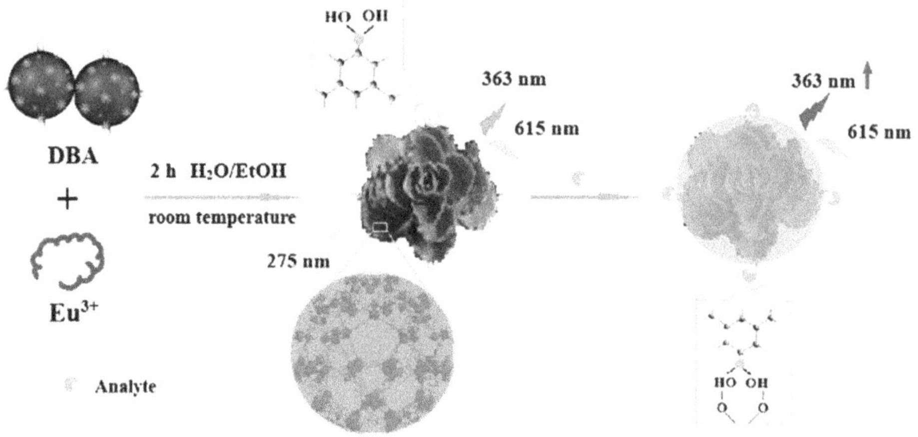

Figure 12. Schematic illustration of the synthetic process of the Eu-MOF detection system. (Reprinted with permission from ref. (123) Copyright, Royal Society of Chemistry).

Formaldehyde Sensing

Formaldehyde (FA), one of the most the simplest aldehyde among all the reactive carbonyl species, is regarded as a major component for sick building syndrome.

Nandi et al. (*124*) reported novel hydrazine-functionalized Al-based MOF, Al-MIL-53-N_2H_3, which possessed MIL-53 framework topology and hydrazine-functionalized H_2BDC-N_2H_3, as a fluorescence turn-on probe for sensitively and selectively detecting FA. Interestingly, the sensor exhibited a rapid response (the 60 s) and highly sensitive for FA with a detection limit of 8.37 μM (0.25 ppm) in HEPES buffer. Note here that, the "turn-on" phenomenon of the probe could be ascribed to the inhibition of PET (from the hydrazine group to the phenyl ring) due to the formation of the hydrazine. Selective response of FA could be assigned to the smallest porous size of MOF and highest electrophilic character of FA.

Compared with nucleic acid, enzyme, and qualitative optical detection of protein macromolecules, MOF-based biomolecular sensing is more inclined to be constructed based on electrochemical quantitative detection. The main mechanism is summarized as the interaction between the fluorescent molecule and biomolecule changes the fluorescent molecular signal or the photoelectric signal, realizing the sensing of the target biomolecules.

Other Sensing

In addition to the sensing aspects mentioned, MOF-based sensing platform is also applied to the sensing other substances. For example, members of the ROS family, ratiometric O_2 (*125*). Other molecular sensors worthy of our attention but it is not covered here, for example, H_2S in live tissue, acetone (*126*), pH (*127*), glutathione (GSH) (*128*), acetylacetone (*129*) sensors, etc.

The sensing mechanisms mainly divided into the following categories: (1) "turn-on" mechanism based on fluorescent signals; (2) photoelectrochemical sensing mechanism based on optical signal changes; (3) chemiluminescent mechanism based on MOF. Among them, the fluorescent "turn-on" is the main mechanism. Electrochemistry plays a very important role in sensing. The following table summarized its biosensing applications. (Table 2)

Table 2. MOF for biosensing applications

Sensors	*Sensing targets*	*Linear range* ($nM \cdot L^{-1}$)	*Detection limits* ($nM \cdot L^{-1}$)	*Refs.*
PHIV/MIL-88B	DNA	0-5 nM	10 pM	(*130*)
{[Ag(L)$_2$]BF$_4$ (1) {[Ag3(L)$_3$](H$_2$O)(CF$_3$SO$_3$)3	l-cysteine d-cysteine	(1) 0.01-0.5 μM (2) 0.03-0.09 μM	(1) 0.0095 μM (2) 0.0057 μM	(*131*)
Hemin@ Cu-MOF-199-SA	H_2O_2.	1.0×10^{-6} -1.0×10^{3} nM	6.9×10^{-7} nM	(*132*)
Cu$_3$(1,3,5-benzenetricarboxylicacid)$_2$/GCE	Dopamine (DA)	5.0×10^{-7}~1.0×10^{-4} M	1.5×10^{-7} M	(*133*)
P-DNA@ ZIF-8	HIV-1 ds-DNA	10 nM-100 nM	1.2 nM	(*134*)
pDNA-Au-Hemin-MIL-88(Fe)-DNAzyme	Prostate specific antigen (PSA)	0.5-500 ng mL^{-1}	0.058 ng·mL^{-1}	(*135*)
NiCo-MOFNs array	Glucose	1-8 mM	0.29 μM	(*109*)
Ni(TPA)-SWCNT-CS/GCE	Glucose	20 μM-4.4 mM	4.6 μM	(*136*)
BCL@MOF nanofibers/chitosan/GCE	Methyl parathion	0.1-38 μM	0.067 μM	(*137*)
AChE-ChO/Pt/MOF/Au	Acetylcholine	0.01-500 μM	0.01 μM	(*138*)
L/AuNP/(Fe-P)n-MOF	T4 polynucleotide kinase (PNK)	1.0 mU·mL^{-1} - 1.0U·mL^{-1}	0.62 mU·mL^{-1}	(*139*)
[Tb-MOF–protamine]/[Hep]	Protamine and Hep	0.5 - 5 μM	0.27 μM	(*140*)
MOF@AuNP@GO	p53 gene/ PSA	0.005 nM-0.01 ngmL^{-1}	-	(*141*)
RhB@MOF1	Land d-lysine	0-0.4 μM	0.024, 0.007 μM	(*142*)
AβO/ZIF-8-ferrocene	amyloid-β oligomer	0.5-100 μM	0.5 μM	(*143*)
FeMOF@AuNP-modified ss-DNA	PNK	0.0005-10 U·mL^{-1}	2.344×10^{-4} U·mL^{-1}	(*144*)

In terms of sense, not only the partial mechanisms of sensing, but also several common sensors are summarized, such as nucleic acid sensing, enzyme/protein biomolecular sensing, small molecule biosensing and other types of sensing. Achievements show promising applications, however, there are still many challenges which can not be ignored in this rapidly developing field. For example, MOF as a hybrid material containing metals and organics, biotoxicity and biocompatibility are the biggest challenges limiting the further *in vivo* clinical application of MOF. Thus, green synthetic strategy should be adopted.

In biosensing, MOF-based researches on biosensors are still in its early stages. The future direction of biosensor development: (1) Low cost, high sensitivity, high stability and long life. The improvement of these characteristics will also accelerate the process of marketization and commercialization of biosensors. To improve the sensitivity, some aspects should be considered: MOFs with high quenching efficiency and good selectivity should be designed. Utilizing nanoparticles, such as MOF-derived porous carbon to develop novel sensors; Oxide enhances the sensing performance, the binding ability and biological activity. The detection in the future should favor the efficiency and accuracy in real samples. Comprehensive quality assessment standard to assess the accuracy and reliability of detection is also needed to be established. The detection of antibodies is still deserves more attention, from the aspect of template elution based on molecular imprinting, the pH-responsive MOF shows great potential for template elution. Meanwhile, the attention should be focused on signal amplification strategy, and the development of other innovative detection methods. The instability of MOF in complex physiological condition should be considered seriously, Water-stable MOFs can be constructed with zwitterionic carboxylate ligands or zwitterionic thiolate and the water-stable MOFs can be obtained by doping metal ions, such as Zn^{2+}, Ca^{2+}, Mg^{2+}, and Mn^{2+} to enhance the stability. (2) Intelligence and integration. Future biosensors should be closely integrated with computers to automatically collect and process data and provide more scientific and accurate results. (3) Functional diversification. The detection should combine with other novel devices, such as lab-on-a-chip, or other technologies, such as electrochemistry, ultrasensitive biosensors with the characteristics of large-scale, automation, high-throughput screening, etc. The combination of MOF with metal oxides, metal particles, or carbon materials for the electrochemical detection is also a promising direction. (4) Miniaturization. The commercialization should be taken into consideration. The high-throughput, automation, miniaturization and low cost are the trend of the future. The sensor disposal, large-scale synthesis, fabrication, cost-effectiveness, and recycling are also the directions.

Additionally, the detection performance and safety of MOF-based biosensing need to be further verified *in vivo*, and the long-term biological toxicity of MOF need to be further evaluated *in vivo*, some biocompatible metal ions, such as Fe^{2+}/Fe^{3+}, Zr^{4+}, Mn^{2+}, Mg^{2+}, Ca^{2+}, Zn^{2+}, biomaterials, such as liposomal and bioconjugates, and green synthetic routes should be considered.

Conclusion

The unique properties of MOF, such as rational aperture design, diversity, easy to functionalize, high BET/Langmuir surface area, etc., have been highlighted. Meanwhile, we also updated and discussed the biosensing applications based on MOF, indicating MOF is a quite promising platform. While, MOF is still in its infancy in biological applications, animal and clinical laboratory studies are limited. How to improve the properties and overcome the weaknesses is the direction that needs to be emphasized in the future.

Acknowledgments

This work was financially supported by the Huxiang Young Talent Support Program of Hunan Province (2018RS3005), the Innovation-Driven Project of Central South University (2020CX048), the Changsha Science and Technology Plan Project (kq2005001, kq2004086), the National Natural Science Foundation of China (81301258), the Natural Science Foundation of Hunan Province (2019JJ60071, 2020JJ4680), the Shenghua Yuying Project of Central South University, and the Open-End Fund for the Valuable and Precision Instruments of Central South University.

References

1. Yaghi, O. M.; Li, G.; Li, H. Selective binding and removal of guests in a microporous metal-organic framework. *Nature* **1995**, *378* (6558), 703–706.
2. Li, H.; Eddaoudi, M.; O'Keeffe, M.; Yaghi, O. M. Design and synthesis of an exceptionally stable and highly porous metal-organic framework. *Nature* **1999**, *402* (6759), 276–279.
3. Rowsell, J. L. C.; Yaghi, O. M. Metal-organic frameworks: a new class of porous materials. *Microporous and Mesoporous Materials* **2004**, *73* (1-2), 3–14.
4. Agostoni, V.; Horcajada, P.; Noiray, M.; Malanga, M.; Aykaç, A.; Jicsinszky, L.; Vargas-Berenguel, A.; Semiramoth, N.; Daoud-Mahammed, S.; Nicolas, V.; Martineau, C.; Taulelle, F.; Vigneron, J.; Etcheberry, A.; Serre, C.; Gref, R. A "green" strategy to construct non-covalent, stable and bioactive coatings on porous MOF nanoparticles. *Sci Rep* **2015**, *5*, 7925.
5. Li, Y.; Xu, Y.; Yang, W.; Shen, W.; Xue, H.; Pang, H. MOF-derived metal oxide composites for advanced electrochemical energy storage. *Small* **2018**, *14* (25), 1704435.
6. Zhang, H.; Liu, X.; Wu, Y.; Guan, C.; Cheetham, A. K.; Wang, J. MOF-derived nanohybrids for electrocatalysis and energy storage: current status and perspectives. *Chem. Commun.* **2018**, *54* (42), 5268–5288.
7. Hong, X. J.; Tang, X. Y.; Wei, Q.; Song, C. L.; Wang, S. Y.; Dong, R. F.; Cai, Y. P.; Si, L. P. Efficient encapsulation of small S2-4 molecules in MOF-derived flowerlike nitrogen-doped microporous carbon nanosheets for high-performance Li-S batteries. *ACS Appl. Mater. Interfaces* **2018**, *10* (11), 9435–9443.
8. Zou, G.; Hou, H.; Ge, P.; Huang, Z.; Zhao, G.; Yin, D.; Ji, X. Metal-organic framework-derived materials for sodium energy storage. *Small* **2018**, *14* (3), e1704435.
9. Qu, C.; Liang, Z.; Jiao, Y.; Zhao, B.; Zhu, B.; Dang, D.; Dai, S.; Chen, Y.; Zou, R.; Liu, M. "One-for-All" strategy in fast energy storage: production of pillared MOF nanorod-templated positive/negative electrodes for the application of high-performance hybrid supercapacitor. *Small* **2018**, *14* (23), e1800285.
10. Hu, J.; Liu, C.; Liu, L.; Li, Q. Thermal energy storage of R1234yf, R1234ze, R134a and R32/MOF-74 nanofluids: a molecular simulation study. *Materials* **2018**, *11* (7), 1164.
11. Cheng, H.; Jiang, X. Y.; Zheng, R. R.; Zuo, S. J.; Zhao, L. P.; Fan, G. L.; Xie, B. R.; Yu, X. Y.; Li, S. Y.; Zhang, X. Z. A biomimetic cascade nanoreactor for tumor targeted starvation therapy-amplified chemotherapy. *Biomaterials* **2019**, *195*, 75–85.
12. Min, H.; Wang, J.; Qi, Y.; Zhang, Y.; Han, X.; Xu, Y.; Xu, J.; Li, Y.; Chen, L.; Cheng, K.; Liu, G.; Yang, N.; Li, Y.; Nie, G. Biomimetic metal-organic framework nanoparticles for cooperative combination of antiangiogenesis and photodynamic therapy for enhanced efficacy. *Adv. Mater.* **2019**, *31* (15), e1808200.

13. Wang, Y.; Yan, J.; Wen, N.; Xiong, H.; Cai, S.; He, Q.; Hu, Y.; Peng, D.; Liu, Z.; Liu, Y. Metal-organic frameworks for stimuli-responsive drug delivery. *Biomaterials* **2019**, 119619.
14. Zhang, X.; Chuah, C. Y.; Dong, P.; Cha, Y. H.; Bae, T. H.; Song, M. K. Hierarchically porous Co-MOF-74 hollow nanorods for enhanced dynamic CO_2 separation. *ACS Appl. Mater. Interfaces* **2018,** *10* (50), 43316–43322.
15. Prasetya, N.; Ladewig, B. P. New Azo-DMOF-1 MOF as a photoresponsive low-energy CO_2 adsorbent and its exceptional CO_2/N_2 separation performance in mixed matrix membranes. *ACS Appl. Mater. Interfaces* **2018,** *10* (40), 34291–34301.
16. Marti, A. M.; Venna, S. R.; Roth, E. A.; Culp, J. T.; Hopkinson, D. P. Simple fabrication method for mixed matrix membranes with *in situ* MOF growth for gas separation. *ACS Appl. Mater. Interfaces* **2018,** *10* (29), 24784–24790.
17. Xu, H.; Cao, J.; Shan, C.; Wang, B.; Xi, P.; Liu, W.; Tang, Y. MOF-derived hollow CoS decorated with CeOx nanoparticles for boosting oxygen evolution reaction electrocatalysis. *Angew. Chem. Int. Ed.* **2018,** *57* (28), 8654–8658.
18. Zhao, W.; Wan, G.; Peng, C.; Sheng, H.; Wen, J.; Chen, H. Key single-atom electrocatalysis in metal-organic framework (MOF)-derived bifunctional catalysts. *ChemSusChem* **2018,** *11* (19), 3473–3479.
19. Liu, X.; Qi, W.; Wang, Y.; Lin, D.; Yang, X.; Su, R.; He, Z. Rational design of mimic multienzyme systems in hierarchically porous biomimetic metal-organic frameworks. *ACS Appl. Mater. Interfaces* **2018,** *10* (39), 33407–33415.
20. Salgaonkar, M.; Nadar, S. S.; Rathod, V. K. Combi-metal-organic framework (Combi-MOF) of alpha-amylase and glucoamylase for one-pot starch hydrolysis. *Int. J. Biol. Macromol.* **2018,** *113*, 464–475.
21. Zhang, F. M.; Sheng, J. L.; Yang, Z. D.; Sun, X. J.; Tang, H. L.; Lu, M.; Dong, H.; Shen, F. C.; Liu, J.; Lan, Y. Q. Rational design of MOF/COF hybrid materials for photocatalytic H_2 evolution in the presence of sacrificial electron donors. *Angew. Chem. Int. Ed.* **2018,** *57* (37), 12106–12110.
22. Han, S. Y.; Pan, D. L.; Chen, H.; Bu, X. B.; Gao, Y. X.; Gao, H.; Tian, Y.; Li, G. S.; Wang, G.; Cao, S. L.; Wan, C. Q.; Guo, G. C. A methylthio-functionalized-MOF photocatalyst with high performance for visible-light-driven H_2 evolution. *Angew. Chem. Int. Ed.* **2018,** *57* (31), 9864–9869.
23. Das, A.; Anbu, N.; Sk, M.; Dhakshinamoorthy, A.; Biswas, S. Highly active urea-functionalized Zr(IV)-UiO-67 metal-organic framework as hydrogen bonding heterogeneous catalyst for friedel-crafts alkylation. *Inorganic Chemistry* **2019,** *58* (8), 5163–5172.
24. Liu, G.; Cadiau, A.; Liu, Y.; Adil, K.; Chernikova, V.; Carja, I. D.; Belmabkhout, Y.; Karunakaran, M.; Shekhah, O.; Zhang, C.; Itta, A. K.; Yi, S.; Eddaoudi, M.; Koros, W. J. Enabling fluorinated MOF-based membranes for simultaneous removal of H_2S and CO_2 from natural gas. *Angew. Chem. Int. Ed. Engl.* **2018,** *57* (45), 14811–14816.
25. Masih, D.; Chernikova, V.; Shekhah, O.; Eddaoudi, M.; Mohammed, O. F. Zeolite-like Metal-organic framework (MOF) encaged Pt(II)-Porphyrin for anion-selective sensing. *ACS Appl. Mater. Interfaces* **2018,** *10* (14), 11399–11405.

26. Lei, B.; Wang, M.; Jiang, Z.; Qi, W.; Su, R.; He, Z. Constructing redox-responsive metal-organic framework nanocarriers for anticancer drug delivery. *ACS Appl. Mater. Interfaces* **2018,** *10* (19), 16698–16706.
27. Liu, L. L.; Yu, Y. Z.; Zhao, X. J.; Wang, Y. R.; Cheng, F. Y.; Zhang, M. K.; Shu, J. J.; Liu, L. A robust Zn(ii)/Na(i)-MOF decorated with $[(OAc)_2(H_2O)_2]n(2n\text{-})$ anions for the luminescence sensing of copper ions based on the inner filter effect. *Dalton Trans.* **2018,** *47* (23), 7787–7794.
28. Yang, X.; Yu, Y. Q.; Peng, L. Z.; Lei, Y. M.; Chai, Y. Q.; Yuan, R.; Zhuo, Y. Strong electrochemiluminescence from MOF accelerator enriched quantum dots for enhanced sensing of Trace cTnI. *Anal. Chem.* **2018,** *90* (6), 3995–4002.
29. Liang, C. C.; Shi, Z. L.; He, C. T.; Tan, J.; Zhou, H. D.; Zhou, H. L.; Lee, Y.; Zhang, Y. B. Engineering of pore geometry for ultrahigh capacity methane storage in mesoporous metal-organic frameworks. *J. Am. Chem. Soc.* **2017,** *139* (38), 13300–13303.
30. Yang, P.; Men, Y.; Tian, Y.; Cao, Y.; Zhang, L.; Yao, X.; Yang, W. Metal-organic framework nanoparticles with near-infrared dye for multimodal imaging and guided phototherapy. *ACS Appl. Mater. Interfaces* **2019,** *11* (12), 11209–11219.
31. Wang, Z.; Jin, C. M.; Shao, T.; Li, Y. Z.; Zhang, K. L.; Zhang, H. T.; You, X. Z. Syntheses, structures, and luminescence properties of a new family of three-dimensional open-framework lanthanide coordination polymers. *Inorganic Chemistry Communications* **2002,** *5* (9), 642–648.
32. Asefa, T.; Shi, Y. L. *Corrugated and nanoporous silica microspheres: Synthesis by controlled etching, and improving their chemical adsorption and application in biosensing. J. Mater. Chem.* **2008,** *18,* 5604–5614.
33. Cui, Y.; Zhu, F.; Chen, B.; Qian, G. Metal-organic frameworks for luminescence thermometry. *Chem Commun (Camb)* **2015,** *51* (35), 7420–31.
34. Della Rocca, J.; Liu, D.; Lin, W. Nanoscale metal-organic frameworks for biomedical imaging and drug delivery. *Accounts of chemical research* **2011,** *44* (10), 957–68.
35. Peller, M.; Böll, K.; Zimpel, A.; Wuttke, S. Metal-organic framework nanoparticles for magnetic resonance imaging. *Inorganic Chemistry Frontiers* **2018,** *5* (8), 1760–1779.
36. Wang, H.-S. Metal-organic frameworks for biosensing and bioimaging applications. *Coordination Chemistry Reviews* **2017,** *349,* 139–155.
37. Eddaoudi, M.; Kim, J.; Rosi, N.; Vodak, D.; Wachter, J.; O'Keeffe, M.; Yaghi, O. M. Systematic design of pore size and functionality in isoreticular MOFs and their application in methane storage. *Science (New York, NY)* **2002,** *295* (5554), 469–472.
38. Deng, H.; Grunder, S.; Cordova, K. E.; Valente, C.; Furukawa, H.; Hmadeh, M.; Gandara, F.; Whalley, A. C.; Liu, Z.; Asahina, S.; Kazumori, H.; O'Keeffe, M.; Terasaki, O.; Stoddart, J. F.; Yaghi, O. M. Large-pore apertures in a series of metal-organic frameworks. *Science (New York, NY)* **2012,** *336* (6084), 1018–23.
39. Park, K. S.; Ni, Z.; Cote, A. P.; Choi, J. Y.; Huang, R.; Uribe-Romo, F. J.; Chae, H. K.; O'Keeffe, M.; Yaghi, O. M. Exceptional chemical and thermal stability of zeolitic imidazolate frameworks. *Proc Natl Acad Sci U S A* **2006,** *103* (27), 10186–10191.
40. Cote, A. P.; Benin, A. I.; Ockwig, N. W.; Keeffe, M.; Matzger, A. J.; Yaghi, O. M. Porous, Crystalline, covalent organic frameworks. *Science (New York, NY)* **2005,** *310* (5751), 1166.

41. Deng, H.; Doonan, C. J.; Furukawa, H.; Ferreira, R. B.; Towne, J.; Knobler, C. B.; Wang, B.; Yaghi, O. M. Multiple functional groups of varying ratios in metal-organic frameworks. *Science (New York, NY)* **2010**, *327* (5967), 846–850.
42. Ferey, G.; Serre, C.; Mellot-Draznieks, C.; Millange, F.; Surble, S.; Dutour, J.; Margiolaki, I. A hybrid solid with giant pores prepared by a combination of targeted chemistry, simulation, and powder diffraction. *Angew. Chem. Int. Ed.* **2004**, *43* (46), 6296–6301.
43. Ferey, G.; Mellot-Draznieks, C.; Serre, C.; Millange, F.; Dutour, J.; Surble, S.; Margiolaki, I. A chromium terephthalate-based solid with unusually large pore volumes and surface area. *Science (New York, NY)* **2005**, *309* (5743), 2040–2042.
44. Bourrelly, S.; Llewellyn, P. L.; Serre, C.; Millange, F.; Loiseau, T.; Ferey, G. Different adsorption behaviors of methane and carbon dioxide in the isotypic nanoporous metal terephthalates MIL-53 and MIL-47. *Journal of the American Chemical Society* **2005**, *127* (39), 13519–13521.
45. Dan-Hardi, M.; Serre, C.; Frot, T.; Rozes, L.; Maurin, G.; Sanchez, C.; Ferey, G. A new photoactive crystalline highly porous titanium(IV) dicarboxylate. *J. Am. Chem. Soc.* **2009**, *131* (31), 10857–10859.
46. Cavka, J. H.; Jakobsen, S.; Olsbye, U.; Guillou, N.; Lamberti, C.; Bordiga, S.; Lillerud, K. P. A new zirconium inorganic building brick forming metal-organic frameworks with exceptional stability. *J. Am. Chem. Soc.* **2008**, *130* (42), 13850–13851.
47. Grunker, R.; Bon, V.; Muller, P.; Stoeck, U.; Krause, S.; Mueller, U.; Senkovska, I.; Kaskel, S. A new metal-organic framework with ultrahigh surface area. *Chem. Commun.* **2014**, *50* (26), 3450–2.
48. Seo, J. S.; Whang, D.; Lee, H.; Jun, S. I.; Oh, J.; Jeon, Y. J.; Kim, K. A homochiral metal-organic porous material for enantioselective separation and catalysis. *Nature* **2000**, *404* (6781), 982–6.
49. Chen, B.; Xiang, S.; Qian, G. Metal-organic frameworks with functional pores for recognition of small molecules. *Accounts of chemical research* **2010**, *43* (8), 1115–24.
50. Garibay, S. J.; Cohen, S. M. Isoreticular synthesis and modification of frameworks with the UiO-66 topology. *Chem Commun (Camb)* **2010**, *46* (41), 7700–2.
51. Furukawa, H.; Ko, N.; Go, Y. B.; Aratani, N.; Choi, S. B.; Choi, E.; Yazaydin, A. O.; Snurr, R. Q.; O'Keeffe, M.; Kim, J.; Yaghi, O. M. Ultrahigh porosity in metal-organic frameworks. *Science (New York, NY)* **2010**, *329* (5990), 424–428.
52. Farha, O. K.; Eryazici, I.; Jeong, N. C.; Hauser, B. G.; Wilmer, C. E.; Sarjeant, A. A.; Snurr, R. Q.; Nguyen, S. T.; Yazaydın, A. Ö.; Hupp, J. T. Metal-Organic Framework Materials with Ultrahigh Surface Areas: Is the Sky the Limit? *Journal of the American Chemical Society* **2012**, *134* (36), 15016–15021.
53. Morris, W.; Doonan, C. J.; Furukawa, H.; Banerjee, R.; Yaghi, O. M. Crystals as molecules: postsynthesis covalent functionalization of zeolitic imidazolate frameworks. *J. Am. Chem. Soc.* **2008**, *130* (38), 12626–12627.
54. Ferey, G.; Serre, C.; Mellot-Draznieks, C.; Millange, F.; Surble, S.; Dutour, J.; Margiolaki, I. A hybrid solid with giant pores prepared by a combination of targeted chemistry, simulation, and powder diffraction. *Angew. Chem. Int. Ed. Engl.* **2004**, *43* (46), 6296–62301.

55. Chae, H. K.; Siberio-Perez, D. Y.; Kim, J.; Go, Y.; Eddaoudi, M.; Matzger, A. J.; O'Keeffe, M.; Yaghi, O. M. A route to high surface area, porosity and inclusion of large molecules in crystals. *Nature* **2004**, *427* (6974), 523–527.
56. Koh, K.; Wong-Foy, A. G.; Matzger, A. J. A Porous Coordination Copolymer with over 5000 m2/g BET Surface Area. *Journal of the American Chemical Society* **2009**, *131* (12), 4184–4185.
57. Su, X.; Chan, C.; Shi, J.; Tsang, M. K.; Pan, Y.; Cheng, C.; Gerile, O.; Yang, M. A graphene quantum dot@Fe3O4@SiO2 based nanoprobe for drug delivery sensing and dual-modal fluorescence and MRI imaging in cancer cells. *Biosens Bioelectron* **2017**, *92*, 489–495.
58. Mars, A.; Hamami, M.; Bechnak, L.; Patra, D.; Raouafi, N. Curcumin-graphene quantum dots for dual mode sensing platform: Electrochemical and fluorescence detection of APOe4, responsible of Alzheimer's disease. *Anal Chim Acta* **2018**, *1036*, 141–146.
59. Lu, C. H.; Yang, H. H.; Zhu, C. L.; Chen, X.; Chen, G. N. A graphene platform for sensing biomolecules. *Angew Chem Int Ed Engl* **2009**, *48* (26), 4785–7.
60. Chang, L.; He, X.; Chen, L.; Zhang, Y. A novel fluorescent turn-on biosensor based on QDs@GSH-GO fluorescence resonance energy transfer for sensitive glutathione S-transferase sensing and cellular imaging. *Nanoscale* **2017**, *9* (11), 3881–3888.
61. Li, Y.; Su, R.; Xu, J.; Bie, J.; Sun, R.; Wang, L.; Liu, X.; Sun, C. Aptamers-based sensing strategy for 17beta-estradiol through fluorescence resonance energy transfer between oppositely charged CdTe quantum dots and gold nanoparticles. *J Nanosci Nanotechnol* **2018**, *18* (3), 1517–1527.
62. Satishkumar, B. C.; Brown, L. O.; Gao, Y.; Wang, C.-C.; Wang, H.-L.; Doorn, S. K. Reversible fluorescence quenching in carbon nanotubes for biomolecular sensing. *Nature Nanotechnology* **2007**, *2*, 560.
63. Yang, R.; Jin, J.; Chen, Y.; Shao, N.; Kang, H.; Xiao, Z.; Tang, Z.; Wu, Y.; Zhu, Z.; Tan, W. Carbon nanotube-quenched fluorescent oligonucleotides: probes that fluoresce upon hybridization. *J Am Chem Soc* **2008**, *130* (26), 8351–8.
64. Kang, E. B.; Choi, C. A.; Mazrad, Z. A. I.; Kim, S. H.; In, I.; Park, S. Y. Determination of cancer cell-based pH-sensitive fluorescent carbon nanoparticles of cross-linked polydopamine by fluorescence sensing of alkaline phosphatase activity on coated surfaces and aqueous solution. *Analytical chemistry* **2017**, *89* (24), 13508–13517.
65. Wu, J.; Liu, W.; Ge, J.; Zhang, H.; Wang, P. New sensing mechanisms for design of fluorescent chemosensors emerging in recent years. *Chem Soc Rev* **2011**, *40* (7), 3483–95.
66. Hu, Z.; Deibert, B. J.; Li, J. Luminescent metal-organic frameworks for chemical sensing and explosive detection. *Chem Soc Rev* **2014**, *43* (16), 5815–40.
67. Jia, J.; Xu, F.; Long, Z.; Hou, X.; Sepaniak, M. J. Metal-organic framework MIL-53(Fe) for highly selective and ultrasensitive direct sensing of $MeHg^+$. *Chem Commun (Camb)* **2013**, *49* (41), 4670–2.
68. Zou, J. P.; Peng, Q.; Wen, Z.; Zeng, G.; Xing, Q. J.; Zheng, F. K. Two novel metal-organic frameworks (MOFs) with (3,6)-connected net topologies: Syntheses, Crystal Structures, Third-Order Nonlinear Optical and Luminescent Properties. *Cryst. Growth Des.* **2010**, *10* (6), 2613–2619.
69. Wang, G. H.; Li, Z. G.; Jia, H. Q.; Hu, N. H.; Xu, J. W. Metal-organic frameworks based on the pyridine-2,3-dicarboxylate and a flexible bispyridyl ligand: Syntheses, structures, and photoluminescence. *CRYSTENGCOMM* **2009**, *11* (2), 292–297.

70. An, J.; Shade, C. M.; Chengelis-Czegan, D. A.; Petoud, S.; Rosi, N. L. Zinc-adeninate metal-organic framework for aqueous encapsulation and sensitization of near-infrared and visible emitting lanthanide cations. *J Am Chem Soc* **2011**, *133* (5), 1220–3.
71. Jung, S.; Kim, Y.; Kim, S. J.; Kwon, T. H.; Huh, S.; Park, S. Bio-functionalization of metal-organic frameworks by covalent protein conjugation. *Chem Commun (Camb)* **2011**, *47* (10), 2904–6.
72. Morris, W.; Briley, W. E.; Auyeung, E.; Cabezas, M. D.; Mirkin, C. A. Nucleic acid-metal-organic framework (MOF) nanoparticle conjugates. *J Am Chem Soc* **2014**, *136* (20), 7261–4.
73. Zhao, D.; Li, J.; Yang, T.; He, Z. "Turn off-on" fluorescent sensor for platinum drugs-DNA interactions based on quantum dots. *Biosens Bioelectron* **2014**, *52*, 29–35.
74. Vaishnavi, E.; Renganathan, R. "Turn-on-off-on" fluorescence switching of quantum dots-cationic porphyrin nanohybrid: a sensor for DNA. *The Analyst* **2014**, *139* (1), 225–34.
75. Yi, F. Y.; Chen, D.; Wu, M. K.; Han, L.; Jiang, H. L. Chemical sensors based on metal-organic frameworks. *ChemPlusChem* **2016**, *81* (8), 675.
76. Wang, Y.; Hu, Y.; He, Q.; Yan, J.; Xiong, H.; Wen, N.; Cai, S.; Peng, D.; Liu, Y.; Liu, Z. Metal-organic frameworks for virus detection. *Biosens Bioelectron* **2020**, *169*, 112604.
77. Zhu, X.; Zheng, H.; Wei, X.; Lin, Z.; Guo, L.; Qiu, B.; Chen, G. Metal-organic framework (MOF): a novel sensing platform for biomolecules. *Chem Commun (Camb)* **2013**, *49* (13), 1276–8.
78. Fang, J. M.; Leng, F.; Zhao, X. J.; Hu, X. L.; Li, Y. F. Metal-organic framework MIL-101 as a low background signal platform for label-free DNA detection. *The Analyst* **2014**, *139* (4), 801–6.
79. Uhlmann, E. Peptide nucleic acids (PNA) and PNA-DNA chimeras: from high binding affinity towards biological function. *Biological chemistry* **1998**, *379* (8-9), 1045–52.
80. Mejia-Ariza, R.; Rosselli, J.; Breukers, C.; Manicardi, A.; Terstappen, L. W.; Corradini, R.; Huskens, J. DNA detection by flow cytometry using PNA-modified metal-organic framework particles. *Chemistry* **2017**, *23* (17), 4180–4186.
81. Qiu, Q.; Chen, H.; Ying, S.; Sharif, S.; You, Z.; Wang, Y.; Ying, Y. Simultaneous fluorometric determination of the DNAs of Salmonella enterica, Listeria monocytogenes and Vibrio parahemolyticus by using an ultrathin metal-organic framework (type Cu-TCPP). *Microchimica Acta* **2019**, *186*, 93.
82. Zhao, H. Q.; Qiu, G. H.; Liang, Z.; Li, M. M.; Sun, B.; Qin, L.; Yang, S. P.; Chen, W. H.; Chen, J. X. A zinc(II)-based two-dimensional MOF for sensitive and selective sensing of HIV-1 ds-DNA sequences. *Anal Chim Acta* **2016**, *922*, 55–63.
83. Sun, B.; Zhao, H. Q.; Xie, B. P.; Bai, L. P.; Jiang, Z. H.; Chen, J. X. Sequence-specific fluorometric recognition of HIV-1 ds-DNA with zwitterionic zinc(II)-carboxylate polymers. *Journal of Inorganic Biochemistry* **2017**, *176*, 17–23.
84. Xie, B. P.; Qiu, G. H.; Sun, B.; Yang, Z. F.; Zhang, W. H.; Chen, J. X.; Jiang, Z. H. Synchronous sensing of three conserved sequences of Zika virus using a DNAs@MOF hybrid: experimental and molecular simulation studies. *Inorganic Chemistry Frontiers* **2019**, *6* (1), 148–152.
85. Xie, B. P.; Qiu, G. H.; Hu, P. P.; Liang, Z.; Liang, Y. M.; Sun, B.; Bai, L. P.; Jiang, Z. H.; Chen, J. X. Simultaneous detection of Dengue and Zika virus RNA sequences with a

three-dimensional Cu-based zwitterionic metal-organic framework, comparison of single and synchronous fluorescence analysis. *Sensors and Actuators B: Chemical* **2018**, *254*, 1133–1140.

86. Qiu, G. H.; Weng, Z. H.; Hu, P. P.; Duan, W. J.; Xie, B. P.; Sun, B.; Tang, X. Y.; Chen, J. X. Synchronous detection of ebolavirus conserved RNA sequences and ebolavirus-encoded miRNA-like fragment based on a zwitterionic copper (II) metal-organic framework. *Talanta* **2018**, *180*, 396–402.
87. Qin, L.; Lin, L. X.; Fang, Z. P.; Yang, S. P.; Qiu, G. H.; Chen, J. X.; Chen, W. H. A water-stable metal-organic framework of a zwitterionic carboxylate with dysprosium: a sensing platform for Ebolavirus RNA sequences. *Chem Commun (Camb)* **2016**, *52* (1), 132–5.
88. Li, Y.; Yu, C.; Yang, B.; Liu, Z.; Xia, P.; Wang, Q. Target-catalyzed hairpin assembly and metal-organic frameworks mediated nonenzymatic co-reaction for multiple signal amplification detection of miR-122 in human serum. *Biosensors and Bioelectronics* **2018**, *102*, 307–315.
89. Chang, J.; Wang, X.; Wang, J.; Li, H.; Li, F. Nucleic acid-functionalized metal-organic framework-based homogeneous electrochemical biosensor for simultaneous detection of multiple tumor biomarkers. *Analytical chemistry* **2019**, *91* (5), 3604–3610.
90. Yang, S. P.; Chen, S. R.; Liu, S. W.; Tang, X. Y.; Qin, L.; Qiu, G. H.; Chen, J. X.; Chen, W. H. Platforms formed from a three-dimensional Cu-based zwitterionic metal-organic framework and probe ss-DNA: selective fluorescent biosensors for human immunodeficiency virus 1 ds-DNA and Sudan virus RNA sequences. *Analytical chemistry* **2015**, *87* (24), 12206–12214.
91. Song, W. J. Intracellular DNA and microRNA sensing based on metal-organic framework nanosheets with enzyme-free signal amplification. *Talanta* **2017**, *170*, 74–80.
92. Wong, L. S.; Khan, F.; Micklefield, J. Selective Covalent Protein Immobilization: Strategies and Applications. *Chemical Reviews* **2009**, *109* (9), 4025–4053.
93. Bommarius, A. S.; Paye, M. F. Stabilizing biocatalysts. *Chemical Society Reviews* **2013**, *42* (15), 6534–6565.
94. Cui, J.; Ren, S.; Sun, B.; Jia, S. Optimization protocols and improved strategies for metal-organic frameworks for immobilizing enzymes: Current development and future challenges. *Coordination Chemistry Reviews* **2018**, *370*, 22–41.
95. Newton, K.; Dugger, D. L.; Wickliffe, K. E.; Kapoor, N.; de Almagro, M. C.; Vucic, D.; Komuves, L.; Ferrando, R. E.; French, D. M.; Webster, J.; Roose-Girma, M.; Warming, S.; Dixit, V. M. Activity of protein kinase RIPK3 determines whether cells die by necroptosis or apoptosis. *Science (New York, NY)* **2014**, *343* (6177), 1357–60.
96. Zhang, G.-Y.; Zhuang, Y.-H.; Shan, D.; Su, G.-F.; Cosnier, S.; Zhang, X.-J. Zirconium-based porphyrinic metal-organic framework (PCN-222): Enhanced photoelectrochemical response and its application for label-free phosphoprotein detection. *Analytical chemistry* **2016**, *88* (22), 11207–11212.
97. Zhou, N.; Su, F.; Guo, C.; He, L.; Jia, Z.; Wang, M.; Jia, Q.; Zhang, Z.; Lu, S. Two-dimensional oriented growth of Zn-MOF-on-Zr-MOF architecture: A highly sensitive and selective platform for detecting cancer markers. *Biosens. Bioelectron.* **2019**, *123*, 51–58.
98. Beristain, A. G.; Molyneux, S. D.; Joshi, P. A.; Pomroy, N. C.; Di Grappa, M. A.; Chang, M. C.; Kirschner, L. S.; Privé, G. G.; Pujana, M. A.; Khokha, R. PKA signaling drives mammary tumorigenesis through Src. *Oncogene* **2014**, *34*, 1160.

99. Myeku, N.; Clelland, C. L.; Emrani, S.; Kukushkin, N. V.; Yu, W. H.; Goldberg, A. L.; Duff, K. E. Tau-driven 26S proteasome impairment and cognitive dysfunction can be prevented early in disease by activating cAMP-PKA signaling. *Nature Medicine* **2016**, *22* (1), 46–53.
100. Wang, Z.; Yan, Z.; Wang, F.; Cai, J.; Guo, L.; Su, J.; Liu, Y. Highly sensitive photoelectrochemical biosensor for kinase activity detection and inhibition based on the surface defect recognition and multiple signal amplification of metal-organic frameworks. *Biosensors and Bioelectronics* **2017**, *97*, 107–114.
101. Wang, C.; Tang, G.; Tan, H. Pyrophosphate ion-triggered competitive displacement of ssDNA from a metal-organic framework and its application in fluorescent sensing of alkaline phosphatase. *Journal of Materials Chemistry B* **2018**, *6* (46), 7614–7620.
102. Ross, S. A.; Gulve, E. A.; Wang, M. Chemistry and biochemistry of type 2 diabetes. *Chem Rev* **2004**, *104* (3), 1255–82.
103. Xu, Y.; Liu, S. Y.; Liu, J.; Zhang, L.; Chen, D.; Chen, J.; Ma, Y.; Zhang, J. P.; Dai, Z.; Zou, X. *In situ* enzyme immobilization with oxygen-sensitive luminescent metal-organic frameworks to realize "All-in-One" multifunctions. *Chemistry* **2019**, *25* (21), 5334–5334.
104. Zhu, N.; Gu, L.; Wang, J.; Li, X.; Liang, G.; Zhou, J.; Zhang, Z. Novel and sensitive chemiluminescence sensors based on 2D-MOF nanosheets for one-step detection of glucose in human urine. *The Journal of Physical Chemistry C* **2019**, *123* (14), 9388–9393.
105. Wen, Y.; Meng, W.; Li, C.; Dai, L.; He, Z.; Wang, L.; Li, M.; Zhu, J. Enhanced glucose sensing based on a novel composite Co(II)-MOF/Acb modified electrode. *Dalton Trans* **2018**, *47* (11), 3872–3879.
106. Meng, W.; Xu, S.; Dai, L.; Li, Y.; Zhu, J.; Wang, L. An enhanced sensitivity towards H_2O_2 reduction based on a novel Cu metal-organic framework and acetylene black modified electrode. *Electrochimica Acta* **2017**, *230*, 324–332.
107. Zhang, Y.; Xu, J.; Xia, J.; Zhang, F.; Wang, Z. MOF-derived porous Ni_2P/graphene composites with enhanced electrochemical properties for sensitive nonenzymatic glucose sensing. *ACS Appl. Mater. Interfaces* **2018**, *10* (45), 39151–39160.
108. Chen, X.; Liu, D.; Cao, G.; Tang, Y.; Wu, C. *In situ* synthesis of a sandwich-like graphene@ZIF-67 heterostructure for highly sensitive nonenzymatic glucose sensing in human serums. *ACS Appl. Mater. Interfaces* **2019**, *11* (9), 9374–9384.
109. Li, W.; Lv, S.; Wang, Y.; Zhang, L.; Cui, X. Nanoporous gold induced vertically standing 2D NiCo bimetal-organic framework nanosheets for non-enzymatic glucose biosensing. *Sensors and Actuators B-Chemical* **2019**, *281*, 652–658.
110. Beckman, K. B.; Ames, B. N. The free radical theory of aging matures. *Physiological Reviews* **1998**, *78* (2), 547–81.
111. Finkel, T.; Serrano, M.; Blasco, M. A. The common biology of cancer and aging. *Nature* **2007**, *448* (7155), 767–74.
112. Houstis, N.; Rosen, E. D.; Lander, E. S. Reactive oxygen species have a causal role in multiple forms of insulin resistance. *Nature* **2006**, *440* (7086), 944–8.
113. Lin, M. T.; Flint Beal, M. Mitochondrial dysfunction and oxidative stress in neurodegenerative diseases. *Nature* **2006**, *443* (7113), 787–95.

114. Yu, H.; Long, D. Highly chemiluminescent metal-organic framework of type MIL-101(Cr) for detection of hydrogen peroxide and pyrophosphate ions. *Microchimica Acta* **2016**, *183* (12), 3151–3157.
115. Mostakim, S. K.; Banesh, S.; Trivedi, V.; Biswas, S. Selective and Sensitive Sensing of hydrogen peroxide by a boronic acid-functionalized metal-organic framework and its application in live-cell imaging. *Inorganic Chemistry* **2018**, *57* (23), 14574–14581.
116. Luo, F.; Lin, Y.; Zheng, L.; Lin, X.; Chi, Y. Encapsulation of hemin in metal-organic frameworks for catalyzing the chemiluminescence reaction of the H_2O_2-luminol system and detecting glucose in the neutral condition. *ACS Appl. Mater. Interfaces* **2015**, *7* (21), 11322–11329.
117. Yi, X.; Dong, W.; Zhang, X.; Xie, J.; Huang, Y. MIL-53(Fe) MOF-mediated catalytic chemiluminescence for sensitive detection of glucose. *Analytical and Bioanalytical Chemistry* **2016**, *408* (30), 8805–8812.
118. Mao, X.; Lu, Y.; Zhang, X.; Huang, Y. β-Cyclodextrin functionalization of metal-organic framework MOF-235 with excellent chemiluminescence activity for sensitive glucose biosensing. *Talanta* **2018**, *188*, 161–167.
119. Droge, W.; Hack, V.; Breitkreutz, R.; Holm, E.; Shubinsky, G.; Schmid, E.; Galter, D. Role of cysteine and glutathione in signal transduction, immunopathology and cachexia. *BioFactors (Oxford, England)* **1998**, *8* (1-2), 97–102.
120. Yang, N.; Song, H.; Wan, X.; Fan, X.; Su, Y.; Lv, Y. A metal (Co)-organic framework-based chemiluminescence system for selective detection of l-cysteine. *The Analyst* **2015**, *140* (8), 2656–2663.
121. He, J.; Wu, X.; Long, Z.; Hou, X. Fast and sensitive fluorescent and visual sensing of cysteine using Hg-metalated PCN-222. *Microchemical Journal* **2019**, *145*, 68–73.
122. Yildirim, A.; Bayindir, M. Turn-on fluorescent dopamine sensing based on *in situ* formation of visible light emitting polydopamine nanoparticles. *Analytical chemistry* **2014**, *86* (11), 5508–12.
123. Du, Q.; Wu, P.; Dramou, P.; Chen, R.; He, H. One-step fabrication of a boric acid-functionalized lanthanide metal–organic framework as a ratiometric fluorescence sensor for the selective recognition of dopamine. *New Journal of Chemistry* **2019**, *43*, 1291–1298.
124. Nandi, S.; Sharma, E.; Trivedi, V.; Biswas, S. Metal-organic framework showing selective and sensitive detection of exogenous and endogenous formaldehyde. *Inorganic Chemistry* **2018**, *57* (24), 15149–15157.
125. Xu, X.-Y.; Yan, B. Nanoscale LnMOF-functionalized nonwoven fibers protected by a polydimethylsiloxane coating layer as a highly sensitive ratiometric oxygen sensor. *Journal of Materials Chemistry C* **2016**, *4* (36), 8514–8521.
126. Zhang, Q.; Wang, J.; Kirillov, A. M.; Dou, W.; Xu, C.; Xu, C.; Yang, L.; Fang, R.; Liu, W. Multifunctional Ln-MOF luminescent probe for efficient sensing of $Fe^{(3+)}$, $Ce^{(3+)}$, and Acetone. *ACS Appl Mater Interfaces* **2018**, *10* (28), 23976–23986.
127. Qi, Z.; Chen, Y. Charge-transfer-based terbium MOF nanoparticles as fluorescent pH sensor for extreme acidity. *Biosensors and Bioelectronics* **2017**, *87*, 236–241.

128. Du, T. Y.; Zhang, H.; Ruan, J.; Jiang, H.; Chen, H. Y.; Wang, X. M. Adjusting the linear range of Au-MOF fluorescent probes for real-time analyzing intracellular GSH in living cells. *ACS Appl. Mater. Interfaces* **2018,** *10* (15), 12417–12423.
129. Yao, S. L.; Liu, S. J.; Tian, X. M.; Zheng, T. F.; Cao, C.; Niu, C. Y.; Chen, Y. Q.; Chen, J. L.; Huang, H.; Wen, H. R. A ZnII-based metal-organic framework with a rare tcj topology as a turn-on fluorescent sensor for acetylacetone. *Inorganic Chemistry* **2019**, *58* (6), 3578–3581.
130. Tian, J.; Liu, Q.; Shi, J.; Hu, J.; Asiri, A. M.; Sun, X.; He, Y. Rapid, sensitive, and selective fluorescent DNA detection using iron-based metal–organic framework nanorods: Synergies of the metal center and organic linker. *Biosensors and Bioelectronics* **2015**, *71*, 1–6.
131. Wang, X. R.; Du, J.; Huang, Z.; Liu, K.; Liu, Y. Y.; Huo, J. Z.; Liu, Z. Y.; Dong, X. Y.; Chen, L. L.; Ding, B. Anion directing self-assembly of 2D and 3D water-stable silver(i) cation metal-organic frameworks and their applications in real-time discriminating cysteine and DNA detection. *Journal of Materials Chemistry B* **2018,** *6* (28), 4569–4574.
132. Cheng, D.; Xiao, X.; Li, X.; Wang, C.; Liang, Y.; Yu, Z.; Jin, C.; Zhou, N.; Chen, M.; Dong, Y.; Lin, Y.; Xie, Z.; Zhang, C. A non-enzymatic electrochemical sensing platform based on Hemin@MOF composites for detecting hydrogen peroxide and DNA. *Journal of the Electrochemical Society* **2018,** *165* (16), B885–B892.
133. Li, J.; Xia, J. F.; Zhang, F. F.; Wang, Z. H.; Liu, Q. Y. A novel electrochemical sensor based on copper-based metal-organic framework for the determination of dopamine. *Journal of the Chinese Chemical Society* **2018,** *65* (6), 743–749.
134. Pan, Y.; Zhan, S.; Xia, F. Zeolitic imidazolate framework-based biosensor for detection of HIV-1 DNA. *Anal Biochem* **2018,** *546*, 5–9.
135. Shao, K.; Wang, B.; Nie, A.; Ye, S.; Ma, J.; Li, Z.; Lv, Z.; Han, H. Target-triggered signal-on ratiometric electrochemiluminescence sensing of PSA based on MOF/Au/G-quadruplex. *Biosens. Bioelectron.* **2018,** *118*, 160–166.
136. Wang, F.; Chen, X.; Chen, L.; Yang, J.; Wang, Q. High-performance non-enzymatic glucose sensor by hierarchical flower-like nickel(II)-based MOF/carbon nanotubes composite. *Mater. Sci. Eng. C-Mater. Biol. Appl.* **2019,** *96*, 41–50.
137. Wang, Z.; Ma, B.; Shen, C.; Cheong, L. Z. Direct, selective and ultrasensitive electrochemical biosensing of methyl parathion in vegetables using Burkholderia cepacia lipase@MOF nanofibers-based biosensor. *Talanta* **2019,** *197*, 356–362.
138. Chauhan, N.; Tiwari, S.; Narayan, T.; Jain, U. Bienzymatic assembly formed @ Pt nano sensing framework detecting acetylcholine in aqueous phase. *Applied Surface Science* **2019,** *474*, 154–160.
139. Song, W.; Yin, W.; Zhang, Z.; He, P.; Yang, X.; Zhang, X. A DNA functionalized porphyrinic metal-organic framework as a peroxidase mimickingcatalyst for amperometric determination of the activity of T4 polynucleotide kinase. *Microchimica Acta* **2019,** *186* (3), 149.
140. Qu, F.; Li, X. N.; Lv, X. X.; You, J. M.; Han, W. L. Highly selective metal-organic framework-based sensor for protamine through photoinduced electron transfer. *J Mater Sci* **2019,** *54* (4), 3144–3155.
141. Huang, X.; He, Z.; Guo, D.; Liu, Y.; Song, J.; Yung, B. C.; Lin, L.; Yu, G.; Zhu, J. J.; Xiong, Y.; Chen, X. "Three-in-one" nanohybrids as synergistic nanoquenchers to enhance no-wash

fluorescence biosensors for ratiometric detection of cancer biomarkers. *Theranostics* **2018**, *8* (13), 3461–3473.

142. Wang, X. R.; Huang, Z.; Du, J.; Wang, X. Z.; Gu, N.; Tian, X.; Li, Y.; Liu, Y. Y.; Huo, J. Z.; Ding, B. Hydrothermal preparation of five rare-earth (Re = Dy, Gd, Ho, Pr, and Sm) luminescent cluster-based coordination materials: The first MOFs-based ratiometric fluorescent sensor for lysine and bifunctional sensing platform for insulin and Al^{3+}. *Inorganic Chemistry* **2018**, 57 (20), 12885–12899.
143. Qin, J.; Cho, M.; Lee, Y. Ferrocene-Encapsulated Zn Zeolitic Imidazole framework (ZIF-8) for optical and electrochemical sensing of amyloid-beta oligomers and for the early diagnosis of Alzheimer's disease. *ACS Appl. Mater. Interfaces* **2019**, *11* (12), 11743–11748.
144. Cui, L.; Hu, J.; Li, C. C.; Wang, C. M.; Zhang, C. Y. An electrochemical biosensor based on the enhanced quasi-reversible redox signal of prussian blue generated by self-sacrificial label of iron metal-organic framework. *Biosens. Bioelectron.* **2018**, *122*, 168–174.

Chapter 2

Mechanistic Advances of Metal–Organic Frameworks Assisted Chemical Sensors

Bidyut Kumar Kundu[*,1]

[1]**Department of Chemistry, School of Applied Sciences, Centurion University of Technology and Management, Bhubaneswar 752050, Odisha, India**

[*]**Email: bidyut.kundu@cutm.ac.in**

Metal-organic frameworks (MOFs) are a mesmerizing set of enormously porous materials constituted of organic linkers and metal ions or clusters that assured excessive potential in many areas. Now a days, the implementation of MOFs as chemical sensors have been widely surveyed because of their exclusive physical and chemical characteristics, crystallinity, ultrahigh surface areas, superior charge/energy transfers ability, structural diversity, and tunable porosity. In addition, the opportunity of postsynthetic modification of its synthetic variability adds an extra dimension. In general, MOFs are assumed one of the most perfect materials for chemical sensing as its properties are primarily controlled by surface reactions. Despite numerous advances, more functional chemical sensors are required to screen ecological conditions, to understand about controlling food quality, quantification of analytes as well as an early diagnosis of disease. Herein, this chapter will be devoted to highlight some of topical exploration in this emerging research field including the backgrounds needing to act MOFs as chemical sensors, numerous luminescent MOF sensors, probable sensing mechanism(s), and diverse sensing targets e.g., temperature, gases, biomolecules (such as protein, DNA, RNA, hormone, antibody, *etc*), explosives, pH, humidity, environmental pollutants, organic molecules, and/or ions.

1. Introduction

A type of porous coordination polymers (PCPs) recognized as Metal-organic frameworks (MOFs), through the past two decades, have practiced explosive development (*1–3*). Due to its special characteristics, like structural diversity, flexibility and adaptability, huge surface area, high porosity, and excellent affinity for adsorption, MOF has shown a broad area of possible applications, which include luminescence, catalysis, separation, gas storage, and drug delivery (*4–6*). In a similar context, zeolite is one of the utmost distinguished recognitions of ordered microporous/nanoporous materials, and has been extensively researched and used in numerous industrial and households' arenas (*7*, *8*). Petrochemical industry are the main targets for them to usually act as catalysts such as molecular sieves for the basis of separation, and ion exchangers in detergents accentuating the universality of the demand for this type of materials (*9*, *10*). Nevertheless, definite boundaries like their structural rigidity or inadequate pore size inspired the progress of another porous materials e.g., MOFs or ordered mesoporous materials (*11*). Apart from the activity toward heterogeneous catalysis, separation, and gas storage ability, zeolites are not sufficient to explore sensing application. Thus, novel frameworks urgently needed to overcome such disputes. The revelation of new structures and on the tailored construction of these materials have been focused extensively during the last decades for a target application *viz.* chemical sensing, which zeolites cannot be able to properly address. In these days, their applications as host materials for chemical sensing or drug delivery has also fascinated rising consideration (*12*).

In various MOF applications, MOF-based chemical sensors undoubtedly show promise as MOF can express various levels of luminescence amplification or extinction in order to mention the interaction between the frameworks and the inserted guest species (*13*). The salient characteristics of MOFs are: (i) Improve the detection sensitivity to concentrate analytes to high levels by its large surface areas; (ii) Definite working sites (e.g., Lewis acid/base sites, open metal positions, and custom pore sizes) can be specific through host-host recognition or size segregation with unique selectivity; (iii) Tunable frameworks or pores, which activate to the substrates for reversible release and uptake to upsurge rejuvenation and recycling.

Figure 1. Schematic demonstration of the key components of a MOF-based chemical sensor.

Numerous approaches for the operation of a precise transduction-recognition method were established and applied, herein providing to our better conception toward molecular sensing. This adding up and recognition also execute the rising toolbox for forthcoming design of chemical sensor. Chemical sensor possesses five key components such as receptor, transducer, amplifier, signal processor followed by a display unit. In chemical sensors, transducer converts the optical/ electronic/ electrochemical activity into corresponding energy signal, which can be displayed as a form of graph in the detection unit e.g., spectrophotometer, voltammetry, circular dichroism, mass spectrometer, and so on (Figure 1).

In this chapter, thus, will be described some of contemporary studies in this emerging research arena including the important literatures reported to act MOFs as chemical sensors, various MOF sensors having luminescent properties, plausible sensing mechanism(s), and miscellaneous sensing targets like temperature, humidity, small molecules, explosives, biomolecules, atmospheric gases/ VOCs, pH, antibiotics, ions, and so on. This starts with designing MOFs with desirable properties to examine the current challenges linking to MOF sensor progresses including, applications of MOF-based chemical sensors, corresponding plausible sensing mechanism(s) followed by numerous important reports published recently. Also, special emphasis has been given to the photoluminescence (more specifically fluorescence) driven MOF sensors for its easy and broad applicability.

2. Design of MOF Sensors

Although it has not been used systematically, the special customization of MOFs structures and properties should provide greater advantages over other types of chemical sensor materials. Selective and sensitive vapor phase and gas phase detection of analytes is required to address a wide range of real-life challenges including environmental monitoring, occupational safety, medical diagnostics, chemical threat detection, industrial process management, and food quality control. Most sensors are based on organic-polymer or inorganic semiconductor films, which absorb or react with analyte species. The variations in mechanical, electrical or photo physical behavior of these films are usually supervised. The magnitudes of the modifications indeed rely strongly on analyte concentration and the physical and chemical properties of analytes, for example, acidity or alkalinity, tendency to donate and accept electrons, and the capability to penetrate and swell the film. Although many chemical sensors have been commercialized positively, they must still be improved. Therefore, the creation of MOF-based chemical sensors requires extremely effective design strategies, which relate to the selection of organic core, metal ions, linker(s) with suitable functionalization, and the exact entities for the post synthetic modifications (Figure 2).

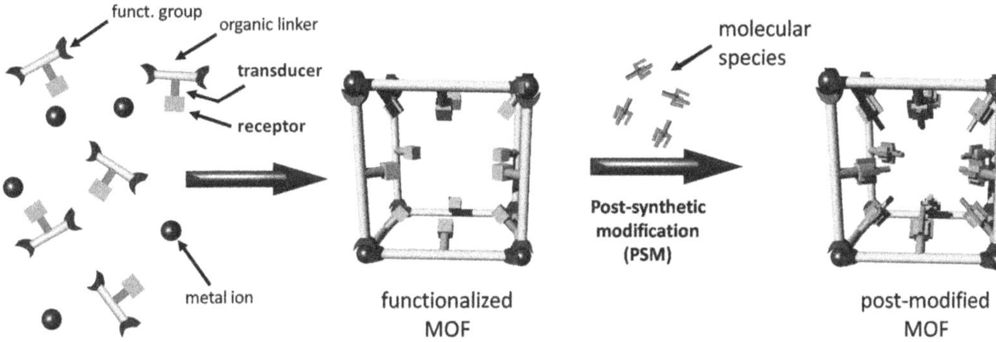

Figure 2. Schematic representation of the modular synthetic route for MOFs followed by post synthetic modifications.

Selectivity, sensitivity, response time, recyclability and material stability are important factors that must be considered to optimize the accomplishment and usefulness of chemical sensors. Highly porous materials (namely MOF) must be intrinsically susceptible to the detection of gases or vapors as these can successfully concentrate analyte substances at a concentration more than that in the outside atmosphere. Indeed, numerous field-portable analytical instruments with state-of-the-art properties, which are utilized by military organizations for the detection of chemical threats include a major step for sample preparation, in which porous absorbent materials are used to concentrate the analyte. These portable systems would simplify by straight analyte identification from the sorbent. Although the sensitivity partly depends on the signal-transduction method, it hugely relies on the kinetics of analyte transportation in the MOF and on the binding strength of the analyte to the MOF (the low limit of detection signifies stronger binding). Very slow transmission will result in long response time, which is tough to differentiate from the baseline drift. The best signal observed should be contingent partly on the capacity of MOF sorption.

3. Applications of MOF-Based Chemical Sensors

The different inherent characteristics of MOF, chemical adaptability and structural diversity make the scope MOF almost unlimited, which is subjected to MOF for in-depth research. In general, the modular structure makes it possible to adjust pore sizes, change the shape of the material and execute various modifications within the frameworks. Therefore, the synergistic effect of the organic and inorganic constituents in the MOF furnishes fascinating properties for this type of porous solids. As a consequence, as described below, this hybrid material is considered an optimistic candidate for countless target applications.

3.1. Ion Sensing

MOFs platforms with diverse structures have developed for their applications into the next generation of multifunctional hybrid materials. One of the most important properties of MOFs is their luminescence response, which may be caused by the accumulation of ions (like metal ions, anions) followed by emission to produce signals. These luminescent characteristics MOF has been utilized as an excellent platform for the development of fluorescence sensors. Its chemically adaptive framework with specific host-host interaction plays a significant role in the selective detection of ions.

3.1.1. Cation Sensing

Many transition-metallic cations are indispensable in living organism, consisting of iron(III) (*14–26*), copper(II) (*27–37*), and zinc(II) (*38–41*), however, a few are poisonous and might even purpose critical harm to health and the environment, e.g., lead(II), mercury(II), and chromium(VI) (*42–49*). The direct, effective, and real-time identification of the trace quantity of transition metal cation is very useful and necessary. CPP-16 MOF particles having pyrene-functionalization can selectively identify Cu^{2+} from a series of metal ions by generating a "*turn-off*" signal (Figure 3). MOFs can behave as an antenna and host to give a fixed scaffold, which can defend cations from being quenched by solvents also moreover to make them sensitive to create tunable fluorescence. The correlation among the of Fe^{3+} concentration and the quenching impact could not be properly geared up with the aid of using the Stern-Volmer (SV) equation that represented the accession of static and dynamic quenching progressions. A EuMOF, $[H_2N(CH_3)_2][Eu(H_2O)_2(BTMIPA)]\cdot 2H_2O$ (H_4BTMIPA=5,5′-methylenebis(2,4,6-trimethyliso phthalic acid)), was built by Sun *et al.*, which could be accomplished to sense Al^{3+} and Fe^{3+} ions selectively among many other ions by the help of the quenching and enrichment of fluorescence, correspondingly, established on the reason of the exchange of cation between metal ions, and $[H_2N(CH_3)_2]^+$ ions (*15*). Again, more luminescence "*turn-on*" signals were detected for the other metal ions, *viz.* Al^{3+} (*50–54*). Mg^{2+} (*53, 55, 56*), Ca^{2+} (*57*), Cr^{3+} (*58*), Mn^{2+} (*54*), Zn^{2+} (*38–41*), Cd^{2+} (*46–49*), Ag^+ (*59–61*), Hg^{2+} (*42–45*), and Pb^{2+} (*62, 63*).

3.1.2. Anion Sensing

MOF-functionalized chemical sensors can proficiently distinguish anions, for example, Cl^-, Br^-, I^-, SCN^-, and N_3^- by the anion-exchange procedure. The free counterions can simply be replaced by target anions, with the corroboration of the representative colors of the analytes (*64, 65*). Zhao *et al.* prepared two cationic heteronuclear MOFs for the identification of I^- ions in water medium with rapid response in luminescence spectra (*66*). Additional researches show that the "*turn-off*" response can be enlightened *via* I^- oxidation through the MOF struts, where the produced I_3^- ions absorb the excitation energy and/or the light to block the ligand-to-metal energy transfer (EnT) of the linker's excited state. The fluorescent "*turn-on*" identification of F^- was accomplished by Lobkovsky *et al.*, by means of a lanthanide-based MOF, [Tb(BTC)].(CH_3OH) (H_3BTC = 1,3,5 benzenetricarboxylic acid), in which the F^- ions along with the terminal CH_3OH molecules, produce strong hydrogen bonds, resulting a high increment of the fluorescence intensity (*67*).

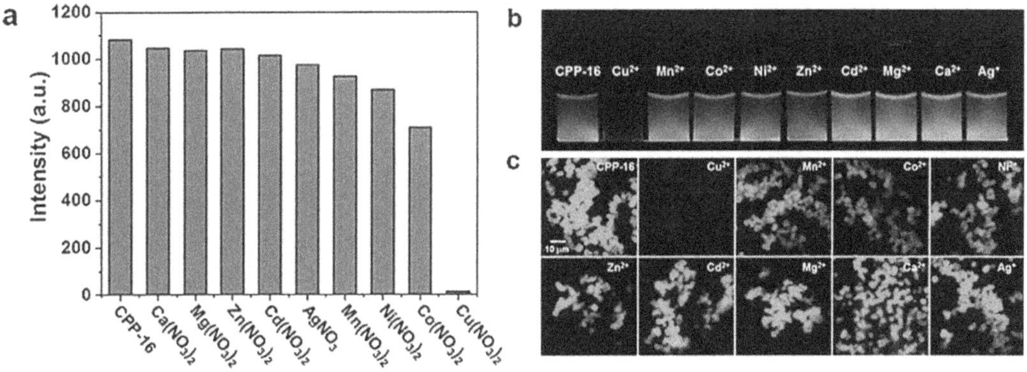

Figure 3. (a) Subsequent reaction with several metal ions, MOF namely CPP-16 changes its fluorescence intensity, (b) UV irradiated digital images of the MOF sensor before and after the reaction with metal ions, and (c) fluorescence microscopic photographs of CPP-16 with and without addition of selected metal ions. Adapted with authorization from Ref. (27). Copyright 2014 American Chemical Society.

The MOF compound NH2-MIL-101 (aluminum, Al) (where, 'MIL' is named as Lavoisier Institute material) initially combines with the dye *viz.* fluorescein in the hole and forms a non-luminous host-guest organization (*68*). After that, the F^- ions coordinate insistently with the Al clusters, which leads to the MOF skeleton decomposition followed by the release of the luminescent dyes. Many other anions, namely, SO_4^{2-}, PO_4^{2-}, $Cr_2O_7^{2-}$, and MnO_4^-, CO_3^{2-} have also been efficaciously sensed by MOFs. A very low limit of detection (1.2 μm/0.03 ppm) could be seen in the presence of toxic CN^- anion in aqueous DMF while incorporating with the heterobimetallic luminescent MOF sensor (Figure 4) (*69*). Moreover, the toxic pollutant anions, for example, OAc^-, ClO_4^-, BF_4^-, CO_3^{2-}, SO_4^{2-}, NO_3^- and halides can be captured by the framework over ion exchange, which result unusually greater luminescent quenching in comparison to the other anions (*70, 71*).

3.2. Explosives Sensing

The identification of explosives (here, analytes) by the help of MOF-mediated sensors are typically executed in a '*turn-off*' style, where the quenching of luminescence of the MOFs can be accomplished *via* photo-induced energy or electron transport of analytes. Most of the luminescent MOFs (LMOFs) having higher-energy conduction bands mould it happen for the transfer of electron to the LUMOs of an electron-inadequate analyte (*72*). The overlay between the fluorescence emission of the MOF sensor and the excitation wavelength from the UV-Vis of analyte indicates the existence of long-distance EnT (*73, 74*). Also, electrostatic interaction plays a significant character in the detection of explosives that is common in the nitrophenol-assisted explosives e.g., 2,4,6-

trinitrophenol (TNP or PA), and/or 2,4,6-trinitrotoluene (TNT) (75). Through SV equation (76–79), the quantitative evaluation of quenching constant (Ksv in M^{-1}), and the limits of detection, which can provide the sensing efficiency of LMOFs for an analyte e.g., explosives (*80*).

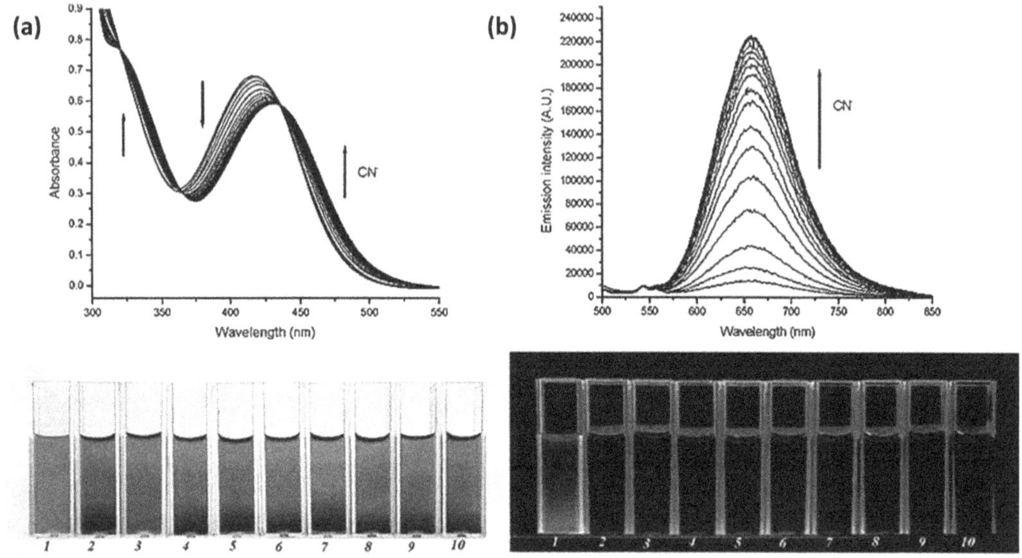

Figure 4. (a) Responses to the UV-Vis spectroscopy, and (b) Fluorescence titration outcomes and luminescent responses. Colorimetric visuals are in the bottom. Modified with agreement from Ref. (69). Copyright 2004 American Chemical Society.

As represented in Figure 5, all nitro-containing compounds were shown to respond in fluorescence quenching to various degrees while unveiled to a series of aromatic analytes (72). The quenching efficiency follows the order: nitrobenzene > m-dinitrobenzene > nitrotoluene ≈ p-dinitrobenzene > DNT, when a LMOFs, namely, $[Zn_2(oba)_2(bpy)]$.DMA [where, DMA = N,N0-dimethylacetamide, bpy = 4,40-bipyridine, H_2oba = 4,40-oxybis(benzoic acid)] was introduced for detection and quantification of nitroaromatics explosives.

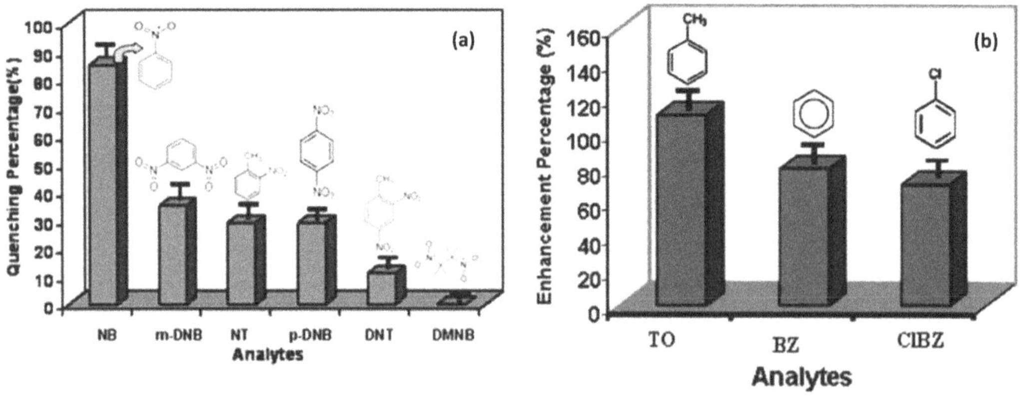

Figure 5. In a luminescent MOF (a) Percentage luminescence quenching by electron-deficient nitro-explosives, and (b) % luminescence enrichment by electron-sufficient analytes. Adapted with consent from Ref. (72). Copyright 2011 American Chemical Society.

Notably, *"turn-on"* recognition of explosives or explosive-like analytes is less frequent however extremely required because of their quick response intrinsic in visible indicators. For example, Wang *et al.* described a panel of MOFs, labeled as TABD-MOF-1, -2, and -3, for the *"turn-on"* identification of active heterocyclic molecules (*81*). This tendency is due to the vapor pressure of the analyte and its electron withdrawing affinity. Dinitrobenzene has high electron deficiency, so it should possess good oxidative quenching ability, but its vapor pressure is low. Nitrobenzene having smaller number of the NO_2 group, gives a higher quenching efficiency probably due to the comparatively more vapor pressure. Nitrotoluenes having deficiency in electrons because of the electron-donating methyl groups, so the quenching efficiency is very low.

3.3. Volatile Organic Compounds (VOCs)

VOCs are the kind of compounds, in which vapor pressure is low enough to evaporate under general atmospheric conditions. Considering the injurious properties of these substances on environment and the human health, it is our unmet prerequisite to detect these substances whenever needed. However, a series of MOFs have been reported as VOC sensors, the arena is still challenging and fascinating because the sensing properties of most MOF frameworks are solution-mediated; i.e., by suspending the MOF in the VOC medium, the luminescence performance can be examined. Although to determine the lower concentration(s) of the analyte in the gaseous phase still very demanding. Cheng *et al.* proposed a 'bimetallic-luminescent center' approach to identify dioxane in glycol (*82*). In addition, to combining the both lanthanide ions in MOF-framework as bimetallic luminescence centers, a self-calibrating sensor can also be realized by capturing the fluorescent dyes in the cages of LMOFs to create a dual emission motif. An emission-fingerprint method was also introduced by Wu *et al.*, with an inter-reference dye@MOF organization for VOCs probing (*83*).

Preferably, the measurements of VOCs are highly accomplished in the gaseous phase, that signifies, a rapid signal to the vapors of analyte, which should be the minimum criteria to act as a practical sensor. Dong *et al.* described a Cu(I)-mediated MOF, *viz.* $[Cu_2(ligand1)_2I_2].(H_2O)_x$ (where, 'ligand1' is 1-benzimidazolyl-3,5-bis(4-pyridyl)benzene), that possesses better capability of sensing with regard to the polar aliphatic VOCs, like halocarbons. Ketones, and alcohols (*84*). Real-time acetonitrile vapor sensing was also reported by Li *et al.*, with preparing an intertwined MOF, $[(CuCN)_3(ligand2)]$.(guests)x (ligand2 = 2,6-bis((3,5-dimethyl-1H-pyrazol-4-yl)methyl)-pyridine) (*85*). The twofold interpenetration of the networks can also be introduced by the help of structural dynamics of the MOF, that permits the entangled frameworks while catching several guest materials in the indication of alteration in emission.

3.4. Gas Sensing

Owed to the significance of atmospheric monitoring, MOF-based chemical sensors for gas molecules, e.g., O_2 (*86–90*), CO_2 (*91, 92*), NH_3 (*93–97*), NO (*98, 99*), H_2S (*100–103*), and between others (*94, 104*), were established. In this regard, three heterobimetallic porous coordinated frameworks were constructed by Lin *et al.*, presenting a ZnMOF in association with $[Ir(ppy)_3]$ (ppy = 2-phenylpyridine) compounds to result very effective phosphorescent analogous (*86*). Likewise, Ho *et al.* reported four iridium(III)-lead(II) heterobimetallic systems for the quick detection of O_2 with 24 s for recovery time of and 50 s as the response time (*90*). A fluorescent material *viz.* the methyl viologen cation (MV^{2+}) was utilized Lu *et al.* to fabricate the CdMOF, that could firmly and specifically sense O_2 within 5 s by developing a detectable color switching by the open eye (*89*).

Although the real-time detection of gaseous CO_2 molecules is very tough in the existence of a competing gases e.g., O_2 or vapor of water. Carrying the reaction between carbon-di-oxide and the selected CDMOF-2 matric, Jeong *et al.* developed a γ-cyclodextrin-derived MOF having many hydroxyl groups (*92*). Further, to understand the real-time identification of H_2S fluctuations in living physiological samples *in situ*, Tang *et al.* offered the porous nanoscaled CuMOF having an excellent selectivity, fast response, and hypotoxicity (*102*). All test outcomes described that the H_2S-responsive biocompatible fluorescent NMOF probe can provide an excellent chance for the progression of probe-derived luminescence imaging methodologies. For example, the malonitrile-based ZIF-90 (MN-ZIF-90) planned and manufactured recently, for the identification of H_2S quantitatively inside HeLa cells by the help of substantial fluorescence augmentation (*101*). Ghosh *et al.* synthesized a nitro group assisted MOF (UiO-66@NO_2), which exhibited a rapid and discriminatory "*turn-on*" response in fluorescence toward H_2S under biological environments, even in the company of potentially impeding chemical substances (*100*).

The same group (*100*) also reported another amine-functionalized fluorescent MOF UiO-66@NH2 (*99*), that displayed a quick response for NO with good sensitivity and selectivity in an bubbly water solution. Dinca *et al.* achieved the fluorescence "*turn-on*" NH_3 detection at an elevated temperature by using MgMOF and ZnMOF sensors, which demonstrated a prominent shift in the fluorescence maxima (*96*). Moreover, a traditional semiconducting was reported by using ZIF-67 to record the fast electrochemical response of formaldehyde with a low limit of detection of 5 ppm at 150 °C (*94*).

3.5. Temperature Sensing

In recent years, mixed metals Ln-derived MOFs have been widely studied as a thermometer that can measure temperature changes in a desired temperature range in proportion. Theoretically, the temperature changes will perturb the EnT in mixed lanthanides and cause variations in photoluminescence characteristics. For example, MOF-thermometer [$Eu_{0.0069}Tb_{0.9931}$-DMBDC] (where, H_2DMBDC = 2,5-dimethoxy-1,4-benzenedicarboxylic acid) was fabricated by Chen *et al.*, which demonstrates an immediate response to temperature from 10-300 K (*105*). A mixed Ln-mediated MOF, namely [$Eu_{0.7}Tb_{0.3}$(D-cam)(Himdc)$_2$(H_2O)$_2$] (where, D-H_2cam = D-camphoric acid), was explored by Du *et al.*, that is proficient of probing the temperature in the range of 100-450 K (*106*). A novel Ln-derived MOF [$Tb_{0.80}Eu_{0.20}$BPDA] (where, H_2BPDA = biphenyl-3,5-dicarboxylic acid) was designed by Qian *et al.*, that is appropriate for the sensing of temperature in the biological width of 298-318 K (*107*). Apart from that, the researchers also establish the impact of the distribution in energy-level on the luminescence properties with the variation of the organic linkers.

Notably, by luminescent guest substances doping into the cavities of the MOF matrix, ratiometric thermometers can also be accomplished. At a glance, Qian *et al.* constructed a new MOF-functionalized thermometer, ZJU-88 by using the changes of the emission of both Eu^{3+} cations and guest species e.g., perylene compounds for the successful sensing of temperatures in physiological region (*108*). The reported ZJU-88-perylene high sensitivity toward minor temperature aberrations of some biological cells such as brain tumor cells, which suggest that this MOF can be potentially applied for such biomedical diagnosis (*109*).

3.6. Humidity Sensing

MOF-functionalized sensors work through changing electrochemical signals or luminescence for the successful measurements of humidity (*110–120*). An open eye colorimetric sensor *viz.* CuMOF was reported by Dong *et al.*, for the quantification of humidity (*110*). Beyond 33% relative humidity (RH), the bright yellow color of these minerals transformed to brown-red and due to the non-radiative energy transition, after different time, the encapsulated water molecules gradually "*turn-off*" the luminescent intensity. The heteroatom-enriched mixed-ligand model of oxalic acid and pyridyl-4,5-imidazole dicarboxylic acid was also used to synthesize another LnMOF to act as the humidity sensor (*111*). Using the concept of OH vibrational quenching, the fluorescence on and off switching of Ln^{3+} emitters can be administered by adsorbing and desorbing the aqueous molecules.

Robinson *et al.* (*112*) and Qiu *et al.* (*113*) prepared a thin film of MOF 'HKUST-1' to be implemented for the water vapor detection. Furthermore, a technique of homogenous nucleation was utilized by Qiu *et al.* to formulate a large-scale copper slice polished steady thin film of HKUST-1 (*113*). The humidity capacitance signal can also be measured by the help of the film sensor unveiled high sensitivity and repeatability having RH values of 11.3, 32.8, 57.6, 75.3, and 84.3%, with a quick retort and salvage period. A MOF-derived electrochemical sensor by increasing a Cu-BTC film was prepared by Robinson *et al.* on a surface acoustic wave (SAW) through a layer-by-layer (LBL) development method to accomplish the identification of vapor of water at concentrations of sub-ppm level (*112*). Besides, a MgMOF has been used as an efficient sensor by Lazarides *et al.*, which is proficient of quickly identifying water traces (0.05-5% v/v) in various organic mediums with the development of fluorescent "*turn-on*" sensing mechanism (*115*).

3.7. Biomolecule and Antibiotics Sensing

Many LMOFs having countless porous assemblies and operationalized linkers, have also been considered to sense antibiotics. One of the key agricultural pollutants is Aflatoxins B1 (AFB1), which directly harm to food such as peanuts or corns, should be detected at low level by introducing a sensor. A limit of detection up to 46 ppb along with the superior efficiency of the luminescence quenching up to 54227 M^{-1} has been demonstrated by MOF-241, $[Zn_2(BPDC)_2(TPPE)]$.(solvents) (where, TPPE = 1,1,2,2-tetrakis(4-(pyridin-4-yl)phenyl)ethane) (*121*), to selectively sense AFB1. Coming to the sensing mechanism, the huge frameworks permitted MOF-241 to capture AFB1 and an efficient interaction to the analytes, that subsequently paved to the outstanding sensing accomplishment. As most antibiotics or pollutants are naturally present in water, the stability of LMOFs in aqueous medium is necessary to follow the sensing mechanism. Zhou *et al.* conveyed two water-stable Zr-functionalized MOFs, which can not only sense explosives and antibiotics in ppb level, but also can effectively bring out the pollutants from the aqueous media (*122*).

More importantly, MOF-based sensors can be successfully implemented to detect biomolecules like DNA and RNA. In a similar context, to sense DNA sequences, a MOF *viz.* UiO- 66-NH_2, also utilized by Jiang *et. al.* (*123*), 5-carboxyfluorescein (5-FAM) fluorophore based free single-stranded DNA (ssDNA) can attribute to the MOF frameworks to produce probe-DNA complex *via* hydrogen bonds and $\pi \rightarrow \pi^*$ stacking interaction in a fluorescence "*turn-off*" manner. While incubated in supporting target DNA, ssDNA breaks out from the MOF channels and produces a double-stranded emissive DNA with the sequence of target, Chen *et al.*, subsequently reported the recently synthesized Dy-functionalized MOF for RNA detection via a similar mechanism (*124*). In addition, the luminescence of the corresponding probe-ssDNA efficaciously gets better by ebolavirus RNA sequences connected to the MOF stage, allowing the limit of detection up to 160 pM.

3.8. pH Sensing

The pH drives an important character in regulatory practice of cells. Luminescence-mediated pH probes are of unmet need because they can quickly track minor changes in pH in the physiological conditions and generally do not damage cells (*106*, *125–134*). MOFs (e.g., $[Eu_3(C_{14}H_6N_2O_4)_4(OH)(H_2O)_4]\cdot 2H_2O$), which demonstrate smooth luminescence responses at different pH values, principally in the biological range of 5 to 7.5, are irreplaceable because of their superior potential as pH probes in living cells or even in aqueous medium (*125*). In the case of these LMOFs, which are appropriate for pH measurement, the mechanism of sensing is mainly related to the protonation/deprotonation processes of the linkers, particularly those that accommodate nitrogen atoms in their struts (*135*). A porphyrinic MOF, *viz.* PCN-225, having high water stablity was by reported Zhou *et al.* to measure the pH values in a wide region of 0-10.2 (*128*). Particularly, the luminescence emission in the pH region of 7-10, displayed a close relationship with the pH changes. In addition, one more water-stable Zr porphyrin MOF, *viz.* PCN-222 was accomplished for visualization pH sensing over the range of 0-6 composed of a discrete reversible colorimetric and luminescent "*turn-on*" or "*turn-off*" pH signal (*127*). Additionally, the postsynthetic modification (PSM) of UiO-66-NH_2UiO-66-N=N-ind was applied to synthesize UiO-66-N=N-ind_{1h} and UiO-66-N=N-ind_{3h} frameworks by a diazotization approach. The UiO-66-N=N-ind_{3h} model revealed a pH-assisted luminescence intensity over a broad range of 1-12, relative to a narrower range of 1-9 for UiO-66-NH_2 (*129*). A NMOF *viz.* FUiO in covalent conjugation with fluorescein isothiocyanate (FITC), was first introduced by Lin *et al.* to provide extraordinary real-time pH sensing (intracellular) response inside living cells (*130*). FUiO also gave an outstanding method to examine NMOF-cell interactions.

4. Plausible Sensing Mechanism

A chemical sensor is defined as an independent analytical device which can deliver evidence regarding the chemical composition and its surrounding, if it is the gas or liquid phase to the outer sphere of the device. Luminescent chemical sensors based on MOFs usually have vaporchromism or solvatochromism, and generate optical signals which are guest-dependent. Therefore, they provide detection of certain chemical entities (for example, gases, VOCs, explosives or ion species), through a more direct and convenient method. By immersing the powdered material in a solution containing the analyte or exposing the material to pure gas or vapor, the detection process can be simplified.

4.1. Solvatochromism

A visible transformation in a material's color is one of the modest, and possibly most prevailing revenues to transduce a sensing signal. In response to an alteration in the identity of the solvent, a large shift to the absorption spectrum of a material is described by solvatochromism. The shifts are usually related to the empirical measurement of solvent polarity, which indicates that the electronic transition accountable for coloring involves in transferring charges which means upon excitation from the ground state of the chromophoric constituent to its excited electronic state of the material a change in dipole moment is taking place. When the excited state possesses more dipole moment in comparison to the ground state, it is favorably sustained by polar solvents, and with increasing solvent polarity red shifts or bathochromic shifts are detected. Again, if the ground state has more dipole moment than that of excited state, then with increasing the solvent polarity hypsochromic shifts (blue shifts) take place. For chromophores embedded in MOFs (like a linker), vapor sorption corresponds to solvate a molecular absorber, and vapochromic drifts should correspond to the solvatochromic effect.

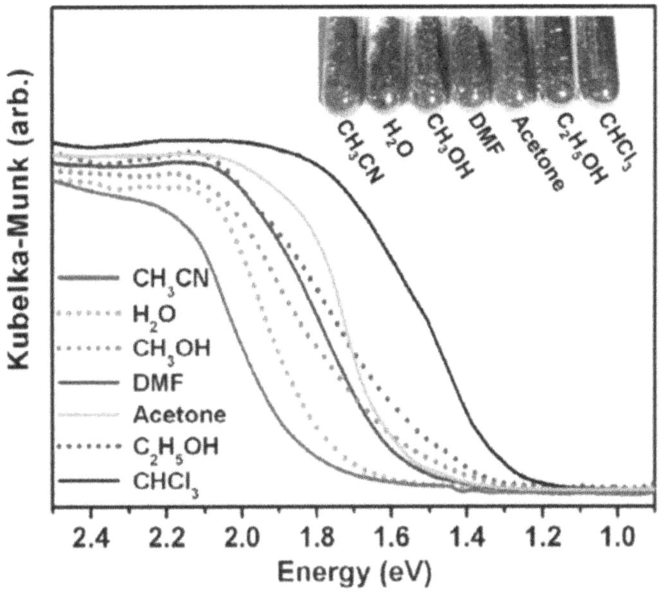

Figure 6. Colorimetric photograph and the visible spectra of MOF crystals in presence of organic as well as aqueous solvents (vapors in condensed phase). Both hydroxylic and nonhydroxylic solvents vapors show a negative vapochromic effect. Reprinted with authorization from Ref. (136). Copyright 2011 American Chemical Society.

At a glance, a 3,6-di(pyridin-4-yl)-1,2,4,5-tetrazine (dptz) ligand based Cu^{2+} MOF was designed and synthesized by Lu *et al.* that demonstrates solvatochromic performance while immersing in solvents in polarity range from chloroform to water (Figure 6) (*136*). Generally, with increasing solvent polarity, a negative solvatochromic consequence is shown by the material having the blue-shifted absorption band. When the calculated band gaps are drawn against the polarity of the solvent, two different groups are observed, one for hydroxy solvents and another for non-hydroxy solvents.

Along with general solvatochromic properties related to the specific solvent-chromophore interactions and solvent polarity, MOFs have showed solvent-associated color variations due to another mechanisms. Moreover, MOFs have features related to unique coordination complexes, as it contains metal ions with their organic components. Also in MOF sensing, the variations in the coordination domain of those metal centers show a vital character. Long *et al.* earlier showed that exposing Co^{2+} MOFs to several vapors could help to move the wavelength maxima across the visible region (*137*). The tetrahedral geometry from octahedral one in coordination environment described a change to this explanation.

4.2. Luminescence

For a long time, luminescent materials have fascinated for researchers and ordinary people. Luminescent materials are such species, which emit energy by means of thermal radiation, also emit light when excited by external stimuli (like UV light, electron beams, X-rays or even mechanical phenomena). Therefore, blocking non-emissive pathways is the key to luminescence in the material, which is usually related to the construction of a rigid scaffold and the prevention of molecular vibrations passing through the material structure (Figure 7). In recent times, luminescent materials have received extensive attention in the field of biomedical science also as chemical sensing agents. Luminescent materials can easily and quickly give a colorimetric response to stimuli, which is its main advantage over other transduction materials. Moreover, the luminescent material can be used in conjunction with a spectrophotometer to detect the sensed substance according to the representative shifts of absorption and radiation, which can be further divided into two parts: radioluminescence and photoluminescence.

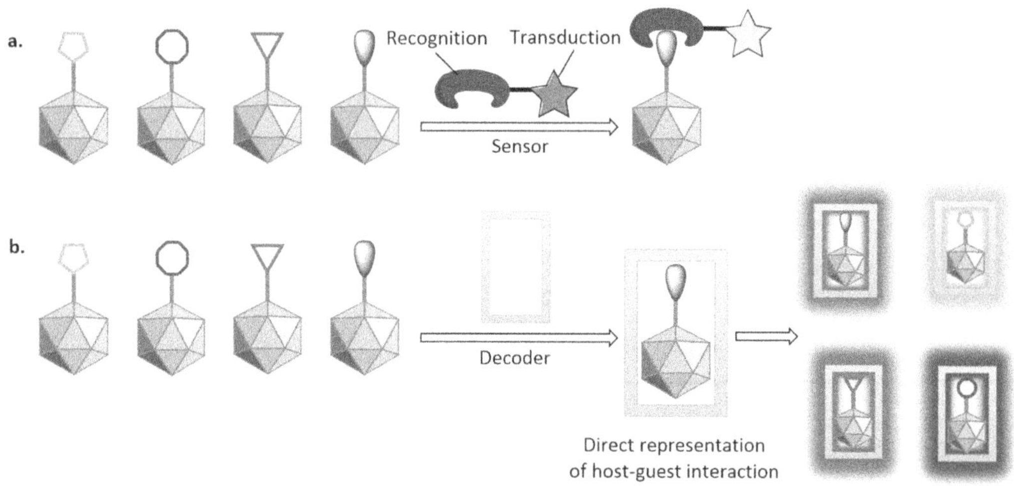

Figure 7. Traditional molecular sensing and molecular decoding conventions. (a) A common molecular detection procedure is illustrated by organic molecules or polymers with two different chemical entities: the recognition segment and the transduction segment. Achieving high sensing capability requires an improvement in the recognition selectivity. (b) Successful molecular decoding of host-guest interacted species leading to the respective signaling responses.

4.2.1. Photoluminescence

The fact that luminescence is more popular than the transduction mechanisms, is the result of some important factors, e.g., producing an indication visible to the naked eye. By using the porosity of MOF, analytes can be selectively detected by size exclusion, where the MOF apparatus are used as a shifter for certain molecules to pass through, by changing the photoluminescence accordingly. With careful selection of ions in MOF, Lewis acid/base sites and functional groups can also be favorably targeted to selective molecules. These mechanisms are frequently observed once sensing ionic particles. Due to its inherent hybrid properties, MOF photoluminescence can be derived from metal ions or organic linkers (Figure 8). The most common sort of luminescence sensing is the linker-based mechanism, that include three subcategories: metal-to-ligand charge transfer (MLCT), ligand-to-metal charge transfer (LMCT), and linker emission. Usually luminescent linkers comprise p-conjugated or aromatic backbones in various existing LMOFs, and these LMOFs emit linker emission when irradiated.

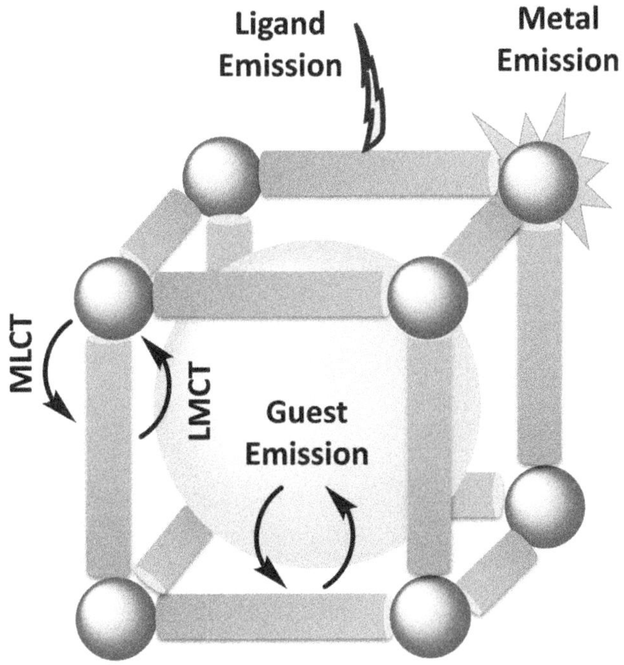

Figure 8. Illustration of probable emission approaches in MOFs. Big sphere inside, guest chromophores; green cylinders, organic linkers; blue spheres, inorganic SBUs.

Fluorescence spectroscopy is recognized for its lower limits of detection that can range the single molecular state. More advantage is the capacity to directly process powdered materials, which is required for film production or other processing. Therefore, orthodox solvothermal synthesis can be used to generate luminescent MOF. Basically, the porous crystal structure of MOF has advantages over other sensors. The ability to customize the adsorption characteristics of the MOF (in general) provides a high mark of molecular selectivity. The size specific signaling in luminescent Zn_3btc_2 (where, btc = benzenetricarboxylate) have been reported by the two different groups in order to achieve the goal regarding the uses of molecular sieving in sensing (*138*, *139*). Coming to the nature of amines in signaling mechanism, easily diffusible small amines such as propylamine, dimethylamine, and ethylamine into the MOF pores to diminish the fluorescence intensity.

Table 1. Sensing applications of selected MOFs

Metal	*MOF*	λ_{ex} *(nm)*	λ_{em} *(nm)*	*Target*	*Ref.*
Cu	$[Cu(pytpy)](NO_3)(CH_3OH)$			Halides	
	[CuL1(H2O)0.5](NO3)2(CH2Cl2)3.25(CH3OH)(H2O)0.5			Halides	
	$[Cu_2(L^7)_2I_2](H_2O)_x$			VOC vapor	
	$[(CuCN)[(CuCN)_3L^8](guests)_x$			Acetonitrile vapor	
Cd	$[Cd_3(NTB)_2(DMA)_3](DMA)_2$			NB, TNP	
	$[CdL[CdL^6(H_2O)](DMF)(H_2O)_2$	371	450	VOCs	*(140)*
	$[Cd_5Cl_6(L^{11})(H^{11})_2](H_2O)_7$	356	470	TNP	*(141)*
	$[Cd_3(BPPA)_3(ABA)_3]$	NS*	514	NP	*(142)*
	$[Cd_3(BPPA)_3(OBA)_3]$	NS	525	PA	*(142)*
	$[Cd(2,6\text{-}NDC)[Cd(2,6\text{-}NDC)_{0.5}(PCA)]$	340	384	TNP	*(143)*
	$[Cd_3(TPT)_2(DMF)_2](H_2O)_{0.5}$	333, 380	390	PA	*(144)*
	Cd(1,4-NDC)(DIA)](MeOH)	370, 368	428	NB, TNP	*(145)*5
	$[Cd(BPDC)(DIA)[Cd(BPDC)(DIA)_{0.5}](DMF)$	NS	434	NB, TNP	*(145)*
	[Cd(BPDC)(DIA)](DEF)		425	NB, TNP	*(145)*
	$[Cd(2,6\text{-}NDC)(H[Cd(2,6\text{-}NDC)(H_2O)]$		380	TNP	*(146)*
	$[Cd_3(CPEIP)_2(DMF)_3](DMF)(EtOH)_2(H_2O)$	342	392	PA	*(147)*
	$[Cd(L[Cd(L^3)(BPDC)](H_2O)_2$	NS	472	Fe^{3+}, $Cr\,O_4^{2-}$, $Cr_2O_7^{2-}$	*(71)*
	$[Cd(L[Cd(L^3)(SDBA)(H_2O)](H_2O)_{0.5}$	NS	458	Metal ions	*(148)*
	$[Cd_3(ETTC)_{1.5}(H_2O)_2(DMF)](DMF)_{12}$	350	472	Organic solvents	*(149)*
	$[Cd(L[Cd(L^{25})(H_2O)](THF)$	380	412	VOCs	*(150)*
	$[Cd_2(TPPE)(BPDC)_2(H_2O)](solvents)_x$		486	$Cr_2O_7^{2-}$	
	$[Cd(TPTZ)(H[Cd(TPTZ)(H_2O)_2(HCOOH)(IPA)_2]$				

Table 1. (Continued). Sensing applications of selected MOFs

Metal	MOF	λ_{ex} (nm)	λ_{em} (nm)	Target	Ref.
Zn	$[Zn_4(DMF)(Ur)_2(2,6\text{-}NDC)_4]$	365	400	TNP	(151)
	$[Zn(L[Zn(L^{19})_2](DMF)_3(H_2O)_5$	NS	602	TNT, NB, DNB, DNT	(152)
	$[Zn_2(BPDC)_2(BPEE)]$	NS	420	DNT, DMNB	(153)
	$[Zn_2(TPTC)(apy)_2x(H_2O)_x](H_2O)$; x 1	319	400, 464	NB	(154)
	$[Zn_2(L^{14})(bpy)(H_2O)_2](H_2O)_3(DMF)_2$	350	476	NB	(155)
	$[Zn_2(OBA)_2(bpy)](DMA)$	280	420	NB	(72)
	$[Zn(2,6\text{-}NDC)(H[Zn(2,6\text{-}NDC)(H_2O)]$		380	TNP, 4-NBA	(146)
	[Zn4O(L15)3(H2O)2]3[Zn4O(L15)3]	333	545	NB, DNB, TNP	(156)
	$[Zn_4O(BPZ)_2(ABDC)](guests)$			Oxygen	
	$[Zn(OBA)(4\text{-}BPDH)[Zn(OBA)(4\text{-}BPDH)_{0.5}](DMF)_{1.5}$		485, 583	PA	(157)
	$[Zn_2L^{16}](Me_2NH_2)(DMF)_{3.5}$	360	463	NB, NT, DNB	(158)
	$[Zn_2L^{16}(H_2O)](Me_2NH_2)(DMF)_2(H_2O)_8$;	360	440	NB, NT	
	$[Zn_2(L^{17})_2(DPYB)]$	382	533	DNB, PA	(159)
	$[Zn(L^{18})(DIBP)](H_2O)_2$	382	523	PA	
	$[Zn_5(HFIPBB,)_4(trz)_2(H_2O)_2]$	260	404, 406	Fe^{3+}	(160)
	$[Zn_2(BDC)_3(DMA)](HDMA)(DMF)_6$	298	578, 593, 614, 648, 687	Acetone, aniline	(161)
	$[Zn_2(TCPE)(H_2O)_2](DMF)_4$	365	514	VOCs	(162)
	$[Zn_2(tcbpe\text{-}F)(4,40\text{-}bpy)](DMA)_x$	365	530	VOCs	(163)
	$[Zn_2(tcbpe\text{-}F)(bpee)](DMA)_x$	305	387, 421, 453	Triethylamine	(164)
	$[Zn_2(TPC4A)(DMF)(H_2O)](H_2O)_3$	330	546	Triethylamine	(165)
	$[Zn(TNC4A)](Me[Zn(TNC4A)](Me_2NH_2)(H_2O)_4$	395	500	I	

Table 1. (Continued). Sensing applications of selected MOFs

Metal	*MOF*	λ_{ex} *(nm)*	λ_{em} *(nm)*	*Target*	*Ref.*
Zn-Ln	$[Ln_2Zn(L^2)_3(H_2O)_4](NO_3)_2(H_2O)_{12}$; Ln = Tb, Eu			NB, NM, NE	
Mg	$[Mg_2(BDC)_2(BPNO)](DMF)_2$			NP	
Ba	$[Ba_5(ADDA)_5(EtOH)_2(H_2O)_3](DMF)_5$			MnO_4	
	$[Ba_3La_{0.5}(l_3\text{-}L^4)_{2.5}(H_2O)_3(DMF)](DMF)_3$			TNP	
Zr	$[Zr_6O_4(OH)_4(L^{12})_6]$				
	$[Zr_6O_4(OH)_4(L13)_6]$	320	440	TNP	*(166)*
	$[Zr_6O_4(OH)_8(H_2O)_4(CTTA)_{8/3}]$	312	372	TNP, NP	*(122)*
	$[Zr_6O_4(OH)_8(H_2O)_4(TTNA)_{8/3}]$	324	410	TNP, NP	
	PCN-224(Pt)	512	655	Dissolved oxygen	*(167)*
	PCN-224-NHPCN-224-NH_2			Hg^{2+}	
	PCN-225			pH	
	UiO-66			pH	
	F-UiO			pH	
	Eu^{3+}/Cu^{2+}@UiO-66-$(COOH)_2$	NS	393, 615	H_2S	*(168)*
	DUT-122	368	508	Solvent vapor	*(169)*
Li	$\{Li_3[Li(DMF)_2](CPMA)_2\}(DMF)_4(H_2O)$	345	430	NB, DNT	*(170)*
Al	MIL-53-L^5	337	567	Cu^{2+}	*(171)*
	NH_2-MIL-101(Al)			F	
	$[Al(OH)(H[Al(OH)(H_2BTEC)](H_2O)$			F, $Cr_2O_7^{2-}$	
In-Ru	$[InRu(DCBPY)_3](Me_2NH_2)(H_2O)_6$	400	657	NB, NT, DMNB, DNT, DNB, TNT	*(172)*

Table 1. (Continued). Sensing applications of selected MOFs

Metal	*MOF*	λ_{ex} *(nm)*	λ_{em} *(nm)*	*Target*	*Ref.*
In	$[In_{10}(TTCA)_{12}](Me_2NH_2)_6(DMF)_{24}(H_2O)_{15}$	410	505	TNP	*(173)*
	$[In_2(OH)_2(H_2TTHA)(H_2O)_2]$			UO_2^{+2}, Ru^{3+}	
In-K	[In0.5K(3-qlc)Cl1.5(H2O)0.5]	NS	455	Temperature	*(174)*
	$[InK(ox)_2](H_2O)_4$	NS	438		
Ti	MIL-125(Ti)-AM-Eu			Anions	
	$[Ti_2(HDOBDC)_2(H_2DOBDC)]$			Fe^{3+}	
Eu	$Eu(BTB)(H_2O)$	350	350, 390, 580, 610, 700	NB, TNP, 2-NT	*(175)*
	$[Eu_2L^9{}_3(H_2O)_4](DMF)_3$			DMF	
	$[Eu_3(MFDA)_4(NO_3)(DMF)_3]$	336	581, 592, 615	DNT, DNB, NT, NM, NB, DMNB	*(176)*
	$[H_3O]_2Eu_{2.5}(BTB)_3(OAc)_{0.5}(H_2O)_3](DMA)_2(H_2O)_{2.5}$	295	538, 592, 614, 653, 696	Al^{3+}	*(177)*
	$[Eu(HPZDC)_2(NO_3)](H_2O)$	394	586, 593, 614, 698	Fe^{3+}, $Cr_2O_7{}^{2-}$	*(178)*
	$[Eu(L^{20})(H_2O)_2](NMP)(H_2O)$	301	590, 612, 650, 700, 618	Fe^{2+}/Fe^{3+}, $Cr_2O_7{}^{2-}$, acetone	*(179)*
	$Eu_2(L^{22})_3(H_2O)_2(DMF)_2](H_2O)_{16}$	280	618	Temperature, humidity	*(180)*
	$[Eu(DBM)_3](Phen)$ $[Eu(DBM)_3](2,20\text{-}bpy)$	373	579, 590, 612, 653, 705	pH, amine vapors	*(181)*
	$[Eu_2(PYDCH)_6](H_2O)_3$	280	589, 612, 646, 690	pH	*(182)*
	$[Eu(L^{24})(OAc)(DMA)]$	330	593, 617, 652, 700	Al^{3+}	*(183)*
	$[Eu_2(BPYDC)_3(H_2O)_3](DMF)_x$	395	580, 592, 614, 653, 700	Oxygen	*(184)*
	Eu–HODA	305	594, 614, 652, 697	Fe^{3+}	*(185)*
	$[Eu(BTB)(phen)](DMF)_{4.5}(H_2O)_2$	300	535, 591, 614, 655, 697	Al^{3+}	*(186)*
	$[Eu_3(phenDCA)_4(OH)(H_2O)_4](H_2O)_2$			pH	
	$[Eu(ox)_2(H_2O)](Me_2NH_2)(H_2O)_3$	395	690, 696, 618	Cu2+	*(187)*

Table 1. (Continued). Sensing applications of selected MOFs

Metal	*MOF*	λ_{ex} *(nm)*	λ_{em} *(nm)*	*Target*	*Ref.*
	$[Eu(HBPTC)(H_2O)_2](DMF)_2$			Carbonate	
	$[Eu_2(QPTCA)(NO_3)_2(DMF)_4](CH_3CH_2OH)_3$	359	594, 615, 699	Temperature	(*108*)
	Ln(BTC); Ln = Tb^{3+}, Eu^{3+}	268	615, 700	Dipicolinic acid	(*188*)
Tb	Fe_3O_4@Tb-BTC	220	440, 545	TNT, PA, DNT	(*189*)
	$[Tb(TTCA)(DMA)(H_2O)](DMA)_7(H_2O)_{9.5}$	370	490, 546, 585, 624	TNT, TNP; 2,4-DNT, 2,6-DNT, NB	(*190*)
	$Tb(BTB)(H_2O)$	350	486, 541, 580, 617, 649, 697	NB, TNP, 2-NT	(*175*)
	$[Tb(BTC)](CH_3OH)$			F^-	
	$[Tb(BPTC)](Me_2NH_2)(solvents)_x$			Fe^{3+}	
	$[Tb_3(TCA)_2(DMA)_{0.5}(OH)_3(H_2O)_{0.5}](H_2O)_3$	395	497, 545, 587, 620	Al^{3+}, Fe^{3+}	(*191*)
	$[Tb_3(L^{23})_2(HCOO)(H_2O)_5](DMF)(H_2O)_4$	327	491, 545, 585, 621	Metal ions, Organic solvent	(*192*)
	$[(CH_3)_2NH_2][Tb(BPTC)](solvents)_x$	310	492, 546, 588, 621	Fe^{3+}	(*193*)
	$[Tb(TATAB)](DMF)_4(H_2O)(MeOH)_{0.5}$ C460(C460 = 7-diethylamino-4-methylcoumarin)	360	545, 350	Temperature	(*194*)
Tb-Eu	[Eu2xTb2(1-x)(FDA)3]			VOCs	
	[Eu0.0069Tb0.9931-DMBDC]			Temperature	
	$[Eu_{0.7}Tb_{0.3}(D\text{-}cam)(HIMDC)_2(H_2O)_2]$			Temperature	
	[Tb0.95Eu0.05HL10]			Temperature	
	[Tb0.80Eu0.20BPDA]			Temperature	
Tb-K	$[TbK(BPDSDC)(DMF)(H_2O)_2]$	360	488, 544, 583, 622	Cr^{3+}	(*195*)
La	$[La(TPT)(DMSO)_2](H_2O)$	342	374	PA	(*196*)
	$[Sm_2(ABTC)_{1.5}(H_2O)_3(DMA)](H_2O)(DMA)$	400	612	Ethanol	(*197*)

Table 1. (Continued). Sensing applications of selected MOFs

Metal	*MOF*	λ_{ex} *(nm)*	λ_{em} *(nm)*	*Target*	*Ref.*
Y	$[Y_9(l_3OH)_8(l_2OH)_3(BTB)_6](DMA)_2$ (solvents)$_x$	290	375	Fe^{3+}	*(198)*
Ag	$[(HMTA)Ag_4I_4](HMTA)$	333	620	Temperature	*(199)*

* NS = excitation wavelength not explicitly specified.

4.2.2. Luminescence Sensing Mechanisms in MOFs

The balanced projection of these architectures is based on countless specimens of separate luminescent coordination compounds. There may be numerous sources of luminescence in the case of molecules as well as MOF. Fluorescence production based on linker(s) is usually formed by the absorption of various MOFs containing conjugated ligands in the UV or visible range along with d^{10} transition metal ions (for example, Cd^{II} or Zn^{II}), which are mainly electronically inert ions). Primarily in MOF containing f-elements, the metal-centered luminescence (mainly phosphorescence) is observed as nodes. In several cases, the extremely absorbing conjugated struts regulated to these centers act as intensifying emission through metal-assisted (spin-orbit coupling expedited) intersystem crossing (i.e., to triplet excited from singlet states of a strut), antennas along with ligand-to-metal EnT.

In luminescent MOFs, the most basic configuration of the signal transduction is to suppress or sometimes increase the photo-induced emission caused by adsorption of the guest. The power of these effects depends on the behavior of the host-guest interactions. Generally, related interactions are those that permit noteworthy overlap between the electron acceptor or the electron donor orbitals, where the MOF or analyte plays certain role. As a result, the most easily detectable analyte is a good electron acceptors (such as nitroaromatic compounds which include conventional simulants or explosives) or donor (*viz.* amines). On the other hand, the guest can be used to change the redox potential of the built-in acceptor or donor, thus reducing and improving erasure and increasing luminescence. In rare cases, the adsorbed material will cause shifts in the MOFs emission frequency. This category of signal transmission is intrinsically more favorable in comparison to quenching in luminescence (*"turn-off"* signaling), due to the analyte uptake can be observed by proper monitoring at shifted single emission wavelength, it is regarded as an increase in luminescence intensity. Similar to the earlier reviews (*73*, *200*, *201*), a comprehensive updated database of LMOFs is thus provided in this chapter and summarized in Table 1 for various target analytes.

The chemosensory material scrutinized also in $Zn_2(bdc)_2(dpNDI)$ (where, bdc = 1,4-benzenedicarboxylate; dpNDI = N,N0-di(4-pyridyl)-1,4,5,8 naphthalenediimide). NDI is recognized to produce exciplex emission, while it interacts with aromatic compounds in solution, and these appearances will be transported to the MOF structure. The MOF was saturated in a variety of pure aromatic molecules encompassing various substances. Analytes such as toluene, benzene, benzonitrile, iodobenzene, xylenes, and anisole. As represented in Figure 9, shifting to longer wavelength has been shown by the systematic exciplex emission by increase in electron-donation competence of the analyte (for example, as determined by the ionization energy of the analyte) having the exception in iodobenzene. Moreover, the effect of heavy atom, promoted by spin-orbit coupling assisted mixing of triplet/singlet excited state, accelerates intersystem crossing and as a consequence it gives phosphorescence mediated radiative decay instead of fluorescence having higher-energy.

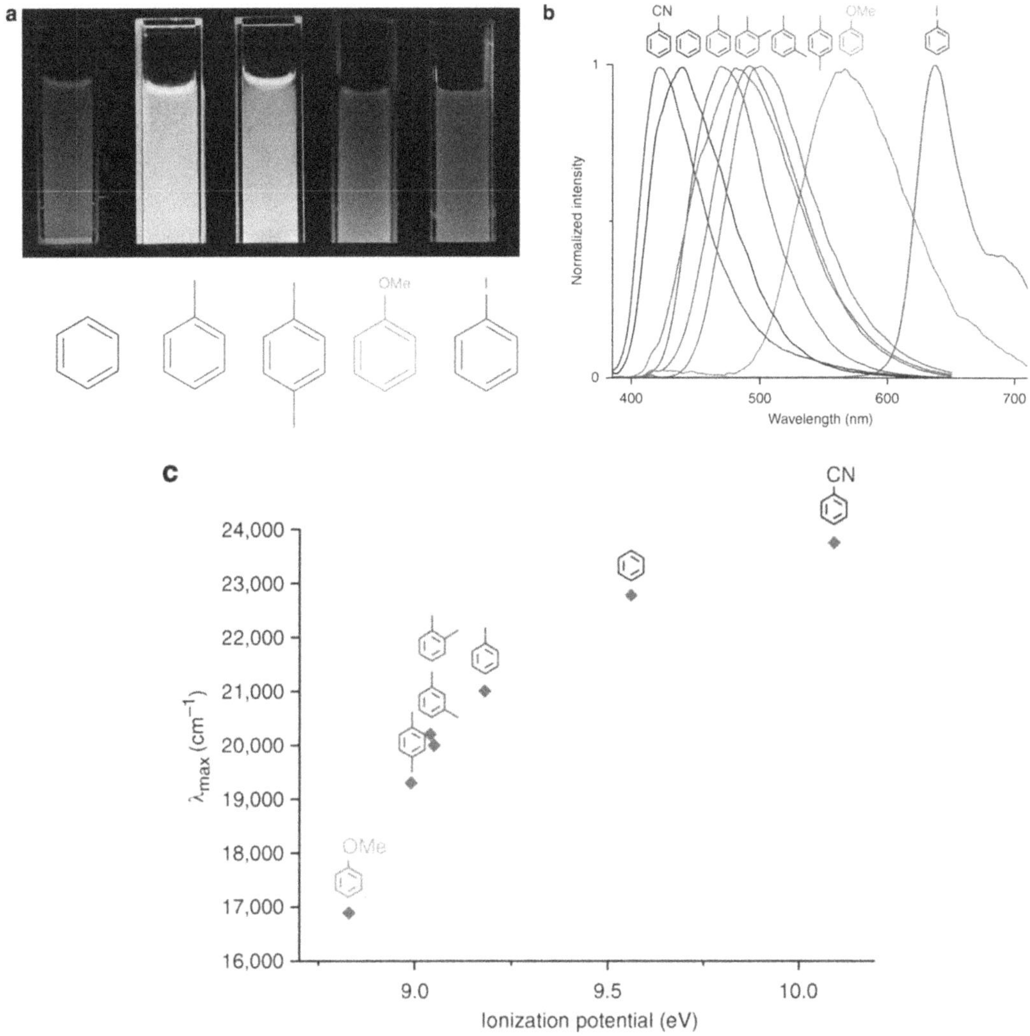

Figure 9. (a) Indication of MOF powders under irradiation at 365 nm in organic solvent. (b) Luminescence spectra of guest-containing MOFs with height-normalization upon excitation at 370 nm. (c) Ionization potential vs. the emission energy of guest-containing MOF. Adapted with permission from Ref. (200). Copyright 2012 ACS Publishing Group.

4.2.3. Radioluminescence

For space exploration, non-proliferation, biological applications and nuclear energy, MOF provides a chance to construct important improvements in the identification of particles in the subatomic level. The flexibility in the synthesis of MOF permits them to be adjusted in ways that standard scintillator materials including plastics, inorganic materials (like sodium iodide) and organic liquids cannot adapt. Moreover, its crystal structure can characterize the interaction between the luminescent linkers with sub-angstrom precision, which create a "nano laboratory" for photophysics research at the level of detail that is unconceivable in non-crystalline materials. Thus, these novel degrees of freedom, that involve both meso and micro length scales along with the balanced projection of both light output nanopore volume and timing can be achieved in MOFs. Scintillation is well-elaborated as the short flashes (ns) emission response to ionizing radiation (e.g., energy typically >10 eV) in MOFs, which initiates from the organic linkers. Furthermore, the organic scintillators based luminescent linker groups, which are comparable to recognized are revealed in Figure 6 of Ref. (*200*), from which it can be seen that the luminescent parts in electronic structures can be varied methodically. Also, on the other hand, the degree of the interaction between emitting assemblies can be changed, either by generating diffused structures, where the orientation and distance among crystal lattices be different from those in the non-interpenetrated structure or by regulating size of the pore. From this point of view, IRMOFs represent a series of particularly interesting isostructures, because there are interpenetrating and non-interpenetrating structures exist that contain the same linker (for example, IRMOF-10 and IRMOF-9, the linker is biphenyl dicarboxylate).

Doty *et al.* first used MOF scintillators (*202*). Bauer *et al.* reported that due to the limitation of the stilbene moiety in the comparatively inflexible *trans* direction, the lifetime of these fluorescent materials meet the criteria of the natural lifetime of emission (*203*). As a result, the non-emissive pathway of *cis-trans* isomerization is blocked in stilbene, which increases its quantum yield. The results led to an assumption that the emission of ionizing radiation will have comparable characteristics. Doty *et al.* evidenced it by detecting the radio luminescence of protons with an energy of 3 MeV (Figure 10) along with the scintillation of alpha particles.

5. Challenges of MOF Sensors

Usually, these types of sensors do not have sufficient chemical selectivity is the main restraint of these sorts of sensors. Another drawback is that generally they involve losses of signals (such as phosphorescence or fluorescence intensity) in retort to sorption of analytes. One of the disadvantages of "*turn-off*" sensors is that this phenomenon can cause loss of signal with analyte capture. By keeping this in mind, this chapter limits the overview of luminescence-mediated chemical sensors to a few current reports in which the porous structure of MOF or its chemical properties conveys specific signaling competences. Scintillation of MOF is also discussed, which luminescence in parallel with the radioactive analytes. Other signal conversion arrangements, which used photons in combination with various types of solvatochromism, variation of local surface plasmon resonance energy by analytes and optical interferometry. The mechanical signal-transduction organizations used with MOF includes micro cantilever device, SAW, and quartz crystal microbalance. Moreover, electrical schemes have been restricted to the impedance spectroscopy so far.

6. Conclusion

Due to the growing research interest of scientists, the wide area of using LMOF as a chemical sensor or probe has developed rapidly in the recent times. Hundreds of LMOF sensors have been carefully designed and established to be very effective in detecting various species, including volatile organic compounds (VOCs), ions, gases, biomaterials and explosives. In addition, there are some excellent fluorescence thermometers that can provide accurate temperature information in a variety of environments. On the basis of the previous review articles (*73*, *142*, *200*, *201*) (in which an extensive LMOF database was created), here this chapter focus to provide a summary of the latest LMOF-mediated chemical sensors, which are classified according to different targeted analytes.

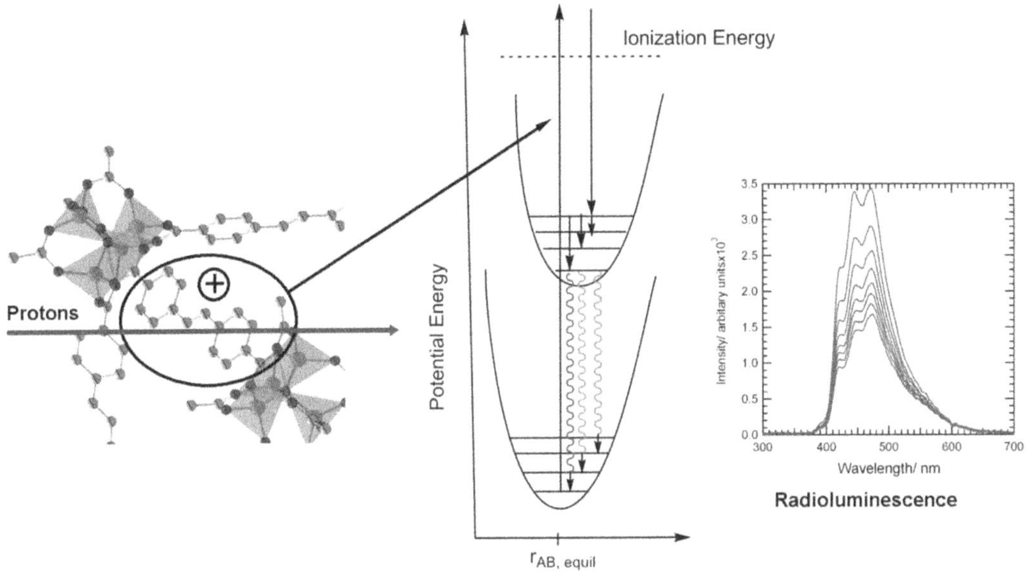

Figure 10. Representation of the ion-beam-induced luminescence (IBIL) progression (by taking an example of the MOF-S1). Showing (left) the interaction of a high-energy proton that ionizes the material. The resulting excitation decays (middle) through radiationless pathways (solid arrows) to the lowest-lying singlet excited state of the fluorescent SDC groups, which then decay by emission of photons (wavy arrows) to various ground-state vibrational levels, producing the IBIL luminescence displaying vibronic structure (right). The series of spectra indicate the effect of increasing proton dose, proceeding from 0.4 to 4 M Gray for the largest to smallest spectra, respectively. Adapted with permission from Ref. (200). Copyright 2012 ACS Publishing Group.

Chemical sensors or probes that display changes in fluorescence emission in response to various analytes are more practical in comparison to the other categories of sensors because the optical signals they produce can easily be predictable through naked eye or by fluorimeter. It has been recently recognized that LMOF makes a tremendous sensor due to its intrinsic advantages (for example, tunable luminescence, crystallinity, diversity of structure and function, and permanent porosity). Through rational design, numerous MOF-mediated sensors with various detection mechanisms have been accomplished. "*Turn-off*" mechanism is usually generated by identifying electron acceptors, where the luminescence quenching is occurred by EnT and/or electron transfer from the sensor to the guest. In order to improve selectivity and sensitivity, MOF sensors also implement a "*turn-on*" detection mode, which usually results in wavelength shifts or luminescent enhancement.

In addition, to detect target compounds functional sites have been successfully exploited. Moreover, embedding multi-luminescent primitives in the frame will produce multiple signal transitions that not only expands the sensing range, but also expands the sensing range. Compared with the other MOF applications (namely, catalysis or separation, gas storage, *etc*), the research of MOF-functionalized chemical sensors or luminescent probes are still in its infancy. Luminescent MOFs with nanoscale dimensions have excellent sensing potential inside living cells, and the impression of installing MOFs in devices like fabricating MOF membrane or thin films is still in the beginning stage. Although the LMOF sensor performs well in the laboratory, a lot of research and development is still needed to make the technology commercially available.

Acknowledgments

Dr. BKK sincerely thanks Centurion University of Technology and Management (CUTM), Odisha for financial support to execute this nice piece of work.

References

1. Horike, S.; Umeyama, D.; Kitagawa, S. Ion conductivity and transport by porous coordination polymers and metal–organic frameworks. *Acc. Chem. Res.* **2013**, *46* (11), 2376–2384.
2. Kumar, P.; Kim, K.-H.; Kwon, E. E.; Szulejko, J. E. Metal–organic frameworks for the control and management of air quality: advances and future direction. *J. Mater. Chem. A* **2016**, *4* (2), 345–361.
3. Matsuyama, K. Supercritical fluid processing for metal–organic frameworks, porous coordination polymers, and covalent organic frameworks. *Journal of Supercritical Fluids* **2018**, *134*, 197–203.
4. Furukawa, H.; Cordova, K. E.; O'Keeffe, M.; Yaghi, O. M. The chemistry and applications of metal-organic frameworks. *Science* **2013**, *341* (6149)
5. Cai, H.; Huang, Y.-L.; Li, D. Biological metal–organic frameworks: structures, host–guest chemistry and bio-applications. *Coord. Chem. Rev.* **2019**, *378*, 207–221.
6. Slater, A. G.; Cooper, A. I. Function-led design of new porous materials. *Science* **2015**, *348* (6238)
7. Kundu, B. K.; Das, M.; Ganguly, R.; Bhobe, P. A.; Mukhopadhyay, S. Role of zeolite encapsulated Cu(II) complexes in electron transfer as well as peroxy radical intermediates formation during oxidation of thioanisole. *J. Catal.* **2020**, *389*, 305–316.
8. Kundu, B. K.; Chhabra, V.; Malviya, N.; Ganguly, R.; Mishra, G. S.; Mukhopadhyay, S. Zeolite encapsulated host-guest Cu(II) Schiff base complexes: Superior activity towards oxidation reactions over homogenous catalytic systems. *Micropor. Mesopor. Mat.* **2018**, *271*, 100–117.
9. Martínez, C.; Corma, A. Inorganic molecular sieves: Preparation, modification and industrial application in catalytic processes. *Coord. Chem. Rev.* **2011**, *255* (13-14), 1558–1580.
10. Clark, J. H.; Macquarrie, D. J. Environmentally friendly catalytic methods. *Chem. Soc. Rev.* **1996**, *25* (5), 303–310.
11. Su, B.-L.; Sanchez, C.; Yang, X.-Y. *Hierarchically Structured Porous Materials: From Nanoscience to Catalysis, Separation, Optics, Energy, and Life Science*; John Wiley & Sons, 2012.
12. Keskin, S.; Kızılel, S. Biomedical applications of metal organic frameworks. *Industrial Engineering Chemistry Research* **2011**, *50* (4), 1799–1812.
13. Cui, Y.; Zhang, J.; He, H.; Qian, G. Photonic functional metal–organic frameworks. *Chem. Soc. Rev.* **2018**, *47* (15), 5740–5785.
14. Xu, H.; Hu, H.-C.; Cao, C.-S.; Zhao, B. Lanthanide organic framework as a regenerable luminescent probe for Fe3+. *Inorg. Chem.* **2015**, *54* (10), 4585–4587.
15. Chen, Z.; Sun, Y.; Zhang, L.; Sun, D.; Liu, F.; Meng, Q.; Wang, R.; Sun, D. A tubular europium–organic framework exhibiting selective sensing of Fe 3+ and Al 3+ over mixed metal ions. *Chem. Commun.* **2013**, *49* (98), 11557–11559.

16. Hou, G.-G.; Liu, Y.; Liu, Q.-K.; Ma, J.-P.; Dong, Y.-B. NbO lattice MOFs based on octahedral M (II) and ditopic pyridyl substituted diketonate ligands: structure, encapsulation and guest-driven luminescent property. *Chem. Commun.* **2011**, *47* (38), 10731–10733.
17. Dang, S.; Ma, E.; Sun, Z.-M.; Zhang, H. A layer-structured Eu-MOF as a highly selective fluorescent probe for Fe3+ detection through a cation-exchange approach. *Journal of Materials Chemistry* **2012**, *22* (33), 16920–16926.
18. Zheng, M.; Tan, H.; Xie, Z.; Zhang, L.; Jing, X.; Sun, Z. Fast response and high sensitivity europium metal organic framework fluorescent probe with chelating terpyridine sites for Fe3+. *ACS Applied Materials Interfaces* **2013**, *5* (3), 1078–1083.
19. Yang, C.-X.; Ren, H.-B.; Yan, X.-P. Fluorescent metal–organic framework MIL-53 (Al) for highly selective and sensitive detection of Fe3+ in aqueous solution. *Anal. Chem.* **2013**, *85* (15), 7441–7446.
20. Tang, Q.; Liu, S.; Liu, Y.; Miao, J.; Li, S.; Zhang, L.; Shi, Z.; Zheng, Z. Cation sensing by a luminescent metal–organic framework with multiple Lewis basic sites. *Inorg. Chem.* **2013**, *52* (6), 2799–2801.
21. Xiang, Z.; Fang, C.; Leng, S.; Cao, D. An amino group functionalized metal–organic framework as a luminescent probe for highly selective sensing of Fe 3+ ions. *J. Mater. Chem. A* **2014**, *2* (21), 7662–7665.
22. Liang, Y.-T.; Yang, G.-P.; Liu, B.; Yan, Y.-T.; Xi, Z.-P.; Wang, Y.-Y. Four super water-stable lanthanide–organic frameworks with active uncoordinated carboxylic and pyridyl groups for selective luminescence sensing of Fe 3+. *Dalton Transactions* **2015**, *44* (29), 13325–13330.
23. Wu, Y.; Yang, G.-P.; Zhou, X.; Li, J.; Ning, Y.; Wang, Y.-Y. Three new luminescent Cd (II)-MOFs by regulating the tetracarboxylate and auxiliary co-ligands, displaying high sensitivity for Fe 3+ in aqueous solution. *Dalton Transactions* **2015**, *44* (22), 10385–10391.
24. Xu, H.; Zhou, S.; Xiao, L.; Wang, H.; Li, S.; Yuan, Q. Fabrication of a nitrogen-doped graphene quantum dot from MOF-derived porous carbon and its application for highly selective fluorescence detection of Fe 3+. *Journal of Materials Chemistry C* **2015**, *3* (2), 291–297.
25. Xu, X.-Y.; Yan, B. Eu (III)-functionalized MIL-124 as fluorescent probe for highly selectively sensing ions and organic small molecules especially for Fe (III) and Fe (II). *ACS Applied Materials Interfaces* **2015**, *7* (1), 721–729.
26. Zhou, Y.; Chen, H.-H.; Yan, B. An Eu 3+ post-functionalized nanosized metal–organic framework for cation exchange-based Fe 3+-sensing in an aqueous environment. *J. Mater. Chem. A* **2014**, *2* (33), 13691–13697.
27. Cho, W.; Lee, H. J.; Choi, G.; Choi, S.; Oh, M. Dual changes in conformation and optical properties of fluorophores within a metal–organic framework during framework construction and associated sensing event. *J. Am. Chem. Soc.* **2014**, *136* (35), 12201–12204.
28. Bhattacharyya, S.; Chakraborty, A.; Jayaramulu, K.; Hazra, A.; Maji, T. K. A bimodal anionic MOF: turn-off sensing of Cu II and specific sensitization of Eu III. *Chem. Commun.* **2014**, *50* (88), 13567–13570.
29. Cai, D.; Guo, H.; Wen, L.; Liu, C. Fabrication of hierarchical architectures of Tb-MOF by a "green coordination modulation method" for the sensing of heavy metal ions. *CrystEngComm* **2013**, *15* (34), 6702–6708.

30. Zhang, L.-N.; Liu, A.-L.; Liu, Y.-X.; Shen, J.-X.; Du, C.-X.; Hou, H.-W. A luminescent europium metal–organic framework with free phenanthroline sites for highly selective and sensitive sensing of Cu2+ in aqueous solution. *Inorg. Chem. Commun.* **2015**, *56*, 137–140.
31. Luo, F.; Batten, S. R. Metal–organic framework (MOF): lanthanide (III)-doped approach for luminescence modulation and luminescent sensing. *Dalton Transactions* **2010**, *39* (19), 4485–4488.
32. Xiao, Y.; Cui, Y.; Zheng, Q.; Xiang, S.; Qian, G.; Chen, B. A microporous luminescent metal–organic framework for highly selective and sensitive sensing of Cu2+ in aqueous solution. *Chem. Commun.* **2010,** *46* (30), 5503–5505.
33. Wang, Z.; Liu, H.; Wang, S.; Rao, Z.; Yang, Y. J. S.; Chemical, A. B. A luminescent Terbium-Succinate MOF thin film fabricated by electrodeposition for sensing of Cu2+ in aqueous environment. *Sensors Actuators B: Chemical* **2015**, *220*, 779–787.
34. Chen, B.; Wang, L.; Xiao, Y.; Fronczek, F. R.; Xue, M.; Cui, Y.; Qian, G. A luminescent metal–organic framework with Lewis basic pyridyl sites for the sensing of metal ions. *Angew. Chem.* **2009**, *121* (3), 508–511.
35. Zhang, P.; Pei, L.; Chen, Y.; Xu, W.; Lin, Q.; Wang, J.; Wu, J.; Shen, Y.; Ji, L.; Chao, H. A Dinuclear Ruthenium (II) Complex as a One-and Two-Photon Luminescent Probe for Biological Cu2+ Detection. *Chemistry–A European Journal* **2013**, *19* (46), 15494–15503.
36. Liu, C.; Yan, B. Highly effective chemosensor of a luminescent silica@ lanthanide complex@ MOF heterostructured composite for metal ion sensing. *RSC advances* **2015**, *5* (123), 101982–101988.
37. Li, Y.-W.; Li, J.-R.; Wang, L.-F.; Zhou, B.-Y.; Chen, Q.; Bu, X.-H. Microporous metal–organic frameworks with open metal sites as sorbents for selective gas adsorption and fluorescence sensors for metal ions. *J. Mater. Chem. A* **2013**, *1* (3), 495–499.
38. Jung, S. H.; Lee, H.; Seo, M. L.; Jung, J. H. Anthracene Dicarboxylate-Based Metal-Organic Framework Gel with Zn 2+ as a TNT Sensor. *Bull. Korean Chem. Soc.* **2013**, *34* (5), 1583–1585.
39. Wei, N.; Zhang, M.-Y.; Zhang, X.-N.; Li, G.-M.; Zhang, X.-D.; Han, Z.-B. Two series of solvent-dependent lanthanide coordination polymers demonstrating tunable luminescence and catalysis properties. *Crystal Growth Design* **2014**, *14* (6), 3002–3009.
40. Xie, G.; Shi, Y.; Hou, F.; Liu, H.; Huang, L.; Xi, P.; Chen, F.; Zeng, Z. *A Highly Selective Fluorescent and Colorimetric Chemosensor for ZnII and Its Application in Cell Imaging*; Wiley Online Library, 2012.
41. Zhao, B.; Chen, X.-Y.; Cheng, P.; Liao, D.-Z.; Yan, S.-P.; Jiang, Z.-H. Coordination polymers containing 1D channels as selective luminescent probes. *J. Am. Chem. Soc.* **2004**, *126* (47), 15394–15395.
42. Tan, H.; Liu, B.; Chen, Y. Lanthanide coordination polymer nanoparticles for sensing of mercury (II) by photoinduced electron transfer. *ACS Nano* **2012**, *6* (12), 10505–10511.
43. Zhu, Y.-M.; Zeng, C.-H.; Chu, T.-S.; Wang, H.-M.; Yang, Y.-Y.; Tong, Y.-X.; Su, C.-Y.; Wong, W.-T. A novel highly luminescent LnMOF film: a convenient sensor for Hg 2+ detecting. *J. Mater. Chem. A* **2013**, *1* (37), 11312–11319.

44. Wu, P.; Liu, Y.; Liu, Y.; Wang, J.; Li, Y.; Liu, W.; Wang, J. Cadmium-based metal–organic framework as a highly selective and sensitive ratiometric luminescent sensor for mercury (II). *Inorg. Chem.* **2015**, *54* (23), 11046–11048.
45. Jia, J.; Xu, F.; Long, Z.; Hou, X.; Sepaniak, M. J. Metal–organic framework MIL-53 (Fe) for highly selective and ultrasensitive direct sensing of MeHg+. *Chem. Commun.* **2013**, *49* (41), 4670–4672.
46. Wei, N.; Zhang, Y.-R.; Han, Z.-B. Thiadiazole-functional porous metal–organic framework as luminescent probe for Cd 2+. *CrystEngComm* **2013**, *15* (44), 8883–8886.
47. Hao, J.-N.; Yan, B. A water-stable lanthanide-functionalized MOF as a highly selective and sensitive fluorescent probe for Cd 2+. *Chem. Commun.* **2015**, *51* (36), 7737–7740.
48. Yang, L.-r.; Song, S.; Zhang, H.-m.; Zhang, W.; Bu, Z.-w.; Ren, T.-g. Synthesis, structure of 3D lanthanide (La (III), Pr (III)) nanoporous coordination polymers containing 1D channels as selective luminescent probes of Pb2+, Ca2+ and Cd2+ ions. *Synth. Met.* **2011**, *161* (21-22), 2230–2240.
49. Liu, C.; Yan, B. Zeolite-type metal organic frameworks immobilized Eu3+ for cation sensing in aqueous environment. *Journal of Colloid Interface Science* **2015**, *459*, 206–211.
50. Shahat, A.; Hassan, H. M.; Azzazy, H. M. Optical metal-organic framework sensor for selective discrimination of some toxic metal ions in water. *Anal. Chim. Acta* **2013**, *793*, 90–98.
51. Cao, L. H.; Shi, F.; Zhang, W. M.; Zang, S. Q.; Mak, T. C. Selective Sensing of Fe3+ and Al3+ Ions and Detection of 2, 4, 6-Trinitrophenol by a Water-Stable Terbium-Based Metal–Organic Framework. *Chemistry–A European Journal* **2015**, *21* (44), 15705–15712.
52. Maity, D.; Govindaraju, T. Naphthaldehyde–Urea/Thiourea Conjugates as Turn-On Fluorescent Probes for Al3+ Based on Restricted C= N Isomerization. *Eur. J. Inorg. Chem.* **2011**.
53. Ma, J. x.; Huang, X. f.; Song, X. q.; Liu, W. s. Assembly of Framework-Isomeric 4 d–4 f Heterometallic Metal–Organic Frameworks with Neutral/Anionic Micropores and Guest-Tuned Luminescence Properties. *Chemistry–A European Journal* **2013**, *19* (11), 3590–3595.
54. Carboni, M.; Lin, Z.; Abney, C. W.; Zhang, T.; Lin, W. A Metal–Organic Framework Containing Unusual Eight-Connected Zr–Oxo Secondary Building Units and Orthogonal Carboxylic Acids for Ultra-sensitive Metal Detection. *Chemistry–A European Journal* **2014**, *20* (46), 14965–14970.
55. Zhao, X.-Q.; Zhao, B.; Shi, W.; Cheng, P. Structures and luminescent properties of a series of Ln–Ag heterometallic coordination polymers. *CrystEngComm* **2009**, *11* (7), 1261–1269.
56. Zhao, B.; Chen, X.-Y.; Chen, Z.; Shi, W.; Cheng, P.; Yan, S.-P.; Liao, D.-Z. A porous 3D heterometal-organic framework containing both lanthanide and high-spin Fe (II) ions. *Chem. Commun.* **2009** (21), 3113–3115.
57. Lu, W.-G.; Jiang, L.; Feng, X.-L.; Lu, T.-B. Three-dimensional lanthanide anionic metal– organic frameworks with tunable luminescent properties induced by cation exchange. *Inorg. Chem.* **2009**, *48* (15), 6997–6999.
58. Zhang, S.-R.; Li, J.; Du, D.-Y.; Qin, J.-S.; Li, S.-L.; He, W.-W.; Su, Z.-M.; Lan, Y.-Q. A multifunctional microporous anionic metal–organic framework for column-chromatographic

dye separation and selective detection and adsorption of Cr 3+. *J. Mater. Chem. A* **2015**, *3* (46), 23426–23434.

59. Hao, J.-N.; Yan, B. Highly sensitive and selective fluorescent probe for Ag+ based on a Eu 3+ post-functionalized metal–organic framework in aqueous media. *J. Mater. Chem. A* **2014**, *2* (42), 18018–18025.
60. Zhang, L.; Jian, Y.; Wang, J.; He, C.; Li, X.; Liu, T.; Duan, C. Post-modification of a MOF through a fluorescent-labeling technology for the selective sensing and adsorption of Ag+ in aqueous solution. *Dalton Transactions* **2012**, *41* (34), 10153–10155.
61. Liu, W.; Jiao, T.; Li, Y.; Liu, Q.; Tan, M.; Wang, H.; Wang, L. Lanthanide coordination polymers and their Ag+-modulated fluorescence. *J. Am. Chem. Soc.* **2004**, *126* (8), 2280–2281.
62. Cui, L.; Wu, J.; Li, J.; Ju, H. Electrochemical sensor for lead cation sensitized with a DNA functionalized porphyrinic metal–organic framework. *Anal. Chem.* **2015**, *87* (20), 10635–10641.
63. Wang, Y.; Wu, Y.; Xie, J.; Hu, X. Metal–organic framework modified carbon paste electrode for lead sensor. *Sensors Actuators B: Chemical* **2013**, *177*, 1161–1166.
64. Ma, J.-P.; Yu, Y.; Dong, Y.-B. Fluorene-based Cu (ii)-MOF: a visual colorimetric anion sensor and separator based on an anion-exchange approach. *Chem. Commun.* **2012**, *48* (24), 2946–2948.
65. Chen, Y.-Q.; Li, G.-R.; Chang, Z.; Qu, Y.-K.; Zhang, Y.-H.; Bu, X.-H. A Cu (I) metal–organic framework with 4-fold helical channels for sensing anions. *Chem. Sci.* **2013**, *4* (9), 3678–3682.
66. Shi, P.-F.; Hu, H.-C.; Zhang, Z.-Y.; Xiong, G.; Zhao, B. Heterometal–organic frameworks as highly sensitive and highly selective luminescent probes to detect I− ions in aqueous solutions. *Chem. Commun.* **2015**, *51* (19), 3985–3988.
67. Chen, B.; Wang, L.; Zapata, F.; Qian, G.; Lobkovsky, E. B. A luminescent microporous metal–organic framework for the recognition and sensing of anions. *J. Am. Chem. Soc.* **2008**, *130* (21), 6718–6719.
68. Hinterholzinger, F. M.; Rühle, B.; Wuttke, S.; Karaghiosoff, K.; Bein, T. Highly sensitive and selective fluoride detection in water through fluorophore release from a metal-organic framework. *Sci Rep* **2013**, *3*, 2562.
69. Chow, C.-F.; Lam, M. H.; Wong, W.-Y. A heterobimetallic ruthenium (II)− copper (II) donor− acceptor complex as a chemodosimetric ensemble for selective cyanide detection. *Inorg. Chem.* **2004**, *43* (26), 8387–8393.
70. Wang, Y.; Cheng, L.; Liu, Z. Y.; Wang, X. G.; Ding, B.; Yin, L.; Zhou, B. B.; Li, M. S.; Wang, J. X.; Zhao, X. J. An Ideal Detector Composed of Two-Dimensional Cd (II)–Triazole Frameworks for Nitro-Compound Explosives and Potassium Dichromate. *Chemistry–A European Journal* **2015**, *21* (40), 14171–14178.
71. Chen, S.; Shi, Z.; Qin, L.; Jia, H.; Zheng, H. Two new luminescent Cd (II)-metal–organic frameworks as bifunctional chemosensors for detection of cations Fe3+, anions CrO42–, and Cr2O72–in aqueous solution. *Crystal Growth Design* **2017**, *17* (1), 67–72.
72. Pramanik, S.; Zheng, C.; Zhang, X.; Emge, T. J.; Li, J. New microporous metal− organic framework demonstrating unique selectivity for detection of high explosives and aromatic compounds. *J. Am. Chem. Soc.* **2011**, *133* (12), 4153–4155.

73. Hu, Z.; Deibert, B. J.; Li, J. Luminescent metal–organic frameworks for chemical sensing and explosive detection. *Chem. Soc. Rev.* **2014**, *43* (16), 5815–5840.
74. Zhang, L.; Kang, Z.; Xin, X.; Sun, D. Metal–organic frameworks based luminescent materials for nitroaromatics sensing. *CrystEngComm* **2016**, *18* (2), 193–206.
75. Kundu, B. K.; Pragti; Reena; Mobin, S. M.; Mukhopadhyay, S. Mechanistic and thermodynamic aspects of a pyrene-based fluorescent probe to detect picric acid. *New J. Chem.* **2019**, *43* (29), 11483–11492.
76. Kundu, B. K.; Singh, R.; Tiwari, R.; Nayak, D.; Mukhopadhyay, S. An amide probe as a selective Al3+ and Fe3+ sensor inside the HeLa and a549 cell lines: Pictet–Spengler reaction for the rapid detection of tryptophan amino acid. *New J. Chem.* **2019**, *43* (12), 4867–4877.
77. Kundu, B. K. Studies on metal complexes of 'N,O-Donor' ligands in modelling metalloenzymes sensing and catalysis. *PhD Thesis, IIT Indore* **2019**.
78. Kundu, B. K.; Pragti; Mobin, S. M.; Mukhopadhyay, S. Studies on the influence of the nuclearity of zinc(ii) hemi-salen complexes on some pivotal biological applications. *Dalton Transactions* **2020**, *49* (43), 15481–15503.
79. Kundu, B. K.; Mandal, P.; Mukhopadhyay, B. G.; Tiwari, R.; Nayak, D.; Ganguly, R.; Mukhopadhyay, S. Substituent dependent sensing behavior of Schiff base chemosensors in detecting Zn2+and Al3+ ions: Drug sample analysis and living cell imaging. *Sensors and Actuators B: Chemical* **2019**, *282*, 347–358.
80. Hasegawa, Y.; Kimura, Y.; Murakoshi, K.; Wada, Y.; Kim, J.-H.; Nakashima, N.; Yamanaka, T.; Yanagida, S. Enhanced emission of deuterated tris (hexafluoroacetylacetonato) neodymium (III) complex in solution by suppression of radiationless transition via vibrational excitation. *The Journal of Physical Chemistry* **1996**, *100* (24), 10201–10205.
81. Guo, Y.; Feng, X.; Han, T.; Wang, S.; Lin, Z.; Dong, Y.; Wang, B. Tuning the luminescence of metal–organic frameworks for detection of energetic heterocyclic compounds. *J. Am. Chem. Soc.* **2014**, *136* (44), 15485–15488.
82. Zhou, J.; Li, H.; Zhang, H.; Li, H.; Shi, W.; Cheng, P. A Bimetallic Lanthanide Metal–Organic Material as a Self-Calibrating Color-Gradient Luminescent Sensor. *Adv. Mater.* **2015**, *27* (44), 7072–7077.
83. Dong, M. J.; Zhao, M.; Ou, S.; Zou, C.; Wu, C. D. A luminescent dye@ MOF platform: emission fingerprint relationships of volatile organic molecules. *Angew. Chem.* **2014**, *126* (6), 1601–1605.
84. Yu, Y.; Ma, J.-P.; Zhao, C.-W.; Yang, J.; Zhang, X.-M.; Liu, Q.-K.; Dong, Y.-B. Copper (I) metal–organic framework: visual sensor for detecting small polar aliphatic volatile organic compounds. *Inorg. Chem.* **2015**, *54* (24), 11590–11592.
85. Wang, J.-H.; Li, M.; Li, D. A dynamic, luminescent and entangled MOF as a qualitative sensor for volatile organic solvents and a quantitative monitor for acetonitrile vapour. *Chem. Sci.* **2013**, *4* (4), 1793–1801.
86. Xie, Z.; Ma, L.; deKrafft, K. E.; Jin, A.; Lin, W. Porous phosphorescent coordination polymers for oxygen sensing. *J. Am. Chem. Soc.* **2010**, *132* (3), 922–923.
87. Ho, M.-L.; Chen, Y.-A.; Chen, T.-C.; Chang, P.-J.; Yu, Y.-P.; Cheng, K.-Y.; Shih, C.-H.; Lee, G.-H.; Sheu, H.-S. Synthesis, structure and oxygen-sensing properties of Iridium (III)-

containing coordination polymers with different cations. *Dalton Transactions* **2012**, *41* (9), 2592–2600.

88. Barrett, S. M.; Wang, C.; Lin, W. Oxygen sensing via phosphorescence quenching of doped metal–organic frameworks. *Journal of Materials Chemistry* **2012**, *22* (20), 10329–10334.
89. Gong, Y.-N.; Lu, T.-B. Fast detection of oxygen by the naked eye using a stable metal–organic framework containing methyl viologen cations. *Chem. Commun.* **2013**, *49* (70), 7711–7713.
90. Chen, Y.-T.; Lin, C.-Y.; Lee, G.-H.; Ho, M.-L. Four new lead (ii)–iridium (iii) heterobimetallic coordination frameworks: synthesis, structures, luminescence and oxygen-sensing properties. *CrystEngComm* **2015**, *17* (10), 2129–2140.
91. Kreno, L. E.; Hupp, J. T.; Van Duyne, R. P. Metal– organic framework thin film for enhanced localized surface plasmon resonance gas sensing. *Anal. Chem.* **2010**, *82* (19), 8042–8046.
92. Gassensmith, J. J.; Kim, J. Y.; Holcroft, J. M.; Farha, O. K.; Stoddart, J. F.; Hupp, J. T.; Jeong, N. C. A metal–organic framework-based material for electrochemical sensing of carbon dioxide. *J. Am. Chem. Soc.* **2014**, *136* (23), 8277–8282.
93. Chen, E.-X.; Yang, H.; Zhang, J. Zeolitic imidazolate framework as formaldehyde gas sensor. *Inorg. Chem.* **2014**, *53* (11), 5411–5413.
94. Achmann, S.; Hagen, G.; Kita, J.; Malkowsky, I. M.; Kiener, C.; Moos, R. Metal-organic frameworks for sensing applications in the gas phase. *Sensors* **2009**, *9* (3), 1574–1589.
95. Travlou, N. A.; Singh, K.; Rodríguez-Castellón, E.; Bandosz, T. J. Cu–BTC MOF–graphene-based hybrid materials as low concentration ammonia sensors. *J. Mater. Chem. A* **2015**, *3* (21), 11417–11429.
96. Shustova, N. B.; Cozzolino, A. F.; Reineke, S.; Baldo, M.; Dincă, M. Selective turn-on ammonia sensing enabled by high-temperature fluorescence in metal–organic frameworks with open metal sites. *J. Am. Chem. Soc.* **2013**, *135* (36), 13326–13329.
97. Zhou, X.-P.; Xu, Z.; He, J.; Zeller, M.; Hunter, A. D.; Clérac, R.; Mathonière, C.; Chui, S. S.-Y.; Che, C.-M. Coordination networks from Cu cations and tetrakis (methylthio) benzenedicarboxylic acid: Tunable bonding patterns and selective sensing for NH3 gas. *Inorg. Chem.* **2010**, *49* (21), 10191–10198.
98. Wu, P.; Wang, J.; He, C.; Zhang, X.; Wang, Y.; Liu, T.; Duan, C. Luminescent metal-organic frameworks for selectively sensing nitric oxide in an aqueous solution and in living cells. *Adv. Funct. Mater.* **2012**, *22* (8), 1698–1703.
99. Desai, A. V.; Samanta, P.; Manna, B.; Ghosh, S. K. Aqueous phase nitric oxide detection by an amine-decorated metal–organic framework. *Chem. Commun.* **2015**, *51* (28), 6111–6114.
100. Nagarkar, S. S.; Desai, A. V.; Ghosh, S. K. A Nitro-Functionalized Metal–Organic Framework as a Reaction-Based Fluorescence Turn-On Probe for Rapid and Selective H2S Detection. *Chemistry–A European Journal* **2015**, *21* (28), 9994–9997.
101. Li, H.; Feng, X.; Guo, Y.; Chen, D.; Li, R.; Ren, X.; Jiang, X.; Dong, Y.; Wang, B. A malonitrile-functionalized metal-organic framework for hydrogen sulfide detection and selective amino acid molecular recognition. *Scientific Reports* **2014**, *4* (1), 1–5.
102. Ma, Y.; Su, H.; Kuang, X.; Li, X.; Zhang, T.; Tang, B. Heterogeneous nano metal–organic framework fluorescence probe for highly selective and sensitive detection of hydrogen sulfide in living cells. *Anal. Chem.* **2014**, *86* (22), 11459–11463.

103. Cui, J.; Wong, Y. L.; Zeller, M.; Hunter, A. D.; Xu, Z. Pd uptake and H2S sensing by an amphoteric metal–organic framework with a soft core and rigid side arms. *Angew. Chem. Int. Ed.* **2014**, *53* (52), 14438–14442.
104. Nickerl, G.; Senkovska, I.; Kaskel, S. Tetrazine functionalized zirconium MOF as an optical sensor for oxidizing gases. *Chem. Commun.* **2015**, *51* (12), 2280–2282.
105. Cui, Y.; Xu, H.; Yue, Y.; Guo, Z.; Yu, J.; Chen, Z.; Gao, J.; Yang, Y.; Qian, G.; Chen, B. A luminescent mixed-lanthanide metal–organic framework thermometer. *J. Am. Chem. Soc.* **2012**, *134* (9), 3979–3982.
106. Han, Y.-H.; Tian, C.-B.; Li, Q.-H.; Du, S.-W. Highly chemical and thermally stable luminescent Eu x Tb 1− x MOF materials for broad-range pH and temperature sensors. *Journal of Materials Chemistry C* **2014**, *2* (38), 8065–8070.
107. Zhao, D.; Rao, X.; Yu, J.; Cui, Y.; Yang, Y.; Qian, G. Design and synthesis of an MOF thermometer with high sensitivity in the physiological temperature range. *Inorg. Chem.* **2015**, *54* (23), 11193–11199.
108. Cui, Y.; Song, R.; Yu, J.; Liu, M.; Wang, Z.; Wu, C.; Yang, Y.; Wang, Z.; Chen, B.; Qian, G. Dual-emitting MOF⊃ dye composite for ratiometric temperature sensing. *Adv. Mater.* **2015**, *27* (8), 1420–1425.
109. Kateb, B.; Yamamoto, V.; Yu, C.; Grundfest, W.; Gruen, J. P. Infrared thermal imaging: a review of the literature and case report. *NeuroImage* **2009**, *47*, T154–T162.
110. Yu, Y.; Zhang, X.-M.; Ma, J.-P.; Liu, Q.-K.; Wang, P.; Dong, Y.-B. Cu (I)-MOF: naked-eye colorimetric sensor for humidity and formaldehyde in single-crystal-to-single-crystal fashion. *Chem. Commun.* **2014**, *50* (12), 1444–1446.
111. Yu, Y.; Ma, J.-P.; Dong, Y.-B. Luminescent humidity sensors based on porous Ln 3+-MOFs. *CrystEngComm* **2012**, *14* (21), 7157–7160.
112. Robinson, A. L.; Stavila, V.; Zeitler, T. R.; White, M. I.; Thornberg, S. M.; Greathouse, J. A.; Allendorf, M. D. Ultrasensitive Humidity Detection Using Metal–Organic Framework-Coated Microsensors. *Anal. Chem.* **2012**, *84* (16), 7043–7051.
113. Liu, J.; Sun, F.; Zhang, F.; Wang, Z.; Zhang, R.; Wang, C.; Qiu, S. In situ growth of continuous thin metal–organic framework film for capacitive humidity sensing. *Journal of Materials Chemistry* **2011**, *21* (11), 3775–3778.
114. Zhang, Y.; Chen, Y.; Zhang, Y.; Cong, H.; Fu, B.; Wen, S.; Ruan, S. A novel humidity sensor based on NH 2-MIL-125 (Ti) metal organic framework with high responsiveness. *J. Nanopart. Res.* **2013**, *15* (10), 1–6.
115. Douvali, A.; Tsipis, A. C.; Eliseeva, S. V.; Petoud, S.; Papaefstathiou, G. S.; Malliakas, C. D.; Papadas, I.; Armatas, G. S.; Margiolaki, I.; Kanatzidis, M. G. Turn-on luminescence sensing and real-time detection of traces of water in organic solvents by a flexible metal–organic framework. *Angew. Chem. Int. Ed.* **2015**, *127* (5), 1671–1676.
116. Zhu, W.-H.; Wang, Z.-M.; Gao, S. Two 3D porous lanthanide– fumarate– oxalate frameworks exhibiting framework dynamics and luminescent change upon reversible de-and rehydration. *Inorg. Chem.* **2007**, *46* (4), 1337–1342.
117. Weiss, A.; Reimer, N.; Stock, N.; Tiemann, M.; Wagner, T. Surface-modified CAU-10 MOF materials as humidity sensors: impedance spectroscopic study on water uptake. *PCCP* **2015**, *17* (33), 21634–21642.

118. Zeng, M.-H.; Tan, Y.-X.; He, Y.-P.; Yin, Z.; Chen, Q.; Kurmoo, M. A porous 4-fold-interpenetrated chiral framework exhibiting vapochromism, single-crystal-to-single-crystal solvent exchange, gas sorption, and a poisoning effect. *Inorg. Chem.* **2013**, *52* (5), 2353–2360.
119. Sapsanis, C.; Omran, H.; Chernikova, V.; Shekhah, O.; Belmabkhout, Y.; Buttner, U.; Eddaoudi, M.; Salama, K. N. Insights on capacitive interdigitated electrodes coated with MOF thin films: Humidity and VOCs sensing as a case study. *Sensors* **2015**, *15* (8), 18153–18166.
120. Ferrando-Soria, J.; Khajavi, H.; Serra-Crespo, P.; Gascon, J.; Kapteijn, F.; Julve, M.; Lloret, F.; Pasán, J.; Ruiz-Pérez, C.; Journaux, Y. Highly selective chemical sensing in a luminescent nanoporous magnet. *Adv. Mater.* **2012**, *24* (41), 5625–5629.
121. Hu, Z.; Lustig, W. P.; Zhang, J.; Zheng, C.; Wang, H.; Teat, S. J.; Gong, Q.; Rudd, N. D.; Li, J. Effective detection of mycotoxins by a highly luminescent metal–organic framework. *J. Am. Chem. Soc.* **2015**, *137* (51), 16209–16215.
122. Wang, B.; Lv, X.-L.; Feng, D.; Xie, L.-H.; Zhang, J.; Li, M.; Xie, Y.; Li, J.-R.; Zhou, H.-C. Highly stable Zr (IV)-based metal–organic frameworks for the detection and removal of antibiotics and organic explosives in water. *J. Am. Chem. Soc.* **2016**, *138* (19), 6204–6216.
123. Zhang, H.-T.; Zhang, J.-W.; Huang, G.; Du, Z.-Y.; Jiang, H.-L. An amine-functionalized metal–organic framework as a sensing platform for DNA detection. *Chem. Commun.* **2014**, *50* (81), 12069–12072.
124. Qin, L.; Lin, L.-X.; Fang, Z.-P.; Yang, S.-P.; Qiu, G.-H.; Chen, J.-X.; Chen, W.-H. A water-stable metal–organic framework of a zwitterionic carboxylate with dysprosium: a sensing platform for Ebolavirus RNA sequences. *Chem. Commun.* **2016**, *52* (1), 132–135.
125. Harbuzaru, B. V.; Corma, A.; Rey, F.; Jordá, J. L.; Ananias, D.; Carlos, L. D.; Rocha, J. A miniaturized linear pH sensor based on a highly photoluminescent self-assembled europium (III) metal–organic framework. *Angew. Chem. Int. Ed.* **2009**, *48* (35), 6476–6479.
126. Harbuzaru, B. V.; Corma, A.; Rey, F.; Jorda, J. L.; Ananias, D.; Carlos, L. D.; Rocha, J. A Miniaturized Linear pH Sensor Based on a Highly Photoluminescent Self-Assembled Europium (III) Metal-Organic Framework We thank the European Network of Excellence FAME and Beamline BM-16 at ESRF in Grenoble for beam time allocation. BVH, AC, JLJ, and FR would like to thank PROMETEO, Fundacion Areces, and Spanish CICYT (MAT2006-08039, M0ZAT2006-14274-C02-01) for financial support. DA, LDC, and JR thank FCT and PTDC. We also thank FJ Martinez-Casado for helpful discussions. *Angew. Chem.* **2009**, *121* (35), 6598.
127. Deibert, B. J.; Li, J. A distinct reversible colorimetric and fluorescent low pH response on a water-stable zirconium–porphyrin metal–organic framework. *Chem. Commun.* **2014**, *50* (68), 9636–9639.
128. Jiang, H.-L.; Feng, D.; Wang, K.; Gu, Z.-Y.; Wei, Z.; Chen, Y.-P.; Zhou, H.-C. An exceptionally stable, porphyrinic Zr metal–organic framework exhibiting pH-dependent fluorescence. *J. Am. Chem. Soc.* **2013**, *135* (37), 13934–13938.
129. Aguilera-Sigalat, J.; Bradshaw, D. A colloidal water-stable MOF as a broad-range fluorescent pH sensor via post-synthetic modification. *Chem. Commun.* **2014**, *50* (36), 4711–4713.
130. He, C.; Lu, K.; Lin, W. Nanoscale metal–organic frameworks for real-time intracellular pH sensing in live cells. *J. Am. Chem. Soc.* **2014**, *136* (35), 12253–12256.

131. Zan, M.; Li, J.; Luo, S.; Ge, Z. Dual pH-triggered multistage drug delivery systems based on host–guest interaction-associated polymeric nanogels. *Chem. Commun.* **2014**, *50* (58), 7824–7827.
132. Lu, Y.; Yan, B. A ratiometric fluorescent pH sensor based on nanoscale metal–organic frameworks (MOFs) modified by europium (III) complexes. *Chem. Commun.* **2014**, *50* (87), 13323–13326.
133. Li, H.-Y.; Wei, Y.-L.; Dong, X.-Y.; Zang, S.-Q.; Mak, T. C. Novel Tb-MOF embedded with viologen species for multi-photofunctionality: photochromism, photomodulated fluorescence, and luminescent pH sensing. *Chem. Mater.* **2015**, *27* (4), 1327–1331.
134. Meng, Q.; Xin, X.; Zhang, L.; Dai, F.; Wang, R.; Sun, D. A multifunctional Eu MOF as a fluorescent pH sensor and exhibiting highly solvent-dependent adsorption and degradation of rhodamine B. *J. Mater. Chem. A* **2015**, *3* (47), 24016–24021.
135. Xu, X.-Y.; Yan, B. An efficient and sensitive fluorescent pH sensor based on amino functional metal–organic frameworks in aqueous environment. *Dalton Transactions* **2016**, *45* (16), 7078–7084.
136. Lu, Z.-Z.; Zhang, R.; Li, Y.-Z.; Guo, Z.-J.; Zheng, H.-G. Solvatochromic behavior of a nanotubular metal– organic framework for sensing small molecules. *J. Am. Chem. Soc.* **2011**, *133* (12), 4172–4174.
137. Beauvais, L. G.; Shores, M. P.; Long, J. R. Cyano-bridged Re6Q8 (Q= S, Se) cluster-cobalt (II) framework materials: versatile solid chemical sensors. *J. Am. Chem. Soc.* **2000**, *122* (12), 2763–2772.
138. Qiu, L.-G.; Li, Z.-Q.; Wu, Y.; Wang, W.; Xu, T.; Jiang, X. Facile synthesis of nanocrystals of a microporous metal–organic framework by an ultrasonic method and selective sensing of organoamines. *Chem. Commun.* **2008** (31), 3642–3644.
139. Zou, X.; Zhu, G.; Hewitt, I. J.; Sun, F.; Qiu, S. Synthesis of a metal–organic framework film by direct conversion technique for VOCs sensing. *Dalton Transactions* **2009** (16), 3009–3013.
140. Hu, X.-L.; Liu, F.-H.; Qin, C.; Shao, K.-Z.; Su, Z.-M. A 2D bilayered metal–organic framework as a fluorescent sensor for highly selective sensing of nitro explosives. *Dalton Transactions* **2015**, *44* (17), 7822–7827.
141. Buragohain, A.; Yousufuddin, M.; Sarma, M.; Biswas, S. 3D luminescent amide-functionalized cadmium tetrazolate framework for selective detection of 2, 4, 6-trinitrophenol. *Crystal Growth Design* **2016**, *16* (2), 842–851.
142. Zhang, Y.; Yuan, S.; Day, G.; Wang, X.; Yang, X.; Zhou, H.-C. Luminescent sensors based on metal-organic frameworks. *Coord. Chem. Rev.* **2018**, *354*, 28–45.
143. Nagarkar, S. S.; Joarder, B.; Chaudhari, A. K.; Mukherjee, S.; Ghosh, S. K. Highly selective detection of nitro explosives by a luminescent metal–organic framework. *Angew. Chem.* **2013**, *125* (10), 2953–2957.
144. Zhang, C.; Sun, L.; Yan, Y.; Li, J.; Song, X.; Liu, Y.; Liang, Z. A luminescent cadmium metal–organic framework for sensing of nitroaromatic explosives. *Dalton Transactions* **2015**, *44* (1), 230–236.
145. Li, X.; Yang, L.; Zhao, L.; Wang, X.-L.; Shao, K.-Z.; Su, Z.-M. Luminescent metal–organic frameworks with anthracene chromophores: small-molecule sensing and highly selective sensing for nitro explosives. *Crystal Growth Design* **2016**, *16* (8), 4374–4382.

146. Ghosh, P.; Saha, S. K.; Roychowdhury, A.; Banerjee, P. Recognition of an explosive and mutagenic water pollutant, 2, 4, 6-trinitrophenol, by cost-effective luminescent MOFs. *Eur. J. Inorg. Chem.* **2015**, *2015* (17), 2851–2857.
147. Sun, L.; Xing, H.; Xu, J.; Liang, Z.; Yu, J.; Xu, R. A novel (3, 3, 6)-connected luminescent metal–organic framework for sensing of nitroaromatic explosives. *Dalton Transactions* **2013**, *42* (15), 5508–5513.
148. Yang, W.; Chang, G.; Wang, H.; Hu, T. L.; Yao, Z.; Alfooty, K.; Xiang, S.; Chen, B. A Three-Dimensional Tetraphenylíethene-Based Metal–Organic Framework for Selective Gas Separation and Luminescence Sensing of Metal Ions. *Eur. J. Inorg. Chem.* **2016**, *2016* (27), 4470–4475.
149. Tomar, K.; Gupta, M.; Gupta, A. K. A novel 2D porous Cd (II) MOF with a (4, 4) connected binodal network: Synthesis and photoluminescence sensing of small molecules. *Inorg. Chem. Commun.* **2016**, *64*, 16–18.
150. Zhao, X.; Li, Y.; Chang, Z.; Chen, L.; Bu, X.-H. A four-fold interpenetrated metal–organic framework as a fluorescent sensor for volatile organic compounds. *Dalton Transactions* **2016**, *45* (38), 14888–14892.
151. Mukherjee, S.; Desai, A. V.; Manna, B.; Inamdar, A. I.; Ghosh, S. K. Exploitation of guest accessible aliphatic amine functionality of a metal–organic framework for selective detection of 2, 4, 6-trinitrophenol (TNP) in water. *Crystal Growth Design* **2015**, *15* (9), 4627–4634.
152. Li, L.; Zhang, S.; Xu, L.; Han, L.; Chen, Z.-N.; Luo, J. An intensely luminescent metal–organic framework based on a highly light-harvesting dyclo-metalated iridium (III) unit showing effective detection of explosives. *Inorg. Chem.* **2013**, *52* (21), 12323–12325.
153. Lan, A.; Li, K.; Wu, H.; Olson, D. H.; Emge, T. J.; Ki, W.; Hong, M.; Li, J. A luminescent microporous metal–organic framework for the fast and reversible detection of high explosives. *Angew. Chem.* **2009**, *121* (13), 2370–2374.
154. Yao, R.-X.; Cui, X.; Jia, X.-X.; Zhang, F.-Q.; Zhang, X.-M. A luminescent zinc (II) metal–organic framework (MOF) with conjugated π-electron ligand for high iodine capture and nitro-explosive detection. *Inorg. Chem.* **2016**, *55* (18), 9270–9275.
155. Wang, G.-Y.; Yang, L.-L.; Li, Y.; Song, H.; Ruan, W.-J.; Chang, Z.; Bu, X.-H. A luminescent 2D coordination polymer for selective sensing of nitrobenzene. *Dalton Transactions* **2013**, *42* (36), 12865–12868.
156. Xie, W.; Zhang, S.-R.; Du, D.-Y.; Qin, J.-S.; Bao, S.-J.; Li, J.; Su, Z.-M.; He, W.-W.; Fu, Q.; Lan, Y.-Q. Stable luminescent metal–organic frameworks as dual-functional materials to encapsulate Ln3+ ions for white-light emission and to detect nitroaromatic explosives. *Inorg. Chem.* **2015**, *54* (7), 3290–3296.
157. Bagheri, M.; Masoomi, M. Y.; Morsali, A.; Schoedel, A. Two dimensional host–guest metal–organic framework sensor with high selectivity and sensitivity to picric acid. *ACS Applied Materials Interfaces* **2016**, *8* (33), 21472–21479.
158. Di, L.; Zhang, J.-J.; Liu, S.-Q.; Ni, J.; Zhou, H.; Sun, Y.-J. Two dynamic ABW-type metal organic frameworks built of pentacarboxylate and Zn2+ as photoluminescent probes of nitroaromatics. *Crystal Growth Design* **2016**, *16* (8), 4539–4546.

159. Shi, Z.-Q.; Guo, Z.-J.; Zheng, H.-G. Two luminescent Zn (II) metal–organic frameworks for exceptionally selective detection of picric acid explosives. *Chem. Commun.* **2015**, *51* (39), 8300–8303.
160. Hou, B.-L.; Tian, D.; Liu, J.; Dong, L.-Z.; Li, S.-L.; Li, D.-S.; Lan, Y.-Q. A water-stable metal–organic framework for highly sensitive and selective sensing of Fe3+ ion. *Inorg. Chem.* **2016**, *55* (20), 10580–10586.
161. Shen, X.; Yan, B. Anionic metal–organic framework hybrids: functionalization with lanthanide ions or cationic dyes and fluorescence sensing of small molecules. *RSC Advances* **2016**, *6* (34), 28165–28170.
162. Wang, F.; Liu, W.; Teat, S. J.; Xu, F.; Wang, H.; Wang, X.; An, L.; Li, J. Chromophore-immobilized luminescent metal–organic frameworks as potential lighting phosphors and chemical sensors. *Chem. Commun.* **2016**, *52* (67), 10249–10252.
163. Liu, D.; Liu, X.; Liu, Y.; Yu, Y.; Chen, F.; Wang, C. Host–guest interaction dictated selective adsorption and fluorescence quenching of a luminescent lightweight metal–organic framework toward liquid explosives. *Dalton Transactions* **2014**, *43* (40), 15237–15244.
164. Wang, R.; Liu, X.; Huang, A.; Wang, W.; Xiao, Z.; Zhang, L.; Dai, F.; Sun, D. J. I. c. Unprecedented solvent-dependent sensitivities in highly efficient detection of metal ions and nitroaromatic compounds by a fluorescent barium metal–organic framework. *Inorg. Chem.* **2016**, *55* (4), 1782–1787.
165. Nagarkar, S. S.; Desai, A. V.; Samanta, P.; Ghosh, S. K. Aqueous phase selective detection of 2, 4, 6-trinitrophenol using a fluorescent metal–organic framework with a pendant recognition site. *Dalton Transactions* **2015**, *44* (34), 15175–15180.
166. Nagarkar, S. S.; Desai, A. V.; Ghosh, S. K. A fluorescent metal–organic framework for highly selective detection of nitro explosives in the aqueous phase. *Chem. Commun.* **2014**, *50* (64), 8915–8918.
167. Yang, J.; Wang, Z.; Li, Y.; Zhuang, Q.; Gu, J. Real-time monitoring of dissolved oxygen with inherent oxygen-sensitive centers in metal–organic frameworks. *Chem. Mater.* **2016**, *28* (8), 2652–2658.
168. Zhang, X.; Hu, Q.; Xia, T.; Zhang, J.; Yang, Y.; Cui, Y.; Chen, B.; Qian, G. Turn-on and ratiometric luminescent sensing of hydrogen sulfide based on metal–organic frameworks. *ACS Applied Materials Interfaces* **2016**, *8* (47), 32259–32265.
169. Drache, F.; Bon, V.; Senkovska, I.; Adam, M.; Eychmüller, A.; Kaskel, S. Vapochromic Luminescence of a Zirconium-Based Metal–Organic Framework for Sensing Applications. *Eur. J. Inorg. Chem.* **2016**, *2016* (27), 4483–4489.
170. Kim, T. K.; Lee, J. H.; Moon, D.; Moon, H. R. Luminescent Li-based metal–organic framework tailored for the selective detection of explosive nitroaromatic compounds: direct observation of interaction sites. *Inorg. Chem.* **2013**, *52* (2), 589–595.
171. Liu, C.; Yan, B. A novel photofunctional hybrid material of pyrene functionalized metal-organic framework with conformation change for fluorescence sensing of Cu2+. *Sensors Actuators B: Chemical* **2016**, *235*, 541–546.
172. Zhang, S.; Han, L.; Li, L.; Cheng, J.; Yuan, D.; Luo, J. A highly symmetric metal–organic framework based on a propeller-like Ru-organic metalloligand for photocatalysis and explosives detection. *Crystal Growth Design* **2013**, *13* (12), 5466–5472.

173. Gong, Y.-N.; Huang, Y.-L.; Jiang, L.; Lu, T.-B. A luminescent microporous metal–organic framework with highly selective CO2 adsorption and sensing of nitro explosives. *Inorg. Chem.* **2014**, *53* (18), 9457–9459.
174. Du, X.; Fan, R.; Wang, X.; Yu, G.; Qiang, L.; Wang, P.; Gao, S.; Yang, Y. Cooperative Crystallization of Chiral Heterometallic Indium (III)–Potassium (I) Metal–Organic Frameworks as Photosensitizers in Luminescence Sensors and Dye-Sensitized Solar Cells. *Crystal Growth Design* **2016**, *16* (3), 1737–1745.
175. Qin, J.-H.; Ma, B.; Liu, X.-F.; Lu, H.-L.; Dong, X.-Y.; Zang, S.-Q.; Hou, H. Ionic liquid directed syntheses of water-stable Eu–and Tb–organic-frameworks for aqueous-phase detection of nitroaromatic explosives. *Dalton Transactions* **2015**, *44* (33), 14594–14603.
176. Zhou, X.; Li, H.; Xiao, H.; Li, L.; Zhao, Q.; Yang, T.; Zuo, J.; Huang, W. A microporous luminescent europium metal–organic framework for nitro explosive sensing. *Dalton Transactions* **2013**, *42* (16), 5718–5723.
177. Xu, H.; Fang, M.; Cao, C.-S.; Qiao, W.-Z.; Zhao, B. Unique (3, 4, 10)-connected lanthanide–organic framework as a recyclable chemical sensor for detecting Al3+. *Inorg. Chem.* **2016**, *55* (10), 4790–4794.
178. Li, G.-P.; Liu, G.; Li, Y.-Z.; Hou, L.; Wang, Y.-Y.; Zhu, Z. Uncommon pyrazoyl-carboxyl bifunctional ligand-based microporous lanthanide systems: sorption and luminescent sensing properties. *Inorg. Chem.* **2016**, *55* (8), 3952–3959.
179. Wen, G.-X.; Wu, Y.-P.; Dong, W.-W.; Zhao, J.; Li, D.-S.; Zhang, J. An ultrastable europium (III)–organic framework with the capacity of discriminating Fe2+/Fe3+ ions in various solutions. *Inorg. Chem.* **2016**, *55* (20), 10114–10117.
180. Wang, D.; Tan, Q.; Liu, J.; Liu, Z. A stable europium metal–organic framework as a dual-functional luminescent sensor for quantitatively detecting temperature and humidity. *Dalton Transactions* **2016**, *45* (46), 18450–18454.
181. Shen, Y.; Xue, P.; Liu, J.; Ding, J.; Sun, J.; Lu, R. High-contrast mechanofluorochromism and acidochromism of D-π-A type quinoline derivatives. *Dyes Pigments* **2019**, *163*, 71–77.
182. Singha, D. K.; Majee, P.; Mondal, S. K.; Mahata, P. pH-Controlled Luminescence Turn-On Behaviour of a Water-Soluble Europium-Based Molecular Complex. *Eur. J. Inorg. Chem.* **2016**, *2016* (28), 4631–4636.
183. Zhai, B.; Li, Z.-Y.; Wu, Z.-L.; Cui, J.-Z. A novel europium metal-organic framework as luminescent probe for detecting Al3+. *Inorg. Chem. Commun.* **2016**, *71*, 23–26.
184. Xu, X.-Y.; Yan, B. Nanoscale LnMOF-functionalized nonwoven fibers protected by a polydimethylsiloxane coating layer as a highly sensitive ratiometric oxygen sensor. *Journal of Materials Chemistry C* **2016**, *4* (36), 8514–8521.
185. Wang, J.; Jiang, M.; Yan, L.; Peng, R.; Huangfu, M.; Guo, X.; Li, Y.; Wu, P. Multifunctional luminescent Eu (III)-based metal–organic framework for sensing methanol and detection and adsorption of Fe (III) ions in aqueous solution. *Inorg. Chem.* **2016**, *55* (24), 12660–12668.
186. Xu, H.; Zhai, B.; Cao, C.-S.; Zhao, B. A bifunctional europium–organic framework with chemical fixation of CO2 and luminescent detection of Al3+. *Inorg. Chem.* **2016**, *55* (19), 9671–9676.
187. Wang, X.; Qin, T.; Bao, S.-S.; Zhang, Y.-C.; Shen, X.; Zheng, L.-M.; Zhu, D. Facile synthesis of a water stable 3D Eu-MOF showing high proton conductivity and its application as a sensitive luminescent sensor for Cu 2+ ions. *J. Mater. Chem. A* **2016**, *4* (42), 16484–16489.

188. Zhang, Y.; Li, B.; Ma, H.; Zhang, L.; Jiang, H.; Song, H.; Zhang, L.; Luo, Y. A nanoscaled lanthanide metal–organic framework as a colorimetric fluorescence sensor for dipicolinic acid based on modulating energy transfer. *Journal of Materials Chemistry C* **2016**, *4* (30), 7294–7301.
189. Cui, Y.; Zou, W.; Song, R.; Yu, J.; Zhang, W.; Yang, Y.; Qian, G. A ratiometric and colorimetric luminescent thermometer over a wide temperature range based on a lanthanide coordination polymer. *Chem. Commun.* **2014**, *50* (6), 719–721.
190. Gong, Y.-N.; Jiang, L.; Lu, T.-B. A highly stable dynamic fluorescent metal–organic framework for selective sensing of nitroaromatic explosives. *Chem. Commun.* **2013**, *49* (94), 11113–11115.
191. Zhang, M.; Han, J.; Wu, H.; Wei, Q.; Xie, G.; Chen, S.; Gao, S. Tb-MOF: a naked-eye and regenerable fluorescent probe for selective and quantitative detection of Fe 3+ and Al 3+ ions. *RSC Advances* **2016**, *6* (97), 94622–94628.
192. Zhao, J.; Wang, Y.-N.; Dong, W.-W.; Wu, Y.-P.; Li, D.-S.; Zhang, Q.-C. A robust luminescent Tb (III)-MOF with Lewis basic pyridyl sites for the highly sensitive detection of metal ions and small molecules. *Inorg. Chem.* **2016**, *55* (7), 3265–3271.
193. Zhao, X.-L.; Tian, D.; Gao, Q.; Sun, H.-W.; Xu, J.; Bu, X.-H. A chiral lanthanide metal–organic framework for selective sensing of Fe (III) ions. *Dalton Transactions* **2016**, *45* (3), 1040–1046.
194. Xia, T.; Song, T.; Cui, Y.; Yang, Y.; Qian, G. A dye encapsulated terbium-based metal–organic framework for ratiometric temperature sensing. *Dalton Transactions* **2016**, *45* (46), 18689–18695.
195. Wang, H.-H.; Zhou, L.-J.; Wang, Y.-L.; Liu, Q.-Y. Terbium-biphenyl-3, 3′-disulfonyl-4,4′-dicarboxylate framework with sulfonate sites for luminescent sensing of Cr3+ ion. *Inorg. Chem. Commun.* **2016**, *73*, 94–97.
196. Zhang, C.; Yan, Y.; Pan, Q.; Sun, L.; He, H.; Liu, Y.; Liang, Z.; Li, J. A microporous lanthanum metal–organic framework as a bi-functional chemosensor for the detection of picric acid and Fe 3+ ions. *Dalton Transactions* **2015**, *44* (29), 13340–13346.
197. Du, P.-Y.; Liao, S.-Y.; Gu, W.; Liu, X. A multifunctional chemical sensor based on a three-dimensional lanthanide metal-organic framework. *J. Solid State Chem.* **2016**, *244*, 31–34.
198. Gao, M.-L.; Wei, N.; Han, Z.-B. Anionic metal–organic framework for high-efficiency pollutant removal and selective sensing of Fe (III) ions. *RSC advances* **2016**, *6* (65), 60940–60944.
199. Zhang, R.-C.; Wang, J.-J.; Zhang, J.-C.; Wang, M.-Q.; Sun, M.; Ding, F.; Zhang, D.-J.; An, Y.-L. Coordination-Induced Syntheses of Two Hybrid Framework Iodides: A Thermochromic Luminescent Thermometer. *Inorg. Chem.* **2016**, *55* (15), 7556–7563.
200. Kreno, L. E.; Leong, K.; Farha, O. K.; Allendorf, M.; Van Duyne, R. P.; Hupp, J. T. Metal–organic framework materials as chemical sensors. *Chem. Rev.* **2012**, *112* (2), 1105–1125.
201. Yi, F. Y.; Chen, D.; Wu, M. K.; Han, L.; Jiang, H. L. Chemical sensors based on metal–organic frameworks. *ChemPlusChem* **2016**, *81* (8), 675–690.

202. Doty, F.; Bauer, C.; Skulan, A.; Grant, P.; Allendorf, M. Scintillating Metal-Organic Frameworks: A New Class of Radiation Detection Materials. *Adv. Mater.* **2009**, *21* (1), 95–101.

203. Bauer, C. A.; Timofeeva, T. V.; Settersten, T. B.; Patterson, B. D.; Liu, V. H.; Simmons, B. A.; Allendorf, M. D. Influence of connectivity and porosity on ligand-based luminescence in zinc metal– organic frameworks. *J. Am. Chem. Soc.* **2007**, *129* (22), 7136–7144.

Chapter 3

Recent Progress in Metal–Organic Frameworks for the Fabrication of Chemical Sensors

Khursheed Ahmad,[1] Mohd Quasim Khan,[2] Waseem Raza,[3,4] and M. A. Gondal[*,5,6]

[1]Department of Chemistry, Indian Institute of Technology Indore, Simrol, Khandwa Road, M.P., 453552 India

[2]Moradabad Muslim Degree College, Katghar, Moradabad, 244001, M.J.P. Rohilkhand University, Bareilly, U.P., India

[3]Department of Chemical Engineering, Indian Institute of Technology Delhi, Hauz Khas New Delhi, 110016 India

[4]Department of Materials Science and Engineering, WW4-LKO, University of Erlangen-Nuremberg, Martensstrasse, 7, 91058 Erlangen, Germany

[5]Laser Research Group, Department of Physics, Center of Research Excellence in Nanotechnology (CENT), King Fahd University of Petroleum and Minerals (KFUPM), P.O. Box 5047, Dhahran 31261, Saudi Arabia

[6]K.A. CARE Energy Research & Innovation Center, King Fahd University of Petroleum and Minerals, Dhahran 31261, Saudi Arabia

[*]Email: magondal@kfupme.edu.sa

Metal-organic frameworks (MOFs) receive much attention because of their highly porous nature and specific surface area. MOFs are crystalline materials with a microporous nature and are comprised of metal ion and multidentate organic linkers attached with coordination bonds. MOF materials also possess good permeability and electrocatalytic properties, which suggests their potential applications in electrochemical and optoelectronic devices. Herein, we summarize the current advances in preparation of MOFs and their hybrid composites for electrochemical sensing applications. The fabrication of electrochemical sensing electrodes and the role therein of MOFs, and their composites as electrode materials, are also reviewed. The current challenges and future perspectives for the development of MOF-based electrochemical sensing are discussed. We believe this chapter is useful to provide information regarding advances in the development of MOF-based electrochemical sensors.

Introduction

Metal-organic framework (MOF) materials have attracted the scientific community because of their excellent properties and potential application in optoelectronics and optoelectrical devices (*1–3*). MOF materials have good porosity and high surface area but suffer from low conductivity, which needs to be addressed (*4–7*). It has been shown that two-dimensional (2D) materials such as reduced graphene oxide (rGO), MXenes (metal carbide or metal nitride), and transition metal oxides possess good electrical properties and conductivity. Previously, rGO has been used as a key material to enhance the conductivity of low-conductive materials. In this regard, MOFs and hybrid composites consisting of MOFs and transition metal oxides or 2D materials have been prepared using novel strategies (*8–11*). MOF materials emerged as highly porous materials in 2013 and were employed as efficient electrode material for the fabrication of electrochemical sensors. MOFs and their hybrid composite materials have been applied in many different applications, such as biomedicine, energy conversion, batteries, supercapacitors, gas/energy storage, photocatalysis, solar cells, and optical/electrochemical sensors (*12–16*). MOFs have also proven to be a key material for the adsorption of different pollutants as well as an important electrode-modifying material for the construction of electrochemical sensors. A glassy carbon electrode (GCE) has been widely used as a working electrode substrate, and its active surface area has to be modified using MOFs as an electrode modifier. This MOF materials–modified GCE has been widely employed for the determination of biomolecules or toxic/hazardous compounds. MOF materials can be used as electrode materials but have some limitations, such as low stability in aqueous solutions and poor electrical conductivity. Hence, it is of great significance to develop strategies or hybrid materials to improve the electrical properties of MOF materials. Therefore, materials scientists have developed novel hybrid composite materials and utilized them as electrode modifiers for the modification of GCEs. The hybrid composite materials consist of MOFs and transition metal oxides or 2D materials such as rGO, and molybdenum disulfide (MoS_2), etc. The preparation of MOF hybrid composites was found to be beneficial — the presence of 2D materials such as rGO not only improves the stability of MOFs, but also enhances electrical conductivity, which provides a better path for electron transportation during electroanalytical measurements (*16*). It has also been observed that MOF-based hybrid composite materials have better electrocatalytic properties compared to the pristine MOFs. Therefore, it is of great importance to develop novel hybrid composite materials for the construction of electrochemical sensor. Previously, numerous efforts and attempts were made by materials scientists to develop highly efficient electrode modifiers (graphene oxide (GO), MnO_2, MgO, TiO_2, SnO_2, and polyaniline) for the fabrication of electrochemical sensors (*17–22*). Some perovskite structures such as $SrTiO_3$, $CaTiO_3$, and $SrZrO_3$ have also been employed as electrode materials (*23–25*). The literature shows that transition metal oxides, 2D materials, layered structured materials, and perovskite materials have been employed as electrode materials, and they can be more efficient electrode modifiers with MOFS (*26*). Thus, MOFs have been prepared with rGO and other metal oxides for electrochemical sensing applications (*27–30*).

In this chapter, we summarize progress in the preparation of MOFs and MOF hybrid composites for electrochemical sensing applications. The current challenges and future prospective are also discussed in the conclusion section. We believe that this chapter can provide deep information on the fabrication and advancements in the fabrication of MOFs and MOF hybrid composite–based electrochemical sensors.

Fabrication of MOF-Based Sensors

In the first step, the GCE needs to be cleaned with alumina slurry and a velvet pad. Further, an appropriate amount of the MOF (electrode modifier) must be dissolved or dispersed in ethanol to be drop casted onto the active area of the GCE. This modified GCE needs to be air-dried. In some cases, binders such as nafion, PVDF, etc. are deposited onto the MOF-modified GCE to improve the adhesiveness of the MOF on the surface of the GCE. Finally, this modified GCE can be used for electroanalytical sensing measurements.

MOFs in Electrochemical Sensing Applications

MOFs as an Electrode Modifier for Hydrazine Sensing

Hydrazine is widely used in aerospace, catalysis, fuels, agriculture, pharmaceutical, and military applications (*31*, *32*). Hydrazine is also utilized as a corrosion inhibitor, antioxidant, and emulsifier product. It is carcinogenic and possesses hazardous effects on humans, animals, and the environment (*33*). Hydrazine can be responsible for damaging of the spinal cord and can also affect the brain, DNA, kidneys, lungs, and liver (*33*, *34*). Thus, the determination of hydrazine is necessary. Previously, conventional determination approaches such as spectroscopic (*35*), luminescence (*36*), and chromatography (*37*) have been used. However, an electrochemical approach has received tremendous attention because of its simplicity, cost-effectiveness, and high sensitivity. In electrochemical sensors, an electrode material plays a vital role and influences the electrochemical performance of the developed sensors. The electrode materials with high surface area and porosity could be more useful for the development of highly efficient hydrazine sensors. In this regard, Rani et al. (*38*) prepared a highly porous Zn-MOF/rGO composite using the solvothermal method. The successful formation of Zn-MOF and Zn-MOF/rGO composite was confirmed by X-ray diffraction (XRD) and Fourier transform infrared spectroscopy (FTIR). The thermogravimetric analysis (TGA) suggested that incorporation of rGO to the Zn-MOF enhanced thermal stability. The XRD of the Zn-MOF, rGO, and Zn-MOF@rGO is depicted in Figure 1. The XRD data confirmed the formation of Zn-MOF, rGO, and Zn-MOF@rGO. The Brunauer–Emmett–Teller (BET) analysis was conducted to investigate the surface area properties of the prepared Zn-MOF and Zn-MOF/rGO. The N_2 adsorption–desorption isotherms of the Zn-MOF and Zn-MOF/rGO was recorded. The BET investigations showed that Zn-MOF has a specific surface area of 42.14 m^2/g whereas an enhanced specific surface area of 112.14 m^2/g was obtained for the Zn-MOF/rGO composite. A mean pore diameter of 11.3 nm and 22.3 nm was observed for Zn-MOF and Zn-MOF/rGO, respectively.

The structural surface morphological features of the Zn-MOF and Zn-MOF/rGO composite were investigated by transmission electron microscopy (TEM). The TEM results showed that Zn-MOF has a hexagonal prismatic-shaped surface morphology. The TEM results for Zn-MOF/rGO confirmed the presence of MOF particles on rGO sheets. The gold (Au) electrode was modified with Zn-MOF/rGO to determine the hydrazine using voltammetric investigations. Nafion film was used as a binder to enhance the adhesiveness of the electrode material. Authors employed electrochemical impedance spectroscopy (EIS) to investigate the electrochemical properties of the modified electrodes. The EIS results showed that the Zn-MOF/rGO modified Au electrode has good electrocatalytic properties compared to the bare Au electrode. Subsequently, cyclic voltammetry (CV) was utilized for further electrochemical investigations. The CV results also showed the better electrocatalytic properties of Zn-MOF/rGO modified Au electrode compared to the bare Au electrode. The CV curves of the Zn-MOF/rGO modified Au electrode were recorded in the presence of hydrazine in 0.1 M PBS at different scan rates. The CV results at different scan rates showed that the electrocatalytic current response of the Zn-MOF/rGO modified Au electrode increases with increasing scan rate. The interference study showed the selective nature of the Zn-MOF/rGO modified Au electrode toward the detection of hydrazine.

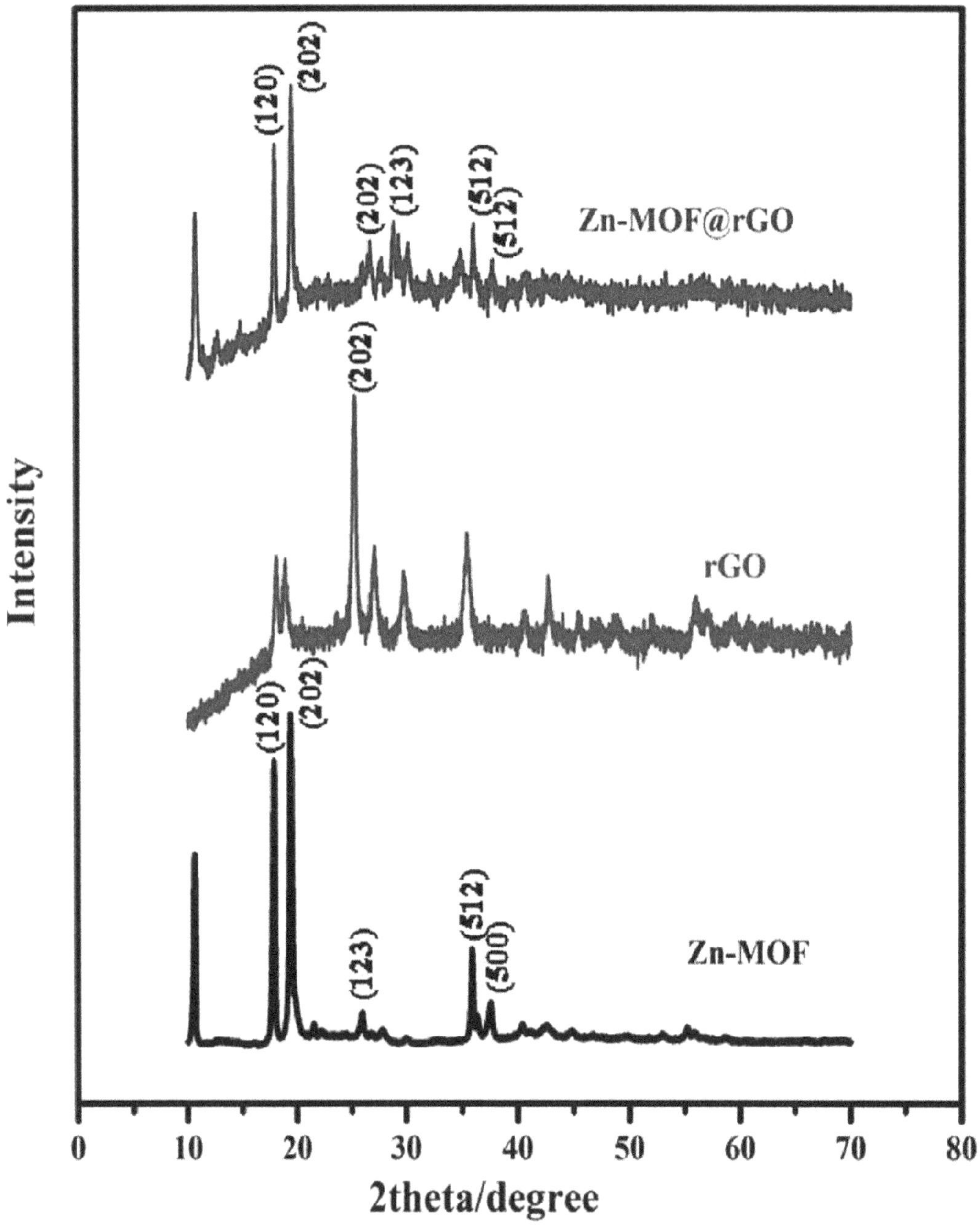

Figure 1. XRD of the Zn-MOF, rGO, and Zn-MOF@rGO. Reprinted with permission from reference (38). Copyright 2020 Elsevier.

The developed Zn-MOF/rGO modified Au electrode also exhibited good reproducibility, sensitivity, and detection limit. This showed that the decoration of Zn-MOF with rGO enhances their electrocatalytic properties. The 2D MOFs may provide more active sites and better electronic communication by increasing metal sites of the active molecules. Thus, Cao et al. (*39*) prepared ultrathin 2D Ni-MOF nanosheets using the solvothermal approach. X-ray photoelectron spectroscopic (XPS) analysis was used to characterize the ultrathin 2D Ni-MOF nanosheets. The XPS results suggested the formation of 2D Ni-MOF. The surface morphological properties of the ultrathin 2D Ni-MOF nanosheets were investigated by scanning electron microscopy (SEM) and TEM. The SEM results showed that Ni-MOF possesses accordion-like surface morphology. This surface morphology was further verified by TEM investigations. The TEM results also showed the presence of 2D Ni-MOF with accordion-like surface morphology. This kind of surface morphology may have surface area that could be useful for better electronic transportation. The GCE was modified with 2D Ni-MOF and electrochemical performance of the 2D Ni-MOF modified GCE was evaluated using CV and amperometric investigations. The 2D Ni-MOF modified GCE exhibited good electrocatalytic properties toward the determination of hydrazine. The amperometric investigations showed that 2D Ni-MOF modified GCE exhibited a good detection limit of 0.23 μM with high sensitivity of 2412 μA/μMcm^2. This indicated the better electrocatalytic properties of the 2D Ni-MOF. The 2D Ni-MOF modified GCE also showed good stability.

MOFs as Electrode Modifiers for Hydrogen Peroxide (H_2O_2) Sensing

H_2O_2 is a small molecule that is widely used in the food, clinics, drugs, environment, and chemical industries (*40, 41*). While H_2O_2 is used in various fields, it does have some negative and toxic effects on human health. Therefore, it is necessary to detect the accurate amount of H_2O_2. Electrochemical sensors developed using MOFs gained much attention for the determination of H_2O_2 using the voltammetric technique. Dang et al. (*42*) prepared Cu-MOF using the ionothermal approach. Further, Cu-MOF was grafted with ammoniated Au nanoparticles (AuNP-NH_2) to obtain AuNP-NH_2/Cu-MOF. The GCE was modified with AuNP-NH_2/Cu-MOF to develop AuNP-NH_2/Cu-MOF/GCE. This developed AuNP-NH_2/Cu-MOF/GCE was further employed as an electrochemical sensor for the sensing of H_2O_2. This developed AuNP-NH_2/Cu-MOF/GCE exhibited a detection limit of 1.2 μM including high sensitivity of 1.71 μA/μMcm^2. Xia et al. (*43*) synthesized Co-MOF on titanium mesh (TM) using facile conditions. The modified electrode Co-MOF/TM was employed as an H_2O_2 sensor. Authors recorded CV curves of the Co-MOF/TM in the presence of H_2O_2 and found that Co-MOF/TM has a higher current response compared to the bare TM. The CV curves of the Co-MOF/TM were also recorded in the presence of different concentrations of H_2O_2. The effect of scan rate was also investigated. It was found that current response increases with increasing scan rates. This increment was found to be linear. This developed Co-MOF/TM exhibited a detection limit of 0.25 μM. The high sensitivity of 98.75 μA/mMcm^2 was obtained using Co-MOF/TM for H_2O_2 sensing. Yang et al. (*44*) synthesized a new Co-MOF (Co(pbda)(4,4-bpy)·$2H_2O]_n$Co-MOF) (where bpy=4,4-bipyridine, H_2pbda=3-(pyridine-3-yloxy)benzene-1,2-dicarboxylic acid) under hydrothermal conditions. The formation of Co-MOF was confirmed by single-crystal X-ray analysis. Further, an H_2O_2 electrochemical sensor was developed using synthesized Co-MOF as an electrode material, whereas GCE was a working electrode. The electrochemical performance of the developed Co-MOF modified GCE was investigated using CV and amperometric investigations. A detection limit of 3.76 μM was obtained using Co-MOF/GCE for H_2O_2 sensing with sensitivity of 83.1 $mA/mMcm^2$. Zhang et al. (*45*) also prepared a new Cu-MOF (Cu(adp)BIB(H_2O)n(BIB=1,4-bisimidazolebenzene) and H_2adp=adipic acid) via the hydrothermal method. The formation of Cu-MOF was evaluated by single-crystal X-ray analysis. The Cu-MOF was deposited onto the active surface area of the GCE (Cu-MOF/GCE) and was utilized as an H_2O_2 sensor. This developed Cu-MOF/GCE exhibited a good detection limit of 0.068 μM for H_2O_2 sensing. Meng et al. (*46*) designed and synthesized a novel Cu-MOF ([Cu_2(bep)(ada).$H_2O]_n$; where beb=1,4-bis(2-ethylbenzimidazol-1-ylmethyl) benzene, H_2ada=1,3-adamantanediacetic acid). The single-crystal X-ray investigations showed the formation of Cu-MOF. The GCE was modified with Cu-MOF using the drop cast approach. The CV and EIS investigations were carried out to examine the electrochemical properties of the modified Cu-MOF/GCE. The CV results showed that Cu-MOF/GCE has good current response compared to the bare GCE. An excellent detection limit of 0.014 μM was obtained using Cu-MOF/GCE for H_2O_2 sensing. This Cu-MOF/GCE also exhibited a good selective nature for H_2O_2 sensing.

In another report, Wang et al. (*47*) prepared 2D Cu-TCPP (TCPP= tetrakis (4-carboxyphenyl) porphyrin) nanosheets. Further, 2D Cu-TCPP was incorporated with multiwalled carbon nanotubes (MWCNTs) and GCE was modified with 2D Cu-TCPP/MWCNTs as electrode material. The electrochemical properties of the fabricated 2D Cu-TCPP/MWCNTs/GCE were evaluated using CV. The modified 2D Cu-TCPP/MWCNTs/GCE showed high current response compared to the bare GCE, 2D Cu-TCPP, and MWCNTs modified GCE. This showed that the 2D Cu-TCPP/MWCNTs has good electrocatalytic properties compared to the MWCNTs and 2D Cu-TCPP. A detection limit of 0.70 μM was obtained for H_2O_2 sensing with sensitivity of 157 $\mu A/mMcm^2$.

Rani et al. (*48*) designed and prepared a new nanocomposite of Sn-MOF and carbon nanotube (CNT). The prepared Sn-MOF@CNT was characterized by XRD, which confirmed the formation of Sn-MOF@CNT nanocomposite. The specific surface area and pore size analysis of the Sn-MOF@CNT were also investigated using BET. The surface morphological features of the Sn-MOF@CNT were also investigated using SEM. The SEM results showed that Sn-MOF is decorated with CNT, which also confirmed the successful formation of Sn-MOF@CNT nanocomposite. The GCE was modified Sn-MOF@CNT as electrode material and electrochemical investigations were carried out to examine the electrocatalytic properties of the modified GCE (Sn-MOF@CNT/GCE). The Sn-MOF@CNT/GCE exhibited good detection limit and sensitivity.

MOFs as Electrode Modifiers for Glucose

Glucose is an important resource for the human body to produce energy molecules (adenosine triphosphate) (*49*). A high glucose level in the human body may cause a metabolic disorder (also called diabetes mellitus). Diabetes cannot be cured, but the accurate determination of glucose levels in the human body may help the patient resume a normal lifestyle. Zhang et al. (*49*) synthesized Ni-MOF and further assembled it with CNT to obtain the Ni-MOF/CNT. The formation of Ni-MOF/CNT was authenticated by XRD and FTIR spectroscopy. The recorded XRD and FTIR patterns of the Ni-MOF and Ni-MOF/CNT are presented in Figure 2a and Figure 2b, respectively. The XRD and FTIR patterns of the Ni-MOF and Ni-MOF/CNT confirmed the formation of Ni-MOF and Ni-MOF/CNT. The surface morphological properties of the Ni-MOF and Ni-MOF/CNT were analyzed by SEM and TEM investigations. The SEM results showed that Ni-MOF has an interlinked spheroidal-shaped structure with lots of pores in the inner surface. The TEM results also showed that the pores in the inner surface of Ni-MOF are irregular and have different depths and diameters. Electrode materials with such morphological properties are beneficial for electrochemical sensing applications (*49*).

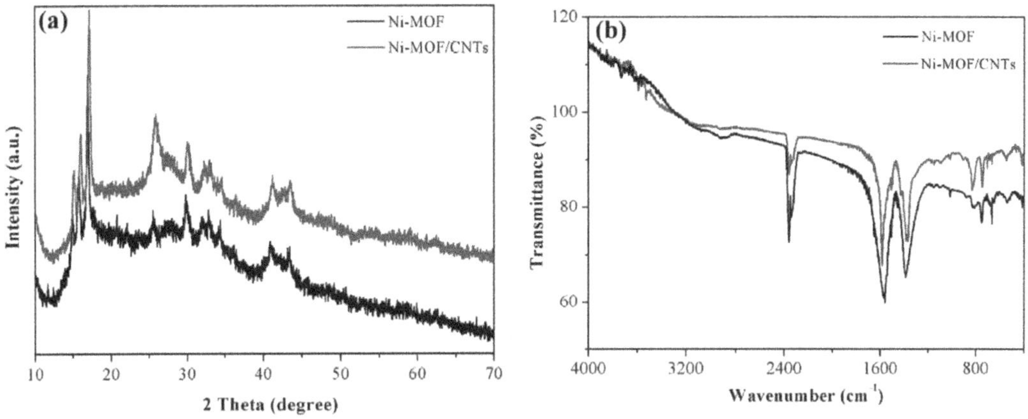

Figure 2. XRD (a) and FTIR (b) patterns of the Ni-MOF and Ni-MOF/CNT. Reprinted with permission from reference (49). Copyright 2018 Elsevier.

However, such electrode materials with improved conductivity would be of more significance for electrochemical sensing applications. Hence, Ni-MOF was assembled to the CNT to prepare the highly efficient electrode modifier for the construction of electrochemical sensing applications. The SEM and TEM results of the Ni-MOF/CNT showed the presence of rough surface, which may provide high specific surface area. The GCE was modified with Ni-MOF/CNT for the construction of glucose sensor (Ni-MOF/CNT/GCE). This Ni-MOF/CNT/GCE exhibited better current response compared to the bare GCE. This enhanced current response may be due to the rough surface and larger surface area of the Ni-MOF/CNT. This developed Ni-MOF/CNT/GCE exhibited a good detection limit of 0.82 μM. This electrode (Ni-MOF/CNT/GCE) was also selective toward the detection of glucose in the presence of other electroactive species. Li et al. (*50*) prepared Co-MOF array on nickel foam (NF). The synthesized Co-MOF/NF was characterized by XRD, SEM, and XPS, all of which confirmed the formation of Co-MOF/NF. Further, this prepared Co-MOF/NF was employed for glucose detection using voltammetric investigations. The obtained CV results showed that the Co-MOF/NF possesses good electrocatalytic properties, and a good detection limit of 1.3 nM was achieved for glucose sensing. The Co-MOF/NF also exhibited excellent sensitivity of 10,886 μA/mMcm^2. In another work, Wang et al. (*51*) have prepared flower-like Ni (II)-terephthalic acid (Ni(TPA))-MOF using simple a solvothermal approach. Further, Ni(TPA)-MOF was assembled with single-walled carbon nanotube (SWCNT) to obtain the Ni(TPA)-SWCNT composite. The synthesized Ni(TPA) and Ni(TPA)-SWCNT was characterized by XRD, which showed the crystalline nature of the Ni(TPA) and Ni(TPA)-SWCNT. The FTIR spectra of the Ni(TPA) and Ni(TPA)-SWCNT also confirmed the formation of Ni(TPA) and Ni(TPA)-SWCNT. The morphological properties of the Ni(TPA) were also investigated by SEM and TEM analysis. The SEM and TEM results of the Ni(TPA) showed the presence of flower-like surface morphology. The TEM image of the Ni(TPA)-SWCNT showed that SWCNT is attached with flower-like Ni-MOF. This confirmed the formation of Ni(TPA)-SWCNT composite. The surface area properties of the Ni(TPA) and Ni(TPA)-SWCNT were analyzed by BET while the thermal stability was checked by TGA. TGA analysis showed that Ni(TPA)-SWCNT has high thermal stability compared to the Ni(TPA). This enhanced thermal stability may be due to the presence of SWCNT. Further, an electrochemical sensor was developed using Ni(TPA)-SWCNT as an electrode material, and CV results showed the better electrocatalytic properties of the Ni(TPA)-SWCNT modified electrode compared to the other electrodes. The constructed electrode exhibited a good detection limit of 4.6 μM for glucose sensing. Zhang et al. (52) also prepared Cu-MOF for glucose sensing applications. The constructed sensor showed a good detection limit of 0.6 μM.

In 2020, Xue et al. (53) prepared Ni@Cu-MOF under facile conditions. The prepared Ni@Cu-MOF was employed as an electrode modifier for the sensing of glucose. The GCE was modified with Ni@Cu-MOF and employed as a working electrode for the sensing of glucose using a voltammetric approach. The developed electrode exhibited a detection limit of 1.67 μM and high sensitivity of 1703 μA/mMcm^2. The developed electrode also exhibited good selective nature toward the sensing of glucose in the presence of electroactive species and decent reproducibility.

Conclusions and Future Perspectives

It has been observed that the sensitive determination of different analytes such as hydrazine, biomolecules, hydrogen peroxide, glucose, and other toxic compounds is a challenging task. Conventional methods have been widely used to determine such analytes, but the electrochemical technique received tremendous attention because of its high sensitivity, cost effectiveness, simple fabrication process, portable instrument, and good detection limit. The performance of any electrochemical sensor is largely influenced by the presence of a working electrode material. Various nanostructured electrode materials have been widely used as electrode materials for electrochemical sensing applications.

MOFs and their hybrid composite materials have been introduced as efficient electrode materials for the development of electrochemical sensors. The highly porous nature and specific surface area of the MOFs provide more active sites for electronic communications. In the case of MOF hybrid composites, it was found that MOFs provide high surface area and active sites for electronic communications, whereas 2D materials such as rGO provide better electron transportation due to their higher electrical conductivity. Moreover, the synergistic interactions between the MOFs and rGO or transition metal oxides enhance the electrochemical performance of the developed sensors. Current challenges for MOF-based sensors include poor electrical conductivity and low stability in aqueous solutions, which restricts their potential applications in aqueous electroanalytical systems. The following key points could be useful in designing and preparing the highly efficient MOFs or MOF hybrid composite-based electrochemical sensors:

a) The novel hybrid composite of MOFs with 2D materials needs to be designed and synthesized; this could be used for electrochemical sensing applications.
b) MOF hybrid composites can also be prepared with doped metal oxides or 2D materials, which may also be other efficient electrode materials.
c) Stability of the MOF materials can also be enhanced by utilizing some hydrophobic or functional groups as well as cationic groups.
d) Some conductive metals can also be incorporated to further enhance the conducting properties of the MOF-based electrode materials.
e) Nanomaterials with unique morphological features and high conductivity can also be incorporated with MOFs to prepare the hybrid composites, which can improve the electron transfer in the redox reactions.

Acknowledgments

M.A. Gondal is thankful to the Deanship of Scientific Research, KFUPM, for supporting this work under project #RG181002. M.A. Gondal is thankful to KACARE for supporting this work under Senior Fellowship Scheme. K.A. acknowledges Discipline of Chemistry, IIT Indore. K.A. thanks Prof. Shaikh M Mobin, IIT Indore, India for his fruitful suggestions.

References

1. Anik, Ü.; Timur, S.; Dursun, Z. Metal Organic Frameworks in Electrochemical and Optical Sensing Platforms: a review. *Microchim. Acta* **2019**, *186*, 196.
2. Wang, H.-S. Metal–Organic Frameworks for Biosensing and Bioimaging Applications. *Coord. Chem. Rev.* **2017**, *349*, 139–155.
3. Giménez-Marqués, M.; Hidalgo, T.; Serre, C.; Horcajada, P. Nanostructured metal–organic frameworks and their bio-related applications. *Coord. Chem. Rev.* **2016**, *307*, 342–360.
4. Zhou, H. C.; Kitagawa, S. Metal-organic frameworks (MOFs). *Chem. Soc. Rev.* **2014**, 3, 5415–5418.
5. Yang, Q.; Xu, Q.; Jiang, H.-L. Metal–Organic Frameworks Meet Metal Nanoparticles: synergistic effect for enhanced catalysis. *Chem. Soc. Rev.* **2017**, *46*, 4774–4808.
6. Eddaoudi, M.; Moler, D.B.; Li, H.; Chen, B.; Reineke, T.M.; O'Keeffe, M.; Yaghi, O.M. Modular chemistry: secondary building units as a basis for the design of highly porous and robust metal—organic carboxylate frameworks. *Acc. Chem. Res.* **2001**, *34*, 319–330.
7. Fe'rey, G.; Mellot-Draznieks, C.; Serre, C.; Millange, F. Crystallized frameworks with giant pores: are there limits to the possible. *Acc. Chem. Res.* **2005**, *38*, 217–225.
8. Meek, ST.; Greathouse, J.A.; Allendorf, MD. Metal-organic frameworks: a rapidly growing class of versatile nanoporous materials. *Adv. Mater.* **2011**, *23*, 249–267.
9. Qiu, S.; Zhu, G. Molecular engineering for synthesizing novel structures of metal–organic frameworks with multifunctional properties. *Coord. Chem. Rev.* **2009**, *253*, 2891–2911.
10. Shaozhou, L.; Fengwei, H. Metal–organic framework composites: from fundamentals to applications. *Nanoscale* **2015**, *7*, 7482–7501.
11. Li, H.; Li, L.; Lin, R.-B.; Zhou, W.; Zhang, Z.; Xiang, S.; Chen, B. Porous metal-organic frameworks for gas storage and separation: Status and challenges. *Energy Chem.* **2019**, *1*, 100006.
12. Hou, C.-C.; Xu, Q. Metal–Organic Frameworks for Energy. *Adv. Energy Mater.* **2019**, *9*, 1801307.
13. Cao, X.; Tan, C.; Sindoro, M.; Zhang, H. Hybrid micro-/nano-structures derived from metal–organic frameworks: preparation and applications in energy storage and conversion. *Chem. Soc. Rev.* **2017**, *46*, 2660–2677.
14. Guan, B. Y.; Yu, X. Y.; Wu, H. B.; Lou, X. W. Complex Nanostructures from Materials based on Metal–Organic Frameworks for Electrochemical Energy Storage and Conversion. *Adv. Mater.* **2017**, *29*, 1703614.
15. Li, D.; Xu, H.-Q.; Jiao, L.; Jiang, H.-L. Porous metal-organic frameworks for gas storage and separation: Status and challenges. *Energy Chem.* **2019**, *1*, 100005.

16. Wu, M.-X.; Yang, Y.-W. Metal–Organic Framework (MOF)-Based Drug/Cargo Delivery and Cancer Therapy. *Adv. Mater.* **2017**, *29*, 1606134.
17. Ahmad, K.; Mobin, S. M. High surface area 3D-MgO flowers as the modifier for the working electrode for efficient detection of 4-chlorophenol. *Nanoscale Adv.* **2019**, *1*, 719–727.
18. Ahmad, K.; Mobin, S. M. Synthesis of MgO microstructures for Congo red dye adsorption and peroxide sensing applications. *J. Environ. Chem. Eng.* **2019**, *7*, 103347.
19. Ahmad, K.; Mohammad, A.; Rajak, R.; Mobin, S. M. Construction of TiO_2 nanosheets modified glassy carbon electrode (GCE/TiO_2) for the detection of hydrazine. *Mater. Res. Express* **2016**, *3*, 074005.
20. Ahmad, K.; Mobin, S. M. Construction of PANI/ITO electrode for electrochemical sensing applications. *Mater. Res. Express* **2019**, *6*, 085508.
21. Ahmad, K.; Mohammad, A.; Mobin, S. M. Hydrothermally grown α-MnO_2 nanorods as highly efficient low cost counter-electrode material for dye-sensitized solar cells and electrochemical sensing applications. *Electrochim. Acta* **2017**, *252*, 549–557.
22. Ahmad, K.; Kumar, P.; Mobin, S. M. Hydrothermally grown SnO_2 flowers as efficient electrode modifier for simultaneous detection of catechol and hydroquinone. *J Electrochem Soc* **2019**, *166*, B1577–B1584.
23. Ahmad, K.; Mohammad, A.; Ansari, SN.; Mobin, S. M. Construction of graphene oxide sheets based modified glassy carbon electrode (GO/GCE) for the highly sensitive detection of nitrobenzene. *Mater. Res. Express* **2018**, *5*, 078005.
24. Raza, W.; Ahmad, K. A highly selective Fe@ZnO modified disposable screen printed electrode based non-enzymatic glucose sensor (SPE/Fe@ZnO). *Mater. Lett.* **2018**, *212*, 231–234.
25. Ahmad, K.; Mohammad, A.; Mathur, P.; Mobin, S. M. Preparation of $SrTiO_3$ perovskite decorated rGO and electrochemical detection of nitroaromatics. *Electrochim. Acta* **2016**, *215*, 435–446.
26. Ahmad, K.; Kumar, P.; Mobin, S. M. A highly sensitive and selective hydroquinone sensor based on a newly designed N-rGO/$SrZrO_3$ composite. *Nanoscale Adv.* **2020**, *2*, 502–511.
27. Ahmad, K.; Kumar, P.; Mobin, S. M. Hydrothermally grown novel pyramids of the $CaTiO_3$ perovskite as an efficient electrode modifier for sensing applications. *Mater. Adv.* **2020**, *1*, 2003–2009.
28. Chidambaram, A.; Stylianou, K. C. Electronic metal–organic framework sensors. *Inorg. Chem. Front.* **2018**, *5*, 979–998.
29. Ma, T.; Li, H.; Ma, J. G.; Cheng, P. Application of MOF-based materials in electrochemical sensing. *Dalton Trans.* **2020**, *49*, 17121–17129.
30. Liu, B. Metal–organic framework-based devices: separation and sensors. *J. Mater. Chem.* **2012**, *22*, 10094–10101.
31. Mehta, S. K.; Singh, K.; Umar, A.; Chaudhary, G. R.; Singh, S. Ultra-high sensitive hydrazine chemical sensor based on low-temperature grown ZnO nanoparticles. *Electrochim. Acta* **2012**, *69*, 128–133.
32. Rahman, M. M.; Alfonso, V. G.; Fabregat-Santiago, F.; Bisquert, J.; Asiri, A. M.; Alshehri, A. A.; Albar, H. A. Hydrazine sensors development based on a glassy carbon electrode modified with a nanostructured TiO2 films by electrochemical approach. *Microchim. Acta* **2017**, *184*, 2123–2129.

33. Gu, X.; Camden, J. P. Surface-Enhanced Raman Spectroscopy-Based Approach for Ultrasensitive and Selective Detection of Hydrazine. *Anal. Chem.* **2015,** *87,* 6460–6464.
34. Garrod, S.; Bollard, M. E.; Nicholls, A. W.; Connor, S. C.; Connelly, J.; Nicholson, J. K.; Holmes, E. Integrated Metabonomic Analysis of the Multiorgan Effects of Hydrazine Toxicity in the Rat. *Chem. Res. Toxicol.* **2005,** *18,* 115–122.
35. Arulraj, A. D.; Vijayan, M.; Vasantha, V. S. Spectrophotometric determination of pico-molar level of hydrazine by using Alizarin red in water and urine samples. *Spectrochim. Acta, Part A* **2015,** *148,* 355–361.
36. Xu, Z.; Pang, M.; Li, C.; Zhu, B. A new colorimetric and far-red fluorescent probe for hydrazine with a large red-shifted absorption spectrum. *Luminescence* **2017,** *32,* 466–470.
37. Oh, J.-A.; Shin, H.-S. Simple and sensitive determination of hydrazine in drinking water by ultra-high-performance liquid chromatography–tandem mass spectrometry after derivatization with naphthalene-2,3-dialdehyde. *J. Chromatogr. A* **2015,** *1395,* 73–78.
38. Rani, S.; Kapoor, S.; Sharma, B.; Kumar, S.; Malhotra, R.; Dilbaghi, N. Fabrication of Zn-MOF@rGO based sensitive nanosensor for the real time monitoring of hydrazine. *J. Alloys Compounds* **2020,** *816,* 152509.
39. Cao, S.; Zheng, S.; Pang, H. Ultrathin nanosheet-assembled accordion-like Ni-MOF for hydrazine hydrate amperometric sensing. *Microchim. Acta* **2020,** *187,* 168.
40. Zhang, R.; Chen, W. Recent advances in graphene-based nanomaterials for fabricating electrochemical hydrogen peroxide sensors. *Biosens. Bioelectron.* **2017,** *89,* 249–268.
41. Teodoro, K. B. R.; Migliorini, F. L.; Christinelli, W. A.; Correa, D. S. Detection of hydrogen peroxide (H2O2) using a colorimetric sensor based on cellulose nanowhiskers and silver nanoparticles. *Carbohydr. Polym.* **2019,** *212,* 235–241.
42. Dang, W.; Sun, Y.; Jiao, H.; Xu, L.; Lin, M. AuNPs-NH_2/Cu-MOF modified glassy carbon electrode as enzyme-free electrochemical sensor detecting H_2O_2. *J. Electroanal. Chem.* **2020,** *856,* 113592.
43. Xia, L.; Luan, X.; Qu, F.; Lu, L. Co-MOF/titanium nanosheet array: An excellent electrocatalyst for non-enzymatic detection of H_2O_2 released from living cells. *J. Electroanal. Chem.* **2020,** *878,* 114553.
44. Yang, L.; Xu, C.; Ye, W.; Liu, W. An electrochemical sensor for H_2O_2 based on a new Co-metal-organic framework modified electrode. *Sens. Actuators B: Chem.* **2015,** *215,* 489–496.
45. Zhang, C.; Wang, M.; Liu, L.; Yang, X.; Xu, X. Electrochemical investigation of a new Cu-MOF and its electrocatalytic activity towards H_2O_2 oxidation in alkaline solution. *Electrochem. Commun.* **2013,** *33,* 131–134.
46. Meng, W.; Xu, S.; Dai, L.; Li, Y.; Zhu, J.; Wang, L. An enhanced sensitivity towards H_2O_2 reduction based on a novel Cu metal–organic framework and acetylene black modified electrode. *Electrochim. Acta* **2017,** *230,* 324–332.
47. Wang, C.; Huang, S.; Luo, L.; Zhou, Y.; Lu, X.; Zhang, G.; Ye, H.; Gu, J.; Cao, F. Ultrathin two-dimension metal-organic framework nanosheets/multi-walled carbon nanotube composite films for the electrochemical detection of H_2O_2. *J. Electroanal. Chem.* **2019,** *835,* 178–185.

48. Rani, S.; Sharma, B.; Malhotra, R.; Kumar, S.; Varma, R. S.; Dilbaghi, N. Sn-MOF@CNT nanocomposite: An efficient electrochemical sensor for detection of hydrogen peroxide. *Environ. Res.* **2020**, *191*, 110005.
49. Zhang, X.; Xu, Y.; Ye, B. An efficient electrochemical glucose sensor based on porous nickel-based metal organic framework/carbon nanotubes composite (Ni-MOF/CNTs). *J. Alloys Compounds* **2018**, *767*, 651–656.
50. Li, Y.; Xie, M.; Zhang, X.; Liu, Q.; Lin, D.; Xu, C.; Xie, F.; Sun, X. Co-MOF nanosheet array: A high-performance electrochemical sensor for non-enzymatic glucose detection. *Sens. Actuators B Chem.* **2019**, *278*, 126–132.
51. Wang, F.; Chen, X.; Chen, L.; Yang, J.; Wang, Q. High-performance non-enzymatic glucose sensor by hierarchical flower-like nickel(II)-based MOF/carbon nanotubes composite. *Mater. Sci. Engin. C* **2019**, *96*, 41–50.
52. Zhang, X.; Luo, J.; Tang, P.; Morante, J. R.; Arbiol, J.; Xu, C.; Li, Q.; Fransaer, J. Ultrasensitive binder-free glucose sensors based on the pyrolysis of in situ grown Cu-MOF. *Sens. Actuators B Chem.* **2018**, *254*, 272–281.
53. Xue, Z.; Jia, L.; Zhu, R. R.; Du, L.; Zhao, Q. H. High-performance non-enzymatic glucose electrochemical sensor constructed by transition nickel modified Ni@Cu-MOF. *J. Electroanal. Chem.* **2020**, *858*, 113783.

Chapter 4

Metal–Organic Framework and Its Nanocomposites as Chemical Sensors

Sushma Rani[1] and Navish Kataria[*,2]

[1]Department of Chemistry, Guru Jambheshwar University of Science and Technology, Hisar 125001, India

[2]Department of Environmental Science and Engineering, J. C. Bose University of Science and Technology, YMCA, Faridabad 121006, Haryana, India

[*]Email: navishkataria08@gmail.com. Phone: +919017125821

Metal-organic frameworks (MOFs) have facilitated the expansion of their potential applications in emerging fields of chemical sensors. MOFs are hybrid inorganic–organic framework materials with unique physiochemical properties in terms of porosity, tunable structure, thermochemical stability, aspect ratio, conductivity, and compatibility for device fabrication. However, traditional MOFs suffer from poor conductivity and low stability, which limits their efficiency in catalysis and other applications that require good electrical conduction. Researchers have started exploring the capability of MOFs as a detection tool for sensing of chemical species and gas molecules. The advantages of MOF structures over other applicants of chemosensory materials still need to be explored. Progress in this field will depend heavily on future generation devices and sensor designing. This chapter is designed to cover the selection of MOFs for chemical sensors applications. MOFs hold enormous potential for chemical sensing owing to their attractive properties, including porosity, superior surface–volume ratio, unsaturated metal sites, and tunable morphology and chemical functionality. The scope of this chapter is further extended to the exploration of MOFs and their composites for chemical sensing. Both ritual and the modern aspects of MOF chemistry illustrate the potential and progressive future in environmental chemical sensing.

Introduction

Metal-organic frameworks (MOFs) are a novel class compound in which metal ions are coordinated to organic ligands to form diverse dimension structures with multidisciplinary field applications. MOFs are a subclass of highly ordered coordination polymers that are viewed as porous. These MOFs have gained significant research interest as hybrid materials that are made from metal ions connected via multitopic organic "linkers" with well-defined coordination geometry (*1*). Because of the availability of different metal ion and organic linkers, more than 20,000 MOFs and assembly forms have been reported, which means limitless possibilities for the structure and function of MOFs. These MOFs possess impressive characteristics such as high structural stability with high surface area, topology properties, good loading ability, and enhanced optical/conductive properties, due to flexible functionalization at the pore sites (*2*). In 1995, Yaghi's research group initiated the exploration of MOFs research. They first synthesized MOF-5 in 1999, which contained Zn-based nodes (*3, 4*). Since then, various type of MOFs — such 1D, 2D, and 3D crystalline structures — have been predominately designed and fabricated. Its exceptional porosity incorporating nano- and meso-sized pores with large internal surface area has been applied in many diverse fields, such as storage of gases and adsorption (*5*), chemical extraction (*6*), photocatalysis (*7*), solar energy harvesting and conversion (*8*), drug release (*9*), sensing, bioimaging, and biosensors (*10*), degradation of chemical warfare agent (*11*), removal of toxic substances from environment (*12*), and loading of (metal and metal oxide) nanoparticles to increase catalytic efficiency (*13*). Many investigators have also started to expend the potential application of MOFs as chemical sensors. Moreover, MOFs can also possess a flexible and rigid structure due to the presence of variable oxidations state of metal and different functional groups on ligands. The selection of different metal nodes, with different functional groups and organic linkers attached by covalent or noncovalent bonding is significantly important in defining the metal-coordination geometry and crystalline structure of MOFs (*14*). These properties of MOFs have significant potential over other available groups of materials for chemical detection.

Here, the comprehensive discussion of MOFs and their emerging applications as chemical sensors are described. The aim of current research is mainly focused on accelerating the working of MOFs and MOF-based frameworks as chemical sensors. Subsequently, many types of MOF sensors were reported and classified by method of signal transduction. Furthermore, some studies revealed that the aromatic or conjugated linker of MOFs can readily exhibit photoluminescence in ultraviolet or visible (typically blue) excitation.

The excitation energy is most typically in the form of photons, the linkers' lowest unoccupied molecular orbital (LUMO) efficiently reduces electrons and generates a hole in the linkers' highest occupied molecular orbital (HOMO). However, several MOFs have the potential to significantly reduce the intensities of photoluminescence once MOFs bound analytes that are either oxidized or reduced (*15–17*). The main limitations of this kind of sensor are the lack of chemical selectivity and loss of signal in response to interaction of analyte. A disadvantage of sensors (turn-off) is loss of signal when other than analyte capture. With these issues in mind, we will focus mainly on the reported porous MOF frameworks to study the sensing abilities. It is thus extensively admitted that with ever-flourishing technology, MOFs have been proposed to be the base of major industrial advancement (*18, 19*). This chapter covers the key features of MOFs including structure, synthesis, and properties of MOFs. Its mainly focused on emerging and future applications predominately as a chemical sensor. It provides an outlook of different types of MOFs and their nanocomposite potential as chemical sensors.

Chemical Sensors

Sensors act as another measurement tool to the identify the chemicals in different environmental matrices. The advantages of sensors include their accuracy, simple use, economical monitoring, and portability, with the potential for real-time detection of chemicals. There are few sensor-based techniques commonly used for chemical detection. The detection approaches for real-world chemicals are categorized into different classes. Sensors can be classified into two steps in terms of recognition and transduction. Every sensor must enclose an immobilized active species that can recognize a target. After the transduction step, it converts the detection species of the target into a measurable change (*20*). These active species in the sensors are either chemical or biological, and their transduction can occur as a result of changes in electronic, optical, and mass properties. Only a few sensor-based methods have been commercialized, such as mass sensor, optical sensor, and electrochemical sensor (*21*).

Materials Used as Chemical Sensors

Due to recent achievements in material science because of the involvement of nanoscience and technology, a large number of 1D, 2D, and 3D nanostructures (including nanoflowers, nanotubes, nanosheet, nanorods, and nanowires) have been published in last few years (*22, 23*). Reported literature on application of fabricated materials such as metal-doped metal oxides, graphene oxide, carbon nanotubes (CNTs), metal oxide, carbon quantum dots (QDs), polymers, MOFs, and hybrid composites has confirmed the potential of materials in diverse fields (*24*). These fabricated materials have been applied as photocatalyst, sensor, adsorbent, energy storage, drug delivery, antibacterial, and so on due to their idiosyncratic properties like high thermal stability, magnetic nature, high volume–surface area, and small size electrical and optical properties (*25*). The development of advanced hybrid materials for the detection or analysis of chemicals in the environment could be a promising tool for future generation equipment designing and field scale applications. Several types of materials have been developed and are being developed for sensing of chemicals. On the bases of time duration, they are classified in conventional materials and advanced hybrid materials or MOFs.

Conventional Materials

Different types of materials can be used for the electrochemical sensing of chemicals. Yang et al. (*26*) synthesized copper hexacyanoferrate coated with polyaniline electrodes for the sensing of sulfite in aqueous medium, which can be widely used as a preservative and organic product to inhibit enzymatic and nonenzymatic pasture using cyclic voltammetry techniques. The polymer-based graphene composites have tremendous electrical and magnetic properties. Gao et al. (*27*) synthesized polypyrrole/graphene composite layered over the glassy carbon electrode (GCE) for the quantitative detection of guanine and adenine. Lee et al. (*28*) developed GCE via electropolymerization of caffeic acid using nanocarbon/CNTs nanocomposite with high concentration of quinine, which can be used for the detection of glutathione via electrochemical methods. Enzyme modified electrodes have also been used as working electrodes for sensing applications. Chen et al. (*29*) developed electrochemical glucose sensors using amperometry. Biosensors for lactate and uric acid were also reported (*30, 31*). Zhao et al. (*32*) synthesized gold-modified Fe_3O_4 nanocomposite for the selective and sensitive analysis of organic pollutants methyl parathion. Cao et al. (*33*) developed bimetallic Pt/Cu nanomaterials for electrochemical sensing of glucose. In a serum sample, detection of glucose performed well by these developed sensors. In recent years, new concepts for novel electrochemical detection based on MOFs were developed.

Advanced Hybrid Nano-MOFs/MOFs

The conventional sensor like metal oxide and doped metal oxide are typically used at high temperatures to increase surface oxygen species reactions. These conventional materials hardly decreased their baseline drift and sensitivity of the sensors. However, MOFs can easily eradicate the challenges of selectivity and sensitivity of the sensors. Most of the MOFs are also recognized as thermally stable, typically not easily degraded at high temperatures ($\geq$ 300 °C) (*10*). These MOF-based sensors or devices overcome the limitations of existing products with respect to cost, production strategies, functionality, and overall performance. In particular, MOFs have prominent research interest as composite materials that are made from metal ion connected with organic ligands with well-defined coordination geometry (*34*). MOFs have gained significant research interest because they contain high structural stability with high surface area, good loading ability, topology properties, and superior optical/conductive properties due to flexible functionalization at pore sites. MOFs are basically three-dimensional crystalline structures that have exceptional porosity incorporating nano and meso-sized pores, with large internal surface area. Depending on the varying geometry, these molecular building blocks control the overall structure of MOFs (*35, 36*).

MOFs and Their Nanocomposites

In 1999, Yaghi et al. first synthesized MOF-5 by bonding metal ions with various ligands that possess properties of both inorganic and organic materials (*37*). Interestingly, MOFs are sometimes defined in three latter codes that explain their overall framework topology. The code that gives the information of the MOF-5 structure is pcu, which stands for primitive cubic net. Precise topological nets can often be targeted when designing MOFs (*4, 38*). In general, it is possible to fabricate MOFs with diverse functionality, different topology, and controlled size through different physical and chemical methods. Several organic linkers are used in synthesis of MOFs, and structures of some of the organic linkers are given in Figure 1.

Moreover, nano-MOFs have been reported in a broad spectrum of applications in the field of coordination chemistry (*39*). Further increment in properties of MOF can be achieved via different surface modification techniques that involve the use of polymers, biomolecules, and ligands (*40*). Due to flexibility and simple synthesis of MOF procedures, an infinite number of combinations is possible and thousands of compounds are being developed every year. Other useful characteristics of MOFs include their often very porous structures with high surface area and easy synthesis that typically gives highly crystalline materials their thermolability (*41*). As a result, MOFs have many applications, perhaps most significantly in sensing of chemicals.

Fabrication/Synthesis Methods of MOFs/Nano-MOFs

MOFs and their nanocomposites are synthesized by several chemical, physical, and biological methods. Their shape and structure are mainly modified by chemical methods. The synthesis routes for different type of crystal structures of MOFs are given in Figure 2. Commonly, nano-MOFs and their composites are prepared by a variety of methods that have been reported to date. Various methods that have been reported for fabrication of MOFs are represented in Figure 3.

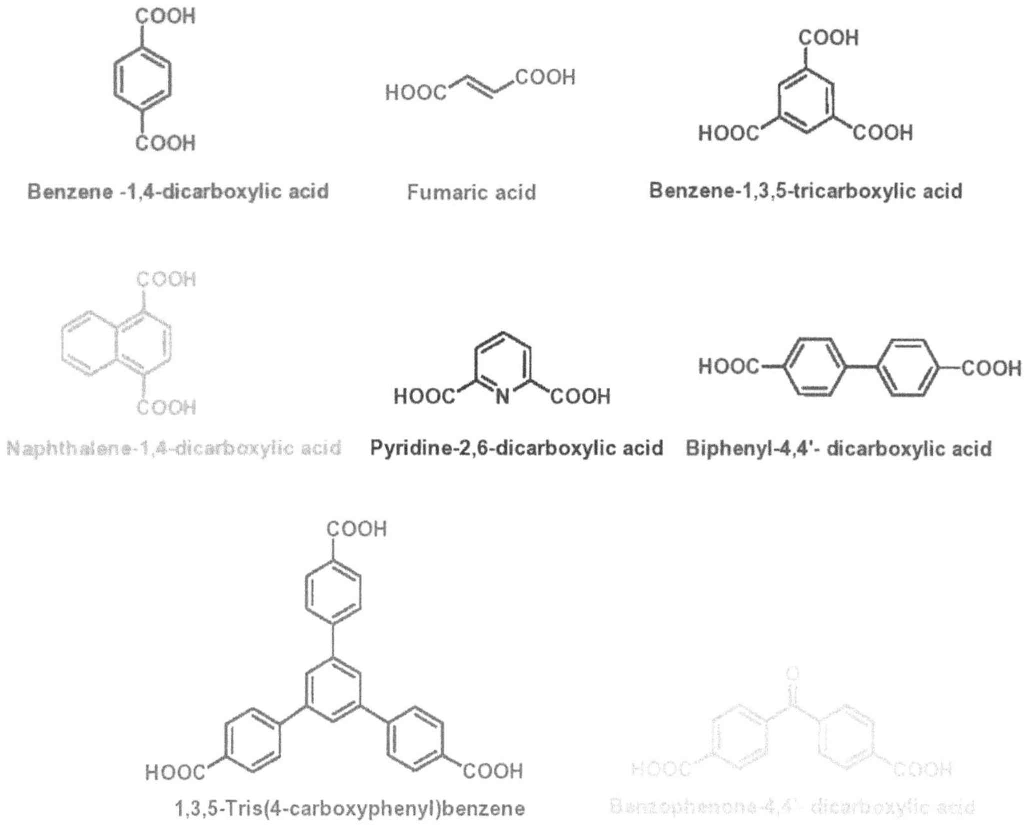

Figure 1. Different types of organic linkers used for the synthesis of MOFs.

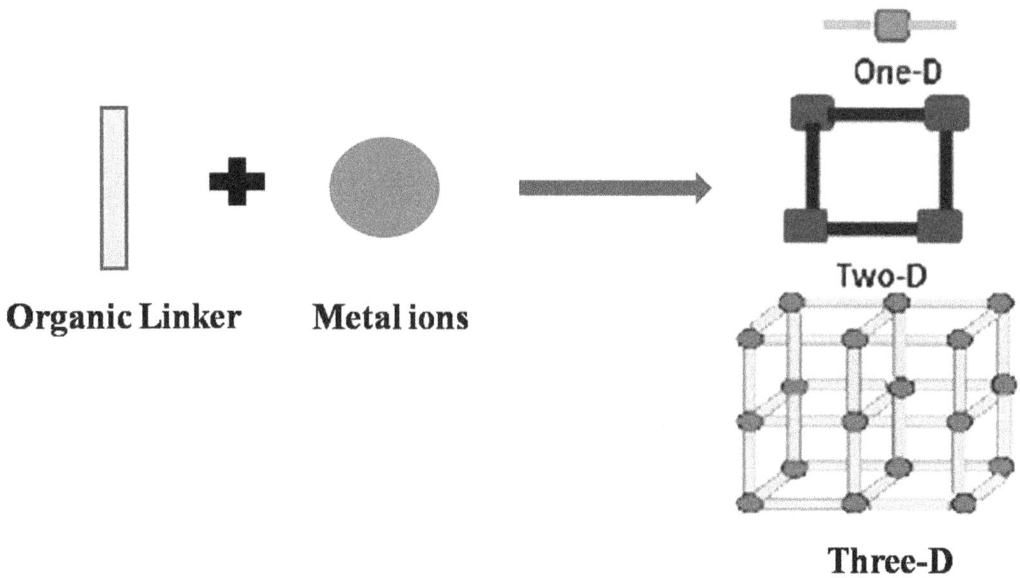

Figure 2. Synthesis routes for different types of crystal structures of MOFs.

Hydrothermal Method

The term "hydrothermal" refers to the synthesis of materials in a sealed heated solution. This method offers several benefits in terms of high reactivity rate, easy control of surface reactions, less air pollution, and low energy consumption. Zhang et al. (*42*) reported ZIF-8 nanofibers possessed highly effective surface area and hierarchical porous features containing ultrathin tellurium nanowires. Hu et al. (*43*) reported a green, mild, and scalable modified hydrothermal development of a series of water-stable UiO-66-type MOFs for huge production and commercialization. Beyond these improvements, hydrothermal synthesis is particularly time consuming (days) in optimization of the size and shape of nano-MOFs and also requires high temperature/pressure conditions. MOF-1, MOF-2, MOF-3, and MOF-4 are different types of Cu-MOFs most common in transport vehicles for drug delivery (*44*). Besides these few disadvantages, large-scale production of MOFs is obtained through hydrothermal methods over other conventional methods. This method needs further exploration in terms of production effectiveness and lower energy costs.

Solvothermal Method

In comparison to other synthesis techniques, the solvothermal method is most common in order to achieve the specific control over morphology of MOFs under mild conditions. This method is carried out at the elevated temperature of the solvent (e.g., dimethylformamide [DMF]; dimethylsulfoxide; and N,N-diethyl-formamide) and entails mixing the reactant in a screw-top vial (*45*). This technique provides higher product yields and better crystallinity of the product over hydrothermal methods (*46*).

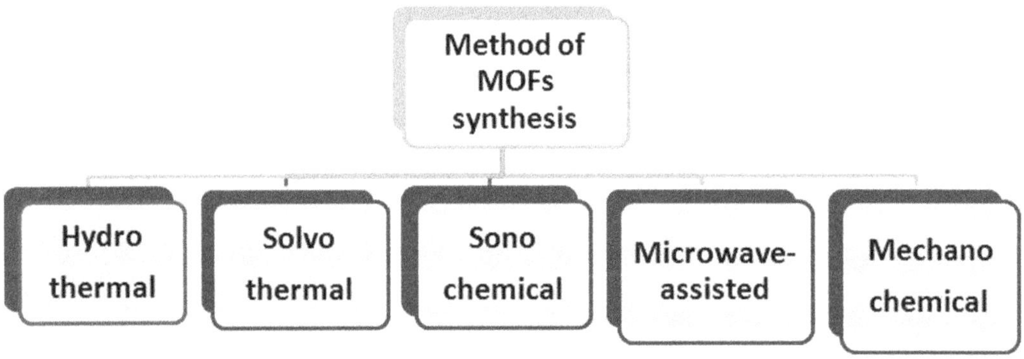

Figure 3. Classification of synthesis methods employed for MOFs.

Sonochemical Method

In sonochemical synthesis, the high-energy ultrasound radiations can be used for a reaction, with frequency between 20 kHz to 10 kHz for the synthesis of MOFs. The effect of ultrasound on liquid and colloid systems is mainly caused by cavitation. In this method, after the irradiation of ultrasound radiation, the formation of bubbles takes place in the solution, which produces fine crystalline (*47*). Compared with conventional synthetic techniques, the ultrasound method is highly permissible, less time consuming, and generates homogeneous nucleation. The nanosized Cu-BTC was synthesized within a short reaction time of 1 min using ultrasound irradiation (*48*). The beneficial effect of ultrasonic irradiation in preparing the small ZIF-8 was also reported (*49*).

Ultrasound-assisted synthesis produced 700 nm ZIF-8. This method has some disadvantages with regard to the environment, such as high-energy sound waves.

Solvent-Free and Mechanochemical Synthesis

Research efforts were focused on solvent-free synthesis approaches that can offer green, fast, high-yield, and less polluting methods. Conventional solvents used for the fabrication of MOFs are often toxic or chlorinated. Guo et al. (*50*) synthesized [$Ni_3(HCOO)_6$] frameworks via the facile method without use of solvent. The solvent-free method has some disadvantages like long duration and problems with isolation in product when the materials are free-flowing powder used in large-scale production.

Microwave-Assisted Synthesis

In this technique, the electromagnetic wavelength lies in the range of 1 m to 1 mm (300 MHz to 300 GHz) and has been used for the fabrication of MOFs. This technique reduced the particle size, phase-selectivity, morphology control, and fast crystallization when compared to conventional heating-based methods. In addition to this, the microwave route also provides an effective way to synthesize porous coordination materials with controllable size and morphology (*51*). In earlier studies, MOFs MIL-100(Cr) and MIL-101 (Cr) were reported by Khan et al. (*47*). Moreover, resultant MOFs have a size of 40–90 nm, which is smaller compared to the particles synthesized by conventional methods. Additionally, Cu-BTC, MOF-74, MIL-53, and ZIF-8 have been successfully fabricated by Chen et al. (*52*) and Thomas-Hillman et al. (*53*) for various potential applications.

Fabrication/Electrodeposition Techniques over the Electrode

The key phase in the development of sensors is the appropriate coating of materials on the electrode surface. Different methods for the coating, namely physical adsorption and covalent bonding, are commonly used. Figure 4 represent the different techniques for the immobilization of synthesized materials over the electrode surface.

Physical Adsorption

This method involves direct adsorption of nanomaterials over the working electrode. While this method is simple and rapid, it faces the major drawbacks of longer response times and leaching of the materials.

Covalent Bonding

This method includes covalent bond formation between functional groups of nanoparticles and binding agent like nafion, PVDF 301F, and PVDF HSV900. Advantages of such methods include stable bond formation between molecules and the matrix, thereby preventing leaching of the nanomaterials.

Crosslinking

Crosslinkers like glutaraldehyde, glyoxal, or other functional materials are well known for the fabrication approach in the development of sensors. Among these, glutaraldehyde is commonly used as it is inexpensive and readily available in commercial quantities. This method decreases the cost but faces the limitations of low mechanical stability and reproducibility.

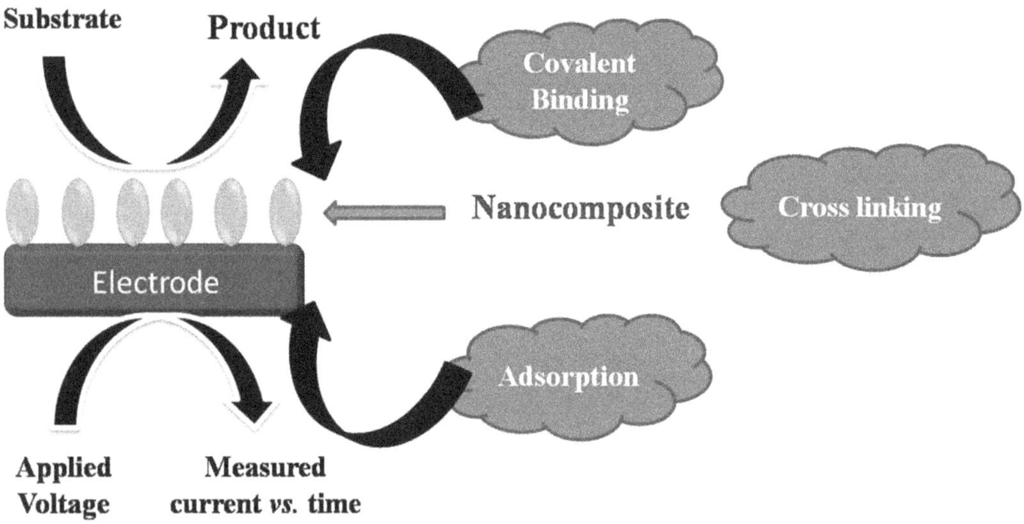

Figure 4. Immobilization of materials over the electrode surface via different methods.

MOF-Based Nanocomposites for Chemical Sensing

A variety of MOF composites with different functionalities, like graphene sheet, CNT, QD, and different metal/metal oxide, have been prepared and explored in various applications. A broad range of nano-MOF composites are being classified into the following main categories. Various properties of each nano-MOF composites materials–based developed sensors along with the applications were discussed in detail. Figure 5 shows a summary of the different methods used for the synthesis of nanocomposites followed by the step involved in development of chemical sensor.

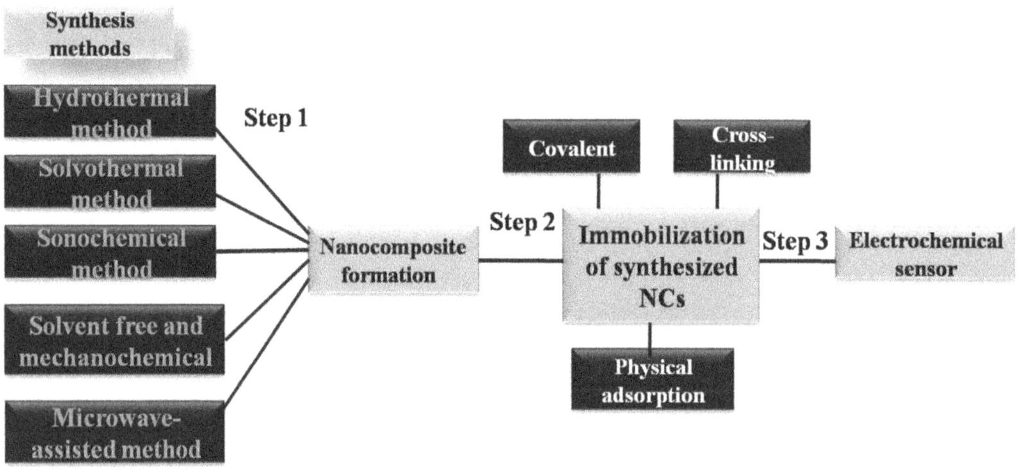

Figure 5. Process of the synthesis of nanocomposites to development of electrochemical sensor.

Graphene-Based MOF Sensors

Graphene-based MOF composites materials have attracted much research in the field of sensing. In this type, composites were made from an assembly of MOFs and graphene-based materials that exhibit the properties of the parent materials with wide-spread applications. Graphene-based composites have introduced a variety of applications because of their low toxicity and unique structure, high electronic as well as thermal, electrochemical, and mechanical properties (*54*). The idea of hybridizing graphene derivatives with MOFs has naturally and rapidly attracted widespread attention (*55*). In graphene, the presence of sp^2 hybridized carbon atoms and ionic groups cannot only function as structural nodes, but also take part in bonding interactions, which is more active in MOFs. The graphene sheets are synthesized by different methods like organic synthesis and chemical synthesis (*56*). The graphene composites with MOFs were widely used for the immobilization of sensors and detection of various analytes. Some selected graphene-based MOF composites for sensing are reported in Table 1.

Carbon Nanotube-Based MOF Sensors

Studies have shown various applications of MOFs/CNT composite for sensing of environmental chemicals. CNTs have been promising materials for sensing due to several intriguing properties. Commonly, CNTs have been fabricated by various methods including arc discharge, laser based, and chemical vapor deposition. CNTs are mainly three types of structures: (1) single-walled carbon nanotubes (SWCNTs), (2) double-walled carbon nanotubes, and (3) multiwalled carbon nanotubes (MWCNTs). CNTs have sp^2-hybridized carbon structure with specific diameter, length, and layers (*65, 66*). These CNTs have various significant chemical and physical properties that make them potentially useful in optics, biosensors, and electrical devices. Considering the advantages of CNTs, the combination of MOFs with CNTs has been developed in order to increase the electrochemical conductivity with electrocatalytically ability (*67, 68*). Similarly, Wang et al. (*69*) synthesized Ni (II)-based MOF/MWCNT composites by using solvothermal methods. MOF nanoparticles dispersed over the conducting CNTs and applied for the electrochemical detection of H_2O_2. MOFs (MWCNTs@$Cu_3(btc)_2$) have been used for the trace detection of lead (limit of detection [LOD] = 0.79 nM) via differential pulse anodic stripping voltammetry (*70*). By introducing the Zn in SWCNTs, a hybrid inorganic–organic framework has been synthesized and employed as adsorbent for sensing of organophosphate compounds (*71*). In another work, Mn-tyr-MWCNT conducting paper was synthesized and fabricated over GCE and demonstrated a significant electroanalytical response toward hydrogen peroxide with good sensitivity of 543 $mAmol^{-1}L\ cm^{-2}$ and an LOD value of $3.8 \times 10^{-6} - 1.02 \times 10^{-4}$ mol/L. This work shows the double pathway response of the working electrode in the presence of Mn and MWCNT electrocatalytic activity. MOF@CNT composite includes sensing system rather than electrochemical sensors in the previous reports. The effect of composite was optimized. As a result, optimized CNTs increased the sensitivity 230% as compared to plain HKUST-1 film (72). Table 2 reports some CNT-based MOFs material for the sensing of analytes.

Table 1. Reported Chemical Sensor Using MOF/G/GO/rGO-Based Composites for the Sensing of Different Analytes

MOFs/G/GO/ rGO formula	*Sensing Methods*	*Target Chemicals*	*Linear Range (μM)*	*LOD (μM)*	*Reference*
Cu(tpa)-EGR	DPV	Dopamine	1–50	0.21	*(57)*
Fe_3O_4@ZIF-8/rGO	DPV	Dopamine	$2.0 \times 10-9$ to $1.0 \times 10-5$ M	6.67×10^{-4}	*(58)*
PDA/ZIF-8@rGO	CV	Glucose	1–1200, 1200–3600	0.33	*(59)*
MOF-ERGO-5	DPV	Catechol (CT)	0.1–566	0.1	*(60)*
Cu-MOFs/CS-RGO	CV	Hydrogen Peroxide (H_2O_2)	0.065–410	0.019	*(61)*
Cu(tpa)-EGR	DPV	Acetaminophen	1–100	0.36	*(57)*
MOF-ERGO-5	DPV	Hydroquinone (HQ)	0.1–476	0.1	*(60)*
Graphene/MOF	DPV	Cd (II)	0.006–1.068	0.0018	*(62)*
Cu-MOF/rGO	CV	Nitrite	33 nm	3–40000	*(63)*
Cu-based MOF-GO	CV	Catechol (CT)	0.1	0.1–476	*(64)*
Ce-MOF/GO	CV	Uric Acid	2.0	50–1050	*(65)*

CV: cyclic voltammetry

Table 2. MOF/CNT-Based Chemical Sensors Reported for the Detection of Various Analytes

MOF/CNT-Based Materials	*Sensing Methods*	*Target Chemicals*	*Linear Range (μM)*	*LOD (μM)*	*Reference*
Ni-MOF/MWCNT	Electrochemical	Urea	3 μM	0.01–1.1 mM	(73)
Cu-MOF-199/SWCNTs	Electrochemical	HQ and CT	0.08–0.1 m mol L^{-1}	0.1–1453 mmol L^{-1}	(74)
Cu(II)-based MOF on MWCNT	Electrochemical	H_2O_2	0.46 μmol L^{-1}	3–70 μmol L^{-1}	(75)
MWCNTs@Cu_3 $(BTC)_2$	DPASV	Pb^{2+}	0.00079 mmol L^{-1}	0.001–0.05 mmol L^{-1}	(76)
SWNT@ZnMOF	Electrochemical	Methyal Parathion	2.3 ng mL	0.034–6.9 μM, 7.9 nM	(77)
MIL-53(Cr-Fe)/Ag/MWCNT	Gas sensing	Ethanol, Methanol Isopropanol	-	30.5–50.5 ppm	(78)
Ce-Zn-MOF/MWCNT	Electrochemical	Bisphenol		7.2 nmol L^{-1}	(79)

DPASV: differential pulse anodic stripping voltammetry

Metal/Metal Oxide-Based MOF Sensors

Based on previous literature, introduction of nanoparticles in sensing systems has a significant enhancement in signals based on the transducers. Moreover, metal/metal oxide nanoparticles have higher conductivity, larger surface area, and good biocompatibility (*80*). Additionally, based on the optical and electrochemical properties, utilized MOFs can also be flexibly used in sensing devices (*81*). MOFs have been used as molecular filters in these systems. On the other hand, MOFs have variable pore size and chemistry that affects sensitivity and absorbance of analytes. In this section, different type of metallic-nanostructure/MOF composites structure in sensing system is explored. For example, silver nanoparticles and other metal oxide nanoparticles attract a lot of attention due to their physical and chemical properties, such as superior electrical conductivity, physicochemical, photocatalytic, better photoelectrochemical activity, and pollutant detection and removal (*24*, *82*, *83*). Vilian et al. (*84*) developed a Pd/NH_2-MIL-101 (Cr) sensor for the electrochemical sensing of nitrite ion. Bagheri et al. (*85*) synthesized MIP-capped AgNPs@Zn-MOF for hydrogen peroxide-terephthalic acid reaction, which results in the formation of high fluorescent product in the range of 0.1-10 μmol L^{-1} with a detection limit of 0.06 μmol L^{-1}. Meng et al. (*86*) developed modified GCE with Ag@ZIF-67 nanocomposite for the catalytically activity toward glucose oxidation. After the loading of Ag nanoparticles, the sensing performance was increased and the LOD value was measured to be 0.66 μM, with a wide linear range between 2–1000 μM. The other selected similar works can be seen in Table 3.

Quantum Dot-Based MOF Sensors

In the last decades, QDs have honed eloquent attention due to their morphology-dependent optical and electronic properties, making them of interest but not limited to medical, analytical, biological, and engineering applications (*66*, *94*). QDs are semiconductor nanocrystals and often range in size from 2 to 10 nm. These semiconductor materials have been doped with elements from groups 12–16, 13–15, or 14–16 of the periodic table, due to their small size and quantum confinement effect (*95*). QDs have multifunctional behavior based on precursor. Coating and encapsulation enhance the optical and electrical properties of these nanoparticles. Advances in the integration of QDs in MOF matrix have enhanced sensing and photocatalysis abilities. Yang et al. (*96*) reported an amine QD doped UiO-66 using a post-synthetic modification approach for the selective fluorescence sensing of 4-nitrotophenol. The 4-nitrophenol has been detected in a concentration range of 0.01–20.0 μM with a detection limit of 3.5 nM. Carbon dots (CDs) as emerging luminescent materials is a hot area of research activity due to their various optical properties, low toxicity, chemical stability, and biocompatibility (*97*, *98*). Reports have clearly demonstrated that the encapsulation approach combines the advantages of MOFs and nanoparticles. CDs@Eu-MOFs is synthesized for fluorescent sensor for the detection of Cr (VI) (*99*). The CDs@Eu-MOFs has been characterized by different techniques like SEM, TEM, FT-IR, and XRD techniques. The synthesized CDs@Eu-MOFs reflected selectivity and high sensitivity with range a from 2 to 100 μM and detection limit of 0.21 μM for Cr (VI). Li et al. (*100*) reported a novel method for homogenous dispersion SnO_x QDs coated on UIO-66 without agglomeration by using vacuum drying, sol-gel, and pyrolysis processing. The electrochemical properties of the ultrafine SnOx@UIO-66 have been studied using cyclic voltammetry and galvanostatic charge/discharge methods. The discharge-specific capacity of the SnOx@Zr-MOF electrode can remain at 994 mA h g^{-1} after 100 cycles, and its columbic efficiencies reached 99%. Saha et al. (*101*) reported cadmium sulfide (CDS) QDs encapsulated into Zn (II)-based metal hydrogel (ZAVA) successfully without any capping agent. The ZAVA hydrogel shows blue color luminescence, but it can change color during CDS QDs encapsulation. After the entrapping, the augmentation can be stopped by changing the gel into xerogel. Stabilization of CDS QDs can be achieved by exclusively exchanging CDS incubated ZAVA gel to CDS incubated MOF (CDS@ZAVXI) crystals. These CDS QDs modified MOFs were used as a catalyst for water splitting in sun light. Meng et al. (*102*) reported CDS quantum modified MIL-101 MOFs by one-step heating method. CD/CDS@MIL-101 composite have been used in photocatalytic hydrogen production. Nguyen et al. (*103*) reported MIL-101(Cr)-CDSe QD composites have been developed by two different methods: (1) loading the CDSe QDs on the surface of MIL-101 and (2) in situ formation of CDSe QDs into the pores of MIL-101. Nature of porous materials depends on the loading amount of CDSe QDs and incorporation of CDSe QDs onto and into to MIL-101, enhancing the optical absorption properties in visible region and, as a result, the composite materials have potential applications in photocatalysis or solar cells. Pan et al. (*104*) synthesized GQD@ZIF-8 composite materials by encapsulation of graphene (edge-sulfonated) QDs as fluorescent base into nanocrystals of the ZIF-8. The synthesized GQD@ZIF-8 materials have good optical properties in water yet can exhibit fluorescence in the solid state due to strong binding with ZIF-8. Some other reported studies on MOF/QDs-based materials synthesis and chemical sensing are summarized in Table 4.

Application of MOFs in Chemical Sensing Techniques

Detection of these chemicals is crucial due to the globalization of terrorist acts and their widespread occurrence, sometimes leaching from soil into the ground and leading to contamination. Their persistence in the environment constitutes a severe threat to human health and the ecosystem. Due to their extraordinary properties, MOFs have been introduced in diverse commercial applications including chemical storage, gas storage, drug loading, sensing, biosensing, environmental remediation, and so on. In this section we will discuss the chemical sensing techniques (e.g., electrochemical, colorimetric, luminescent, and fluorescent sensor) of MOFs/nano-MOFs (see Figure 6).

MOFs as Electrochemical Sensors

In recent years, plenty of attention in the industrial sector has been paid to MOFs, among their other applications. Their applications have been further premediated as an outstanding candidate for sensing of metal ions, solvents, explosives, and chemicals. Figure 7 represents the principle of electrochemical sensors. Nowadays, chemical sensors are widely used for sensing due to their wide linear range and lower detection limit. Moreover, scientists have also explored electrochemical-based MOF sensors (*70*, *113*). Cu-MOF was mixed with Au-SH-SiO_2 nanoparticles casting on glassy carbon electrode for electrocatalytic oxidation of hydrazine and L-cysteine. MOF composite-modified electrode showed efficient catalytic activity and decreased over potential. This particular sensor provides a promising platform for the analysis of hydrazine and L-cysteine in water and biological samples with accuracy and precision (*70*, *114*).

Table 3. MOF/Metal/Metal Oxide-Based Chemical Sensors for the Detection of Different Analytes

Metal/Metal Oxide Nanomaterial/ MOFs	*Sensing Methods*	*Target Chemicals*	*Linear Range (μM)*	*LOD (μM)*	*Reference*
Pt NPs@UiO-66-2 core-shell	Electrochemical sensor	H_2O_2	5–14.75 μM	3.06 μM	*(49)*
Ag@MOF(Ag) nanocomposite	Fluorescence quenching	TNP	0.04 μM	1–29 μM	*(87)*
CuNPs/ZIF-8	Electrochemical sensor	Glucose	2.76 μM	0-0.7 mM	*(88)*
ZnO@ZIF−8	Gas sensor	Hydrogen gas	-	-	*(89)*
MIP-capped AgNPs@ZnMOF	Florescent sensing	Hydrogen peroxide	0.1–10 μmol L^{-1}	0.06 μmol L^{-1}	*(90)*
Ag@ZIF-67	Electrochemical sensing	Glucose	2–1000 μM	0.66 μM	*(86)*
Pd@NH2-MIL-101 (Cr)	Electrochemical sensing	Nitrite	5–150 nM	1.3 nM	*(84)*
PtNP/MIL-101 (Cr)	Electrochemical sensing	Dopamine, uric acid, hypoxanthine	0.5–162 μM	0.42 μM	*(91)*
CoNPs/NH2-MOFs-derived MC	Colorimetric sensing	Glucose	0.25–30 μM	156 nM	*(92)*
ZnO@ZIF-8	Photoelectrochemical sensor	Hydrogen peroxide	-	-	*(93)*
Ag@ZnMOF	Colorimetric sensor	Peroxide	0.4–15 mg L^{-1}	0.1 mg L^{-1}	*(85)*

Table 4. MOF/QDs-Based Chemical Sensors Reported for the Sensing of Different Analytes

Quantum Dots-Based/MOFs	*Sensing Methods*	*Target Chemicals*	*Linear Range (μM)*	*LOD (μM)*	*Reference*
Mn^{2+}:ZnS/ZIF-8	Fluorescence	Co^{2+}	-	-	*(105)*
ZnOQDs/MOFs	-	Phosphate ions	0.512	53 nM	*(106)*
IRMOF-3/CdTeQDs	ECL iluminescence	Cardiac troponin antigen	1.1 fg ML^{-1} to 11 ng mL^{-1}	0.46 fg mL^{-1}	*(107)*
Zn-MOF/CdSe/ZnSQDs	Fluorescence	Pyrraline	5×10^{-6}–1×10-3 mol L^{-1}	3.9×10^{-6}	*(108)*
MOF-545-Zn@MQDs	ECL immunosensors	Carcino-embryonic antigen	0.18–1000 ng m L^{-1}	3.9×10^{-6} pg m L^{-1}	*(109)*
Amine-CQDs@UiO-66	Fluorescence	4-nitrophenol	0.001–20.0 μM	3.5 Nm	*(96)*
BPEI-CQDs/ZIF-8	-	Cu^{2+}	2–10000 nM	80 pM	*(110)*
CDs@Eu-DPAMOFs	-	Cu^{+2}	50 nM–10 μM	26.3 nM	*(111)*
Zn-MOF-8@AgQDs	Electrochemical	2,4-DNT	0.0002 μM to 0.9 μM	0.041 μM	*(112)*

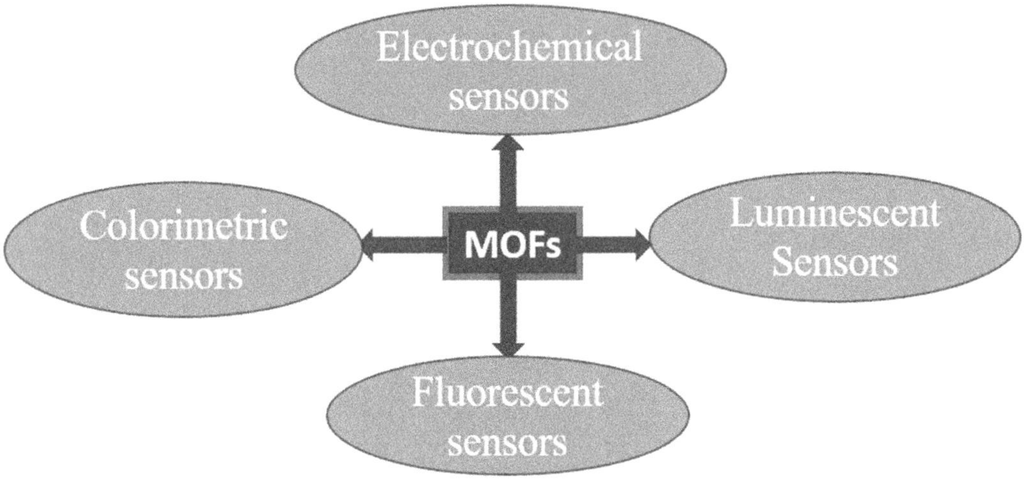

Figure 6. Graphical representation of MOFs applied in different chemical sensing techniques.

Electrocatalytic and electrochemical sensing ability of MIL-101 showed good oxidation potentials 218 mV and 336 mV for dopamine and uric acid, respectively. They were found to be a better liner range from MIL-CPE (carbon paste electrode) than that of the previous results (*115*). Zn_4O $(BDC)_3$ (MOF-5) modified CPE has been synthesized at different pH levels, deposition time, and electrolyte solutions for the determination of lead ions. MOF-5 has been reported for determination of lead in real water samples (*116*). NH_2-MIL-125(Ti) modified CPE showed photooxidation activity toward Mn^{2+}, present in tea samples at pH 8. The stripping peak currents showed linear dependance with concentration of Mn^{2+} in the range from 1.0×10^{-8} to 1.0×10^{-5} M (*117*). Gassensmith et al. (*117*) reported the cubic structure of having an efficient binding capability with CO_2. In pore-filling methanolic medium, ionic conductivity of CDMOF-2 decreased (~550fold) in the presence of CO_2. This has been exploited for measuring CO_2 concentrations quantitatively even in the availability of other gases in environment. These MOFs provide a better future in the field of sensing than sensitive current methods (*118*) reported nano-MOFCu(II)-{CuL- $[AlOH]_2$}n (PAC; H_6L = meso-tetrakis (4-carboxylphenyl)porphyrin) as heterogeneous host for detection of H_2S molecule concerned in human pathological and physiological processes. H_2S is highly reactive and diffusible and thus detection of H_2S is very difficult in real samples.

This nano-MOF explores rapid response, high selectivity, and hypotoxicity for analysis of H_2S in real water samples at optimized pH. Zhang et al.'s (*119*) synthesized 2D MOFs are effective tools for the analysis of nitroaromatic explosives through fluorescence quenching. By using a shift of photoluminescence (PL) spectra first NENU-503 MOFs can be used for nitroaromatic molecules contacting NO_2 groups. High stability and recyclability of NENU-503 make it a promising candidate for the detection of explosives. Roushani et al. (*62*)] reported TMU-16-NH_2 ([$Zn_2(NH_2\text{-}BDC)_2$(4-bpdh)]·3DMF) MOFs for electrochemical sensing of cadmium ion in solutions due to soft-soft interactions between nitrogen-containing groups of TMU-16-NH_2 and Cd^{2+}. This process has great potential for rapidly recognizing cadmium ions using differential pulse voltammetry (DPV). The $Cu_3(btc)_2$ (btc = benzene-1,3,5-tricarboxylic acid) MOF has been coordinated with chitosan (CS) electrode fabricated on the surface of graphene oxide (ERGO) (*120*). It provides electrochemical sensing stands for detection of dihydroxy benzene isomers of catechol (CT), resorcinol (RS), and hydroquinone (HQ). Figure 7 represents the principle of electrochemical sensors. The pioneering research outlining the potential of electrochemical MOFs and their composites for analytes detection are summarized in Table 5.

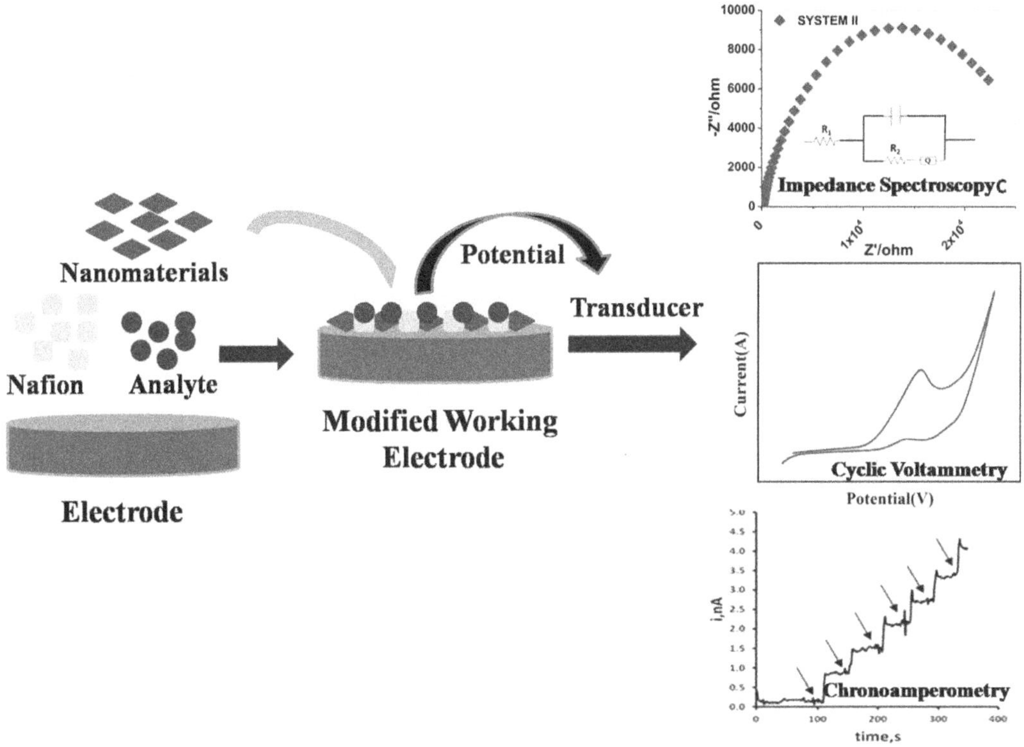

Figure 7. Working principle of the electrochemical sensor.

MOFs as Fluorescent Sensors

Fluorescent-based sensing is a promising tool for rapid detection of selective organic and inorganic analytes for clinical applications (*135*). During the last few decades, the field of fluorescent chemosensors or probes has blossomed because of the development of fluorescent sensors for various analytes and applications of these probes to the environmental sciences, biology, physiology, and disease diagnosis areas. The presence of specific functional sites on MOF surfaces has great importance in the area of biosensors, and functionalized MOFs can be used as fluorescence quenchers for fluorescence resonance energy transfer (FRET)-based biosensing. Researchers have found that a few MOFs possess good fluorescence-quenching abilities, causing a change in the fluorescence of the fluorophore (*136*). Besides, MIL-101 (*137*, *138*), MIL-88A (*139*), UiO-66-NH_2 (*140*), and $[Gd(TIA)(HCOO)]_n$ (*141*) were used as fluorescent sensors for DNA/RNA with good sensitivity and selectivity.

Table 5. MOFs as Electrochemical Sensors for Detection of Chemical Analytes

MOFs	*Electrode*	*Chemical Analytes*	*Linear Range(μm)*	*Detection Limit (μm)*	*Reference*
Cu-MOF	CPE	NO_3^{2-}	40 nM–1.1 mM	40 nM	*(115)*
MOF-525	FTO	NO^{3-}	20–800	2.1	*(121)*
Au-SH-SiO_2@Cu-MOF	GCE	Hydrazine	0.04–500μM	0.01μM	*(122)*
Au-SH-SiO_2@Cu-MOF	GCE	L-cysteine	0.02–300μM	0.008 μM	*(112)*
Cu(tpa)GO	GCE	Acetaminophen, dopamine	1–100, 1–50 μM	0.36 and 0.21μM	*(57)*
Co-MOCP	CPE	Reduced glutathione	2.5–0.95 μM	2.5 μM	*(123)*
GDH@ZIF-70	GCE	Glucose	100–2000 μM	---	*(113)*
Cu-MOF	CPE	$NO_3$2-	50.0 nM–717.2 μM	30.0 nM	*(124)*
MIL-101 CPE	CPE	Dopamine Uric acid	5–250	NA	*(114)*
H_2N-Fe-MIL-88B@OMC	GCE	Hydrazine p-NT	0.006–0.061, 20–225	5.3 nM, 8 μM	*(125)*
Cu-BTC/ERGO	GCE	Trinitrophenol	0.2–10	0.1	*(126)*
UiO-66/MC	GCE	HQ, CT and RS	0.5–100 μM, 0.4–100 μM, 30–400 μM	0.056 μM, 0.072 μM, 3.51 μM	*(127)*
ZIF-8	SPE	Nitroaromatic	$(1–460 \times 10^{-9}$ m)	346×10^{-12} M	*(128)*
AgNPs-Zn-MOF	GCE	H_2O_2	1 μM–5 Mm	67 nM	*(129)*
Co-MOF	GCE	H_2O_2	5 μM–9.0 mM	3.76 μM	*(120)*
Ag@ZIF-67	GCE	H_2O_2	5.0 μM–275 mM	1.5 μM	*(130)*
2D Cu-TCPP/MWCNT	GCE	H_2O_2	0.001–8.159 mM	0.70 Mm	*(131)*
Ni(II)-MOFs/CNTs/	GCE	H_2O_2	0.01–51.6 mM	2.1 μmol L^{-1}	*(69)*
$Cu(BTC)_2$	GCE	Hydroquinone	-	0.44 μM	*(132)*
Cu-MOF/rGO	GCE	NO^{2-}	3–40000 μM	33 nM	*(63)*

Table 5. (Continued). MOFs as Electrochemical Sensors for Detection of Chemical Analytes

MOFs	*Electrode*	*Chemical Analytes*	*Linear Range(μm)*	*Detection Limit (μm)*	*Reference*
Cu-MOF-199/SWCTs	GCE	Hydroquinone, Catechol	0.1−1453 $\mu mol \cdot L^{-1}$	0.08 μM, 1 μM	(*133*)
AuPd/UiO-66-NH_2	GCE	NO^{2-}	0.05–5666 μM	0.01 μM	(*132*)
Au/Zn-MOF	GCE	NO^{2-}	5−65000 μM	1 μM	(*134*)
Au/Cu-MOF	CPE	NO^{2-}	0.05−717.2μM	0.03 μM	(*124*)
ZIF-8	CPE	TNT	1−460 nM	346 nM	(*128*)
Zn-MOF@rGO	AuE	Hydrazine	0.001−100	8.7×10^{-3} μM	(*112*)
Zn-MOF-8@Ag	AuE	2,4-DNT	0.002−0.9	0.041 μM	(*112*)

FTO: fluorine-doped tin-oxide glass; SPE: screen-printed electrodes; AuE: gold electrode; MOCP: metal-organic coordination polymer

Wang and coworkers synthesized ultrathin lanthanum@ MOF nanosheets labeled with fluorescent dyes. These nanosheets have better fluorescence-quenching abilities and selectivity for DNA (*142*). The fluorescence quenching occurred in the presence of dye on $\{[La_2(TDA)_3]2H_2O\}_n$ and depends on the transfer of charge from dye molecules to the lanthanum ions. During the aptameric sensing, the negatively charged dye experiences fluorescence "turn-down" process while the positively charged dye follows fluorescence "turn-down followed by turn-up" process. Several composites of QDs and MOFs have shown great promise for enzyme activity sensing (*26*, *143*). Tan and coworkers synthesized Tb^{3+}/Eu^{3+} doped MOFs for radiometric sensing of dipicolinic acid (DPA) (*144*). DPA is a major component for bacterial spores. Addition of DPA into Tb^{3+} results in a transfer of energy from Tb^{3+} to Eu^{3+}. By regulating transfer of energy in Tb/Eu@bio-MO, visual colorimetric sensing of DPA in MOF has been analyzed for the first time. Tb^{3+} and adenosine monophosphate have been used as sensors for detection of DPA (anthrax) in aqueous solutions. In the presence of the fluorescent, the color of Tb/Eu@bio-MOF-1 changed to green.

The porous nature of MOFs is an effective and reliable tool for detection of antibiotics. However, a small portion of work has been reported in the sensing/removal of antibiotics based on MOFs (*145*). SiO_2-Au nanocomposite-based fluorescence biosensor was made by self-organizing 5-FAM-based peptides via cation with magnetic Fe_3O_4@SiO_2-Au nanomaterials (MNCPs), which significantly decreased the fluorescence intensity of FAM (*107*). In another report, four ruthenium (II) complexes were explored for the fluorescence anisotropy and polyacrylamide gel electrophoresis experiments (PAGE)-based detection of miR-185. The developed platform demonstrated a detection time of 1 min (*146*). There are highly selective "turn on" fluorescent sensors for Pb^{2+} ions based on a $NaY(MoO_4)_2$ material exhibiting either downshifting or up conversion luminescence. The selective sensing ability of the $NaY(MoO_4)_2$:Eu^{3+} and $NaY(MoO_4)_2$:Sm^{3+} downshifting materials is not as distinguished as for the $NaY(MoO_4)_2$:Tm^{3+}, Ho^{3+}, Yb^{3+} up conversion material (*147*). Some amine-grafted MOFs have been used as potential sensors for choline (Cho) and acetylcholine (ACh). Among different amine-functionalized MOFs, N,N,N′,N′-tetramethyl-1,4-butanediamine (TMBDA) modified with MIL-100(Fe) demonstrated the highest activity toward peroxidase (*148*). Zn^{2+} and Cd^{2+} have been widely used as host metal for the formation of matrix coordination-induced emission effect (MCIE) (*149*). The selective and reversible sensor use porous MOFs, NSU-1@ zinc oxide, for fluorescent sensing of ammonia (*150*). Some other potential fluorescent MOFs and their composites for analytes detection are summarized in Table 6.

In another report, an MOF/DNA hybrid system was employed for the fluorescent-based detection of mercury (II) ions. UiO-66@NH_2 and FAM labeled with DNA has been used as a sensor of Hg^{2+} sensitively and selectively. Fluorescence emission of DNA is quenched effectively by UiO-66-NH_2, but it can be recovered with Hg^{2+}due to T–Hg^{2+} –T interactions (*159*). A novel opto-sensor has been developed for the analysis of pyrroline in milk powders using the QDs modified with CdSe/ZnS as fluorescent nanoparticles (*147*, *160*). Although different kinds of detection methods are given as a consequence, fluorescence-based tools are selective and sensitive for various analytes. But it is a challenging task to construct simple, fast, and low-cost methods for detection of Hg^{2+} ions in water samples. Also, fluorescence-based detection techniques are more sensitive than others due to their good selectivity to specific targets, measuring of fluorescence, and absorbance in different manners as compared to other techniques.

Table 6. MOFs as Fluorescent Sensors for Detection of Chemical Analytes

Sensing Material	*Techniques*	*LOD*	*Analytes Sample*	*Reference*
$[Cr_3F(H_2O)_2(BDC)_3nH_2O]$ n (MIL-101)	Fluorescence-quenching platform	73 pM	DNA	(*137*)
[Gd (TIA)(HCOO)] n	Fluorescence-quenching platform	-	DNA and Hg^{2+}	(*141*)
$[Fe_3F(H_2O)_2O(BTC)_2nH_2O]$ n [MIL-100(Fe)]	Immobilization matrices	5 μM	Glucose	(*151*)
$[Zn(MeIM)_2]$n (ZIF-8)	Mimetic multienzyme	-	Glucose Oxidase	(*104*)
{Al(OH)[H_2BTEC]$_5$ (H_2O/H_4BTEC)}n (MIL-121)	Fluorescent probe	(0.05 ~ 8.0 mg/mL)	Hippuric acid	(*152*)
$[Cu_3(BTC)_2(H_2O)_3]$n (HKUST-1)	Fluorescence-quenching platform	-	DNA	(*153*)
Eu^{3+}/CDs@**1**	Fluorescence	0.034 μM, 0.0096 μM	Fe^{3+} ions, *p*-nitrotoluene (*p*-NT) and	(*154*)
Eu^{3+}/CDs^{+2}	Fluorescence	0.062 μM, 0.0111 μM	2,4-dinitrophenol (2,4-DNP)	(*154*)
[Cd(L)2]·(DMF)0.92 (1) (L = L−1), 4-(4-carboxyphenyl)-1,2,4-triazole (HL), DMF = N,N′-dimethylformamide)	Fluorescence	3.9 ppm	Cu^{2+} ions	(*155*)
UiO-66@NO_2 188	Fluorescence	188	for H_2S	(*156*)
UiO-66@N_3	Fluorescence	118	for H_2S	(*157*)
Ce-UiO-66-N_3	Fluorescence	12.2 μM for **1′-N$_3$** and	for H_2S	(*158*)
Ce-UiO-66-NO_2		34.8 μM for **2′-NO$_2$**)		
Eu^{3+}/Cu^{2+}@UiO-66-$(COOH)_2$				
ammonium cerium(IV) nitrate and H_2BDC-X (BDC = 1,4-benzenedicarboxylate; X = – N_3 for **1-N$_3$** and –NO_2 for **2-NO$_2$**) linkers in DMF/H_2O (DMF = *N*,*N*-dimethylformamide)				

MOFs as Colorimetric Sensors

Colorimetric sensing has created intense consideration in chemistry because of agreements on its advantages, such as rapidity, simplicity, economical, and direct visual readout (*74*). These methods are sensitive and highly specific, but the necessity of costly and sophisticated instruments bounds their applications. Colorimetric detection would be a more desirable technique, in which analyte signal interface reaction concludes a change in color is unquestionably crucial for its successful execution. MOFs have been used as fascinating materials because of their tunable nature. The boundless choice of metals and organic compounds affords results into various possible combinations for different applications. Color change during 3,3′,5,5′-tetramethylbenzidine (TMB) oxidation in the presence of H_2O_2 and its inhibition effect on AA examined in human blood serum is shown in Figure 8 (*161*).

The binding of Fe-MIL-88A with target/aptamer led to the generation of blue color due to its high peroxidase-like catalytic efficiency toward TMB. The mimetic activity of Fe-MIL-88A can be strongly repressed in presence of target molecule and consistent aptamer. These biomolecules specific characteristics of MOF have been used directly to understand the colorimetric detection of molecules (*59*). Colorimetric assay for sensing of alkaline phosphatase (ALP) has been developed by Cu-MOFs. The high binding efficiency of Cu^{2+} with PPi greatly repressed the catalytic activity of Cu-MOFs, which resulted in no colorimetric signal. However, when ALP is induced, the negative behavior of PPi can be decreased as the hydrolysis of PPi into orthophosphates (Pi) with low attraction to Cu^{2+}. The quantitative analysis of ALP activity with a detection limit is found to be 0.19 U/L and shows good sensing presentation in human serum samples (*126*).

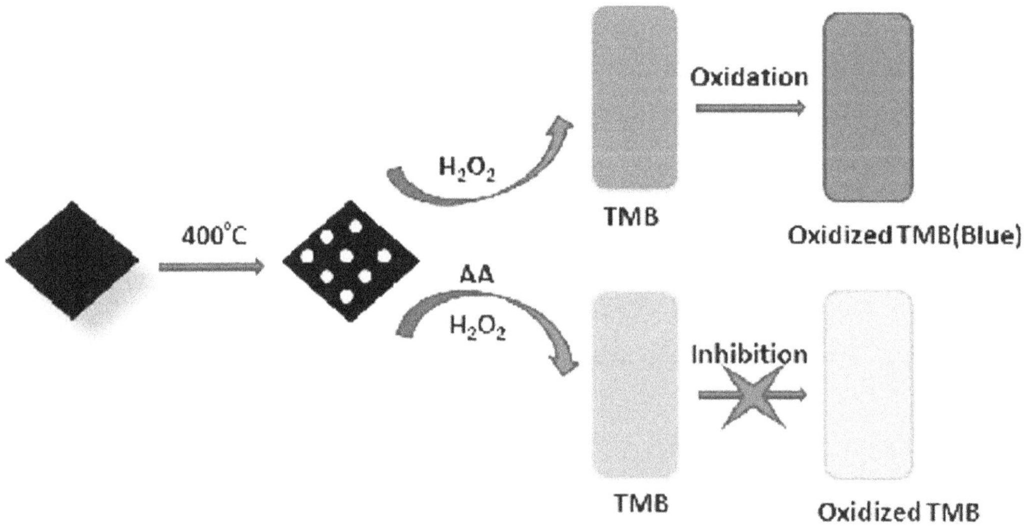

Figure 8. Oxidation and inhibition effect of TMB is demonstrated by the color change.

The MnO_2 and TMB have been used for the selective colorimetric analysis of glutathione (GSH) in human serum. The LOD for GSH determination was observed as low as 0.11 μM (*162*). On the other hand, the paper-based Fe-MIL-101 MOF microfluidic colorimetric platform was used for the detection of glucose in urine and serum (*163*). The MOF-based glucose colorimetric sensor was worked on the HRP mimic principle. μPADs are a very stimulating podium in sensing applications due to the fact that they are affordable, nonrefundable, swift and technically simple, and have high quantity and low sample ingesting (*164*). The Fe-MIL-101 MOF mimics peroxidase on profitable cellulose paper. This method is modest to set up and perform an analysis of glucose in urine and serum due to its low cost and gives accurate results. The glucose oxidase (GOx) reacted with glucose and led to the formation of H_2O_2. Further, the produced H_2O_2 was used for the oxidation of TMB (*165*). The MOF was used for the determination of the resulted product, which gives a blue–green color in the detection zone (*166*). Moreover, the MOF-based colorimetric method can be used for the sensing of glucose in real water samples, attaining good percentage recovery. ZIF-8@BHb hybrid composites have been used for detection of glucose with 423% improvement in catalytic activity over free BHb in aqueous solution. (*167*). As described earlier, Fe (III)@MIL-68 MOF has been found with intrinsic peroxidase-like action for colorimetric sensing (*168*). The previously reported Fe@MOFs, MIL-68 and MIL-100, hold inherent peroxidase-like action for biosensing with significant color change in solution in occurrence of H_2O_2 (*169*). For the determination of cholesterol, Fe_3O_4@MIL-100 core-shell nanoparticles were synthesized consisting of two Fe-based catalytic materials that increase the catalytic effect and its catalytic efficiency. A colorimetric detection method of cholesterol using Fe_3O_4@MIL-100 has been used for the analysis of cholesterol levels in human serum. Mixed-valence-state cobalt nanomaterials have tremendous properties due to their extra ordinarily high catalytic activities and their electron mobilities. This proposed sensitive and colorimetric sensor for the analysis of antioxidant ability of GSH in eye drops and healthcare products (*127*). The three-dimensional porous structure is highly sensitive with efficient mimetic peroxidase activity of FePor-TFPA-COP and explored its application for quantitate detection of alpha fetoprotein through the enzyme-linked immunosorbent assay (ELISA) routine, making it a good candidate for biochemical immunoassay and a promising diagnostic tool for clinical use. Polyoxometalates (POMs) are an important class of metal oxide clusters, having adjustable compositions with a high concentration of oxygen surface and high binding efficiency, constituting of multifunctional materials. A novel and stable MOF polyoxometalate Cu_6 $(Trz)_{10}(H_2O)_4$ $[H_2SiW_{12}O_{40}]\cdot 8H_2O$ has been developed for sensitive colorimetric assessment for detection of hydrogen peroxide (*74*). Some reported studies related to colorimetric sensing are given in Table 7.

Table 7. Summarized Studies of MOFs for Colorimetric Sensing of Various Chemicals

Sensing Material	*Method*	*LOD*	*Analytes Sample*	*Reference*
$Cu_6(Trz)_{10}(H_2O)_4[H_2SiW_{12}O_{40}]$	Colorimetric	detection limit of 1.37 mm	H_2O_2	(*74*)
$Co_3(BTC)_2·12H_2O$	Colorimetric detection	LOD of 0.03 M	Detection of glutathione	(*126*)
RuUiO-67	Colorimetric	0.5 lM	Hg_2?	(*126*)
$Tb1.7Eu0.3(BDC)_3(H_2O)_4$	Colorimetric	0.25 mM	Cd^2? lead-polluted water samples	(*55*)
$[Cd(L)_2]·(DMF)0.92$ (1) ($L = L^{-1}$, ligand 4-(4-carboxyphenyl)-1,2,4-triazole (HL), DMF = N,N′-dimethylformamide)	Colorimetric	LOD of 3.9 ppm	Cu^{2+} ions	(*155*)
Fe_3O_4@MIL-100(Fe)	Colorimetric	0.8 µM	Cholesterol	(*162*)
Fe(III)- based MOFs, MIL-68 and MIL-100	Colorimetric	detection limit of 0.256 mM and 0.155 mM	H_2O_2	(*167*)
Fe-MIL-101 MOF	Colorimetric	2.5 µmol L^{-1}	Glucose	(*163*)

MOFs as Photoluminescent Sensors

The luminescence properties of MOFs have been subject to a lot of research using them as sensing elements for chemical sensors to date (*77*, *170*). The prominence of luminescence over other transduction processes is due to a number of factors, including the ability to produce a signal that can be observed with the naked eye. Another benefit is the capability to work directly with solid and liquid samples, eliminating the requirement for film manufacturing or additional processing. As a result, luminous MOFs can be synthesized using traditional solvothermal methods. Two sets of researchers have demonstrated how molecular sieving has been employed for morphological selectivity and sensitivity in order to achieve this goal (*171*, *172*). Aniline and butylamine, on the other hand, did not show any quenching, owing to size exclusion. A Cd^{2+} based MOF has been found to have a similar size-selective effect. Only nitrite quenches the luminescence when various anions are introduced. While the researchers hypothesized on transfer of energy to NO_2 through strut-based C-H electron density that is reportedly involved in H-bonding with nitrite, we believe quenching of excited state by nitrite is a more probable process. Finally, the protracted 3D structure may permit dipolar interaction between nearby ligands, foster transfer of energy on a shorter time than the singlet excited-state lifetime. As a result, a single molecule interaction event could disturb many signaling middles, similar to what was exposed with polymers with alternative double bonds. The many examples of discrete luminous coordination complexes are used to rationally construct these designs. Multiple sources of light can exist in both the molecular and MOF instances. The presence of conjugation in MOFs leads to absorption of light UV spectrum, as well as prevalent d-block metal ions linker-based fluorescence. Due to guest adsorption, the most predominant form of analysis transduction in MOFs is extinguishing, or infrequent amplification, of photoinduced emission. The extension of these effects can be disturbed by the nature of the bonding between the MOFs and ligands. The important interactions are often those that permit for considerable electron donor–acceptor orbital overlapping with the host and guest system. As a result, samples that have electron donors or acceptors groups are the most easily detected (e.g., nitroaromatics, including common explosives or simulants). Moreover, numerous authors have described sensing applications, particularly those based on luminescent MOFs. A Cd@MOF developed for selective sensing of nitrate ion in the presence of anionic species (such as I^-, Cl^-, ClO^{4-}, and NO^{3-}) has also been proposed (*173*). Kaur et al. (*174*) developed Eu@AuNPs nanocomposites for sensitive detection of trinitrotoluene (TNT) using PL spectroscopy. Xie et al. (*175*) used phosphorescent iridium (III) complexes as struts in an MOF. PL intensity is suppressed due to transfer of energy to O_2. Yb^{3+} ions were synthetically confined in the sites of a Zn-adenitate MOF has been also quenched by O_2, according to An et al. (*176*). Lan and colleagues published two Zn@MOFs that can sense nitro-containing compounds, which could be useful in explosive detection. The different types of MOF-based PL sensors for chemicals that have been reported are shown in Table 8.

Table 8. Reported Studies of MOFs as Photoluminescent Sensors for Detection of Chemicals

Luminescent MOF	*Ligand*	*Analyte (Metal Ion)*	*Medium*	*Detection Limit (Ksv, M−1)*	*Reference*
[Zn(4,4′-AP)(5-AIA)].(DMF)0.5]n, 4,4′-AP = 4,4′-azopyridine 5-AIA = 5-amino	Isophthalic acid	Hg^{2+}	Water	10−11 M (1.011×10^9)	*(177)*
$[Zn_2(BDC)_2(DABCO)]$	H_2BDC = 1,4-benzene dicarboxylic acid DABCO is 1,4-diazabicyclo[2.2.2]octane	Zn_{2+}	Water	0.7 μM	*(178)*
$[Zn_4O(TPA)_3]$ (MOF-5)	TPA: terephthalic acid	Pb^{2+}	Water	2 nM	*(179)*
$\{[Zn_3(bpydb)_2(atz)_2(DMF)](DMF)_6\}n$ $bpydbH_2$ = 4,4′-(4,4′-bipyridine-2,6-diyl)-dibenzoic acid	Hatz = 3-amino-1,2,4-triazole	Fe^{3+}	Water	1.15×10^4	*(146)*
TMU-17-NH_2 Zn-based	amine/azine-	Fe^{3+}	DMF	0.7 μM (>41000)	*(180)*
([Cd (Ligand)]·$_2$DMF)	Ligand = 5,5′,5′-((1,3,5-triazine-2,4,triyl)-tris(azanediyl))tris(3-methylbenzoic acid)	Cu^{2+}	DMF	3 ppm	*(181)*
$[CdL(H_2O)]\cdot 1DMF\cdot 1.5H_2O$	L is 1-(4-carboxyphenyl)-5-methyl-2,5-dihydro-1 Hpyrazole-3-carboxylic acid,	Fe^{3+}	DMF	3.5×10^4	*(182)*
$[Eu_2L_2(DMF)(H_2O)_2]\cdot 2DMF\cdot H_2O$	[H_3L is 5-(1-carboxynaphthalen-4-yl)benzene-1,3-dioic acid]	Cu^{2+} Fe^{3+}	Water	18.3 ppm, 12.2 ppm	*(146)*
KFUPM-3 2,5-Bis(allyloxy)	terephthalic acid	Pd^{2+}	Water	44 ppb	*(183)*
Eu^{+3} doped UIO-66-type MOF	1,4-benzodicarboxylic acid	Hg^{2+}	Water	8.26 nM	*(184)*
$[Tb(DBB)(H_2O)_2]$, DBB = 4-(3,5-dicarboxylatobenzyloxy)	benzoic acid	Fe^{3+}	Water	10−9 M	*(185)*
[Tb(TCBPA)·DMF]n	H_3TCBPA is tris(4′-carboxybiphenyl) amine	Fe^{3+}	Water	3.56 ppb	*(186)*

Table 8. (Continued). Reported Studies of MOFs as Photoluminescent Sensors for Detection of Chemicals

Luminescent MOF	*Ligand*	*Analyte (Metal Ion)*	*Medium*	*Detection Limit (Ksv, M−1)*	*Reference*
{[Mn2(Bript)$_2$(4,4 bpy).5(DMF)]·(H$_2$O)$_2$O)}n	H$_2$ Bript = 4-Bromoisophthalic acid 4,4 -bpy = 4,4 -bipyridine	Hg^{2+}	Water	48 μM (1390.5)	*(187)*
{[Co$_3$(phen) $_2$ (HL) $_2$]·(H$_2$O)$_2$}n	H4L = 3,3′,5,5′-azoxybenzenetetracarboxylic acid Phen = Phenanthroline	Fe^{3+}	Water	1.79 × 10−6 M (8.5 × 10^3)	*(188)*
{[Cd$_2$(adp)(nic) $_2$ (H$_2$O)$_3$(3-ampyr)]·5H$_2$O}n	(adp)(nic) $_2$ (H$_2$O)$_3$(3- ampyr)]	nitro compound	Liquid	18 ppb	*(189)*
Co$_3$[Co(CN)$_6$]$_2$	CN^-	ethanol	Liquid		*(190)*
[Cd(atc)(H$_2$O) $_2$]n	Act	bovine serum albumin (BSA)	Liquid	100 ppt	*(157)*
Tb(HL)(H$_2$O)$_4$]H$_2$O	HL	Nitrobenzene	Liquid	1 ppb	*(191)*
Ln$_2$ (C$_{17}$H$_8$F$_6$O$_4$)$_3$	(C$_{17}$H$_8$F$_6$O$_4$)$_3$	water, ethanol	Liquid	-	*(40)*

MOFs as Radio-Luminescent Sensors

MOFs have the potential to make substantial advancements in the detection and analysis of particles (atomic and subatomic), which could be useful in nuclear power, space exploration, and nonproliferation. MOFs' synthetic elasticity permits them to be altered in ways other than traditional scintillator materials such as polymers, organic liquids, and inorganic. Furthermore, their crystalline structure permits sub angstrom precise characterization of interactions among luminous linkers, resulting in a "nanolaboratory" for discovering photo-physics at a degree of feature inconceivable with noncrystalline nanoparticles. From this perspective of organic linker, IRMOFs are a particularly intriguing isostructural series since they contain both interpenetrated and noninterpenetrated structures including the biphenyl dicarboxylate. Bauer et al. (*192*) previously demonstrated those fluorescence lifetimes of stilbene groups contain MOFs being constrained in a *trans* alignment that is relatively stiff. The nonradiative *cis-trans* pathway in stilbene has been prevented, resulting in an increase in quantum yield. As a result, it was hypothesized that ionizing radiation's emission would have comparable qualities. In comparison to crystalline organics, they are also highly unaffected by radiations, suggesting that the arrangement of coordination bonding and relatively wide interlinked size decreases harmful radical processes. Unpredictably, the radioluminescence of the two stilbene@MOFs varies from their PL. Feng et al. (*193*) developed MOF structures to modify optical properties by structural changes and via doping. The charge transfer effects produced by photons and protons have been studied. At 400 nm, the tidy MOF exhibits photoluminescence based on linkers ("monomer"). However, reaction with protons having energy 3 MeV results in a broad peak at 475 nm, indicating that ionization-induced structural alteration causes excimer production between neighboring NDC linkers. When the electron donor diethyl aniline (DEA) is infiltrated into Zn (II)@NDC, a third form of luminescence is produced. The presence of an exciplex between DEA and NDC results in broad emission. In the infiltrating IRMOF-8 complex's proton-IBIL, all three forms of emission (monomer, excimer, and exciplex) may be seen. These examples highlight MOFs' extraordinary potential for "crystal engineering," in which structural property associations can be leveraged to create new sparkling materials.

Challenges and Future Perspectives

The superior characteristics of MOFs demonstrate excellent applications in the biological and biomedical fields. However, a few challenges in enlargement of new solids and change in the construction and processing of the frameworks are associated with MOF-based systems, which should be addressed before their practical use in biological systems. A major challenge associated with MOFs is their toxicity to living organisms. Important parameter considerations associated with size, nature, thermal stability, and chemical stability of nanomaterials is optimized for toxicological assessment of MOFs. Moreover, existence of toxic heavy metal ions, organic ligands, and functionalized materials in MOFs can cause damage to the ecosystem. The important progress of nano-MOFs for loading and release of cargos is investigated by using computational approaches, molecular dynamics method. Moreover, there are still many challenges that need to be measured. National and international agencies must safely manage the use of MOFs based on their toxicity and further establish regulations. However, this attention should not be to degree that MOFs are seen as equally toxic to persistent organic pollutants endocrine disrupting chemicals, and other emerging pollutants the US Environmental Protection Agency has indicated have a high impact on human health and the environment (*194*). Research needs to focus on development of low/no toxicity MOF-based systems for industrial, biomedical, and drug delivery application. Moreover, different challenges of MOFs relating to their particular solvent-based dispersion and toxicity are also discussed, with a main emphasis on upcoming directions for this developing field of research. Both traditional and modern aspects of MOFs and nano-MOF chemistry are illustrated for their continually expanding research interest and bright future in the biomedical field.

Conclusion

This chapter has described the current status of MOF-based nanocomposites and their potential application as chemical sensor. Scientist have reported and are currently exploring hybrid advance MOFs integrated material for diverse applications. MOFs as chemical sensors have exhibited excellent potential for sensing of organic species, metal ions, radiation, explosives, and gas molecules in various environmental matrices. Reported literatures confirmed that MOF-based materials have expressed fast response, high sensitivity, and high selectivity toward chemicals as compare to conventional materials. Pure MOFs fabricated with other materials or nanocomposites, such as CNTs, reduce graphene oxide, polymers, metal oxides, QDs, etc. to enhance response and sensitivity toward detection of analytes in water system. These MOFs could be significantly applied in the fabrication and development of sensing probes, electronic electrodes, chip sensors, and portable sensors for field scale application. Dissemination of the MOF thin film coating approach is important in designing future generation devices. These new engineered devices with MOF thin films build a productive interface between MOF-supported surface and analytes. While, the assumption will be found everywhere, designing MOF materials that respond specifically to a single analyte will be difficult for the near future. Studies indicate that MOFs could significantly interact with analyst and collectively work as a chemical nose in different matrices.

References

1. Auffan, M.; Rose, J.; Bottero, J. Y.; Lowry, G. V.; Jolivet, J. P.; Wiesner, M. R. Towards a definition of inorganic nanoparticles from an environmental, health and safety perspective. *Nature Nanotechnology* **2009**, *4*, 634–641.
2. Yi, F. Y.; Lin, Q. P.; Zhou, T. H.; Mao, J. G. A Series of New Manganese (II) Sulfonate-Arsonates with 2D Layer, 1D Chain, and 0D Clusters Structures. *Inorganic Chemistry* **2010**, *49*, 3489–3500.
3. Li, H.; Eddaoudi, M.; O'Keeffe, M.; Yaghi, O. M. Design and synthesis of an exceptionally stable and highly porous metal-organic framework. *Nature* **1999**, *402*, 276–279.
4. Yaghi, O. M.; O'Keeffe, M.; Ockwig, N. W.; Chae, H. K.; Eddaoudi, M.; Kim, J. Reticular synthesis and the design of new materials. *Nature.* **2003**, *423*, 705–714.
5. Bao, C.; Zhou, L.; Shao, Y.; Wu, Q.; Zhu, H.; Li, K. A novel Au-loaded magnetic metal–organic framework/graphene multifunctional composite: Green synthesis and catalytic application. *Journal of Industrial and Engineering Chemistry* **2016**, *38*, 132–140.
6. Mason, J. A.; Veenstra, M.; Long, J. R. Evaluating metal–organic frameworks for natural gas storage. *Chemical Science* **2014**, *5*, 32–51.
7. Lee, J.; Farha, O. K.; Roberts, J.; Scheidt, K. A.; Nguyen, S. T.; Hupp, J. T. Metal–organic framework materials as catalysts. *Chemical Society Review* **2009**, *38*, 1450–1459.
8. So, M. C.; Wiederrecht, G. P.; Mondloch, J. E.; Hupp, J. T.; Farha, O. K. Metal–organic framework materials for light-harvesting and energy transfer. *Chemical Communications* **2015**, *51*, 3501–3510.

9. Horcajada, P.; Serre, C.; Vallet-Regí, M.; Sebban, M.; Taulelle, F.; Férey, G. Metal-Organic Frameworks as Efficient Materials for Drug Delivery. *Angew. Chem. Int. Ed. Engl.* **2006**, *45* (36), 5974–5978.
10. Kreno, L. E.; Leong, K.; Farha, O. K.; Allendorf, M.; Van Duyne, R. P.; Hupp, J. T. Metal–organic framework materials as chemical sensors. *Chemical Reviews* **2011**, *112*, 1105–1125.
11. Mondloch, J. E.; Katz, M. J.; Isley, W. C., III; Ghosh, P.; Liao, P.; Bury, W.; Snurr, R. Q. Destruction of chemical warfare agents using metal–organic frameworks. *Nature Materials* **2015**, *14* (5), 512–516.
12. Decoste, J. B.; Peterson, G. W. Metal–organic frameworks for air purification of toxic chemicals. *Chemical reviews.* **2014**, *114*, 5695–5727.
13. Wu, Z. S.; Yang, S.; Sun, Y.; Parvez, K.; Feng, X.; Müllen, K. 3D nitrogen-doped graphene aerogel-supported Fe_3O_4 nanoparticles as efficient electrocatalysts for the oxygen reduction reaction. *Journal of the American Chemical Society* **2012**, *134*, 9082–9085.
14. Deng, H.; Doonan, C. J.; Furukawa, H.; Ferreira, R. B.; Towne, J.; Knobler, C. B.; Yaghi, O. M. Multiple functional groups of varying ratios in metal-organic frameworks. *Science* **2010**, *327*, 846–850.
15. Allendorf, M. D.; Bauer, C. A.; Bhakta, R. K.; Houk, R. J. T. *Chem. Soc. Rev.* **2009**, *38*, 1330.
16. Jiao, L.; Seow, J. Y. R.; Skinner, W. S.; Wang, Z. U.; Jiang, H. L. Metal–organic frameworks: Structures and functional applications. *Materials Today* **2019**, *27*, 43–68.
17. Rocha, J.; Carlos, L. D.; Paz, F. A. A.; Ananias, D. *Chem. Soc. Rev.* **2011**, *40*, 926.
18. Wriedt, M. Systematic investigations on magneto-structural correlations of copper (II) coordination polymers based on organic ligands with mixed carboxylic and nitrogen-based moieties. *Dalton Transactions* **2012**, *41*, 4207–4216.
19. Hoffart, D. J.; Loeb, S. J. Metal–Organic Rotaxane Frameworks: Three-Dimensional Polyrotaxanes from Lanthanide-Ion Nodes, Pyridinium N-Oxide Axles, and Crown-Ether Wheels. *Angewandte Chemie International Edition* **2005**, *44*, 901–904.
20. Caygill, J. S.; Davis, F.; Higson, S. P. J. Current trends in explosive detection techniques. *Talanta* **2012**, *88*, 14–29.
21. Gares, K. L.; Hufziger, K. T.; Bykov, S. V.; Asher, S. A. Review of explosive detection methodologies and the emergence of standoff deep UV resonance Raman. *J. Raman Spectrosc.* **2016**, *47*, 124–141.
22. Huang, X. J.; Choi, Y. K. Chemical sensors based on nanostructured materials. *Sensors and Actuators B: Chemical* **2007**, *122*, 659–671.
23. Joshi, S.; Kataria, N.; Garg, V. K.; Kadirvelu, K. Pb^{2+} and Cd^{2+} recovery from water using residual tea waste and SiO_2@ TW nanocomposites. *Chemosphere* **2020**, *257*, 127277.
24. Kumar, S. S.; Ghosh, P.; Kataria, N.; Kumar, D.; Thakur, S.; Pathania, D.; Singh, L. The role of conductive nanoparticles in anaerobic digestion: Mechanism, current status and future perspectives. *Chemosphere* **2021**, *280*, 130601.
25. Yu, X.; Cheng, H.; Zhang, M.; Zhao, Y.; Qu, L.; Shi, G. Graphene-based smart materials. *Nature Reviews Materials* **2017**, *2*, 1–13.

26. Yang, W.; Zhang, G.; Weng, W.; Qiu, B.; Guo, L.; Lin, Z.; Chen, G. Signal on fluorescence biosensor for MMP-2 based on FRET between semiconducting polymer dots and a metal–organic framework. *Rsc Advances* **2014**, *4*, 58852–58857.
27. Gao, Y. S.; Xu, J. K.; Lu, L. M.; Wu, L. P.; Zhang, K. X.; Nie, T.; Wu, Y. Overoxidized polypyrrole/graphene nanocomposite with good electrochemical performance as novel electrode material for the detection of adenine and guanine. *Biosensors and Bioelectronics* **2014**, *62*, 261–267.
28. Lee, P. T.; Ward, K. R.; Tschulik, K.; Chapman, G.; Compton, R. G. Electrochemical detection of glutathione using a poly (caffeic acid) nanocarbon composite modified electrode. *Electroanalysis* **2014**, *26* (2), 366–373.
29. Chen, C.; Xie, Q.; Yang, D.; Xiao, H.; Fu, Y.; Tan, Y.; Yao, S. Recent advances in electrochemical glucose biosensors: A review. *RSC Advances* **2013**, *3*, 4473–4491.
30. Rassaei, L.; Olthuis, W.; Tsujimura, S.; Sudhölter, E. J.; Vandenberg, A. Lactate biosensors: current status and outlook. *Analytical and Bioanalytical Chemistry* **2014**, *406*, 123–137.
31. Erden, P. E.; Kılıç, E. A review of enzymatic uric acid biosensors based on amperometric detection. *Talanta* **2013**, *107*, 312–323.
32. Zhao, Y.; Zhang, W.; Lin, Y.; Du, D. The vital function of Fe 3 O 4@ Au nanocomposites for hydrolase biosensor design and its application in detection of methyl parathion. *Nanoscale* **2013**, *5*, 1121–1126.
33. Cao, X.; Wang, N.; Jia, S.; Shao, Y. Detection of glucose based on bimetallic PtCu nanochains modified electrodes. *Analytical Chemistry* **2013**, *85*, 5040–5046.
34. Drexler, K. E. Nanotechnology: from Feynman to funding. *Bulletin of Science, Technology & Society* **2004**, *24*, 21–7.
35. Chakraborty, O.; Park, H.; Medishetty, R.; Jagadese, J. V. Two-Dimensional Metal-Organic Framework Materials: Synthesis, Structures, Properties and Applications. *Chemical Reviews* **2021**.
36. Cui, Y.; Yue, Y.; Qian, G.; Chen, B. Luminescent functional metal–organic frameworks. *Chemical Reviews* **2012**, *112* (2), 1126–1162.
37. Eddaoudi, M.; Moler, D. B.; Li, H.; Chen, B.; Reineke, T. M.; O'Keeffe, M.; Yaghi, O. M. Modular chemistry: secondary building units as a basis for the design of highly porous and robust metal-organic carboxylate frameworks. *Acc. Chem. Res.* **2001**, *34*, 319–330.
38. Furukawa, H.; Cordova, K. E.; O'Keeffe, M.; Yaghi, O. M. The chemistry and applications of metal-organic frameworks. *Science* **2013**, *341*, 1230444.
39. Hu, Z.; Deibert, B. J.; Li, J. Luminescent metal–organic frameworks for chemical sensing and explosive detection. *Chem. Soc. Rev.* **2014**, *43*, 5815–5840.
40. Nagarkar, S. S.; Joarder, B.; Chaudhari, A. K.; Mukherjee, S.; Ghosh, S. K. Highly selective detection of nitro explosives by a luminescent metal–organic framework. *Angewandte Chemie International Edition* **2013**, *52* (10), 2881–2885.
41. Zhou, H.; Long, J. R.; Yaghi, O. M. Introduction to metal–organic frameworks. *Chem. Rev.* **2012**, *112*, 673–674.
42. Zhang, W.; Wu, Z. Y.; Jiang, H. L.; Yu, S. H. Nanowire-directed templating synthesis of metal–organic framework nanofibers and their derived porous doped carbon nanofibers for enhanced electrocatalysis. *Journal of the American Chemical Society* **2014**, *136*, 14385–14388.

43. Hu, Z.; Peng, Y.; Kang, Z.; Qian, Y.; Zhao, D. A modulated hydrothermal (MHT) approach for the facile synthesis of UiO-66-type MOFs. *Inorganic Chemistry* **2015**, *54*, 4862–8.
44. Sun, K.; Li, L.; Yu, X.; Liu, L.; Meng, Q.; Wang, F.; Zhang, R. Functionalization of mixed ligand metal-organic frameworks as the transport vehicles for drugs. *Journal of Colloid and Interface Science* **2017**, *486*, 128–35.
45. Hammond, O. S.; Edler, K. J.; Bowron, D. T.; Torrente-Murciano, L. Deep eutectic-solvothermal synthesis of nanostructured ceria. *Nature Communications* **2017**, *8*, 14150.
46. Tzitzios, V.; Kostoglou, N.; Giannouri, M.; Basina, G.; Tampaxis, C.; Charalambopoulou, G.; Rebholz, C. Solvothermal synthesis, nanostructural characterization and gas cryo-adsorption studies in a metal– organic framework (IRMOF-1) material. *International Journal of Hydrogen Energy* **2017**, *42*, 23899–23907.
47. Khan, N. A.; Jhung, S. H. Synthesis of metal-organic frameworks (MOFs) with microwave or ultrasound: Rapid reaction, phase-selectivity, and size reduction. *Coordination Chemistry Reviews* **2015**, *285*, 11–23.
48. Chevreau, H.; Permyakova, A.; Nouar, F.; Fabry, P.; Livage, C.; Ragon, F.; Garcia-Marquez, A.; Devic, T.; Steunou, N.; Serre, C.; Horcajada, P. Synthesis of the biocompatible and highly stable MIL-127 (Fe): from large scale synthesis to particle size control. *Cryst Eng Comm.* **2016**, *18*, 4094–101.
49. Yang, X.; Xue, H.; Yang, Q.; Yuan, R.; Kang, W.; Lee, C. S. Preparation of porous ZnO/ZnFe 2 O 4 composite from metal–organic frameworks and its applications for lithium-ion batteries. *Chemical Engineering Journal* **2017**, *308*, 340–346.
50. Guo, Y.; Hu, J.; Liu, X.; Sun, T.; Zhao, S.; Wang, S. Scalable solvent-free preparation of [Ni3 (HCOO) 6] frameworks for highly efficient separation of CH4 from N2. *Chemical Engineering Journal* **2017**, *327*, 564–572.
51. Zhang, L.; Li, S.; Xin, J.; Ma, H.; Pang, H.; Tan, L.; Wang, X. A nonenzymatic voltammetric xanthine sensor based on the use of platinum nanoparticles loaded with a metal-organic framework of type MIL-101(Cr). Application to simultaneous detection of dopamine, uric acid, xanthine and hypoxanthine. *Microchim. Acta* **2019**, *186*, 9.
52. Chen, C.; Feng, X.; Zhu, Q.; Dong, R.; Yang, R.; Cheng, Y.; He, C. Microwave-assisted rapid synthesis of well-shaped MOF-74 (Ni) for CO2 efficient capture. *Inorganic Chemistry* **2019**, *58*, 2717–2728.
53. Thomas-Hillman, I.; Laybourn, A.; Dodds, C.; Kingman, S. W. Realising the Environmental Benefits of Metal-Organic Frameworks: Recent Advances in Microwave Synthesis. *J. Mater. Chem. A* **2018**, *6*, 11564–11581.
54. Biswal, B. P.; Shinde, D. B.; Pillai, V. K.; Banerjee, R. Stabilization of graphene quantum dots (GQDs) by encapsulation inside zeolitic imidazolate framework nanocrystals for photoluminescence tuning. *Nanoscale* **2013**, *5*, 10556–10561.
55. Zheng, S.; Li, X.; Yan, B.; Hu, Q.; Xu, Y.; Xiao, X.; Pang, H. Transition-metal (Fe, Co, Ni) based metal organic frameworks for electrochemical energy storage. *Advanced Energy Materials* **2017**, 7 (18), 1602733.
56. Koehler, F. M.; Stark, W. J. Organic synthesis on graphene. *Accounts of chemical research.* **2013**, *46*, 2297–2306.

57. Wang, X.; Wang, Q.; Wang, Q.; Gao, F.; Gao, F.; Yang, Y.; Guo, H. Highly dispersible and stable copper terephthalate metal–organic framework–graphene oxide nanocomposite for an electrochemical sensing application. *ACS Appl. Mater. Interfaces* **2014**, *6*, 11573–11580.
58. Wang, Y.; Zhang, Y.; Hou, C.; Liu, M. Magnetic Fe 3 O 4@ MOFs decorated graphene nanocomposites as novel electrochemical sensor for ultrasensitive detection of dopamine. *RSC Advances* **2015**, *5* (119), 98260–98268.
59. Wang, Y.; Zhu, Y.; Binyam, A.; Liu, M.; Wu, Y.; Li, F. Discovering the enzyme mimetic activity of metal-organic framework (MOF) for label-free and colorimetric sensing of biomolecules. *Biosensors and Bioelectronics* **2016**, *86*, 432–438.
60. Chen, Q.; Li, X.; Min, X.; Cheng, D.; Zhou, J.; Li, Y.; Zhang, C. Determination of catechol and hydroquinone with high sensitivity using MOF-graphene composites modified electrode. *Journal of Electroanalytical Chemistry* **2017**, *789*, 114–122.
61. Qi, Z.; Wang, L.; You, Q.; Chen, Y. PA-Tb-Cu MOF as luminescent nanoenzyme for catalytic assay of hydrogen peroxide. *Biosensors and Bioelectronics* **2017**, *96*, 227–232.
62. Roushani, M.; Valipour, A.; Saedi, Z. Electroanalytical sensing of Cd2+ based on metal–organic framework modified on metal–organic framework modified carbon paste electrode. *Sensors and Actuators B: Chemical* **2016**, *233*, 419–425.
63. Saraf, M.; Rajak, R.; Mobin, S. M. A fascinating multitasking Cu-MOF/rGO hybrid for high performance supercapacitors and highly sensitive and selective electrochemical nitrite sensors. *Journal of Materials Chemistry A* **2016**, *4* (42), 16432–16445.
64. Peng, B.; Cui, J.; Wang, Y.; Liu, J.; Zheng, H.; Jin, L.; Wu, Y. CeO 2− x/C/rGO nanocomposites derived from Ce-MOF and graphene oxide as a robust platform for highly sensitive uric acid detection. *Nanoscale* **2018**, *10* (4), 1939–1945.
65. Cevik, S.; Anik, U. Double-walled carbon nanotube-based carbon paste electrode as xanthine biosensor. *Microchim. Acta* **2009**, *166*, 209–213.
66. Timur, S.; Anik, U.; Odaci, D.; Gorton, L. Development of a microbial biosensor based on carbon nanotube (CNT) modified electrodes. *Electrochem Commun.* **2007**, *9*, 1810–1815.
67. Uka, D.; Bunzen, H.; Berger, J.; Kieslich, G.; Fischer, R. Charge-Transfer-Induced Electrical Conductivity in a Tetrathiafulvalene-Based Metal–Organic Framework. *Chemistry of Materials* **2021**, *33* (7), 2532–2542.
68. Wen, P.; Gong, P.; Sun, J.; Wang, J.; Yang, S. Design and synthesis of NiMOF/CNT composites and rGO/carbon nitride composites for an asymmetric supercapacitor with high energy and power density. *J Mater Chem A.* **2015**, *3*, 13874–13883.
69. Wang, M. Q.; Zhang, Y.; Bao, S. J.; Yu, Y. N.; Ye, C. Ni (II)-based metalorganic framework anchored on carbon nanotubes for highly sensitive nonenzymatic hydrogen peroxide sensing. *Electrochimica Acta* **2016**, *190*, 365–370.
70. Wang, Y.; Wu, Y.; Xie, J.; Ge, H.; Hu, X. Multi-walled carbon nanotubes and metal–organic framework nanocomposites as novel hybrid electrode materials for the determination of nano-molar levels of lead in a lab-on-valve format. *Analyst* **2013**, *138* (17), 5113–20.
71. Merkoci, A. Carbon nanotubes in analytical sciences. *Microchim. Acta* **2006**, *152*, 157–174.
72. Chappanda, K. N.; Shekhah, O.; Yassine, O.; Patole, S. P.; Eddaoudi, M.; Salama, K. N. The quest for highly sensitive QCM humidity sensors: the coating of CNT/MOF composite sensing films as case study. *Sensors Actuators B* **2018**, *257*, 609–619.

73. Tran, T. Q. N.; Das, G.; Yoon, H. H. Nickel-metal–organic framework/MWCNT composite electrode for non-enzymatic urea detection. *Sensors and Actuators B: Chemical* **2017**, *243*, 78–83.
74. Zhou, E. L.; Qin, C.; Huang, P.; Wang, X. L.; Chen, W. C.; Shao, K. Z.; Su, Z. M. A Stable Polyoxometalate Pillared Metal–Organic Framework for Proton Conducting and Colorimetric Biosensing. *Chemistry-A European Journal* **2015**, *21* (33), 11894–11898.
75. Zhou, E.; Zhang, Y.; Li, Y.; He, X. Cu (II)-based MOF immobilized on multiwalled carbon nanotubes: synthesis and application for non-enzymatic detection of hydrogen peroxide with high sensitivity. *Electroanalysis* **2014**, *26*, 2526–2533.
76. Rezapour, M. Nano-Composite Carbon Paste Electrochemical Sensors for Monitoring of Lead Ions in Real Samples. *Current Analytical Chemistry* **2017**, *13* (1), 31–39.
77. Wang, F.; Zhao, J.; Gong, J.; Wen, L.; Zhou, L.; Li, D. New Multifunctional Porous Materials Based on Inorganic–Organic Hybrid Single Walled Carbon Nanotubes: Gas Storage and HighSensitive Detection of Pesticides. *Chemistry–A European Journal* **2012**, *18* (37), 11804–11810.
78. Ghanbarian, M.; Seinali, S.; Mostafavi, A. A novel MIL-53(Cr-Fe)/Ag/CNT nanocomposite based resistive sensor for sensing of volatile organic compounds. *Sensors Actuators B.* **2018**, *267*, 381–391.
79. Chen, D. H. J.; Ye, R.; Huang, Q. L. X.; Chen, S. An Electrochemical Bisphenol: a Sensor Based on Bimetallic Ce-Zn-MOF. *Electrocatalysis* **2021**, *12*, 456–468.
80. Kalyoncu, D.; Tepeli, Y.; Kırgöz, U.; Buyrac, A.; Anik, U. Electro-nano diagnostic platforms for simultaneous detection of multiple Cancer biomarkers. *Electroanalysis* **2017**, *29*, 2832–2838.
81. Falcaro, P.; Ricco, R.; Yazdi, A.; Imaz, I.; Furukawa, S.; Maspoch, D.; Ameloote, R.; Evan, J. D.; Doonan, C. J. Application of metal and metal oxide nanoparticles@MOFs. *Coord Chem Rev.* **2016**, *307*, 237–254.
82. Shanmugaraj, K.; Ilanchelian, M. Colorimetric determination of sulfide using chitosan-capped silver nanoparticles. *Microchim. Acta* **2016**, *183*, 1721–1728.
83. Kataria, N.; Garg, V. K. Application of EDTA modified Fe_3O_4/sawdust carbon nanocomposites to ameliorate methylene blue and brilliant green dye laden water. *Environmental Research* **2019**, *172*, 43–54.
84. Vilian, A. E.; Dinesh, B.; Muruganantham, R.; Choe, S. R.; Kang, S. M.; Huh, Y. S.; Han, Y. K. A screen printed carbon electrode modified with an aminofunctionalized metal–organic framework of type MIL-101(Cr) and with palladium nanoparticles for voltammetric sensing of nitrite. *Microchim. Acta* **2017**, *184* (12), 4793–4801.
85. Bagheri, N.; Khataee, A.; Hassanzadeh, J.; Habibi, B. Visual detection of peroxide-based explosives using novel mimetic Ag nanoparticle/ZnMOF nanocomposite. *Journal of Hazardous Materials* **2018**, *360*, 233–242.
86. Meng, W.; Wen, Y.; Dai, L.; He, Z.; Wang, L. A novel electrochemical sensor for glucose detection based on Ag@ZIF-67 nanocomposite. *Sensors Actuators B Chem.* **2018**, *260*, 852–860.

87. Guo, M.; Liu, S.; Guo, H.; Sun, Y.; Guo, X.; Deng, R. The mixed-ligand strategy to assemble a microporous anionic metal–organic framework: Ln 3+ post-functionalization, sensors and selective adsorption of dyes. *Dalton Transactions* **2017**, *46*, 14988–94.
88. Shi, L.; Zhu, X.; Liu, T.; Zhao, H.; Lan, M. Encapsulating Cu nanoparticles into metal-organic frameworks for nonenzymatic glucose sensing. *Sensors Actuators B Chem.* **2016**, *227*, 583–590.
89. Wu, X.; Xiong, S.; Mao, Z.; Hu, S.; Long, X. A designed ZnO@ ZIF-8 core-shell nanorod film as a gas sensor with excellent selectivity for H2 over CO. *Chem. Eur. J.* **2017**, *23* (33), 7969–7975.
90. Bagheri, N.; Khatae, A.; Habibi, B.; Hassanzadeh, J. Mimetic Ag nanoparticle/Zn-based MOF nanocomposite (AgNPs@ZnMOF) capped with molecularly imprinted polymer for the selective detection of patulin. *Talanta* **2018**, *179*, 710–718.
91. Zhang, L.; Li, S.; Xin, J.; Ma, H.; Pang, H.; Tan, L.; Wang, X. A nonenzymatic voltammetric xanthine sensor based on the use of platinum nanoparticles loaded with a metal-organic framework of type MIL-101(Cr). Application to simultaneous detection of dopamine, uric acid, xanthine and hypoxanthine. *Microchim. Acta* **2019**, *186* (1), 9.
92. Dong, W.; Zhuang, Y.; Li, S.; Zhang, X.; Chai, H.; Huang, Y. High peroxidase-like activity of metallic cobalt nanoparticles encapsulated in metal– organic frameworks derived carbon for biosensing. *Sensors Actuators B Chem.* **2018**, *255*, 2050–2057.
93. Zhan, W. W.; Kuang, Q.; Zhou, J. Z.; Kong, X. J.; Xie, Z. X.; Zheng, L. S. Semiconductor@metal–organic framework Core–Shell heterostructures: a case of ZnO@ZIF-8 Nanorods with selective Photoelectrochemical response. *J. Am. Chem. Soc.* **2013**, *135* (5), 1926–1933.
94. Michalet, X. Thoughtful peer review is worth the time it takes. *Nature* **2005**, *435* (7046), 1160.
95. Bawendi, M. G.; Steigerwald, M. L.; Brus, L. E. The quantum mechanics of larger semiconductor clusters ("quantum dots"). *Annual Review of Physical Chemistry* **1990**, *41* (1), 477–496.
96. Yang, J. M.; Hu, X. W.; Liu, Y. X.; Zhang, W. Fabrication of a carbon quantum dots-immobilized zirconium-based metal-organic framework composite fluorescence sensor for highly sensitive detection of 4-nitrophenol. *Microporous and Mesoporous Materials* **2019**, *274*, 149–154.
97. Sun, Y. P.; Zhou, B.; Lin, Y.; Wang, W.; Fernando, K. S.; Pathak, P.; Luo, P. G. Quantum-sized carbon dots for bright and colorful photoluminescence. *Journal of the American Chemical Society* **2006**, *128* (24), 7756–7757.
98. Li, Z.; Bassett, W. P.; Askim, J. R.; Suslick, K. S. Differentiation Among Peroxide Explosives with an Optoelectronic Nose. *Chem. Commun.* **2015**, *51*, 15312–15315.
99. Wang, Y.; He, J.; Zheng, M.; Qin, M.; Wei, W. Dual-emission of Eu based metal-organic frameworks hybrids with carbon dots for ratiometric fluorescent detection of Cr (VI). *Talanta* **2019**, *191*, 519–525.
100. Li, W.; Li, Z.; Yang, F.; Fang, X.; Tang, B. Synthesis and Electrochemical Performance of SnO x Quantum Dots@ UiO-66 Hybrid for Lithium Ion Battery Applications. *ACS Applied Materials & Interfaces* **2017**, *9* (40), 35030–35039.
101. Saha, S.; Das, G.; Thote, J.; Banerjee, R. Photocatalytic metal–organic framework from CdS quantum dot incubated luminescent metallohydrogel. *Journal of the American Chemical Society* **2014**, *136* (42), 14845–14851.

102. Meng, X. B.; Sheng, J. L.; Tang, H. L.; Sun, X. J.; Dong, H.; Zhang, F. M. Metal-organic framework as nanoreactors to co-incorporate carbon nanodots and CdS quantum dots into the pores for improved H2 evolution without noblemetal cocatalyst. *Applied Catalysis B: Environmental* **2019**, *244*, 340–346.
103. Nguyen, P. T.; Le, H. V.; Nguyen, K. T. T.; Hoang, V. A. T.; Dinh, T. T.; Quach, B. V. M.; Phung, X. B. T. Chromium-Based Metal-Organic Framework MIL101 (Cr)–CdSe Quantum Dot Composites: Synthesis, Morphology, Gas Adsorption and Photoluminescent Properties. *Journal of Electronic Materials* **2019**, *48* (1), 662–668.
104. Pan, D.; Wang, L.; Li, Z.; Geng, B.; Zhang, C.; Zhan, J.; Wang, L. Synthesis of graphene quantum dot/metal–organic framework nanocomposites as yellow phosphors for white light-emitting diodes. *New Journal of Chemistry* **2018**, *42* (7), 5083–5089.
105. Fu, X.; Li, H.; Le, R.; Hong, D.; Yang, B.; Gu, W.; Liu, X. Synthesis of Mn^{2+} doped ZnS quantum dots/ZIF-8 composite and its applications as a fluorescent probe for sensing Co^{2+} and dichromate. *Journal of Solid State Chemistry* **2018**, *264*, 35–41.
106. Zhao, D.; Wan, X.; Song, H.; Hao, L.; Su, Y.; Lv, Y. Metal–organic frameworks (MOFs) combined with ZnO quantum dots as a fluorescent sensing platform for phosphate. *Sensors and Actuators B: Chemical* **2014**, *197*, 50–57.
107. Yang, X.; Yu, Y. Q.; Peng, L. Z.; Lei, Y. M.; Chai, Y. Q.; Yuan, R.; Zhuo, Y. Strong electrochemiluminescence from MOF accelerator enriched quantum dots for enhanced sensing of trace TnI. *Analytical Chemistry* **2018**, *90* (6), 3995–4002.
108. Liu, H.; Ni, T.; Mu, L.; Zhang, D.; Wang, J.; Wang, S.; Sun, B. Sensitive detection of pyrraline with a molecularly imprinted sensor based on metalorganic frameworks and quantum dots. *Sensors and Actuators B: Chemical* **2018**, *256*, 1038–1044.
109. Xin, W. L.; Jiang, L. F.; Zong, L. P.; Zeng, H. B.; Shu, G. F.; Marks, R.; Shan, D. MoS_2 quantum dots-combined zirconium-metalloporphyrin frameworks: Synergistic effect on electron transfer and application for bioassay. *Sensors and Actuators B: Chemical* **2018**, *273*, 566–573.
110. Lin, X.; Gao, G.; Zheng, L.; Chi, Y.; Chen, G. Encapsulation of strongly fluorescent carbon quantum dots in metal–organic frameworks for enhancing chemical sensing. *Analytical Chemistry* **2014**, *86*, 1223–1228.
111. Hao, J.; Liu, F.; Liu, N.; Zeng, M.; Song, Y.; Wang, L. Ratiometric fluorescent detection of Cu2+ with carbon dots chelated Eu-based metal-organic frameworks. *Sensors and Actuators B: Chemical* **2017**, *245*, 641–647.
112. Rani, S.; Sharma, B.; Kapoor, S.; Malhotra, R.; Varma, R. S.; Dilbaghi, N. Construction of Silver Quantum Dot Immobilized Zn-MOF-8 Composite for Electrochemical Sensing of 2, 4-Dinitrotoluene. *Applied Sciences* **2019**, *9* (22), 495.
113. Ma, W.; Jiang, Q.; Yu, P.; Yang, L.; Mao, L. Zeolitic imidazolate framework-based electrochemical biosensor for in vivo electrochemical measurements. *Analytical Chemistry* **2013**, *85* (15), 7550–7.
114. Li, Y.; Huangfu, C.; Du, H.; Liu, W.; Li, Y.; Ye, J. Electrochemical behavior of metal–organic framework MIL-101 modified carbon paste electrode: An excellent candidate for electroanalysis. *Journal of Electroanalytical Chemistry* **2013**, *709*, 65–69.

115. Yuan, B. Q.; Zhang, J. C.; Zhang, R. C.; Shi, H. Z.; Guo, X. L.; Guo, Y. Y.; Guo, X. Y.; Cai, S. S.; Zhang, D. J. Electrochemical and electrocatalytic properties of a stable Cu-based metal–organic framework. *Int. J. Electrochem. Sci.* **2015**, *104*, 899–910.
116. Xu, Q.; Wang, Y.; Jin, G.; Jin, D.; Li, K.; Mao, A.; Hu, X. Photooxidation assisted sensitive detection of trace Mn 2+ in tea by NH 2-MIL-125 (Ti) modified carbon paste electrode. *Sensors and Actuators B: Chemical* **2014**, *201*, 274–280.
117. Gassensmith, J. J.; Kim, J. Y.; Holcroft, J. M.; Farha, O. K.; Stoddart, J. F.; Hupp, J. T.; Jeong, N. C. A Metal–Organic Framework-Based Material for Electrochemical Sensing of Carbon Dioxide. *Journal of the American Chemical Society* **2014**, *136* (23), 8277–8282.
118. Ma, Y.; Su, H.; Kuang, X.; Li, X.; Zhang, T.; Tang, B. Heterogeneous nano metal–organic framework fluorescence probe for highly selective and sensitive detection of hydrogen sulfide in living cells. *Analytical Chemistry* **2014**, *86* (22), 11459–11463.
119. Zhang, Y.; Li, B.; Ma, H.; Zhang, L.; Zheng, Y. Rapid and facile ratiometric detection of an anthrax biomarker by regulating energy transfer process in bio-metal-organic framework. *Biosensors and Bioelectronics* **2016**, *85*, 287–293.
120. Yang, L.; Xu, C.; Ye, W.; Liu, W. An electrochemical sensor for H2O2 based on a new Co-metal-organic framework modified electrode. *Sensor Actuator B Chem.* **2015**, *215*, 489–496.
121. Kung, C. W.; Li, Y. S.; Lee, M. H.; Wang, S. Y.; Chiang, W. H.; Ho, K. C. In situ growth of porphyrinic metal–organic framework nanocrystals on graphene nanoribbons for the electrocatalytic oxidation of nitrite. *Journal of Materials Chemistry A* **2016**, *4* (27), 10673–10682.
122. Hosseini, H.; Ahmar, H.; Dehghani, A.; Bagheri, A.; Fakhari, A. R.; Amini, M. M. Au-SH-SiO2 nanoparticles supported on metal-organic framework (Au-SH-SiO2@ Cu-MOF) as a sensor for electrocatalytic oxidation and determination of hydrazine. *Electrochimica Acta.* **2013**, *88*, 301–309.
123. Yuan, B.; Zhang, R.; Jiao, X.; Li, J.; Shi, H.; Zhang, D. Amperometric determination of reduced glutathione with a new Co-based metal-organic coordination polymer modified electrode. *Electrochemistry Communications* **2014**, *40*, 92–95.
124. Yuan, B.; Zhang, J.; Zhang, R.; Shi, H.; Wang, N.; Li, J.; Zhang, D. Cu-based metal–organic framework as a novel sensing platform for the enhanced electrooxidation of nitrite. *Sensors and Actuators B: Chemical* **2016**, *222*, 632–637.
125. Yuan, S.; Bo, X.; Guo, L. In-situ growth of iron-based metal-organic framework crystal on ordered mesoporous carbon for efficient electrocatalysis of p-nitrotoluene and hydrazine. *Analytica Chimica Acta* **2018**, *1024*, 73–83.
126. Wang, Z.; Yang, J.; Li, Y.; Zhuang, Q.; Gu, J. Zr-Based MOFs integrated with a chromophoric ruthenium complex for specific and reversible Hg 2+ sensing. *Dalton Transactions* **2018**, *47* (16), 5570–5574.
127. Deng, X.; Fang, Y.; Lin, S.; Cheng, Q.; Liu, Q.; Zhang, X. Porphyrin-Based Porous Organic Frameworks as a Biomimetic Catalyst for Highly Efficient Colorimetric Immunoassay. *ACS Applied Materials & Interfaces* **2017**, *9* (4), 3514–3523.
128. Prathap, M. U. A.; Gunasekaran, S. Rapid and Scalable Synthesis of Zeolitic Imidazole Framework (ZIF8) and its Use for the Detection of Trace Levels of Nitroaromatic Explosives. *Advanced Sustainable Systems* **2018**, *2* (10), 1800053.

129. Arul, P.; John, S. A. Silver nanoparticles built-in zinc metal–organic framework modified electrode for the selective non-enzymatic determination of H2O2. *Electrochimica Acta* **2017**, *235*, 680–689.
130. Dong, Y.; Duan, C.; Sheng, Q.; Zheng, J. Preparation of Ag@ zeolitic imidazolate framework-67 at room temperature for electrochemical sensing of hydrogen peroxide. *Analys.* **2019**, *144* (2), 521–529.
131. Wang, C.; Huang, S.; Luo, L.; Zhou, Y.; Lu, X.; Zhang, G.; Cao, F. Ultrathin two-dimension metal-organic framework nanosheets/multi-walled carbon nanotube composite films for the electrochemical detection of H2O2. *J. Electroanal. Chem.* **2019**, *835*, 178–185.
132. Yang, J.; Yang, L.; Ye, H.; Zhao, F.; Zeng, B. Highly dispersed Au-Pd alloy nanoparticles immobilized on UiO-66-NH2 metal-organic framework for the detection of nitrite. *Electrochimica Acta* **2016**, *219*, 647–654.
133. Zhou, J.; Li, X.; Yang, L.; Yan, S.; Wang, M.; Cheng, D.; Zhang, C. The CuMOF-199/single-walled carbon nanotubes modified electrode for simultaneous determination of hydroquinone and catechol with extended linear ranges and lower detection limits. *Analytica Chimica Acta* **2015**, *899*, 57–65.
134. Yadav, D. K.; Ganesan, V.; Sonkar, P. K.; Gupta, R.; Rastogi, P. K. Electrochemical investigation of gold nanoparticles incorporated zinc based metal-organic framework for selective recognition of nitrite and nitrobenzene. *Electrochimica Acta* **2016**, *200*, 276–282.
135. Yoon, J. Recent Progress on Fluorescent Probes. In *Coordination Chemistry Reviews*; Elsevier, 2018, p 1.
136. Yang, S. P.; Chen, S. R.; Liu, S. W.; Tang, X. Y.; Qin, L.; Qiu, G. H.; Chen, W. H. Platforms formed from a three-dimensional Cu-based zwitterionic metal–organic framework and probe-DNA: selective fluorescent biosensors for human immunodeficiency virus 1 ds-DNA and Sudan virus RNA sequences. *Analytical Chemistry* **2015**, *87* (24), 12206–12214.
137. Fang, J. M.; Leng, F.; Zhao, X. J.; Hu, X. L.; Li, Y. F. Metal–organic framework MIL-101 as a low background signal platform for label-free DNA detection. *Analyst* **2014**, *139* (4), 801–806.
138. Ferey, G.; Mellot-Draznieks, C.; Serre, C.; Millange, F.; Dutour, J.; Surble, S.; Margiolaki, I. A chromium terephthalate-based solid with unusually large pore volumes and surface area. *Science* **2005**, *309* (5743), 2040–2042.
139. Mejia Ariza, R.; Rosselli, J.; Breukers, C.; Manicardi, A.; Terstappen, L. W.; Corradini, R.; Huskens, J. DNA Detection by Flow Cytometry using PNA-Modified Metal–Organic Framework Particles. *Chemistry-A European Journal* **2017**, *23* (17), 4180–4186.
140. Chen, M.; Gan, N.; Zhou, Y.; Li, T.; Xu, Q.; Cao, Y.; Chen, Y. A novel aptamer-metal ions-nanoscale MOF based electrochemical biocodes for multiple antibiotics detection and signal amplification. *Sensors and Actuators B: Chemical* **2017**, *242*, 1201–1209.
141. Zhao, S. N.; Wu, L. L.; Feng, J.; Song, S. Y.; Zhang, H. J. An ideal detector composed of a 3D Gd-based coordination polymer for DNA and Hg^{2+} ion. *Inorganic Chemistry Frontiers* **2016**, *3* (3), 376–380.
142. Wang, H. S.; Li, J.; Li, J. Y.; Wang, K.; Ding, Y.; Xia, X. H. Lanthanide-based metal-organic framework nanosheets with unique fluorescence quenching properties for two-color intracellular adenosine imaging in living cells. *NPG Asia Materials* **2017**, *9* (3), e354.

143. Wang, K.; Li, N.; Zhang, J.; Zhang, Z.; Dang, F. Size-selective QD@ MOF core-shell nanocomposites for the highly sensitive monitoring of oxidase activities. *Biosensors and Bioelectronics* **2017**, *87*, 339–344.

144. Tan, H.; Ma, C.; Chen, L.; Xu, F.; Chen, S.; Wang, L. Nanoscaled lanthanide/nucleotide coordination polymer for detection of an anthrax biomarker. *Sensors and Actuators B: Chemical* **2014**, *190*, 621–626.

145. Jung, B. K.; Hasan, Z.; Jhung, S. H. Adsorptive removal of 2, 4-dichlorophenoxyacetic acid (2, 4-D) from water with a metal–organic framework. *Chemical Engineering Journal* **2013**, *234*, 99–105.

146. Sun, S.; Wang, F.; Sun, Y.; Guo, X.; Ma, R.; Zhang, M.; Guo, H.; Xie, Y.; Hu, T. Construction of a Dual-function Metal–organic Framework: Detection of Fe^{3+}, Cu^{2+}, Nitroaromatic Explosives, and a High Second-harmonic Generation Response. *Ind. Eng. Chem. Res.* **2019**, *58*, 17784–17791.

147. Liu, J.; Kaczmarek, A. M.; Van Deun, R. Downshifting/upconversion NaY (MoO4) 2 luminescent materials as highly sensitive fluorescent sensors for Pb2+ ions detection. *Sensors and Actuators B: Chemical* **2018**, 255, 2163–2169.

148. Valekar, A. H.; Batule, B. S.; Kim, M. I.; Cho, K. H.; Hong, D. Y.; Lee, U. H.; Hwang, Y. K. Novel amine-functionalized iron trimesates with enhanced peroxidase-like activity and their applications for the fluorescent assay of choline and acetylcholine. *Biosensors and Bioelectronics* **2018**, *100*, 161–168.

149. Shustova, N. B.; McCarthy, B. D.; Dinca, M. Turn-on fluorescence in tetraphenylethylene-based metal–organic frameworks: an alternative to aggregation-induced emission. *Journal of the American Chemical Society* **2011**, *133* (50), 20126–20129.

150. Dolgopolova, E. A.; Rice, A. M.; Martin, C. R.; Shustova, N. B. Photochemistry and photophysics of MOFs: steps towards MOF-based sensing enhancements. *Chemical Society Reviews* **2018**, *47* (13), 4710–4728.

151. Huo, S. H.; Yan, X. P. Metal–organic framework MIL-100 (Fe) for the adsorption of malachite green from aqueous solution. *Journal of Materials Chemistry* **2012**, *22* (15), 7449–7455.

152. Volkringer, C.; Loiseau, T.; Guillou, N.; Férey, G.; Haouas, M.; Taulelle, F.; Stock, N. High-Throughput Aided Synthesis of the Porous Metal– Organic Framework-Type Aluminum Pyromellitate, MIL-121, with Extra Carboxylic Acid Functionalization. *Inorganic Chemistry* **2010**, *49* (21), 9852–9862.

153. Wang, H. Metal–organic frameworks for biosensing and bioimaging applications. *Coordination Chemistry Reviews* **2017**, *349*, 139–155.

154. Tao, Y.; Jiang, Y.; Huang, Y.; Liu, J.; Zhang, P.; Chen, X.; Xu, J. Carbon dots@ metal–organic frameworks as dual-functional fluorescent sensors for Fe^{3+} ions and nitro explosives. *CrystEngComm* **2021**.

155. Senthilkumar, S.; Goswami, R.; Smith, V. J.; Bajaj, H. C.; Neogi, S. Pore wall-functionalized luminescent Cd (II) framework for selective CO_2 adsorption, highly specific 2, 4, 6-trinitrophenol detection, and colorimetric sensing of Cu^{2+} ions. *ACS Sustainable Chemistry & Engineering* **2018**, *6* (8), 10295–10306.

156. Nagarkar, S. S.; Desai, A. V.; Ghosh, S. K. A Nitro-Functionalized Metal–Organic Framework as a Reaction-Based Fluorescence Turn On Probe for Rapid and Selective H2S Detection. *Chemistry–A European Journal* **2015**, *21* (28), 9994–9997.
157. Nagarkar, S. S.; Saha, T.; Desai, A. V.; Talukdar, P.; Ghosh, S. K. Metal-organic framework based highly selective fluorescence turn-on probe for hydrogen sulphide. *Scientific Reports* **2014**, *4* (1), 1–6.
158. Buragohain, A.; Biswas, S. Cerium-based azide-and nitro-functionalized UiO-66 frameworks as turn-on fluorescent probes for the sensing of hydrogen sulphide. *CrystEngComm* **2016**, *18* (23), 4374–4381.
159. Zhang, M.; Feng, G.; Song, Z.; Zhou, Y. P.; Chao, H. Y.; Yuan, D.; Liu, B. Two-dimensional metal–organic framework with wide channels and responsive turn-on fluorescence for the chemical sensing of volatile organic compounds. *Journal of the American Chemical Society* **2014**, *136* (20), 7241–7244.
160. Razavi, S. A. A.; Masoomi, M. Y.; Morsali, A. Double Solvent Sensing Method for Improving Sensitivity and Accuracy of Hg (II) Detection Based on Different Signal Transduction of a Tetrazine-Functionalized Pillared Metal–Organic Framework. *Inorganic Chemistry* **2017**, *56* (16), 9646–9652.
161. Gao, L.; Zhuang, J.; Nie, L.; Zhang, J.; Zhang, Y.; Gu, N.; Yan, X. Intrinsic peroxidase-like activity of ferromagnetic nanoparticles. *Nature Nanotechnology* **2007**, *2* (9), 577–583.
162. Liu, X.; Wang, Q.; Zhang, Y.; Zhang, L.; Su, Y.; Lv, Y. Colorimetric detection of glutathione in human blood serum based on the reduction of oxidized TMB. *New Journal of Chemistry* **2013**, *37* (7), 2174–2178.
163. Ortiz-Gómez, I.; Salinas-Castillo, A.; García, A. G.; Álvarez-Bermejo, J. A.; de Orbe-Payá, I.; Rodríguez-Diéguez, A.; Capitán-Vallvey, L. F. Microfluidic paper-based device for colorimetric determination of glucose based on a metal-organic framework acting as peroxidase mimetic. *Microchimica Acta* **2017**, *185* (1), 1–8.
164. Jin, L.; Meng, Z.; Zhang, Y.; Cai, S.; Zhang, Z.; Li, C.; Shen, Y. Ultrasmall Pt nanoclusters as robust peroxidase mimics for colorimetric detection of glucose in human serum. *ACS Applied Materials & Interfaces* **2017**, *9* (11), 10027–10033.
165. Chen, W.; Chen, J.; Feng, Y. B.; Hong, L.; Chen, Q. Y.; Wu, L. F.; Xia, X. H. Peroxidase-like activity of water-soluble cupric oxide nanoparticles and its analytical application for detection of hydrogen peroxide and glucose. *Analyst* **2012**, *137* (7), 1706–1712.
166. Yin, Y.; Gao, C.; Xiao, Q.; Lin, G.; Lin, Z.; Cai, Z.; Yang, H. Protein-Metal–organic Framework Hybrid Composites with Intrinsic Peroxidase-like Activity as a Colorimetric Biosensing Platform. *ACS Applied Materials & Interfaces* **2016**, *8* (42), 29052–29061.
167. Zhang, J. W.; Zhang, H. T.; Du, Z. Y.; Wang, X.; Yu, S. H.; Jiang, H. L. Water-stable metal–organic frameworks with intrinsic peroxidase-like catalytic activity as a colorimetric biosensing platform. *Chemical Communications* **2014**, *50* (9), 1092–1094.
168. Fateeva, A.; Horcajada, P.; Devic, T.; Serre, C.; Marrot, J.; Grenèche, J. M.; Férey, G. Synthesis, Structure, Characterization, and Redox Properties of the Porous MIL68 (Fe) Solid. *European Journal of Inorganic Chemistry* **2010**, *24*, 3789–3794.
169. Wu, Y.; Ma, Y.; Xu, G.; Wei, F.; Ma, Y.; Song, Q.; Xu, X. Metal-organic framework coated Fe_3O_4 magnetic nanoparticles with peroxidase-like activity for colorimetric sensing of cholesterol. *Sensors and Actuators B: Chemical* **2017**, *249*, 195–202.

170. Tranchemontagne, D. J.; Mendoza-Cortes, J. L.; O'Keeffe, M.; Yaghi, O. M. *Chem. Soc. Rev.* **2009**, *38*, 1257.
171. Qiu, Y.; Deng, H.; Mou, J.; Yang, S.; Zeller, M.; Batten, S. R.; Li, J. In situ tetrazole ligand synthesis leading to a microporous cadmium–organic framework for selective ion sensing. *Chemical Communications* **2009**, *36*, 5415–5417.
172. Zou, X.; Zhu, G.; Hewitt, I. J.; Sun, F.; Qiu, S. Synthesis of a metal–organic framework film by direct conversion technique for VOCs sensing. *Dalton Trans.* **2009**, *16*, 3009–3013.
173. Yang, S.; Karve, V. V.; Justin, A.; Kochetygov, I.; Espin, J.; Asgari, M.; Trukhina, O.; Sun, D. T.; Peng, L.; Queena, W. L. Enhancing MOF performance through the introduction of polymer guests. *Coordination Chemistry Reviews* **2021**, *427*, 213525.
174. Kaur, R.; Paul, A. K.; Deep, A. Nanocomposite of europium organic framework and quantum dots for highly sensitive chemosensing of trinitrotoluene. *Forensic Science International* **2014**, *242*, 88–93.
175. Xie, Z.; Ma, L.; deKrafft, K. E.; Jin, A.; Lin, W. Porous Phosphorescent Coordination Polymers for Oxygen Sensing. *J. Am. Chem. Soc.* **2009**, *132*, 922–923.
176. An, J. Y.; Shade, C. M.; Chengelis-Czegan, D. A.; Petoud, S.; Rosi, N. L. Zinc-Adeninate Metal—Organic Framework for Aqueous Encapsulation and Sensitization of Near-infrared and Visible Emitting Lanthanide Cations. *J. Am. Chem. Soc.* **2011**, *133*, 1220–1223.
177. Pankajakshan, A.; Kuznetsov, D.; Mandal, S. Ultrasensitive Detection of Hg (II) Ions in Aqueous Medium Using Zinc-based Metal–organic Framework. *Inorg. Chem.* **2019**, *58*, 1377–1381.
178. Shayegan, H.; Farahani, Y. D.; Safarifard, V. A Pillar-layer Metal-organic Framework as A Turn-on Luminescent Sensor for Highly Selective and Sensitive Detection of Zn(II) Ion. *J. Solid State Chem.* **2019**, *279*, 120968.
179. Xu, S.; Zhan, L.; Hong, C.; Chen, X.; Chen, X.; Oyama, M. Metal–organic Framework-5 as a Novel Phosphorescent Probe for the Highly Selective and Sensitive Detection of Pb (II) in Mussels. *Sensors Actuators B Chem.* **2020**, *308*, 127733.
180. Farahani, Y. D.; Safarifard, V. A Luminescent Metal-organic Framework with Pre-designed Functionalized Ligands as an Efficient Fluorescence Sensing for Fe^{3+} Ions. *J. Solid State Chem.* **2019**, *270*, 428–435.
181. Ding, M.; Flaig, R. W.; Jiang, L.; Yaghi, O. M. Carbon capture and conversion using metal-organic framework and MOF- based materials. *Chem. Soc. Rev.* **2019**, *48*, 2783.
182. Ge, Y.; Yao, S.; Sun, X.; Yu, C.; Li, G.; Liu, Y. A Luminescent Metal-organic Framework with Helical SBUs for Highly Effective Detection of Fe^{3+} Ions. *Inorg. Chem. Commun.* **2018**, *93*, 52–55.
183. Helal, A.; Nguyen, H. L.; Al-Ahmed, A.; Cordova, K. E.; Yamani, Z. H. An Ultrasensitive and Selective Metal–organic Framework Chemosensor for Palladium Detection in Water. *Inorg. Chem.* **2019**, *58*, 1738–1741.
184. Xiaoxiong, Z.; Wenjun, Z.; Cuiliu, L.; Xiaohong, Q.; Chengyu, Z. Eu^{3+}-postdoped Uio66-type Metal–organic Framework as a Luminescent Sensor for Hg^{2+} Detection in Aqueous Media. *Inorg. Chem.* **2019**, *58*, 3910–3915.

185. Chai, H.; Ren, Y.; He, H.; Wang, Z.; Zhang, Y.; Gao, L. A Microporous Tb-MOF Luminescent Sensor Based on A Flexible Tricarboxylate for Highly Sensitive Detection of Acetone and Fe^{3+} Ions in Aqueous and Isopropanol. *J. Solid State Chem.* **2019**, *271*, 162–167.
186. Xia, Z.; Ren, C.; Xu, W.; Li, F.; Qiao, C.; Wei, Q.; Zhou, C.; Chen, S.; Gao, S. Ultrasensitive Fe^{3+} Luminescence Sensing and Supercapacitor Performances of a Triphenylamine-based TbIII-MOF. *J. Solid State Chem.* **2020**, *282*, 121083.
187. Li, Y.; Ma, D.; Chen, C.; Chen, M.; Li, Z.; Wu, Y.; Zhu, S.; Peng, G. A Hydrostable and Bromine-functionalized Manganese-organic Framework with Luminescence Sensing of Hg^{2+} and Antiferromagnetic Properties. *J. Solid State Chem.* **2019**, *269*, 257–263.
188. Liu, Y.; Liu, C.; Zhang, X.; Liu, L.; Ge, C.; Zhuang, X.; Zhang, N.; Yang, Q.; Huang, Y.; Zhang, Z. Highly Selective and Sensitive Detection of Fe^{3+}, Al^{3+} and Picric Acid by a Water-stable Luminescent MOF. *J. Solid State Chem.* **2019**, *272*, 1–8.
189. Tu, M.; Wannapaiboon, S.; Fischer, R. A. Liquid phase stepwise growth of surface mounted metal–organic frameworks for exploratory research and development of applications. *Inorganic Chemistry Frontiers* **2014**, *1* (6), 442–463.
190. Wang, Z.; Chen, Q. Molecular and Biomolecular Spectroscopy and Biomolecular Spectroscopy Vapochromic behavior of MOF for selective sensing of ethanol. *Spectrochimica Acta Part A* **2018**, *194*, 158–162.
191. Hazra, A.; Bonakala, S.; Reddy, S. K.; Balasubramanian, S.; Maji, T. K. Effect of Pillar Modules and Their Stoichiometry in 3D Porous Frameworks of Zn (II) with [Fe (CN) 6] 3–: High CO2/N2 and CO2/CH4 Selectivity. *Inorganic Chemistry* **2013**, *52* (19), 11385–11397.
192. Bauer, C. A.; Timofeeva, T. V.; Settersten, T. B.; Patterson, B. D.; Liu, V. H.; Simmons, B. A.; Allendorf, M. D. *J. Am. Chem. Soc.* **2007**, *129*, 7136.
193. Feng, L.; Dong, C.; Li, M.; Li, L.; Jiang, X.; Gao, R.; Wang, R.; Zhang, L.; Ning, Z.; Gao, D.; Bi, J. Terbium-based metal-organic frameworks: highly selective and fast respond sensor for styrene detection and construction of molecular logic gate. *Journal of Hazardous Materials* **2020**, *388*, 121816.
194. Sajid, M. Toxicity of nanoscale metal organic frameworks: a perspective. *Environ. Sci. Pollut. Res.* **2016**, *23*, 14805–14807.

Chapter 5

Metal–Organic Frameworks as Sensors

Bhawna Uttam* and Ravi Kumar*

Department of Chemistry, J.C. Bose University of Science and Technology, YMCA, Faridabad, Haryana 121006, India

***Email: bhawnauttamchem@gmail.com**

***Email: ravi.dhamija@rediffmail.com**

There is an increasing demand for the development of chemical sensors for the detection of harmful pollutants in the environment as evident from enormous publications in this area. Metal-organic frameworks (MOFs) have been widely and rapidly growing as chemical sensors for such pollutants. The various special features of MOFs such as large surface area, structural flexibility, high porosity, high adsorption efficiency are the main reasons for the development of these frameworks as sensors. Luminescent properties of the MOFs have been explored for various applications such as gas storage and separation, drug delivery, sensing of various ions and biomolecules. The luminescent MOFs detect the analyte species *via* various mechanisms such as cation exchange, structure collapse, overlap mechanism and analyte-linker interactions. The present chapter highlights the recent advances in the area of metal-organic frameworks as cation, anion, pH and temperature sensors.

Introduction

During the last two decades, there has been rapid and explosive growth in the field of metal-organic frameworks (*1–4*). Metal-organic frameworks (MOFs) are mainly crystalline structures in which the metal cation and their metal clusters are coordinated to multi-topic organic linker molecules or ions *via* coordinate bonding. Infinite number of combinations are possible by using these metal ions and linkers to form MOFs. The flexible coordination environment with different geometries *viz.*, tetrahedral, pyramidal, octahedral, bipyamidal has been offered by commonly used metal cations such as Zn^{2+}, Ln^{2+}, Cu^{2+}, Al^{3+}, Cd^{2+} etc. during the synthesis of MOFs (*5–7*). MOFs possess the highly ordered network structure due to ability of the metal – ligand bonds to reversibly allows the arrangement of metal ions and organic linkers during the polymerization process (*8*, *9*). The carboxylates, amines, phosphates, nitriles, sulfonates groups in organic linkers play a vital role in the formation of MOFs (*10–13*). The common techniques used for the synthesis of metal-organic frameworks include solvothermal, electrochemical, hydrothermal, mechanochemical and layer- by-layer growth (*14–18*). The MOFs have been utilized and explored thoroughly for various applications such as catalysis, luminescence, drug delivery, sensing and gas separation etc. (*19–25*) The specific silent features, namely, structural diversity, chemical and thermal stability, flexibility, functional tunability, greater porosity, high surface area and large adsorption capacity of the metal-organic frameworks make them suitable candidate for such diverse applications. Chemo- and bio- sensing are among promising applications of these porous frameworks due to different degree of luminescence response (luminescence enhancement/quenching) obtained upon addition of guest species to the host MOFs, non toxicity and biodegradability (*26, 27*). MOFs are used as potential chemical sensors due to some silent features, such as, 1) specific recognition with exceptionally high selectivity of the guest species *via* host guest interaction; 2) facile recognition of analytes due to the large surface area of MOFs; 3) regeneration and recyclability of the MOFs due to flexible porosity. In last few years, the optical properties of the MOFs have also been explored to serve as optoelectronic and electronic sensors (*28, 29*). In general, MOFs based materials as sensors are categorized in to four different categories namely, crystalline materials, functional composites, nanoscale materials and membranes (*30–33*). The present chapter focus on the development of MOF-based sensors for the detection of cations, anions, pH and temperature.

Mechanism of Analyte Detection by Luminescent MOF

The photoluminescence signal of luminescent MOFs changes in presence of specific analytes and this change in the signal is referred as 'response'. Upon addition of the analytes, change appears in the intensity and the wavelength of the photoluminescence spectra of the host MOF. There are different mechanisms by which the analytes are detected by MOFs and categorized as overlap based mechanism, structural collapse mechanism, multi-interaction, linker-analyte interaction and cation exchange (*34–37*).

Overlap Mechanism

Overlap mechanism is the common mechanism for analyte detection. In this mechanism, photoluminescence excitation and emission band of luminescent MOFs overlay with the absorption spectra of the guest ion. In case of non-emissive analyte, process of energy transfer through resonance occurs from fluorescent MOF to the guest species *via* various interactions such as dipole-dipole or ion-dipole interactions. This process is dependent on the distance between the guest ion and the fluorescent MOF as host and, leads to the enhancement in the quenching efficiency and sensitivity. Another possible method is the competitive absorption of light by analyte and fluorophore resulting in turn-off or fluorescence quenching of the emission spectra. This method is generally applied for the detection of the metal ions such as Fe^{3+} and Cr^{3+} (*38–41*).

Structure Collapse Mechanism

Structure collapse mechanism refers to the transformation and collapse of the frameworks of crystalline luminescent metal-organic framework that results in to free organic linkers and metal ions. The resulting free organic molecule/linker based fluorescence emission appears at the cost of metal ligand charge transition emission. In this mechanism, there is quenching in the photoluminescence emission band, along with the reduction of intensity of band. In case of d- block, charge transfer takes place from ligand to the metal ion whereas for f-block, luminescent MOFs resonance energy transfers from organic linker to metal (*42–44*).

Cation Exchange Mechanism

Cation exchange mechanism mainly explains the interchange of the guest species/metal ions with the metal present in the metal-organic framework. To facilitate this process, the analyte cations should be able to diffuse in the pores by replacing the cationic counter parts. During this process, there will be cleavage of the bond between metal ions and linker groups. In case of f- block lanthanide MOFs, the guest ions are exchanged with the lanthanides and hence released in the solution. As a result, there will be a decrease in the intensity of characteristic peaks of lanthanide MOF (*45, 46*).

Linker-Analyte Interaction

Linker-Analyte Interaction is another useful mechanism for the analyte detection by MOFs. In this mechanism, coordination interaction between the organic linker and metal ions leads to the change in the conformation and the electron density of the organic linker. Therefore, the photoluminescence enhancement or quenching both can be obtained based on the structural and conformational changes through these ion-linker interaction. The electrostatic, coordination and chelation interactions leads to photo transfer of electron or energy transfer from luminescent MOF to the guest ion center. This results in either the photoluminescence quenching or enhancement (in case of radiative relaxation) (*46–49*).

MOFs as Cation Sensors

MOFs have been widely found to act as potential sensor for the detection of various types of metal ions. Metal ions from various sources are released in environment that impacts severe hazardous effects. Detection of the cations by MOFs has become an important area of concern. The porosity with an accessible free space present within the pores of MOFs makes them excellent host for the guest molecules and these guest molecules can be easily captured. Therefore, these frameworks are either decorated with suitable functional groups or are used in the form of hybrid materials. The interactions between metal ions and the host molecule, such as MOFs, can induce perturbation in photoluminescence properties such as changes in absorption and emission spectra. Specific metal ions can be easily detected due to change in the emission signal of MOF through optimized host guest interactions between the MOF and metal ion. This area has been reviewed from different points of view as evident from some recent examples in this field (*50–52*). The astounding progress and developments made in metal ions detection using MOFs are discussed in this section.

Hg^{2+}

Inorganic, organic and elemental mercury are found to be very toxic for humans depending on their dose and exposure rate. Even the acute exposure to mercury vapors can lead to failure of respiratory system whereas the chronic exposure results in the fatigue, weakness, gastrointestinal disturbance etc. Therefore, there is dire need for the development of selective and efficient Hg^{2+} sensors. Ma and co-workers successfully synthesized a water-stable indium-based anionic MOF having formula $[(CH_3)_2NH_2][In(TNB)_{4/3}].(2DMF)(3H_2O)$ with organic linker as H_3TNB (4,4′,4″-nitrilobenzoic acid). A composite of indium MOF has been prepared by the cationic dye encapsulation by following the ion exchange mechanism. The as-prepared composite was found to have sensitive quenching response toward Hg^{2+} with 1.75 ppb limit of detection (*53*). In another report by Xu and Yan, a colorimetric fluorescent probe composed of Eu-MOF (Eu^{3+}/CDs@MOF-

253) with encapsulated carbon dots for the detection of Hg^{2+} has been developed. The MOF-253 was developed using Al(OH)(bpydc) and, pyridine-dicarboxylic acid derivative as organic linker. The optical properties of CD and Eu^{3+} leads to dual emission fluorescent-functionalized MOFs. These MOFs also show good stability in aqueous samples. The (I_{Eu}/I_{CD}) ratio shows a linear relationship (0.065–150 μM range) with increasing concentration of Hg^{2+} with a limit of detection as 13ppb as shown in Figure 1 (54).

Chemosensors, having nitrogen centers or amine groups and sulfur-rich probes are associated with reasonable selectivity. These sensors, however, have two major drawbacks, (i) ariel oxidation in case of both amines and sulfides upon long time storage, (ii) no suitable response signal in sulfur-rich environment in case of sulfur based probes. In order to avoid these limitations, an excellent chemodosimeter based butyne-functionalized MOF, UiO-66@Butyne, was recently developed *via* ligand modulation by Ghosh et al. and, used for the selective sensing of Hg^{2+} ions via oxymercuration reaction. The detection limit for Hg^{2+} ions is found to be 10.9 nM (Figure 1) (55). A solution and solid state zinc-based MOF bearing an azo group was synthesized by Mandal et al. keeping in view that the Hg^{2+} ions has a higher affinity toward bases having nitrogen center. This Zn-based MOF has various components as linker like 4,4'azopyridine, 5-aminoisophthalic acid and N,N-dimethylformamide (56). The presence of the nitrogen containing ligands directs the Hg^{2+} ions for specific interaction that results alteration in the organic molecule based luminescence properties. This water-stable MOF interacts with Hg^{2+} ion due to –N=N– group and, can detect the picomolar concentrations of Hg^{2+} ions without interference from other ions present in the aqueous medium.

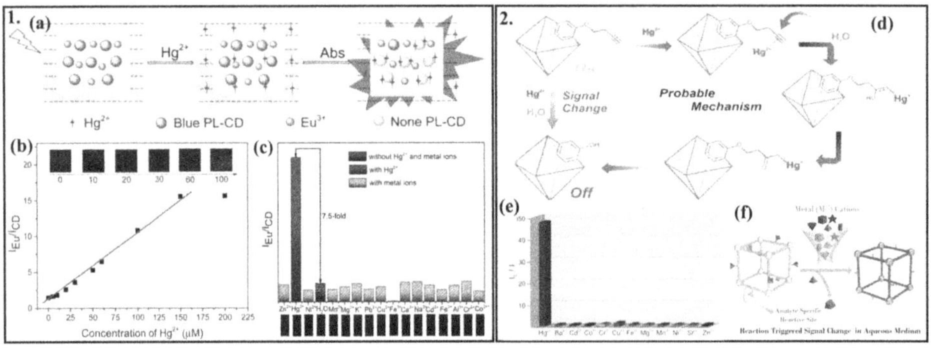

Figure 1. Box 1 shows the Eu^{3+} MOF system for Hg^{2+} detection, (a) schematic representation of the Hg^{2+} detection by Eu^{3+}/CDs@MOF-253; (b) Plot of I_{Eu}/I_{CD} vs. concentration of Hg^{2+} at λ_{ex} = 360 nm, Inset shows photographs of test plates treated by varying conc. of Hg^{2+}; (c) The bar graph depicting the comparison of the photoluminescence intensity of ratio in aqueous medium containing different metal ions having concentration 100 mM and corresponding photographs are taken by irradiation of UV light. Reproduced with the permission from reference (54) copyright 2016, Royal Society of Chemistry. Box 2, UiO-66@Butyne MOFs: (d and f) shows the mechanism of Hg^{2+} sensing by MOF; (e) The relative fluorescence response of UiO66@Butyne toward various metal ions. Reproduced with the permission from reference (55) copyright 2018, American Chemical Society.

On the basis of the reported values, it is concluded that the detection limit using luminescent MOFs for Hg^{2+} ions is in the order of 10^{-9} M with the lowest limit of the order 10^{-12} M by using Zn-based MOFs bearing N-donor sites. However, as the major challenge is the stability of luminescent

MOFs in aqueous solutions, the selection of the functional ligands plays a vital role in designing these MOFs so that the framework retains its photo-stability as well water-stability.

Al^{3+}

The high concentration of the presence of Al^{3+} ions in human body have an severe effect on health and, its detection is also much needed. The urgency for its detection leads to the development of various efforts using luminescent MOFs and various methodologies have been applied in this direction. These MOFs are usually designed on the basis of (i) the interaction between the Al^{3+} ions and the organic linkers present in the framework (*57–62*); (ii) framework collapse, known as 'Sacrificial Sensors' (*63–66*).

Construction of versatile Co- and Cd-based MOFs was reported for the Al^{3+} detection. A 1,2,4-triazole based MOF with the formula, $[Co(OBA)(DATZ)_{0.5}(H_2O)]$, where H_2OBA= 4,4′-oxybis(benzoic acid) and DATZ= 3,5-diamino-1,2,4-triazole, was designed by Mahata and co-workers with detection limit of 57.6 ppb. The Al^{3+} ions have selective interactions with nitrogen atoms of the triazole moiety as evident from 6.3 times signal enhancement along with ~24 nm blue shift in the emission spectra. The reason for the blue shift in the emission profile is mainly due to the interaction between the Al^{3+} and the triazole moiety (*57*). In another report, a Cd-based LMOF bearing pyridyl ring was developed by Lang et al. This MOF was synthesized *via* post-synthetic modification of a 2-D framework, and found to be a potential sensor having good stability, selectivity and sensitivity toward Al^{3+} ions (*61*).

The most practical and easiest method for the quantitative diagnosis of contaminant metal ion is the development of naked-eye sensor to detect and identify the pollutants. Morsali et al. presented a unique way to detect the Al^{3+} ions using a Zn-based MOF. This MOF has the formula, (TMU-60: [Zn(OBA)(L)].DMF, having 'L' as organic linker namely 5,6-dipyridin-4-yl-1,2,3,4-tetrahydropyrazine. This Zn-based MOF exhibited interaction with Al^{3+} ions having detection limit greater than 500 ppb, leading to quantitative degradation of these ions along with color change. The exact detection of the lower values (upto 100 ppb) is possible due to the fluorescence enhancement of TMU-60. This is due to the collapse framework mechanism and therefore the release of the interacting organic ligand take place (*65*). A water-stable terbium-based metal-organic framework with formula, $\{Tb(L_1)_{1.5}(H_2O)].3H_2O\}_n$(Tb-MOF), where H_2L_1= 2-(2-hydroxy-propionylamino)-terephthalic acid), with 1D rod-shaped $[Tb(CO_2)_3]_n$ secondary building units and honey-comb tubular channels has been reported to sense Al^{3+} and Fe^{3+}. · The different metal ions, Cu^{2+}, Al^{3+} and Fe^{3+} showed different extent of quenching behavior of the photoluminescence spectra of Tb^{3+}. In case of Al^{3+}, the collapse of Tb-MOF structure to Tb^{3+} and organic ligand has been reported that resulted into bright blue light. However, the emission color is not reversible, which limits the recyclability of this MOF (*63*). Sun and co-workers reported the synthesis and utilization of europium-based MOF having tubular channels due to the presence of carboxylate ligand. This Eu-MOF has the formula, $(H_2N(CH_3)_2][Eu(H_2O)_2(BTMIPA)].2H_2O$, where H_4 BTMIPA= 5,5′-methylenebis(2,4,6-trimethylisophthalic acid). This MOF works through fluorescence quenching and enhancement respectively *via* a cation exchange. The outercations in the channels are exchanged with the Al^{3+} and Fe^{3+} ions. A detailed study using acetylacetone (acac) as chelating agent indicated the different response of Al^{3+} (dynamic mechanism) and Fe^{3+} ions (static mechanism) on luminescent intensity (Figure 2) (*66*).

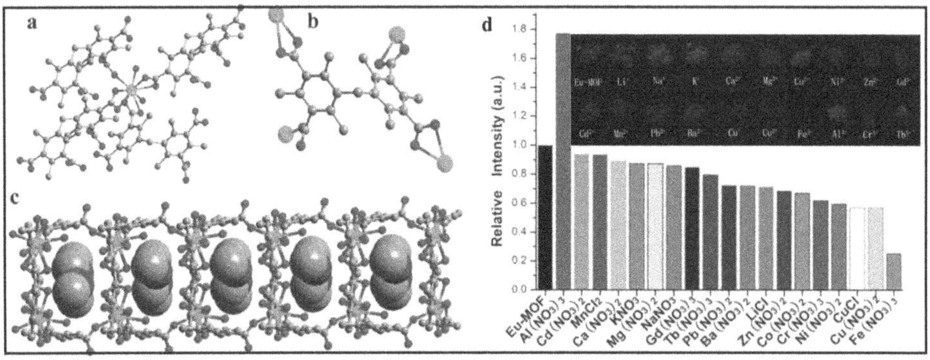

Figure 2. (a) The coordination environment of Eu^{3+} ions in Eu-MOF; (b) The BTMIPA organic linker linking mode; (c) The space filling mode of two dimensional tubular framework of Eu-MOF with the counterions of $[H_2N(CH_3)_2]^+$ (The color of atoms: Oxygen-red, carbon-gray, europium-green and nitrogen-blue; (d) Photoluminescence intensity of band at 617 nm due to $^5D_0$- $^7F_2$ transition of Eu-MOF treated in presence of different metal ions. Inset shows the color variation in the samples upon addition of metal ions under the UV light irradiation. The figure is reproduced with the permission from reference (66), copyright 2013, Royal Society of Chemistry.

On the basis of previous reported methods, it is concluded that both d-block MOFs and f-block MOFs can effectively sense the Al^{3+} ions. However, designing and utilization of (d)-LMOFs having suitable functional linkers with hard acidic character of the metal ions, seems better way toward selective and sensitive detection (*63–65*).

Cu^{2+}

Cu^{2+} ions are necessary for the human body but a high dose of these metal ions can lead to some serious health issues, like, Alzhemier's disease. These metal ions can be easily detected using MOFs, usually through metal-linker interaction mechanism. Besides this, composites of MOF-based materials can also be applied as effective sensors. Till date, the detection limit for the Cu^{2+} ions using MOFs is in the order of 10^{-3} to 10^{-8} M. Cu^{2+} ions can be easily sensed through O, N atom-Cu^{2+} interactions using charge and energy transfer via different mechanism (*67–74*). Chen and co-workers developed MOF by encapsulating the functionalized carbon dots having high fluorescence quantum yield into the zeolitic imidazolates (Figure 3b) (*75*). The as-synthesized MOF was found to be an efficient sensor for the Cu^{2+} ions detection with a range from 2-1000 nM. Harnessing the property of ligand interactions (free O, N atom) with Cu^{2+} ions, MOF with pyrene-functionalized ligand was developed as a fluorescent chemosensor (Figure 3a) (*76*). This MOF contains three important parts: Construction of MOF structure using terephthalic acid, imine moiety for Cu^{2+} binding sites and pyrene moiety acting as signaling unit. The coordination between the Cu^{2+} ions and the imine binding sites results in the conformational change in pyrene units from the excimer state to the monomer form. Similarly in another report, the pyridyl group of Tb-based MOF leads to the perturbation of Tb^{3+} emission and enhancement in the ligand based emission (*77*).

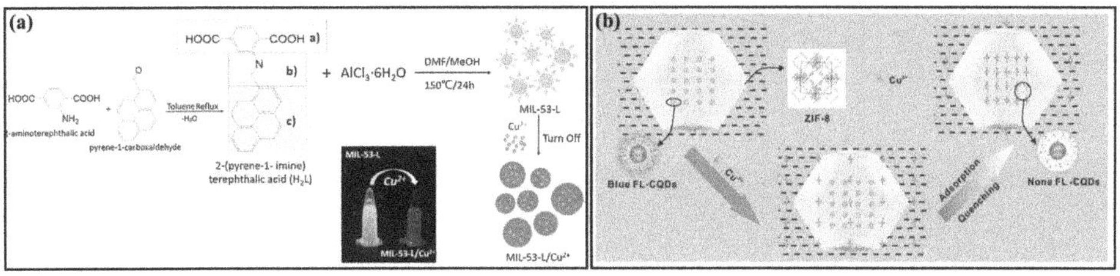

Figure 3. (a) Schematic illustration of for MIL-53-L fluorescent MOF for selective detection of Cu^{2+} showing the organic ligand, MOF MIL-53, detection unit and the fluorescence change, Reproduced with permission from reference (76), copyright 2016, Elsevier; (b) Schematic representation of the process for Sensing Cu^{2+} using quantum dots encapsulated fluorescent zeolitic imidazolates. Reproduced with permission from reference (75), copyright 2013, American Chemical Society.

Huang and co-workers developed anionic coordination polymer $[(CH_3)_2NH_2].[Cu_2(CN)_3]$ and utilized for the detection of Cu^{2+} through exchange of Cu^{2+} ions with guest cations, $[(CH_3)_2NH_2]^+$. FT-IR spectra shows an intense shift of peak for –CN group indicating strong interaction between Cu^{2+} ions and –CN group (*78*). Similar type of cation exchange mechanism between Cu^{2+} ions and $[(CH_3)_2NH_2]^+$ or $[NH_4]^+$ inside the pores of anionic MOF is being followed in other examples (*79, 80*). Certain other reports include 1D enhancement based on (i) Cu^{2+}-$[(CH_3)_2NH_2]^+$ cation exchange (*81*), (ii) Cu^{2+} and Pb^{2+} cation exchange using N atom of the porphyrin-metal ion interaction (*82*). Cu^{2+} ions detection has also been reported through interaction using other sites in MOF leading to two dimensional response (*83–85*).

Cr^{3+}

Cr^{3+} ions are important in trace amounts for human body as these ions play a vital role in the metabolism of carbohydrates, however, an excessive amount is quite toxic and harmful (*86*). The detection limit for Cr^{3+} ions is of the order of 10^{-8} M (*87, 88*), and, the outstanding value of K_{sv} is found to be about 10^4 M^{-1} (*89*). For the detection of Cr^{3+} ions, various MOFs have been designed and follows mechanisms usually the resonance energy transfer due to MOF excitation band and Cr^{3+} absorption band overlap (*90–92*), Cr^{3+}-MOF linker interactions (*93–98*).

Zhang and co-workers developed a 3D anionic MOF with $Zn_4(trz)_2$ chain for the visual detection of Cr^{3+} ions. The MOF with formula, $[Me_2NH_2]_4[Zn_6(qptc)_3(trz)_4].6H_2O$, where ($H_4qptc$ = terphenyl-2,5,2′,5′-teracarboxylic acid and trz= 1,2,4-triazole) was synthesized using triazole based organic ligand to π-conjugated aromatic carboxylate system. The MOF features $Zn_4(trz)_2$ chain-based three dimensional anionic structure with channels filled with $[Me_2NH_2]^+$ as can be noticed from Figure 4 (*87*). A possible mechanism involving the interactions between free carboxylate oxygen atoms with Cr^{3+} ions leading to transfer of charge from MOF to Cr^{3+} ions was speculated. Another, Zn-based MOFs, $[Zn_2(TPOM)(NH_2\text{-}BDC)_2].H_2O$, for the detection of Cr^{3+} ions were developed using TPOM= tetrakis(4-pyridyloxy methylene)methane and NH_2-BDC = 2-aminoterephthalic acid as organic linkers (Figure 4) (*95*). The modified MOF containing free NH_2

and carboxylic functional groups results in the formation of hexatomic ring due to the coordination between free N and O atoms of the MOF that provides more stability, and higher luminescence efficiency.

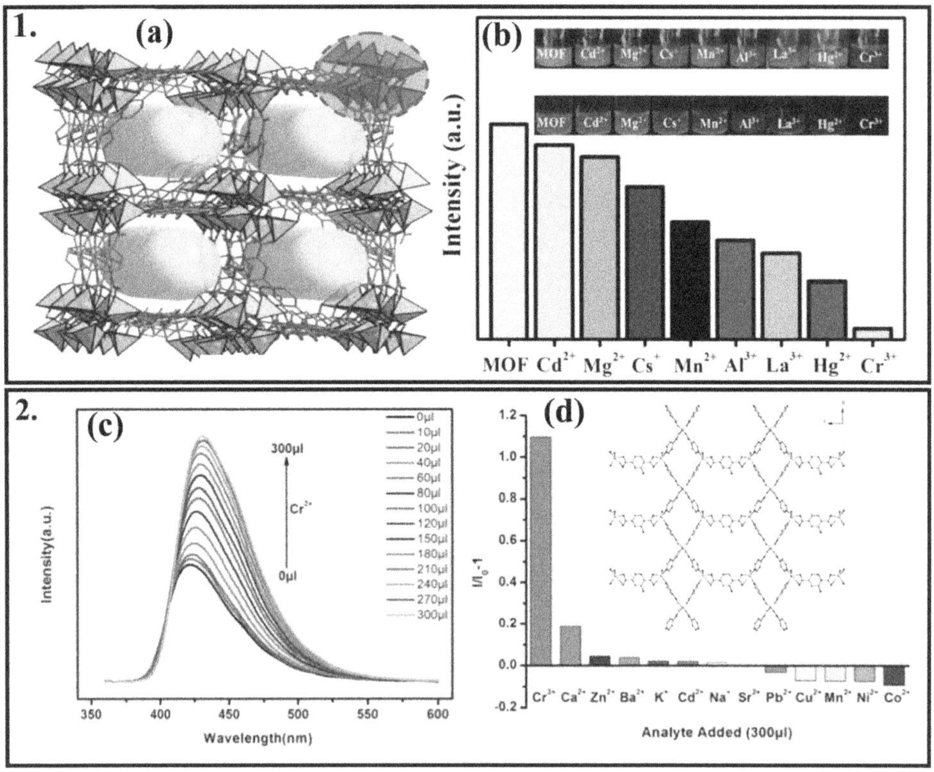

Figure 4. Box 1, three dimensional Zn-anionic MOF: (a) view of 3D anionic MOF with $Zn_4(trz)_2$ chain along c axis; (b) Bar graph depicting the Luminescent intensity change Zn-MOF upon addition of various metal ions and inset shows the photographs of the solutions in presence of metal ions under sunlight and UV light. The figures are reproduced with permission from reference (87), copyright 2017, American Chemical Society. Box 2, NH_2-Zn-MOF: (c) Titration of Zn-MOF with increasing concentration of Cr^{3+}; (d) The bar graph depicting the changes in the fluorescence intensity upon addition of different metal ions (Inset shows the 3-D structure of the NH_2-Zn-MOF. The figures are reproduced with permission from reference (95), copyright 2016, Royal Society of Chemistry.

Fe^{3+}

Fe^{3+} ions have been detected using MOFs to a large extent through different mechanisms, (i) Fe^{3+}-linker interaction (*64*, *94*, *96–98*); (ii) cation exchange mechanism (*99–101*); (iii) collapse of the framework (*102*, *103*); (iv) through multiple mechanisms (*104*).. The K_{sv} values related to Fe^{3+} ions have been reported in the range of 10^3 – 10^5 M^{-1}, while it reaches to 10^6 – 10^7 M^{-1} in case of some Ln^{3+} MOFs. Moreover, the detection limits for Fe^{3+} ions usually lie in range of 10^{-3}-10^{-7} M, however, in some cases, the detection limit is up to the order of 10^{-8} or 10^{-9} M.

Li and co-workers developed two robust lanthanide MOFs $\{[Ln(L)(H_2O)]\cdot 7H_2O\}_n$ having Eu^{3+} and Tb^{3+} as metal center and H_3L = 5-(3′,5′-dicarboxylphenyl) picolinic acid as organic ligand and, used these MOFs for the selective detection of Fe^{3+} ions (*105*). These MOFs contain a three dimensional structure and have been associated with good luminescence response, and excellent stability in water. The detection limits toward Fe^{3+} are calculated to be 0.67 μM for Eu-based MOF and 1.26 μM for Tb-based MOF, respectively. Furthermore, the fluorescence quenching obtained is due to the synergetic effect of energy transfer and competitive absorption between MOF and metal ions. Similarly, Ma and co-workers has also designed Zn-based MOFs utilizing the aromatic moieties as fluorescent tags. The two MOFs, $[Zn_2(TPC4A)\text{-}(DMF)(H_2O)_4].3H_2O$, where TPC4A =2,8,14,20-tetra-phenyl-6,12,18,24-tetramethoxy-4,10,16,22-tetra-carboxymethoxy-resorcin[4]arene and, $[(CH_3)_2NH_2]_2[Zn\ (TNC4A)].4H_2O$, where TNC4A = 2,8,14,20-tetra-1-naphthyl-6,12,18,24-tetra-methoxy-4,10,16,22-tetracarboxymethoxy-resorcin[4]arene were synthesized and effectively used for the detection of Fe^{3+} ions through fluorescent quenching (Figure 5) (*106*). In another report, development of Cd^{2+}-based MOF containing free sulfonate groups was achieved for the detection of various metal ions including Fe^{3+} ions. The Fe^{3+} ions show high electron withdrawing power in sulfonate---Fe^{3+} interactions (*104*).

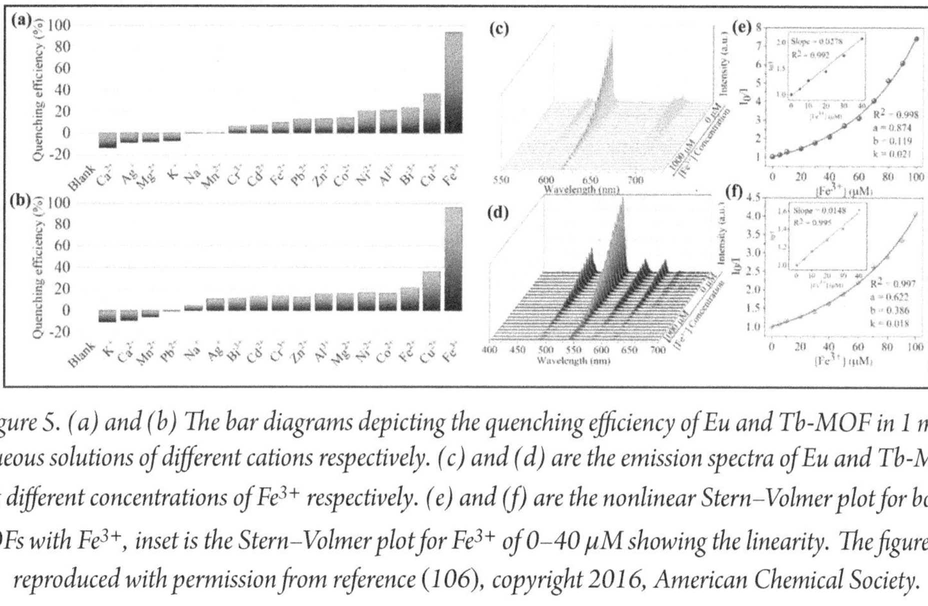

Figure 5. (a) and (b) The bar diagrams depicting the quenching efficiency of Eu and Tb-MOF in 1 mM aqueous solutions of different cations respectively. (c) and (d) are the emission spectra of Eu and Tb-MOF at different concentrations of Fe^{3+} respectively. (e) and (f) are the nonlinear Stern–Volmer plot for both MOFs with Fe^{3+}, inset is the Stern–Volmer plot for Fe^{3+} of 0–40 μM showing the linearity. The figures are reproduced with permission from reference (106), copyright 2016, American Chemical Society.

Lanthanide Metal Ions

Lanthanide metal ions have gained much attention in last few years and are of great scientific interest due to their diverse applications such as semiconducting materials, fluorescent sensors and, as excellent catalysts. The lanthanide metal ions, Eu^{3+}, Tb^{3+}, Dy^{3+} and Sm^{3+} can be detected using metal-organic frameworks usually through Ln^{3+} encapsulation into the porous MOFs, exchange of cations in the pores of anionic MOF and, coordination interaction of functional organic ligands (*107–112*).

In case of (f)-MOFs as sensors, one dimensional fluorescence quenching or enhancement is obtained in the emission profile. The 1D turn-on response is generally associated with colorimetric detection of the metal ions that can be easily visualized by naked-eye through a change in the color. For example, color of the sensor is changed from normal to red, green, blue, orange for Eu^{3+}, Tb^{3+}, Dy^{3+} and Sm^{3+} respectively.

For example, Liu and co-workers synthesized a thermostable magnesium metal-organic framework (Mg-MOF) and used it as a selective, sensitive, rapid colorimetric sensor for the detection of Eu^{3+} ions. This Mg-MOF contains holes of the nanoscale dimensions along with some non-coordinating -N- atoms inside the walls. This makes it a potential candidate for the guest metal ions (Figure 6a) (*112*). In another report, a terbium MOF, having 5-(4-carboxyphenyl)picolinic acid as organic linker, has been developed and demonstrated as a turn-on fluorescent sensor for the detection of Eu^{3+} and Dy^{3+}, with a detection limit as 5×10^{-8} and 1×10^{-4} M respectively (Figure 6b,c) (*109*). Besides, a 3D microporous Zn-based MOF with formula, $[Zn_3(Httca)_2(4,4'\text{-bpy})(H_2O)_2]_n$ (MOF–COOH), where H_4ttca = 1,1′:2′,1″-terphenyl-4,4′-4″, 5′ tetra carboxylic acid was prepared by Liu et al. This MOF contains uncoordinated carbonyl groups pointing to the pores in channels and acts as potential sites for the exchange of cations. This MOF has been shown to effectively sensitize Tb^{3+} cation (*113*).

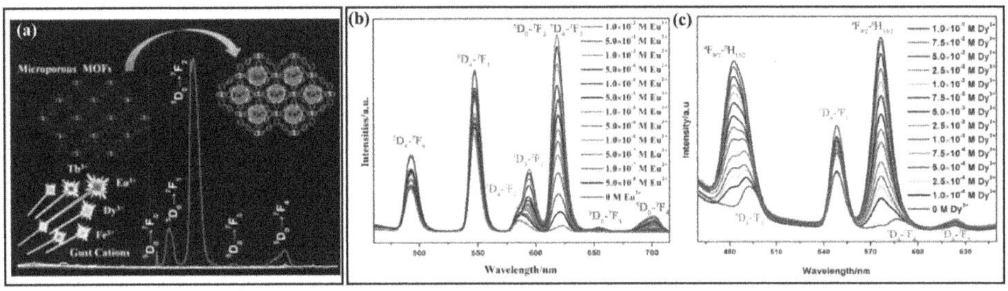

Figure 6. (a) Schematic representation of detection of Eu^{3+} by Mg-MOF system. The figure is reproduced with permission from reference (112), copyright 2015, Royal Society of Chemistry; (b) and (c) are the luminescence spectra of CTGU-1 with varying concentration of Eu^{3+} and Dy^{3+} respectively. The figures are reproduced with permission from reference (109), copyright 2017, Wiley-VCH.

Detection of several other metal ions, such as Ag^{+}, Ba^{2+}, Mn^{2+}, Pb^{2+}, Ca^{2+}, Ga^{3+}, Pd^{2+}, Ru^{3+}, Zn^{2+} etc. have been reported in the literature using MOFs albeit to a much limited extent. Majority of these metal ions have been detected *via* linker-metal ion mechanism. More investigations are much needed in this area due to the potential risks and environmental issues related to these metal ions (*107–112*).

MOFs as Anion Sensors

MOFs have been explored as anion sensors over the last two decades. The MOFs have been found to be promising candidates as efficient anion sensors due to their lager surface area, large pore volume, active sites for binding and tuneable structure. Anionic pollutants have several negative effects on the human health and environment. Therefore, the MOFs have attracted much attention for the development of accurate and rapid detection of anions such as CO_3^{2-}, PO_4^{3-}, F^-, $Cr_2O_7^{2-}$, CrO_4^-, etc. (*114–117*)

CO_3^{2-} Detection

Large amount of carbonates are very toxic as these have severe effects on the human health especially on gastrointestinal track leading to vomiting, abdominal pain etc. In literature, variety of MOFs are available for the selective sensing of carbonate ions (*118*, *119*). Yang and co-workers have reported Eu-MOF-based luminescent films to detect the CO_3^{2-} ions in the aqueous media. The luminescent films, $[\{[Eu(HBPTC)(H_2O)_2].2DMF\}_n]$, have been fabricated using electrodeposition in an anhydride system having benzophenone 3,3′,4,4′-tetracarboxylate (BPTC) as a linker. These luminescent films made up of MOF have been demonstrated as a highly selective colorimetric sensor for the CO_3^{2-} ion in biological systems. (*20*). Hu et al. has developed terbium MOF as sensitive, selective and recyclable sensor for the Al^{3+} and CO_3^{2-} ions. Using solvothermal conditions, a terbium MOF $[Tb(ppda)(ox)_{0.5}(H_2O)_2]_n$ has been developed using the two mixed organic linkers, phthalic acid derivative and oxalic acid. Four photoluminescence emission bands are obtained from terbium MOF and they are attributed to its 4f- characteristic peaks for electronic transitions of the Tb^{3+} ion at 492, 545, 587, and 623 nm respectively. The 2D layer structure of the terbium MOF has been found as excellent sensor for carbonate ion with limit of detection 3.76×10^{-7} M (Figure 7) (*121*).

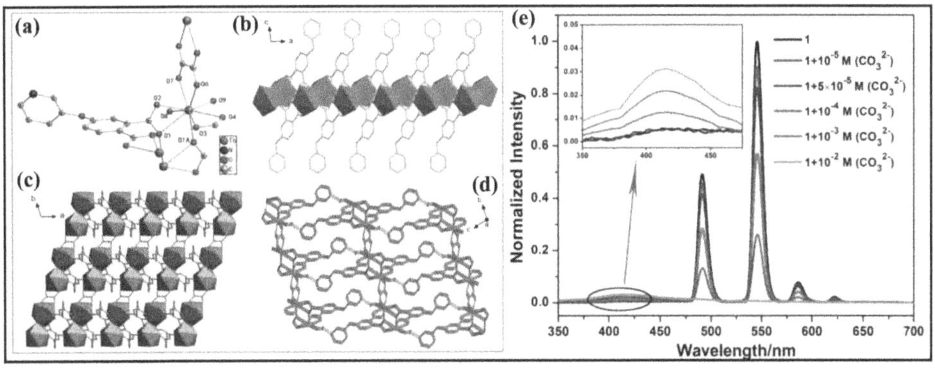

Figure 7. (a) The coordination site of the Tb^{3+} ion; (b) The one dimensional chain along the b axis; (c) Two dimensional layer of Tb-MOF along the c axis in 1; (d) H-bonding interactions in three dimensional supramolecular architecture; (e) The emission spectra of Tb-MOF with varying concentration of CO_3^{2-} ions. The figures are reproduced with permission from reference (121), copyright 2019, Royal Society of Chemistry.

PO_4^{3-} Detection

Phosphates are involved in various physiological processes such as bone mineralization, cellular signal transduction and membrane integrity maintenance. Various inorganic/lipid phosphates are found in biological system and play an important role in maintaining the balance of the phosphate in the body as well as homeostasis (*122, 123*). Phosphates have been used in large amount in the modern industry and agriculture, however it may lead to various problems such as eutrophication, excessive growth of the algae, depletion of dissolved oxygen and reduction in the quality of water (*124, 125*). Conventional methods such as colorimetry, enzymatic biosensors, electrochemical methods for the detection of the PO_4^{3-} have lot of limitations namely tedious process, time consuming, expensive instruments etc. Therefore, there is an increasing demand for the development of the facile, efficient and selective detection methods for these ions. There are some reports based on the MOF-based sensors for the phosphate detection and there is still more development going on in this field. Ding and co-workers has developed a unique fluorescent Eu-MOF for the detection of the phosphate under physiological conditions (*126*). The Eu-MOF containing terphenyl tricarboxylate ligands, as organic linkers, exhibits excellent fluorescence and can be used as self-calibrating probe for the detection of PO_4^{3-} (Figure 8). Dual emission response in two emission band was obtained upon addition of PO_4^{3-} with continuous change in the fluorescence color. Upon addition of PO_4^{3-}, a hydrogen bond formation was observed between the ligand and PO_4^{3-}, resulting in enhanced emission intensity of the ligand. On the other hand, the coordination of PO_4^{3-} and metal clusters, partly interrupt the energy transfer process from H_3L_3 to Eu_2O_2 resulted in decrease in the Eu^{3+} ions emission intensity. Tang et al. has developed a simple method for the detection of hyperthyroid liver injury by monitoring the blood phosphorous and glutathione levels by using the simple MOFs. A novel fluorescent core-shell composite of UiO-66$(OH)_2$ as core and CuMOFs as shell, has been synthesized and found to be simultaneous sensor for the glutathione and phosphate through the specific interaction between the Zr^{4+} and the phosphate. Different stages of mice models having hyperthyroidism-induced liver injury were established by administering different doses of levothyroxin sodium tablets. The MOFs based nanosensors have been used for evaluating the same by monitoring the blood phosphorous and glutathione levels in serum, liver, kidney etc. based on fluorescence detection and in situ fluorescence imaging. The changes in the phosphate and GSH were positively correlated in the liver and kidney with the degree of hyperthyroid liver injury. This nanosensor has shown good sensing properties for the early diagnosis of liver injury caused by hyperthyroidism (Figure 8) (*127*). Lv and co-workers has designed an efficient sensing platform composed of Zn-MOF combined with ZnO quantum dots for the detection of the phosphate (*128*). This sensing platform exhibited good sensitivity toward phosphate in a linear range of 0.5-12 μM and the limit of detection as 53 nM.

Chen et al. reported two isomorphous Ni^{2+}/Zn^{2+}-triazolate cluster based MOFs $\{[M_2(\mu_3\text{-}OH)(trz)(sdba)(H_2O)]\cdot 3H_2O\}_\infty$ having 1,2,4-triazole (Htrz) and 4,4′-sulfonyl- dibenzoate (H_2sdba) as organic likers by using mixed ligand strategy (*129*). Among a variety of anions studied, the Zn^{2+} based MOF resulted in the fluorescent enhancement upon addition of PO_4^{3-} ions and found to be an unusual material for the selective sensing of phosphate in aqueous solution.

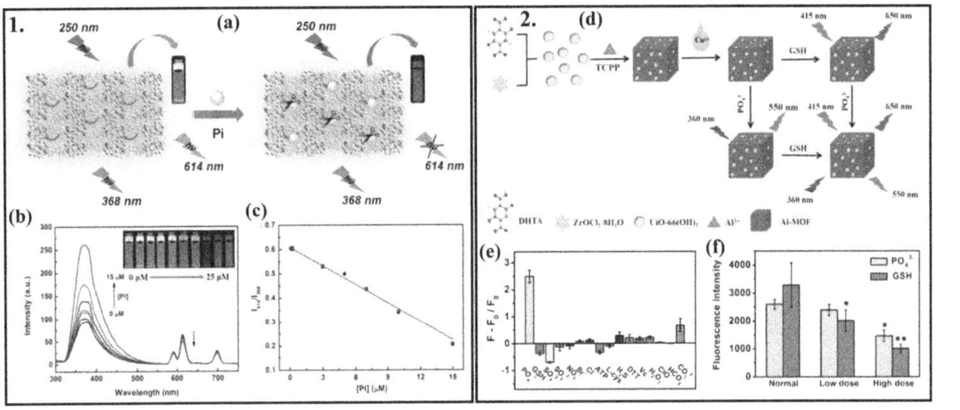

Figure 8. Box 1 shows fluorescent Eu-MOF, (a) The schematic representation of the dual-emissive Eu-MOF for visual detection of Pi; (b) The fluorescence emission spectral titration of Eu-MOF with increasing concentration of Pi. Inset shows the photographs of varying concentration of phosphate under 302 nm UV light; (c) The linear plot of fluorescence intensity ratio (I_{614}/I_{368}) vs. P_i concentration. The figures are reproduced with permission from reference (126), copyright 2018, Royal Society of Chemistry. Box 2 shows the nanoscale metal-organic framework (NMOF), (d) Schematic representation for the preparation and recognition mechanism for phosphate and glutathione by core-shell nanosensor; (e) Bar diagram depicting the relative fluorescence intensity change upon addition of different anions at λ_{max} at 550 nm; (f) The bar diagram showing the fluorescence emission in mice modal with normal, low and high dosed of phosphate and glutathione. The figures are reproduced with permission from reference (127), copyright 2020, American Chemical Society.

F^- Detection

Among the various anions, the fluoride recognition has become one of the attractive targets due to their importance in various biological and medical processes. Low concentration of the fluoride has been added to the drinking water to have a beneficial effect on the dental health and used for the treatment of osteoporosis (*130*). The exposure to the elevated levels of fluoride ions either in drinking water or excess intake from other sources can lead to various health issues and problems such as dental fluorosis, urolithiasis, acute stomach, kidney problems and DNA damage etc. (*131, 132*) The fluoride ions are easily absorbed by the body and excreted very slowly, thus the traditional methods such as ion chromatography and ion selective electrodes are effectively used for the fluoride detection. Equipment free and inexpensive methods such as luminescent chemosensors have gained much attention in the recent years and showed great significance for the environmental protection and human health. MOFs have been explored as luminescent chemosensors for various analytes including fluoride. Hou and co-workers has synthesized mixed lanthanide MOFs using Tb^{3+} and Eu^{3+} metal ions and a triazine based planar organic linker ligand, 4,4′,4″-s-triazine-2,4,6-triyltribenzoate (TATB) by fine-tuning the ratio of Tb^{3+}/Eu^{3+} during synthesis (*133*). This water-stable mixed MOF system has demonstrated as a ratiometric luminescent sensor for the F^-. With increasing concentration of fluoride ion, there is decrease in the emission intensity ratio (I_{545}/I_{617} nm) along with the color change from green to red. This sensing strategy by mixed MOF chemosensors is very simple, selective, and reliable with limit of detection for fluoride much lower than 2 ppm and has been used for the visual detection of fluoride in water with a smartphone by identifying the RGB values (Figure 9). Campbell et al. also developed a simple and robust method using redox-active naphthalene diimide MOF for reversible detection of fluoride in water. Incorporation of napthalene diimide in the mesoporous MOF leads to the development of solid state sensor devices (*134*). Tian and co-workers developed Eu-MOF using 2-aminoterephthalic acid as ligand and it emits intense blue-violet fluorescence at 425 nm. The fluorescence of Eu-MOF is quenched upon addition of Eu^{3+} with detection limit of Eu^{3+} as $(2.481 \pm 0.014) \times^{-7}$ M. The addition of fluoride to the mixture of Eu-MOF and Eu^{3+}, replaces the Eu^{3+} with detection limit of $(1.145 \pm 0.009) \times 0^{-6}$ M (Figure 9) (*135*). There is an increasing demand for the portable analytical devices, so paper based sensor has also been developed using these Eu-MOFs fabricated on polyvinyl alcohol (PVA-1788). These paper based sensors are used for the development of visual detection of fluoride in ground water.

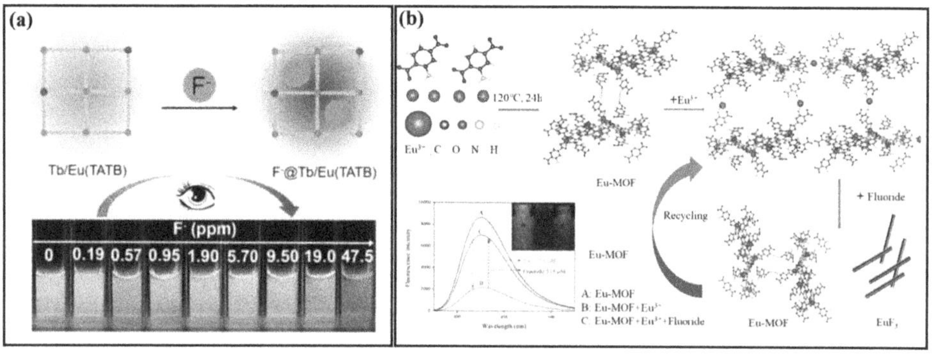

Figure 9. (a) Schematic representation of visual detection of fluoride by using Tb/Eu fluorescent MOF, the figures are reproduced with permission from reference (133), copyright 2020, American Chemical Society; (b) The strategy for the Eu-MOF synthesis, mechanism of fluoride detection and its recyclability. the figures are reproduced with permission from reference (135), copyright 2020, Elsevier.

$Cr_2O_7^{2-}/CrO_4^{2-}$ Detection

In the aqueous solution, hexavalent chromium typically exists in oxoanionic species such as $Cr_2O_7^{2-}$, CrO_4^{2-}, $HCrO_4^-$ depending on the pH and the environmental conditions. $Cr_2O_7^{2-}/CrO_4^{2-}$ are among the major industrial pollutants generated from various processes such as metal plating, leather tanneries, wood preservation and dye/paper industries (*136–138*). Excess exposure or intake of $Cr_2O_7^{2-}/CrO_4^{2-}$ ions can lead to noticeable carcinogenic and mutagenic changes, DNA damage, nasal epithelia, disruption of protein in the humans (*139–142*). Non biodegradability and bioaccumulation of the $Cr_2O_7^{2-}$ and CrO_4^{2-} anions leads to serious toxicity to organisms and humans. Thus, there is a dire need of the rapid, sensitive and accurate chemosensors for the detection of $Cr_2O_7^{2-}/CrO_4^{2-}$ ions.

Wang et al. developed highly stable luminescent MOF having formula $Eu_7(mtb)_5(H_2O)_{16}]\cdot NO_3\cdot 8DMA\cdot 18H_2O$ using 4-[tris(4-carboxyphenyl) methyl] benzoic acid as organic linker (*143*). The Eu-MOF shows the high selectivity, sensitivity and instant luminescence quenching toward the chromate ions in aqueous solutions and even in the high ionic strength solution such as sea water (Figure 10a). The limit of detection of the previously mentioned MOFs toward the chromate ions in deionized water, lake water and sea water are 0.56, 2.88, and 1.75 ppb respectively. Cao and co-workers developed a water-stable MOF having large surface area to remove the $Cr_2O_7^{2-}$ ions from aqueous solutions. Among a variety of MOFs, NU-1000 (Zr-MOF) shows the highest absorption capacity toward chromate ions having sorption capacity 76.8 mg/g. Due to the presence of several active sites, large pore size and high surface area, NU-1000 can selectively and efficiently detect the $Cr_2O_7^{2-}$ with a limit of detection 1.8 μM (Figure 10b, c) (*144*).

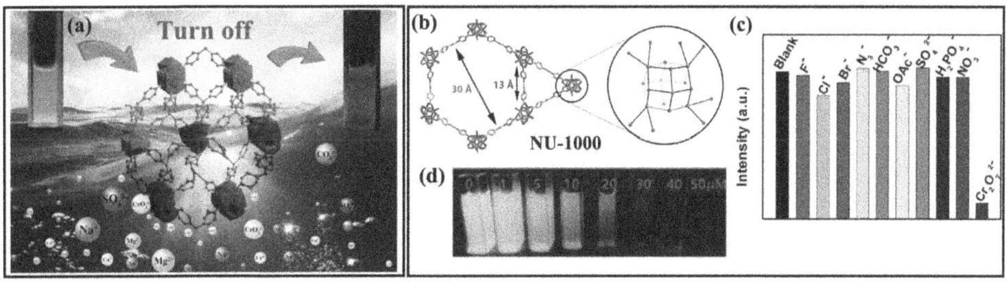

Figure 10. (a) Schematic representation showing the detection of $Cr_2O_7^{2-}$ using Eu-MOF. The figure is reproduced with permission from reference (143), copyright 2017, American Chemical Society. (b) The magnified view of 8-connected $Zr_6(\mu_3\text{-}OH)_8(OH)_8$ cluster of NU-1000; (c) The bar diagram depicting the fluorescent intensities of NU-1000 at 518 nm in 0.5 mM solution of different anions; (d) The photographs showing the fluorescence quenching obtained with different concentration of $Cr_2O_7^{2-}$ taken under UV light (365 nm) using NU-1000 MOF. The figures are reproduced with permission from reference (144), copyright 2017, American Chemical Society.

MOFs as Efficient pH Sensors

In many fields such as biological diagnostics, bioprocessing and environmental monitoring, pH is very critical phenomenon. The inappropriate pH leads to various diseases in the biological systems (*145*). Treatments to various diseases can be achieved by monitoring the small changes in the tissue pH. Traditional pH sensors have various limitations, such as, long time response, low accuracy of measurement, high impedance etc. MOFs have shown remarkable responses toward different pH values (*146*, *147*). In general, MOFs suffers from certain limitation such as stability in aqueous media, non resistance to harsh acid and alkaline solutions. In a recent report by Pan and co-workers, two non-luminescent MOFs based on the heterometallic clusters (tetranuclear and hexanuclear) have been demonstrated as efficient pH sensors. In this study, Li^+, Co^{2+} and semirigid 5-[bis(4-carboxybenzyl) amino] isophthalic acid (H_4L) are assembled together using solvent thermal reaction that results in the formation of two MOFs, CoLi(HL)·4DMA (**HNU-23**) and $CoLi_2(L)(DMF)_2\cdot DMF\cdot H_2O$ (**HNU-24**). The XRD data suggests that the HNU-23 porous framework consists of Co_2Li_2 clusters and HL_3^- ligands. Both the HNU-23 and HNU-24 have been demonstrated as H^+ sensors and, these MOFs showed fluorescence enhancement and linear H^+ response with a wide range of 0.02 to 0.2 M for HNU-23 and 0.1 to 0.2 M for HNU-24. (Figure 11) (*148*). The fluorescence enhancement and the mechanism of sensing is explained by the disruption of the assembly of both MOFs under strongly acidic conditions by gradual release of the fluorophore ligand (H_4L). Li et al. also synthesized two isoreticular, highly porous, chemical stable, flexible Zr-MOFs for the wavelength based luminescence sensing of aqueous pH. The two

Zirconium-MOFs, namely, BUT-62 and BUT-63 were synthesized from two trapezoids ligands having strong fluorescence *viz*., carbazole tertayl tetrabenzoic acid derivative (H_4CTTB) and carbazole tertayl naphthoic acid derivative (H_4CTNA) as shown in Figure 11. The fluorescence properties of BUT-62 and BUT-63 have been explored for their potential applications in sensing the pH ranging from 0-9 of aqueous systems. Upon gradual increase in the acidity of the solution from pH 9 to 0, the fluorescence spectra gradually blue shifted with both the Zr-MOFs. Due to structure dependency, the fluorescence spectral changes are more sensitive in case of BUT-63 as compared to BUT-62. Overall, a linear relationship was obtained in the fluorescence λ_{max} of both MOFs with change in the pH of the aqueous systems and hence found potential pH sensor for the environmental and biological systems (*149*).

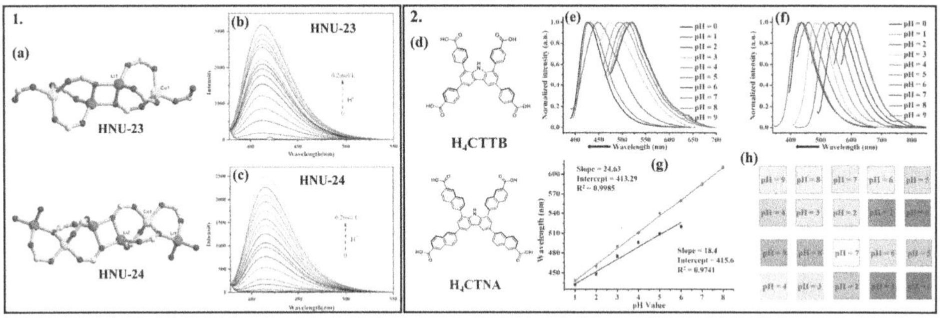

Figure 11. Box 1. shows the luminescent MOFs based on heterometallic clusters, (a) coordination environment of Co^{2+} and Li^+ in Co_2Li_2 clusters (HNU-23) and Co_2Li_4 clusters (HNU-24); (b) and (c) are the fluorescence emission spectra with different concentration of H^+ concentration for HNU-23 and HNU-24 respectively. The figures are reproduced with permission from reference (148), copyright 2019, Elsevier. Box 2. Shows the isoreticular Zr-MOF, (d) Structure of the organic linker of both MOFs BUT-62 (H_4CTTB) and BUT-63 (H_4CTNA); (e) and (f) are the normalized emission spectra of BUT-62 and BUT-63 MOFs at varying pH from 1-9 at room temperature; (g) plot of λ_{max} vs. pH for BUT-62 (Black) and BUT-63 (red); (h) photographs of the carboxymethyl plate coated with BUT-62 (upper row) and BUT-63 (lower row) at different pH values. The figures are reproduced with permission from reference (149), copyright 2018, Royal Society of Chemistry.

Bradshaw and co-workers reported pH dependent fluorescence response of colloidal dispersion of UiO-66-NH_2 having Zr(6) cluster and 2-aminoterephtalic acid as linker across a wide range of pH from 1 to 9. The increased stability and applicability of these MOFs increases up to pH 12 upon post modification of the pendant amine groups with the indole moieties (*150*). Zhou et al. has also synthesized MOF with (4, 8) connected **sqc** net by the reaction of Zr^{4+} salt with porphyrinic teracarboxylic acid. The zirconium porphyrin frameworks, PCN-225 shows exceptional stability in the aqueous systems, and found to be a promising sensors with a wide range of pH from 1 to 11 (*151*). Lin and co-workers has synthesized and developed a NMOF for real time application of the intracellular pH measurement in live cells. The fluorescein isothiocyanate (FITC) has been covalently attached to the UiO-NMOF resulted in the formation of F-UiO-NMOF carrying the efficient fluorescence, excellent sensing and high FITC loading. The UiO-NMOF consists of zirconium metal center and amino-triphenyldicarboxylic acid as a bridging ligand. The fluorescent F-UiO-NMOF has been explored to reveal the endosome acidification process (endocytosis) and intracellular trafficking process of NMOF (*152*). In a recent report by Zang and co-workers, Eu-

MOF has been prepared by using tetrazole modified carboxylic linkers (as proton hopping sites) with not only hydrothermal stability but also acid and alkali resistance. Proton conducting membranes were prepared by using the Eu-MOFs in combination with the Nafion. The emission of the hybrid membranes exhibits the luminescence pH sensing due to the tetrazolium unit present in the backbone of the Eu-MOF. Due to the protonation and deprotonation of the tetrazoilum units at low and high pH, change occurs in emission intensity of the Eu-MOF (*153*). The Eu-MOF hybrid membranes have been demonstrated as efficient pH sensors and showed a linear response over a wide range of pH as shown in Figure 12.

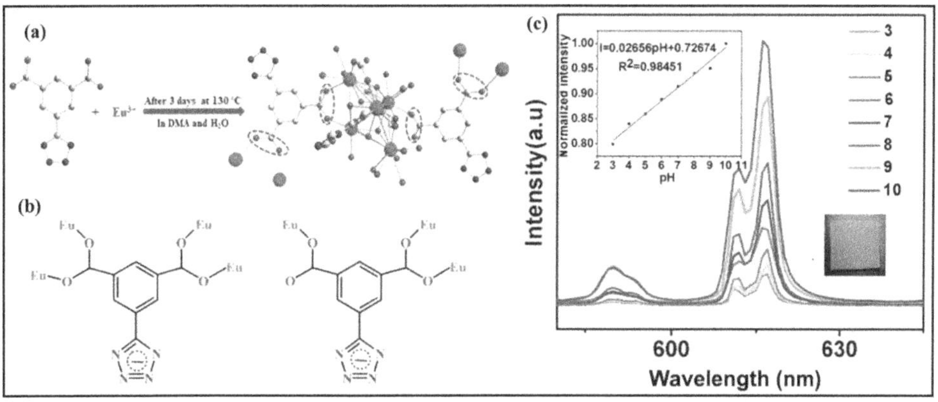

Figure 12. The synthesis scheme of the Eu-MOF; (b) different coordination modes of the tetrazoium based ligand with Eu ions; (c) The emission spectra of the Eu-MOF/Nafion hybrid membrane at varying pH from 3 to 10, the inset shows the linear graph of normalized intensity vs. pH at λ_{max} 616 nm. The figures are reproduced with permission from reference (153), copyright 2020, American Chemical Society.

MOFs as Temperature Sensors

In the past few years, the accurate measurement of temperature in the area of microelectronics, micro-optics, photonics and nanomedicine is found to be an emerging field. Moreover, in the field of pathology and physiology research, accurate temperature measurements play a crucial role. Many biological processes such as protein synthesis, cellular respiration, gene expression, ion transportation, cell division, determination of growth of the tissue are affected by the variation in temperature within the cell. Due to non applicability at molecular scale, the traditional temperature measurement probes such as thermistors and thermocouples suffer from various limitations, whereas the luminescence based temperature probes have higher accuracy and wide applicability even at micro and nanoscales. In this regard, the mixed lanthanide metal-organic frameworks, having instantaneous response and good sensitivity/selectivity even at nanoscales, have been explored and attracted much attention (*154–156*). In the past few years, a lot of such lanthanides MOFs based thermometers have been explored for temperature sensing in many temperature regions such as physiological (298-398 K), medium (100-300 K) and ascryogenic (< 100 K) with very high performances. However, the high temperature sensing is rare and still a challenging emerging field.

Chen et al. reported $Eu_{0.0069}Tb_{0.9931}$-DMBDC luminescent MOF using 2,5-dimethoxy-1,4-benzenedicarboxylate (DMBDC) as an organic ligand and found its potential application for the temperature sensing based on the luminescent intensity of the Tb^{3+} and Eu^{3+} metal center (*156*). Recently, Qian and co-workers developed the mixed lanthanide MOFs to serve as thermometers to operate in high temperature range using Tb^{3+} and Eu^{3+} ions. Eu/Tb mixed MOF labeled as

$\mathbf{Eu_{0.37}Tb_{0.63}}$**-BTCa** has been synthesized *via* solvothermal reaction using 1,3,5-benzenetricarboxylate (H_3BTC) as a linker in DMF-water system. These mixed MOFs are thermostable and thermosensitive, as their high triplet excited states energy can efficiently sensitize the Tb^{3+} and Eu^{3+} ions at elevated temperatures. $\mathbf{Eu_{0.37}Tb_{0.63}}$**-BTCa** has been explored for its application as a luminescent thermometer for the sensing of temperature in the range of 313-473 K with high sensitivity, excellent repeatability and accurate temperature measurement as shown in Figure 13 (*157*). This group also reported mixed lanthanide MOF, $Tb_{0.8}Eu_{0.2}$BPDA using biphenyl-3,5-dicarboxylic acid, with higher thermostability and rigidity. $Tb_{0.8}Eu_{0.2}$BPDA works as a ratiometric luminescent thermometer operating in the physiological temperature range from 303 to 328 K. The intensity ratio of two emissions (I_{Tb}/I_{Eu}) as a function of temperature can be used as the ratiometric thermometric parameter and possess higher sensitivity than previously reported M'LnMOFs (*158*). In another recent report, Liu et al. has constructed a lanthanide MOF with dual-emissive properties. In this work, 4-aminonapthalimide has been introduced in the pores of two lanthanide MOFs having Eu and Gd metal center considering the size effects and explored its temperature sensing properties. Temperature sensing of dye encapsulated MOFs has been performed in the range of 35-85°C and resulted in the development of reversible ratiometric luminescent temperature sensor (Figure 13) (*159*). This co-doped approach of the lanthanide MOFs has been found to give excellent results.

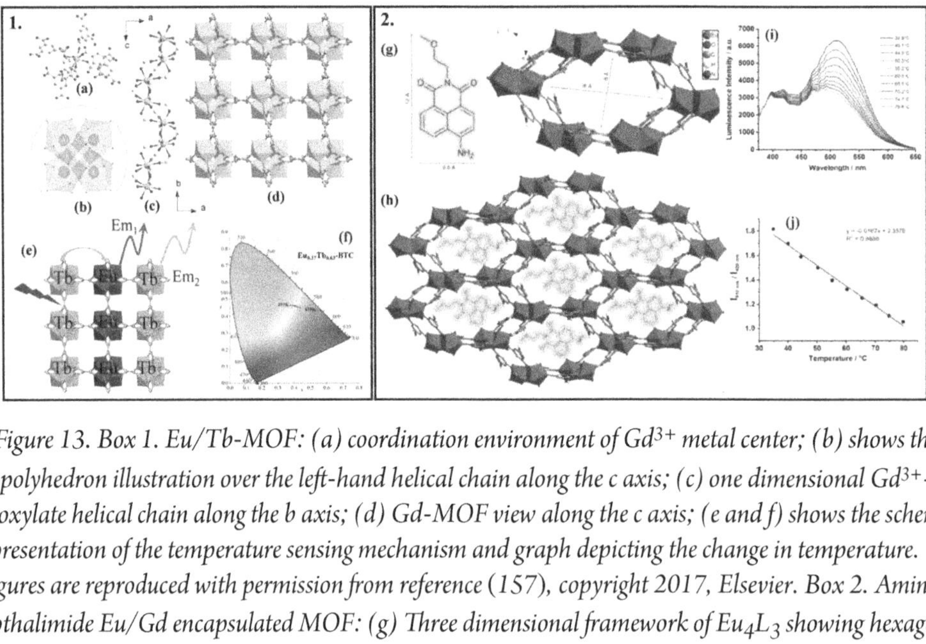

Figure 13. Box 1. Eu/Tb-MOF: (a) coordination environment of Gd^{3+} metal center; (b) shows the polyhedron illustration over the left-hand helical chain along the c axis; (c) one dimensional Gd^{3+}-carboxylate helical chain along the b axis; (d) Gd-MOF view along the c axis; (e and f) shows the schematic representation of the temperature sensing mechanism and graph depicting the change in temperature. The figures are reproduced with permission from reference (157), copyright 2017, Elsevier. Box 2. Amino-napthalimide Eu/Gd encapsulated MOF: (g) Three dimensional framework of Eu_4L_3 showing hexagonal channels and molecular structure and size of Eu/Gd MOF; (h) Schematic representation of napthalimide encapsulated MOF; (i) The luminescence spectra of ANA-Gd MOF at λ_{ex} =364 nm at different temperatures; (j) Plot of I_{510}/I_{420} at different temperature. The figures are reproduced with permission from reference (159), copyright 2019, Wiley-VCH.

Liu and co-workers synthesized three isostructural mixed lanthanide metal-organic frameworks $\mathbf{Tb_{1-x}Eu_xL}$ (x=0.05, 0.1 and 0.2) by using 3-bis (3 carboxyphenyl) imidazolium (H_2L) as organic ligand using the hydrothermal method. These co-doped MOFs have been reported to show more sensitivity and accuracy due to the energy transfer from ($Tb^{3+} \rightarrow Eu^{3+}$) with increase in temperature. The characteristics peaks of Eu^{3+} and Tb^{3+} at 617 and 547 nm respectively in the luminescent spectra belongs to $^5D_0 \rightarrow {}^7F_2$ and $^5D_4 \rightarrow {}^7F_5$ transitions of $^5D_0 \rightarrow {}^7F_2$, revealing the energy transfer process of Tb^{3+} to Eu^{3+} (*160*). The excellent stability of $\mathbf{Tb_{0.9}Eu_{0.1}L}$ under physiological conditions and sensing over a wide temperature range of 303 to 423 K suggests the applicability of the co-doped MOFs for detection of temperature in biological systems. Shi and co-workers has reported Zn-based MOFs Using solvothermal reaction of $Zn(NO_3)_2(H_2O)_6$ and H_6TDPAT (2,4,6-tris(3,5-dicarboxyl phenylamino)-1,3,5-triazine) in DMF. Luminescent Zn-TDPAT MOFs have been found to be effective for the quantitative detection of temperature from 164 to 276 K (*161*).

Overall, the luminescent MOFs have been demonstrated as excellent sensors for the metal ions, anions, pH and temperature. Other than these applications, MOFs have also been explored as potential sensors for bio-molecules, and other small organic molecules like nitro aromatics (picric acid and 2,4,6-trinitrotoluene, etc).

Conclusion

There is a rapid growth in the research area pertaining to the development of luminescent metal-organic frameworks (MOFs). These MOFs have been widely explored as excellent sensors and, their sensing applications are attributed to their structural diversity, flexibility, high pore size, tuneable luminescent properties, high adsorption capacity, easy identification of the analyte, naked-eye detection, etc. The sensing applications of these MOFs has been explained via different mechanisms, such as cation exchange, structural collapse, ligand-metal interaction and overlap mechanism. In this chapter, applications of MOFs as cation, anion, pH and temperature sensors have been discussed with some specific comments related to selectivity, sensitivity, stability, response time and reusability of these MOFs toward various analytes.

Acknowledgments

Bhawna Uttam and Ravi Kumar acknowledges the Department of Chemistry, J.C. Bose University of Science and Technology, YMCA, Faridabad, India. BU is also thankful to CSIR, Delhi for the financial support.

References

1. Tranchemontagne, D. J.; Mendoza-Cortes, J. L.; O'Keeffe, M.; Yaghi, O. M. Secondary building units, nets and bonding in the chemistry of metal-organic frameworks. *Chem. Soc. Rev.* **2009**, *38*, 1257.
2. Batten, S. R.; Neville, S. M.; Turner, D. R. Porous coordination polymers. In *Coordination Polymers: Design*; The Royal Society of Chemistry: Cambridge, UK, 2009; Chapter 10, pp 313–344.
3. Uemura, K.; Matsuda, R.; Kitagawa, S. Flexible microporous coordination polymers. *J. Solid State Chem.* **2005**, *178*, 2420–2429.
4. Farha, O. K.; Yazaydin, A. O.; Eryazici, I.; Malliakas, C. D.; Hauser, B. G.; Kanatzidis, M. G.; Nguyen, S. T.; Snurr, R. Q.; Hupp, J. T. De novo synthesis of a metal-organic framework material featuring ultrahigh surface area and gas storage capacities. *Nat. Chem.* **2010**, *2*, 944–948.

5. Li, H.; Eddaoudi, M.; O'Keeffe, M.; Yaghi, O. M. Design and synthesis of an exceptionally stable and highly porous metal-organic framework. *Nature* **1999**, *402*, 276–279.
6. Burnett, B. J.; Barron, P. M.; Choe, W. Recent advances in porphyrinic metal-organic frameworks: materials design, synthetic strategies, and emerging applications. *CrystEngComm* **2012**, *14*, 3839–3846.
7. Diamantis, S. A.; Margariti, A.; Pournara, A. D.; Papaefstathiou, G. S.; Manos, M. J.; Lazarides, T. Luminescent metal-organic frameworks as chemical sensors: common pitfalls and proposed best practices. *Inorg. Chem. Front.* **2018**, *5*, 1493–1511.
8. Furukawa, H.; Ko, N.; Go, Y. B.; Aratani, N.; Choi, S. B.; Choi, E. Ultrahigh porosity in metal-organic frameworks. *Science* **2010**, *329*, 424–428.
9. Gutschke, S. O. H.; Price, D. J.; Powell, A. K.; Wood, P. T. Hydrothermal synthesis, structure, and magnetism of $[Co_2(OH)\{1,2,3\text{-}(O_2C)_3C6H3\}(H_2O)] \cdot H_2O$ and $[Co_2(OH)\{1,2,3\text{-}(O_2C)_3C_6H_3\}]$: magnetic Δ-chains with mixed cobalt geometries. *Angew. Chem. Int. Ed.* **2001**, *40*, 1920–1923.
10. Eddaoudi, M.; Moler, D. B.; Li, H.; Chen, B.; Reineke, T. M.; Keeffe, M. O. Modular chemistry: secondary building units as a basis for the design of highly porous and robust metal-organic carboxylate frameworks. *Acc. Chem. Res.* **2001**, *34*, 319–330.
11. Kumagai, H.; Kepert, C. J.; Kurmoo, M. Construction of hydrogen-bonded and coordination-bonded networks of cobalt (II) with pyromellitate: synthesis, structures, and magnetic properties. *Inorg. Chem.* **2002**, *41*, 3410–3422.
12. Prior, T. J.; Rosseinsky, M. J. Crystal engineering of a 3-D coordination polymer from 2-D building blocks. *Chem. Commun.* **2001**, *5*, 495–496.
13. Yaghi, O. M.; Li, H.; Groy, T. L. Construction of porous solids from hydrogen bonded metal complexes of 1,3,5-Benzenetricarboxylic acid. *J. Am. Chem. Soc.* **1996**, *118*, 9096–9101.
14. Keeffe, M. O.; Yaghi, O. M. Deconstructing the crystal structures of metal-organic frameworks and related materials into their underlying nets. *Chem. Rev.* **2012**, *112*, 675–702.
15. Robin, A. Y.; Fromm, K. M. Coordination polymer networks with O- and N-donor what they are, why and how they are made. *Coord. Chem. Rev.* **2006**, *250* (15–16), 2127–2215.
16. Cook, T. R.; Zheng, Y.-R.; Stang, P. J. Metal-organic frameworks and self-assembled supramolecular coordination complexes: comparing and contrasting the design, synthesis, and functionality of metal-organic materials. *Chem. Rev.* **2012**, *113*, 734–777.
17. Long, J. R.; Yaghi, O. M. The pervasive chemistry of metal-organic frameworks. *Chem. Soc. Rev.* **2009**, *38* (5), 1213–1214.
18. Stock, N.; Biswas, S. Synthesis of metal-organic frameworks (MOFs): routes to various MOF topologies, morphologies, and composites. *Chem. Rev.* **2012**, *112*, 933–969.
19. Kumar, P.; Deep, A.; Kim, K.-H. Metal-organic frameworks for sensing applications. *Trends in Analytical Chemistry* **2015**, *73*, 39–53.
20. Collins, D. J.; Zhou, H.-C. Hydrogen storage in metal-organic frameworks. *J. Mater. Chem.* **2007**, *17*, 3154.
21. Li, J.-R.; Kuppler, R. J.; Zhou, H.-C. Selective gas adsorption and separation in metal-organic frameworks. *Chem. Soc. Rev.* **2009**, *38*, 1477.

22. Li, J.-R.; Ma, Y.; McCarthy, M. C.; Sculley, J.; Yu, J.; Jeong, H.-K.; Balbuena, P. B.; Zhou, H.-C. Carbon dioxide capture-related gas adsorption and separation in metal-organic frameworks. *Coord. Chem. Rev.* **2011**, 255, 1791.
23. Burtch, N. C.; Jasuja, H.; Walton, K. S. Water Stability and Adsorption in Metal-Organic Frameworks. *Chem. Rev.* **2014**, *114*, 10575–10612.
24. Cui, Y.; Yue, Y.; Qian, G.; Chen, B. Luminescent Functional Metal-Organic Frameworks. *Chem. Rev.* **2012**, *112*, 1126–1162.
25. Horcajada, P.; Gref, R.; Baati, T.; Allan, P. K.; Maurin, G.; Couvreur, P.; Flrey, G.; Morris, R. E.; Serre, C. Metal-Organic Frameworks in Biomedicine. *Chem. Rev.* **2012**, *112*, 1232–1268.
26. McKinlay, A. C.; Horcajada, R. E.; Ferey, P.; Gref, G.; Couvreur, R. P.; Serre, C. BioMOFs: metal-organic frameworks for biological and medical applications. *Angew. Chem. Int. Ed.* **2010**, *49*, 6260–6266.
27. Wriedt, M.; Zhou, H. Systematic investigations on magneto-structuralcorrelations of copper (II) coordination polymers based on organic ligands with mixed carboxylic and nitrogen-based moieties. *Dalton Trans.* **2012**, *41*, 4207–4216.
28. Stavila, V.; Talin, A. A.; Allendorf, M. D. MOF-based electronic and opto-electronic devices. *Chem. Soc. Rev.* **2014**, *43*, 5994–6010.
29. Zhang, X.; Wang, W.; Hu, Z.; Wang, G.; Uvdal, K. Coordination polymers for energy transfer: Preparations, properties, sensing applications, and perspectives. *Coord. Chem. Rev.* **2015**, *284*, 206–235.
30. Heinke, L.; Tu, M.; Wannapaiboon, S.; Fischer, R. A.; Wcll, C. Surface-mounted metal-organic frameworks for applications in sensing and separation. *Microporous Mesoporous Mater.* **2015**, *216*, 200–215.
31. Masoomi, M. Y.; Morsali, A. Applications of metal–organic coordination polymers as precursors for preparation of nano-materials. *Coord. Chem. Rev.* **2012**, *256*, 2921–2943.
32. Shekhah, O.; Liu, J.; Fischer, R. A.; Wcll, C. MOF thin films: existing and future applications. *Chem. Soc. Rev.* **2011**, *40*, 1081–1106.
33. Li, W.; Zhang, Y.; Li, Q.; Zhang, G. Metal—organic framework composite membranes: Synthesis and separation applications. *Chem. Eng. Sci.* **2015**, *135*, 232–257.
34. Hu, Z.; Deibert, B. J.; Li, J. Luminescent metal-organic frameworks for chemical sensing and explosive detection. *Chem. Soc. Rev.* **2014**, *43*, 5815–5840.
35. Meyer, L. V.; Schçnfeld, F.; Mìller-Buschbaum, K. Lanthanide based tuning of luminescence in MOFs and dense frameworks – from mono- and multimetal systems to sensors and films. *Chem. Commun.* **2014**, *50*, 8093–8108.
36. Sharma, S.; Desai, A. V.; Joarder, B.; Ghosh, S. K. A Water-Stable Ionic MOF for the Selective Capture of Toxic Oxoanions of Se^{VI} and As^{V} and Crystallographic Insight into the Ion-Exchange Mechanism. *Angew. Chem., Int. Ed.* **2020**, 59, 7788–7792.
37. Zhang, X. J.; Wang, W. J.; Hu, Z. J.; Wang, G. N.; Uvdal, K. Coordination polymers for energy transfer: Preparations, properties, sensing applications, and perspectives. *Coord. Chem. Rev.* **2015**, *284*, 206–235.
38. Wang, M.; Guo, L.; Cao, D. Metal-organic framework as luminescence turn-on sensor for selective detection of metal ions: Absorbance caused enhancement mechanism. *Sens. Actuators, B.* **2018**, *256*, 839–845.

39. Han, L.-J.; Yan, W.; Chen, S.-G.; Shi, Z.-Z.; Zheng, H.-G. Exploring the Detection of Metal Ions by Tailoring the Coordination Mode of V-Shaped Thienylpyridyl Ligand in Three MOFs. *Inorg. Chem.* **2017**, *56*, 2936–2940.
40. Jin, H.; Xu, J.; Zhang, L.; Ma, B.; Shi, X.; Fan, Y.; Wang, L. Multi-responsive luminescent sensor based on Zn (II) metal-organic framework for selective sensing of Cr(III), Cr(VI) ions and *p*-nitrotolune. *J. Solid State Chem.* **2018**, *268*, 168–174.
41. Wen, G.-X.; Wu, Y.-P.; Dong, W.-W.; Zhao, J.; Li, D.-S.; Zhang, J. An Ultrastable Europium(III)–Organic Framework with the Capacity of Discriminating Fe^{2+}/Fe^{3+} Ions in Various Solutions. *Inorg. Chem.* **2016**, *55*, 10114–10117.
42. Tuergong, M.; Nizamidin, P.; Yimit, A.; Simayi, R. In situ growth and optical gas adsorption performance of Zn(ii) metal-organic framework membranes at room temperature. *Analyst* **2019**, *144*, 4887–4896.
43. Liu, W.; Huang, X.; Xu, C.; Chen, C.; Yang, L.; Dou, W.; Chen, W.; Yang, H.; Liu, W. A Multi-responsive Regenerable Europium–Organic Framework Luminescent Sensor for Fe^{3+}, Cr^{VI} Anions, and Picric Acid. *Chem. – Eur. J.* **2016**, *22*, 18769–18776.
44. Li, Q.; Wu, X.; Huang, X.; Deng, Y.; Chen, N.; Jiang, D.; Zhao, L.; Lin, Z.; Zhao, Y. Tailoring the Fluorescence of AIE-Active Metal-Organic Frameworks for Aqueous Sensing of Metal Ions. *ACS Appl, Mater. Interfaces* **2018**, *10*, 3801–3809.
45. Xu, X.-Y.; Yan, B. Eu(III) functionalized Zr-based metal-organic framework as excellent fluorescent probe for Cd^{2+} detection in aqueous environment. *Sens. Actuators, B* **2016**, *222*, 347–353.
46. Xu, X.-Y.; Yan, B. Eu(III)-Functionalized MIL-124 as Fluorescent Probe for Highly Selectively Sensing Ions and Organic Small Molecules Especially for Fe(III) and Fe(II). *ACS Appl. Mater. Interfaces* **2015**, *7*, 721–729.
47. Wei, N.; Zhang, Y.-R.; Han, Z.-B. Thiadiazole-functional porous metal-organic framework as luminescent probe for Cd^{2+}. *CrystEngComm* **2013**, *15*, 8883–8886.
48. Lu, Y.; Yan, B.; Liu, J.-L. Nanoscale metal-organic frameworks as highly sensitive luminescent sensors for Fe^{2+} in aqueous solution and living cells. *Chem. Commun.* **2014**, *50*, 9969–9972.
49. Hao, J.-N.; Yan, B. Highly sensitive and selective fluorescent probe for Ag^{+} based on a Eu^{3+} post-functionalized metal-organic framework in aqueous media. *J. Mater. Chem. A* **2014**, *2*, 18018–18025.
50. Pamei, M.; Puzari, A. Luminescent transition metal-organic frameworks: An emerging sensor for detecting biologically essential metal ions. *Nano-Structures & Nano-Objects* **2019**, *19*, 100364.
51. Razavi, S. A. A.; Morsali, A. Metal ion detection using luminescent-MOFs: Principles, strategies and roadmap. *Coordination Chem Rev.* **2020**, *415*, 213299.
52. Kanan, S. M.; Malkawi, A. Recent Advances in Nanocomposite Luminescent Metal-Organic Framework Sensors for Detecting Metal Ions. *Comments on Inorganic Chemistry* **2021**, *41* (1), 1–66.
53. Fu, H.-R.; Zhao, Y.; Xie, T.; Han, M.-L.; Ma, L.-F.; Zang, S. Q. Stable dye-encapsulated indium–organic framework as dual-emitting sensor for the detection of $Hg^{2+}/Cr_2O_7^{2-}$ and a wide range of nitro-compounds. *J. Mater. Chem. C* **2018**, *6*, 6440–6448.

54. Xu, X.-Y.; Yan, B. Fabrication and application of a ratiometric and colorimetric fluorescent probe for Hg^{2+} based on dual-emissive metal-organic framework hybrids with carbon dots and Eu^{3+}. *J. Mater. Chem. C* **2016**, *4*, 1543–1549.
55. Samanta, P.; Desai, A. V.; Sharma, S.; Chandra, P.; Ghosh, S. K. Selective Recognition of Hg^{2+} ion in Water by a Functionalized Metal-Organic Framework (MOF) Based Chemodosimeter. *Inorg. Chem.* **2018**, *57*, 2360–2364.
56. Pankajakshan, A.; Kuznetsov, D.; Mondal, S. Ultrasensitive Detection of Hg(II) Ions in Aqueous Medium Using Zinc-Based Metal-Organic Framework. *Inorg. Chem.* **2019**, *58*, 1377–1381.
57. Singha, D. K.; Mahata, P. Highly Selective and Sensitive Luminescence Turn-On-Based Sensing of Al^{3+} Ions in Aqueous Medium Using a MOF with Free Functional Sites. *Inorg. Chem.* **2015**, *54*, 6373–6379.
58. Zhou, X.; Cheng, J.; Li, L.; Chen, Q.; You, Y.; Xiao, H.; Huang, W. A europium (III) metal-organic framework as ratiometric turn-on luminescent sensor for Al^{3+} ions. *Sci. China Mater.* **2018**, *61*, 752–757.
59. Chand, S.; Mondal, M.; Pal, S. C.; Pal, A.; Maji, S.; Mandal, D.; Das, M. C. Two azo-functionalized luminescent 3D Cd(II) MOFs for highly selective detection of Fe^{3+} and Al^{3+}. *New J. Chem.* **2018**, *42*, 12865–12871.
60. Razavi, S. A. A.; Masoomi, M. Y.; Morsali, A. Morphology-dependent sensing performance of dihydro-tetrazine functionalized MOF toward Al(III). *Ultrason. Sonochem.* **2018**, *41*, 17–26.
61. Li, W.-X.; Gu, J.-H.; Li, H.-X.; Dai, M.; Young, D. J.; Li, H.-Y.; Lang, J.-P. Post-synthetic Modification of a Two-Dimensional Metal-Organic Framework via Photodimerization Enables Highly Selective Luminescent Sensing of Aluminum(III). *Inorg. Chem.* **2018**, *57*, 13453–13460.
62. Zhang, J.; Gong, L.; Feng, J.; Wu, J.; Zhang, C. Two luminescent Zn(II)/Cd(II) metal-organic frameworks as rare multifunctional sensors. *New J. Chem.* **2017**, *41*, 8107–8117.
63. Cao, L. H.; Shi, F.; Zhang, W. M.; Zang, S. Q.; Mak, T. C. Selective Sensing of Fe^{3+} and Al^{3+} Ions and Detection of 2,4,6-Trinitrophenol by a Water-Stable Terbium-Based Metal-Organic Framework. *Chem. Eur. J.* **2015**, *21*, 15705–15712.
64. Lv, R.; Chen, Z.; Fu, X.; Yang, B.; Li, H.; Su, J.; Gu, W.; Liu, X. A highly selective and fast-response fluorescent probe based on Cd-MOF for the visual detection of Al^{3+} ion and quantitative detection of Fe^{3+} ion. *J. Solid State Chem.* **2018**, *259*, 67–72.
65. Rouhani, F.; Rafizadeh-Masuleh, F.; Morsali, A. Selective sacrificial metal-organic frameworks: a highly quantitative colorimetric naked-eye detector for aluminium ions in aqueous solutions. *J. Mater. Chem. A* **2019**, 7, 18634–18641.
66. Chen, Z.; Sun, Y.; Zhang, L.; Sun, D.; Liu, F.; Meng, Q.; Wang, R.; Sun, D. A tubular europium–organic framework exhibiting selective sensing of Fe^{3+} and Al^{3+} over mixed metal ions. *Chem. Commun.* **2013**, *49*, 11557–11559.
67. Senthilkumar, S.; Goswami, R.; Smith, V. J.; Bajaj, H. C.; Neogi, S. Pore Wall-Functionalized Luminescent Cd(II) Framework for Selective CO_2 Adsorption, Highly Specific 2,4,6-Trinitrophenol Detection, and Colorimetric Sensing of Cu^{2+} Ions. *ACS Sustain. Chem. Eng.* **2018**, *6*, 10295–10306.

68. Li, L.; Shen, S.; Lin, R.; Bai, Y.; Liu, H. Rapid and specific luminescence sensing of Cu(II) ions with a porphyrinic metal-organic framework. *Chem. Commun.* **2017**, *53*, 9986–9989.
69. Du, J.-L.; Gao, J.-P.; Li, C.-P.; Zhang, X.-Y.; Hou, J.-X.; Jing, X.; Mu, Y.-J.; Li, L.-J. A stable 3D Cd(ii) metal-organic framework for highly sensitive detection of Cu2+ ions and nitroaromatic explosives. *RSC Adv.* **2017**, *7*, 49618–49625.
70. Dan, W.; Liu, X.; Deng, M.; Ling, Y.; Chen, Z.; Zhou, Y. A highly stable indium phosphonocarboxylate framework as a multi-functional sensor for Cu2+ and methylviologen ions. *Dalton Trans.* **2015**, *44*, 3794–3800.
71. Liu, B.; Wu, W.-P.; Hou, L.; Wang, Y.-Y. Rapid and specific luminescence sensing of Cu(II) ions with a porphyrinic metal-organic framework. *Chem. Commun.* **2014**, *50*, 8731–8734.
72. Zhao, J.; Wang, Y.-N.; Dong, W.-W.; Wu, Y.-P.; Li, D.-S.; Zhang, Q.-C. A Robust Luminescent Tb(III)-MOF with Lewis Basic Pyridyl Sites for the Highly Sensitive Detection of Metal Ions and Small Molecules. *Inorg. Chem.* **2016**, *55*, 3265–3271.
73. Ma, H.; Wang, L.; Chen, J.; Zhang, X.; Wang, L.; Xu, N.; Yang, G.; Cheng, P. A multi-responsive luminescent sensor for organic small-molecule pollutants and metal ions based on a 4d–4f metal-organic framework. *Dalton Trans.* **2017**, *46*, 3526–3534.
74. Zhang, H.; Fan, R.; Chen, W.; Fan, J.; Dong, Y.; Song, Y.; Du, X.; Wang, P.; Yang, Y. 3D Lanthanide Metal-Organic Frameworks Based on Mono-, Tri-, and Heterometallic Tetranuclear Clusters as Highly Selective and Sensitive Luminescent Sensor for Fe^{3+} and Cu^{2+} Ions. *Cryst. Growth Des.* **2016**, *16*, 5429–5440.
75. Lin, X.; Gao, G.; Zheng, L.; Chi, Y.; Chen, G. Encapsulation of Strongly Fluorescent Carbon Quantum Dots in Metal-Organic Frameworks for Enhancing Chemical Sensing. *Anal. Chem.* **2013**, *86*, 1223–1228.
76. Liu, C.; Yan, B. A novel photofunctional hybrid material of pyrene functionalized metal-organic framework with conformation change for fluorescence sensing of Cu^{2+}. *Sens. Actuators, B.* **2016**, *235*, 541–546.
77. Feng, X.; Feng, Y.; Guo, N.; Sun, Y.; Zhang, T.; Ma, L.; Wang, L. Series d–f Heteronuclear Metal-Organic Frameworks: Color Tunability and Luminescent Probe with Switchable Properties. *Inorg. Chem.* **2017**, *56*, 1713–1721.
78. Xu, X.-Y.; Chen, Q.-C.; Yu, Y.-D.; Huang, X.-C. Ligand Induced Anionic Cuprous Cyanide Framework for Cupric Ion Turn on Luminescence Sensing and Photocatalytic Degradation of Organic Dyes. *Inorg. Chem.* **2015**, *55*, 75–82.
79. Li, Y.-W.; Li, J.-R.; Wang, L.-F.; Zhou, B.-Y.; Chen, Q.; Bu, X.-H. Microporous metal-organic frameworks with open metal sites as sorbents for selective gas adsorption and fluorescence sensors for metal ions. *J. Mater. Chem. A 1* **2013**, 495–499.
80. Liu, S.; Li, J.; Luo, F. The first transition-metal metal-organic framework showing cation exchange for highly selectively sensing of aqueous Cu(II) ions. *Inorg. Chem. Commun.* **2010**, *13*, 870–872.
81. Xu, X.-Y.; Chen, Q.-C.; Yu, Y.-D.; Huang, X.-C. Ligand Induced Anionic Cuprous Cyanide Framework for Cupric Ion Turn on Luminescence Sensing and Photocatalytic Degradation of Organic Dyes. *Inorg. Chem.* **2016**, *55*, 75–82.
82. Chen, Y.-Z.; Jiang, H.-L. Porphyrinic Metal-Organic Framework Catalyzed Heck-Reaction: Fluorescence "Turn-On" Sensing of Cu(II) Ion. *Chem. Mater.* **2016**, *28*, 6698–6704.

83. Ma, Y.; Xu, G.; Wei, F.; Cen, Y.; Ma, Y.; Song, Y.; Xu, X.; Shi, M.; Muhammad, S.; Hu, Q. A dual-emissive fluorescent sensor fabricated by encapsulating quantum dots and carbon dots into metal-organic frameworks for the ratiometric detection of Cu^{2+} in tap water. *J. Mater. Chem. C* **2017**, *5*, 8566–8571.
84. Weng, H.; Yan, B. A Eu(III) doped metal-organic framework conjugated with fluorescein-labeled single-stranded DNA for detection of Cu(II) and sulfide. *Anal. Chim. Acta* **2017**, *988*, 89–95.
85. Hao, J.; Liu, F.; Liu, N.; Zeng, M.; Song, Y.; Wang, L. Ratiometric fluorescent detection of Cu^{2+} with carbon dots chelated Eu-based metal-organic frameworks. *Sens. Actuators B* **2017**, *245*, 641–647.
86. Gu, T.-Y.; Dai, M.; Young, D. J.; Ren, Z.-G.; Lang, J.-P. Luminescent Zn(II) Coordination Polymers for Highly Selective Sensing of Cr(III) and Cr(VI) in Water. *Inorg. Chem.* **2017**, *56*, 4668–4678.
87. Jia, X.-X.; Yao, R.-X.; Zhang, F.-Q.; Zhang, X.-M. A Fluorescent Anionic MOF with $Zn_4(trz)_2$ Chain for Highly Selective Visual Sensing of Contaminants: Cr(III, Ion and TNP. *Inorg. Chem.* **2017**, *56*, 2690–2696.
88. Guo, X.-Y.; Zhao, F.; Liu, J.-J.; Liu, Z.-L.; Wang, Y.-Q. An ultrastable zinc(II)–organic framework as a recyclable multi-responsive luminescent sensor for Cr(III), Cr(VI) and 4-nitrophenol in the aqueous phase with high selectivity and sensitivity. *J. Mater. Chem. A* **2017**, *5*, 20035–20043.
89. Jin, H.; Xu, J.; Zhang, L.; Ma, B.; Shi, X.; Fan, Y.; Wang, L. Multi-responsive luminescent sensor based on Zn (II) metal-organic framework for selective sensing of Cr(III), Cr(VI) ions and *p*-nitrotolune. *J. Solid State Chem.* **2018**, *268*, 168–174.
90. Hu, J.-S.; Dong, S.-J.; Wu, K.; Zhang, X.-L.; Jiang, J.; Yuan, J.; Zheng, M.-D. An ultrastable magnesium-organic framework as multi-responsive luminescent sensor for detecting trinitrotoluene and metal ions with high selectivity and sensitivity. *Sens. Actuators B* **2019**, *283*, 255–261.
91. Guo, X.-Y.; Zhao, F.; Liu, J.-J.; Liu, Z.-L.; Wang, Y.-Q. An ultrastable zinc(II)–organic framework as a recyclable multi-responsive luminescent sensor for Cr(III), Cr(VI) and 4-nitrophenol in the aqueous phase with high selectivity and sensitivity. *J. Mater. Chem. A* **2017**, *5*, 20035–20043.
92. Sun, Z.; Yang, M.; Ma, Y.; Li, L. Multi-Responsive Luminescent Sensors Based on Two-Dimensional Lanthanide–Metal-Organic Frameworks for Highly Selective and Sensitive Detection of Cr(III) and Cr(VI, Ions and Benzaldehyde. *Cryst. Growth Des.* **2017**, *17*, 4326–4335.
93. Wang, H.-H.; Zhou, L.-J.; Wang, Y.-L.; Liu, Q.-Y. Terbium-biphenyl-3,3′-disulfonyl-4,4′-dicarboxylate framework with sulfonate sites for luminescent sensing of Cr^{3+} ion. *Inorg. Chem. Commun.* **2016**, *73*, 94–97.
94. Wang, H.; Qin, J.; Huang, C.; Han, Y.; Xu, W.; Hou, H. Mono/bimetallic water-stable lanthanide coordination polymers as luminescent probes for detecting cations, anions and organic solvent molecules. *Dalton Trans.* **2016**, *45*, 12710–12716.

95. Lv, R.; Wang, J.; Zhang, Y.; Li, H.; Yang, L.; Liao, S.; Gu, W.; Liu, X. An amino-decorated dual-functional metal-organic framework for highly selective sensing of Cr(III) and Cr(VI) ions and detection of nitroaromatic explosives. *J. Mater. Chem. A* **2016**, *4*, 15494–15500.
96. Rubin, H. N.; Reynolds, M. M. Amino-incorporated Tricarboxylate Metal-Organic Framework for the sensitive Fluorescence Detection of Heavy Metal Ions with insights into the origin of a photoluminescence response. *Inorg. Chem.* **2019**, *58*, 10671–10679.
97. Farahani, Y. D.; Safarifard, V. A luminescent metal-organic framework with pre-designed functionalized ligands as an efficient fluorescence sensing for Fe3+ ions. *J. Solid State Chem.* **2019**, *270*, 428–435.
98. Jin, J.; Yang, G.; Liu, Y.; Cheng, S.; Liu, J.; Wu, D.; Wang, Y.-Y. *Inorg. Chem.* **2019**, *58*, 339–348.
99. Xu, N.; Zhang, Q.; Hou, B.; Cheng, Q.; Zhang, G. A Novel Magnesium Metal-Organic Framework as a Multiresponsive Luminescent Sensor for Fe(III) Ions, Pesticides, and Antibiotics with High Selectivity and Sensitivity. *Inorg. Chem.* **2018**, *57*, 13330–13340.
100. Xiao, Z.-Z.; Han, L.-J.; Wang, Z.-J.; Zheng, H.-G. Three Zn(II)-based MOFs for luminescence sensing of Fe^{3+} and $Cr_2O_7^{2-}$ ions. *Dalton Trans.* **2018**, *47*, 3298–3302.
101. Jing, T.; Chen, L.; Jiang, F.; Yang, Y.; Zhou, K.; Yu, M.; Cao, Z.; Li, S.; Hong, M. Fabrication of a Robust Lanthanide Metal-Organic Framework as a Multifunctional Material for Fe(III) Detection, CO_2 Capture, and Utilization. *Cryst. Growth Des.* **2018**, *18*, 2956–2963.
102. Wang, R.; Dong, X.-Y.; Xu, H.; Pei, R.-B.; Ma, M.-L.; Zang, S.-Q.; Hou, H.-W.; Mak, T. C. W. A super water-stable europium–organic framework: guests inducing low-humidity proton conduction and sensing of metal ions. *Chem. Commun.* **2014**, *50*, 9153–9156.
103. Kang, Y.; Zheng, X.-J.; Jin, L.-P. A microscale multi-functional metal-organic framework as a fluorescence chemosensor for Fe(III), Al(III) and 2-hydroxy-1-naphthaldehyde. *J. Colloid Interface Sci.* **2016**, *471*, 1–6.
104. Yang, Y.-J.; Wang, M.-J.; Zhang, K.-L. A novel photoluminescent Cd(II)–organic framework exhibiting rapid and efficient multi-responsive fluorescence sensing for trace amounts of Fe^{3+} ions and some NACs, especially for 4-nitroaniline and 2-methyl-4-nitroaniline. *J. Mater. Chem. C* **2016**, *4*, 11404–11418.
105. Dong, J. P.; Li, B.; Jin, Y.-J.; Wang, L.-Y. Efficient detection of Fe(iii) and chromate ions in water using two robust lanthanide metal-organic frameworks. *CrystEnggComm* **2021**, *23*, 1677–1683.
106. Zhao, S.-S.; Yang, J.; Liu, Y.-Y.; Ma, J.-F. Fluorescent Aromatic Tag-Functionalized MOFs for Highly Selective Sensing of Metal Ions and Small Organic Molecules. *Inorg. Chem.* **2016**, *55*, 2261–2273.
107. Qin, J. S.; Zhang, S. R.; Du, D. Y.; Shen, P.; Bao, S. J.; Lan, Y. Q.; Su, Z. M. A Microporous Anionic Metal-Organic Framework for Sensing Luminescence of Lanthanide(III) Ions and Selective Absorption of Dyes by Ionic Exchange. *Chem. Eur. J.* **2014**, *20*, 5625–5630.
108. Bhattacharyya, S.; Chakraborty, A.; Jayaramulu, K.; Hazra, A.; Maji, T. K. A bimodal anionic MOF: turn-off sensing of Cu(II) and specific sensitization of Eu(III). *Chem. Commun.* **2014**, *50*, 13567–13570.

109. Xu, G. W.; Wu, Y. P.; Dong, W. W.; Zhao, J.; Wu, X. Q.; Li, D. S.; Zhang, Q. A Multifunctional Tb-MOF for Highly Discriminative Sensing of Eu^{3+}/Dy^{3+} and as a Catalyst Support of Ag Nanoparticles. *Small* **2017**, *13*, 1602996.
110. Qin, J. S.; Bao, S. J.; Li, P.; Xie, W.; Du, D. Y.; Zhao, L.; Lan, Y. Q.; Su, Z. M. A Stable Porous Anionic Metal-Organic Framework for Luminescence Sensing of Ln^{3+} Ions and Detection of Nitrobenzene. *Chem. Asian. J.* **2014**, *9*, 749–753.
111. Zhao, S. N.; Song, X. Z.; Zhu, M.; Meng, X.; Wu, L. L.; Feng, J.; Song, S. Y.; Zhang, H. J. Encapsulation of Ln^{III} Ions/Dyes within a Microporous Anionic MOF by Post-synthetic Ionic Exchange Serving as a Ln^{III} Ion Probe and Two-Color Luminescent Sensors. *Chem. Eur. J.* **2015**, *21*, 9748–9752.
112. Gao, Y.; Zhang, X.; Sun, W.; Liu, Z. A robust microporous metal-organic framework as a highly selective and sensitive, instantaneous and colorimetric sensor for Eu^{3+} ions. *Dalton Trans.* **2015**, *44*, 1845–1849.
113. Cao, J.; Gao, Y.; Wang, Y.; Du, C.; Liu, Z. A microporous metal-organic open framework containing uncoordinated carbonyl groups as postsynthetic modification sites for cation exchange and Tb^{3+} sensing. *Chem. Commun.* **2013**, *49*, 6897–6899.
114. Choi, S. J.; Yoon, B.; Lin, S.; Swager, T. M. Functional Single-Walled Carbon Nanotubes for Anion Sensing. *ACS Appl. Mater. Interfaces* **2020**, *12*, 28375–28382.
115. Dixit, S.; Agarwal, N. Synthesis of imidazoaryl-BODIPY derivatives for anion sensing applications. *J. Photochem. Photobiol. A.* **2017**, *343*, 66–71.
116. Gale, P. A.; Caltagirone, C. Fluorescent and colorimetric sensors for anionic species. *Coord. Chem. Rev.* **2018**, *354*, 2–27.
117. Singha, D. K.; Majee, P.; Mondal, S. K.; Mahata, P. A luminescent cadmium based MOF as selective and sensitive iodide sensor in aqueous medium. *J. Photochem. Photobiol. A.* **2018**, *356*, 389–396.
118. Zhan, Z. Y.; Jia, Y. J.; Li, D. H.; Zhang, X. L.; Hu, M. A water-stable terbium-MOF sensor for the selective, sensitive, and recyclable detection of Al^{3+} and CO_3^{2-} ion. *Dalton Trans.* **2019**, *48*, 15255–15262.
119. Jain, A. K.; Gupta, V. K.; Raisoni, J. R. Anion recognition using newly synthesized hydrogen bonding diamide receptors: PVC based sensors for carbonate. *Electrochim. Acta* **2006**, *52*, 951–957.
120. Liu, H.; Wang, H.; Chu, T.; Yua, M.; Yang, Y. An electrodeposited lanthanide MOF thin film as a luminescent sensor for carbonate detection in aqueous solution. *J. Mater. Chem. C* **2014**, *2*, 8683–8690.
121. Zhan, Z.; Jia, Y.; Li, D.; Zhang, X.; Hu, M. A water-stable terbium-MOF sensor for the selective, sensitive, and recyclable detection of Al^{3+} and CO_3^{2-} ions. *Dalton Trans.* **2019**, *48*, 15255–15262.
122. Yi, K.; Zhang, L. Embedding dual fluoroprobe in metal-organic frameworks for continuous visual recognition of Pb^{2+} and PO_4^{3-} via fluorescence 'turn-off-on' response: Agar test paper and fingerprint. *J. Hazard. Mater.* **2020**, *389*, 122141–122148.

123. Qin, W.; Zhang, X. Y.; Zhang, J. P. A New Multifunctional Zinc–Organic Framework with Rare Interpenetrated Tripillared Bilayers as a Luminescent Probe for Detecting Ni^{2+} and PO_4^{3-} in Water. *Cryst. Growth Des.* **2020**, *20*, 5120–5128.
124. Warwick, C.; Guerreiro, A.; Soares, A. Sensing and analysis of soluble phosphatesin environmental samples: A review. *Biosens. Bioelectron.* **2013**, *41*, 1–11.
125. Law al, A. T.; Adeloju, S. B. Progress and recent advances in phosphate sensors: a review. *Talanta* **2013**, *114*, 191–203.
126. Cheng, Y.; Zhang, H.; Yang, B.; Wu, J.; Wang, Y.; Ding, B.; Huo, J.; Li, Y. Highly efficient fluorescence sensing of phosphate by dual-emissive lanthanide MOFs. *Dalton Trans.* **2018**, *47*, 12273.
127. Zhang, W.; Wang, X.; Li, P.; Zhang, W.; Wang, H.; Tang, B. Evaluating Hyperthyroidism-Induced Liver Injury Based on In Situ Fluorescence Imaging of Glutathione and Phosphate via Nano MOFs Sensor. *Anal. Chem.* **2020**, *92*, 8952–8958.
128. Zhaoa, D.; Wana, X.; Songa, H.; Haoa, L.; Sub, Y.; Lv, Y. Metal-organic frame works (MOFs) combined with ZnO quantum dots as a fluorescent sensing platform for phosphate. *Sensors and Actuators B* **2014**, *197*, 50–57.
129. Chena, Y-Q.; Tiana, Y.; Lib, N.; Liuc, S.-J. Ni(II)/Zn(II)-triazolate clusters based MOFs constructed from a V-shaped dicarboxylate ligand: Magnetic properties and phosphate sensing. *Journal of Solid State Chemistry* **2018**, *262*, 100–105.
130. Liu, H. P.; Wang, H. M.; Chu, T. S.; Yu, M. H.; Yang, Y. Y. An electrodeposited lanthanide MOF thin film as a luminescent sensor for carbonate detection in aqueous solution. *J. Mater. Chem. C* **2014**, *2*, 8683–8690.
131. Ayoob, S.; Gupta, A. K. Fluoride in drinking water: a review on the status and stress effects. *Environ. Sci. Technol.* **2006**, *36*, 433–487.
132. Mandinic, Z.; Curcic, M.; Antonijevic, B.; Carevic, M.; Mandic, J.; Djukic-Cosic, D.; Lekic, C. P. *Sci. Total Environ.* **2010**, *408*, 3507–3512.
133. Zeng, X.; Hu, J.; Zhang, M.; Wang, F.; Wu, L.; Hou, X. Visual Detection of Fluoride Anions Using Mixed Lanthanide Metal—Organic Frameworks with a Smartphone. *Anal. Chem.* **2020**, *92*, 2097–2102.
134. Wentz, H. C.; Campbell, M. G. Fluoride detection with a redox-active naphthalene diimide metal-organic framework. *Polyhedron* **2018**, *154*, 309–313.
135. Che, H.; Li, Y.; Zhang, S.; Chen, W.; Tian, X.; Yang, C.; Lu, L.; Zhou, Z.; Nie, Y. A portable logic detector based on Eu-MOF for multi-target, on-site, visual detection of Eu^{3+} and fluoride in groundwater. *Sensors & Actuators: B. Chemical* **2020**, *324*, 128641.
136. Xiao, Q. Q.; Li, Y. H.; Liu, D.; Cufi, G. H. A water-stable luminescent Co(II) coordination polymer as probe for efficient detection of Cr(VI)-anions ($Cr_2O_7^{2-}$ and CrO_4^{2-}) in aqueous solution. *Inorg. Chem. Commun.* **2020**, *111*, 107665–107668.
137. Katayev, E. A.; Ustynyuk, Y. A.; Sessler, J. L. Receptors for tetrahedral oxyanions. *Coord. Chem. Rev.* **2006**, *250*, 3004.
138. Keith, L. H.; Teillard, W. A. Special report: Priority pollutants: perspective view. *Environ. Sci. Technol.* **1979**, *13*, 416–23.

139. Zhitkovich, A. Importance of chromium-DNA adducts in mutagenicity and toxicity of chromium(VI). *Chem. Res. Toxicol.* **2005**, *18*, 3–11.
140. Vonburg, R.; Liu, D. Chromium and Hexavalent Chromium. *J. Appl. Toxicol.* **1993**, *13*, 225–230.
141. Bagchi, D.; Stohs, S. J.; Downs, B. W.; Bagchi, M.; Preuss, H. G. Cytotoxicity and oxidative mechanisms of different forms of chromium. *Toxicology* **2002**, *180*, 5–22.
142. Rapti, S.; Sarma, D.; Diamantis, S. A.; Skliri, E.; Armatas, G. S.; Tsipis, A. C.; Hassan, Y. S.; Alkordi, M.; Malliakas, C. D.; Kanatzidis, M. G.; Lazarides, T.; Plakatouras, J. C.; Manos, M. J. All in one porous material: exceptional sorption and selective sensing of hexavalent chromium by using a Zr^{4+} MOF. *J. Mater. Chem. A* **2017**, *5*, 14707.
143. Liu, W.; Wang, Y.; Bai, Z.; Li, Y.; Wang, Y.; Chen, L.; Xu, L.; Diwu, J.; Chai, Z.; Wang, S. Hydrolytically Stable Luminescent Cationic Metal-Organic Framework for Highly Sensitive and Selective Sensing of Chromate Anions in Natural Water Systems. *ACS Appl. Mater. Interfaces* **2017**, *9*, 16448–16457.
144. Lin, Z.-J.; Zheng, H.-Q.; Zheng, H.-Y.; Lin, L.-P.; Xin, Q.; Cao, R. Efficient Capture and Effective Sensing of $Cr_2O_7^{2-}$ from Water Using a Zirconium Metal–Organic Framework. *Inorg. Chem.* **2017**, *56*, 14178–14188.
145. Harbuzaru, B. V.; Corma, A.; Rey, F.; Jorda, J. L.; Ananias, D.; Carlos, L. D.; Rocha, J. A. Miniaturized Linear pH Sensor Based on a Highly Photoluminescent Self-Assembled Europium(III) Metal-Organic Framework. *Angew. Chem., Int. Ed.* **2009**, *48*, 6476–6479.
146. Zhao, Y.; Wan, M.-Y.; Bai, J.-P.; Zeng, H.; Lu, W.; Li, D. pH Modulated Luminescence Switching in a Eu-MOF: Rapid Detection of Acidic Amino Acids. *J. Mater. Chem. A* **2019**, *7*, 11127–11133.
147. Bai, Z.; Liu, S.; Cheng, G.; Wu, G.; Liu, Y. High Proton Conductivity of MOFs-Polymer Composite Membranes by Phosphoric Acid Impregnation. *Microporous Mesoporous Mater.* **2020**, *292*, 109763–109785.
148. Li, M.; Yanga, W.; Qiu, P.; Ren, G.; Li, C.; Chen, Z.; Wang, Y.; Pan, Q. Two efficient pH sensors based on heteronuclear metal-organic frameworks. *Journal of Luminescence* **2019**, *205*, 380–384.
149. Lv, X.-L.; Xie, L.-H.; Wang, B.; Zhao, M.; Cui, Y.; Li, J.-R. Flexible metal-organic frameworks for the wavelength-based luminescence sensing of aqueous Ph. *J. Mater. Chem. C* **2018**, *6*, 10628.
150. Aguilera-Sigalat, J.; Bradshaw, D. A colloidal water-stable MOF as a broad-range fluorescent pH sensor via post-synthetic modification. *Chem. Commun.* **2014**, *50*, 4711.
151. Jiang, H.-L.; Feng, D.; Wang, K.; Gu, Z.-Y.; Wei, Z.; Chen, Y.-P.; Zhou, H.-C. An Exceptionally Stable, Porphyrinic Zr Metal–Organic Framework Exhibiting pH-Dependent Fluorescence. *J. Am. Chem. Soc.* **2013**, *135*, 13934–13938.
152. He, C.; Lu, K.; Lin, W. Nanoscale Metal–Organic Frameworks for Real-Time Intracellular pH Sensing in Live Cells. *J. Am. Chem. Soc.* **2014**, *136*, 12253–12256.
153. Huang, S.-Z.; Liu, S.-S.; Zhang, H.-j.; Han, Z.; Zhao, G.; Dong, X-Y.; Zang, S.-Q. Dual-Functional Proton-Conducting and pH-Sensing Polymer Membrane Benefiting from a Eu-MOF. *ACS Appl. Mater. Interfaces* **2020**, *12*, 28720–28726.

154. Cui, Y.; Xu, H.; Yue, Y.; Guo, Z.; Yu, J.; Chen, Z. A luminescent mixed-lanthanide metal-organic framework thermometer. *J. Am. Chem. Soc.* **2012**, *134*, 3979–3982.
155. Cui, Y.; Zhu, F.; Chen, B.; Qian, G. Metal-organic frameworks for luminescence thermometry. *Chem. Commun.* **2015**, *51*, 7420–7431.
156. Cui, Y.; Xu, H.; Yue, Y.; Guo, Z.; Yu, J.; Chen, Z.; Gao, J.; Yang, Y.; Qian, G.; Chen, B. A Luminescent Mixed-Lanthanide Metal-Organic Framework Thermometer. *J. Am. Chem. Soc.* **2012**, *134*, 3979–3982.
157. Wang, H.; Zhao, D.; Cui, Y.; Yang, Y.; Qian, G. A Eu/Tb-mixed MOF for luminescent high-temperature sensing. *Journal of Solid State Chemistry* **2017**, *246*, 341–345.
158. Zhao, D.; Rao, X.; Yu, J.; Cui, Y.; Yang, Y.; Qian, G. Design and Synthesis of an MOF Thermometer with High Sensitivity in the Physiological Temperature Range. *Inorg. Chem.* **2015**, *54*, 11193–11199.
159. Liu, W.; Chen, C.; Huang, X.; Xie, E.; Liu, W. Functional Construction of Dual-Emitting 4-Aminonaphthalimide Encapsulated Lanthanide MOFs Composite for Ratiometric Temperature Sensing. *Chem. Eur. J.* **2019**, *25*, 10054–10058.
160. Yang, Y.; Wang, Y.; Feng, Y.; Song, X.; Cao, C.; Zhang, G.; Liu, W. Three isostructural Eu^{3+}/Tb^{3+} co-doped MOFs for wide-range ratiometric temperature sensing. *Talanta* **2020**, *208*, 120354.
161. Ma, D.; Li, B.; Zhou, X.; Zhou, Q.; Liu, K.; Zeng, G.; Li, G.; Shi, Z.; Feng, S. A dual functional MOF as a luminescent sensor for quantitatively detecting the concentration of nitrobenzene and temperature. *Chem. Commun.* **2013**, *49*, 8964.

Chapter 6

Metal–Organic Framework-Based Artificial Antibodies for Sensing Applications

Trupti R. Das,[1] Prama Bhattacherjee,[2] Santanu Patra,[*,3,4]
Penny P. Govender,[5] Bindu Mangala,[*,6] and Sudheesh K. Shukla[*,5,7]

[1]CIPET, Institute of Petrochemicals Technology (IPT) – Bhubaneswar, B-25 CNI Complex, Patia, Bhubaneswar 751024, India

[2]Department of Chemistry, Indian Institute of Technology (ISM) Dhanbad, Jharkhand 826004, India

[3]Institute of Advanced Materials, IAAM, Gammalkilsvägen 18, Ulrika 590 53, Sweden

[4]VBRI, 7/16, Kalkaji Extension, New Delhi 110019, India

[5]Department of Chemical Sciences, University of Johannesburg, P.O. Box 17011, Doornfontein Campus, 2028 Johannesburg, South Africa

[6]Department of Chemistry, J.C. Bose University of Science and Technology YMCA, Faridabad 121006, India

[7]Department of Biomedical Engineering, School of Biological Engineering and Life Science, Shobhit Institute of Engineering & Technology (Deemed-to-be University), Modipuram, Meerut 250110, India

[*]Email: spatra.ism@gmail.com

[*]Email: bindumangla@gmail.com

[*]Email: sudheeshkshukla010@gmail.com

The United Nations Sustainable Development Goals provided 17 guidelines for all counties to ensure health and safety for all by 2030. Health disasters like COVID-19 risk human lives, and there is a great need for heightened awareness. Routine and early-phase monitoring is turning out to be a vital requirement for sustaining a standard fitness level and evaluating one's physical strength. Various new age technologies and methods such as the selective artificial antibody-based metal-organic frameworks (MOFs), which have been used in practical applications. MOFs have been used for creating sensing devices, but they suffer from a low selectivity in complex matrixes. To increase the selectivity of sensor, artificial antibodies such as molecularly imprinted polymers (MIPs) come into the limelight. This chapter illustrates developments in the model of molecular imprinting utilizing MOFs to create chemical sensors. It presents key strategies in fabrication procedures, implementation, and applications in sensing devices. It also includes commentary on future developments in this field.

Introduction

The porous coordination polymers called the metal-organic frameworks (MOFs) have gained special attention because of their wide application ranges because of their porosity. Research in this field has been rapidly increasing. In 1999, the abbreviation "MOF" was coined by Yaghi et al. To illustrate the volume of free pores, they picked a "yellow sphere" in their diagram (*1*). These are the classes of materials that have mixed crystalline structures and a combination of metal centers and natural linkers. They offer a novel compound flexibility along with a designable structure, an extraordinarily large perpetually inward porosity. Thus, MOFs comprise their own group of porous materials that surpass the impediments of previously recognized porous materials (initiated carbon, zeolites, and mesoporous silica) (*2*). First, most endeavors were coordinated for development of new frameworks, which included innovative geographies and designs with the largest conceivable surface zones. The objective was to discover materials that have outstandingly excellent gas storage capacity, particularly for putting away hydrogen. Reactive structures have become incorporated more often as well. MOF structures have elevated researchers' expectations because of their adaptability and expanded compound functionalization in comparison with zeolites. In the past, researchers have tended to focus on reticular frameworks (*3*), post-synthesis functionalization (*4*), protection (*5*), solid solution or multivariate constructions (copolymers) (*6*), and mixtures of precious stone heterostructure methods (*7, 8*). It is possible to switch the natural linker to fuse responsive groups inside a MOF. While endeavors progress, observing the mechanical operations and preparation of MOF materials are contributing to new fields such as handling of MOFs as films, which has some significant applications, such as compound sensors (*9*) and membranes (Figure 1) (*10*).

Motivation to produce MOF films repeatedly originates from two strictly associated fields: zeolite films, especially the instant blend and cultivated growth strategies, and coordination polymer films in the setting of Langmuir Blodgett and layer-by-layer dainty film readiness methods (*11, 12*). We characterize MOFs as half-breed, organic–inorganic, strong materials with metal centers connected by natural moieties. MOFs include perpetual inward porosities, reversible desorption, and adsorption of guest particles, and an all-encompassing three-dimensional (3D) structure with a translucent construction that can be surveyed essentially by X-ray diffraction. Consequently, coordination of polymer-type films containing of organic–inorganic crossover materials that are porous, however, need long-range request (e.g., films made of Hofmann clathrates) are past the extent of this survey. Metal cyanide materials such as Prussian blue analogs, will not be considered here on the grounds that the plausibility of linker functionalization or isoreticular development does not exist at all or is very restricted—as opposed to common MOFs, which permit gatherings at the natural linkers. Further data on Prussian blue analogs can be observed elsewhere (*13, 14*). Also, we preclude all one-dimensional and two-dimensional coordination polymer types without 3D translational order. Moreover, films created from powder MOF materials that are stored by, for instance, dropping or screen-printing approaches are not reviewed here because such coatings are not viewed as "thin" films (≤ 50 μm).

Methods of Synthesis for MOFs

Direct Synthesis

Here, we assembled the strategies of utilizing a substrate and exposed (or altered by natural particles) along with a suitable development arrangement.

Development of MOFs happens at the surface and is sometimes arranged simultaneously. This development prompts the growth of crystal associated with the substrate surface. Other methods include solvothermal growth (Figure 2).

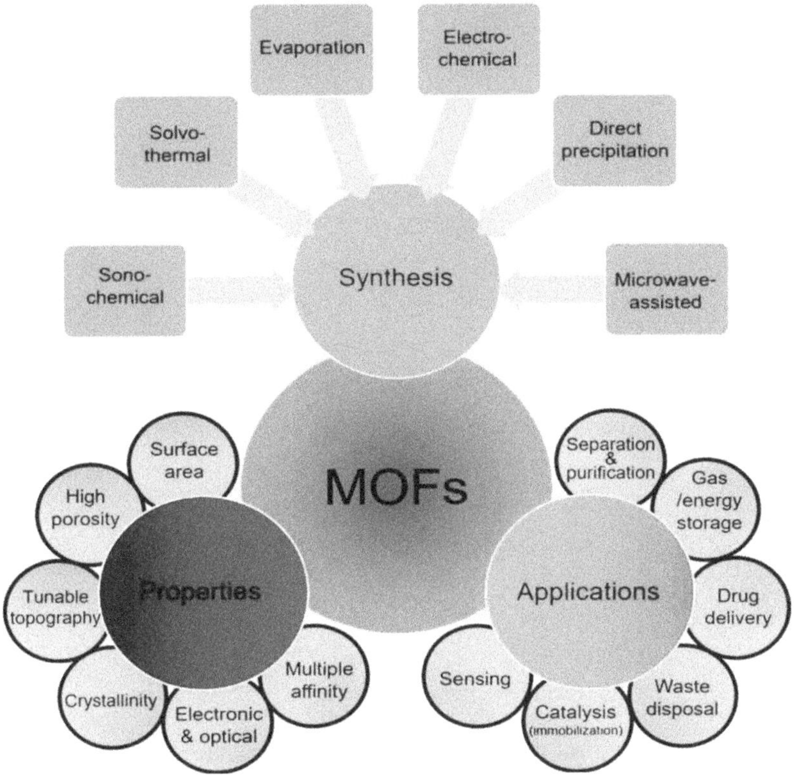

Figure 1. Graphical diagram showing the synthesis, properties, and applications of MOFs. Reproduced with permission from reference (6). Copyright 2020 MDPI.

Seeded Growth

Secondary growth is also known as seeded growth. In the case of zeolite films, this is a popular process. Because of the heterogeneous nucleation density, fabrication is difficult (*11*, *12*). Concentration of many MOFs on porous ceramic supports is low, so the method of thin-film fabrication has become more widespread. There are two steps to this method: seed preparation and seed placement. After that, the MOF develops into a film. Seeds of various kinds can be utilized for this process, as will be discussed later in this chapter. In the second stage of development, solvothermal conditions are often used.

Electrochemical Methods

Scientists at Badische Anilin- und SodaFabrik were the first to implement electrochemical methods to synthesize MOFs (*15*, *16*). In this method, a metal electrode undergoes oxidation in a solution containing the ligand, producing the metal ions. Ameloot et al. (*17*) demonstrated that metal electrodes can be coated with a MOF film in the absence of stirring by changing the conditions. The disadvantage of this process is that metallic nanoparticles of silver can defile the film despite thorough washing.

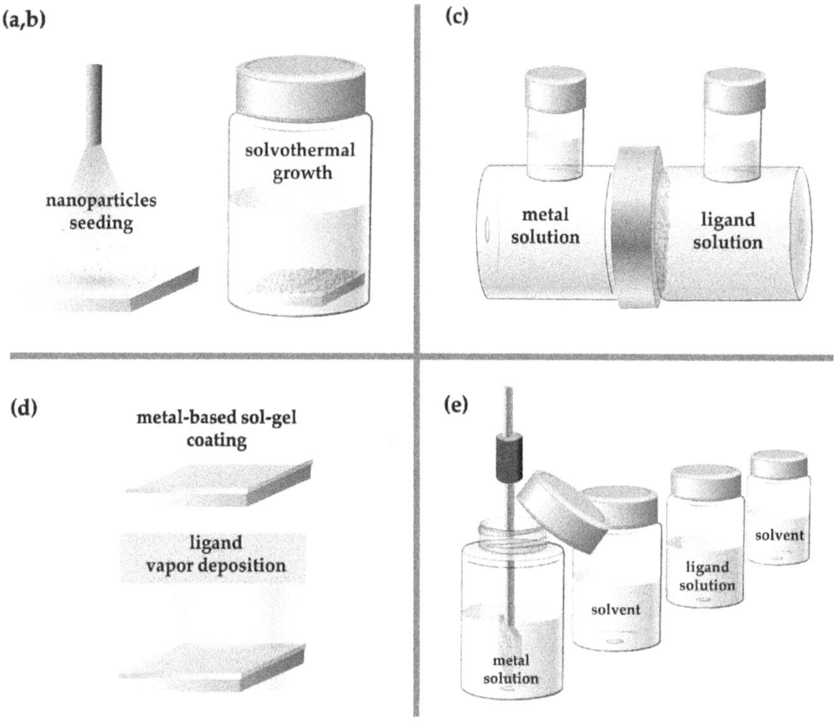

Figure 2. Schematic representation of the MOF membrane preparation methods, like solvothermal growth method and seeding (a, b), counter diffusion (c), gel-vapor deposition (d), and layer-by-layer (e). Reproduced with permission from reference (11). Copyright 2018 MDPI.

Assembly of Preformed MOF Nanocrystals

Demessence et al. have researched the construction of MOF films by assembling preformed artifacts, which are typically colloids (*18*). This method involves preparation of well-arranged MOF particles and dip coating them onto a surface. The authors used three separate MOF structures to demonstrate the concept's generality: MIL-89 (MIL=Matériaux de l'Institut Lavoisier) (*18*), which is an iron muconate. Two advantages of this method are the existence of intergrain mesoporosity and precise regulation of particle size. When these two advantages are combined, they allow for easier analyte diffusion into the film, which can be beneficial in some applications.

Stepwise Dosing of Reagents

Use of reagents comes under the deposition technique, which is closely associated with the use of chemicals. The method of liquid-phase epitaxy is applicable to depositions in which conditions must be met for linear layer-by-layer growth, particularly a gold substrate that has been S-adenosyl methionine modified and manually or in situ automated. Continuous flowing does not permit the reunification.

Characterization

Because it is not as thick, the surface-anchored metal-organic frameworks have characterization problems, including morphology, porosity, and structural characterization. Subsequent sections will look at some of these issues in detail, as well as give several examples of how they have been used in the past.

Applications of MOFs

Gas Absorption and Storage

One of the direst environmental issues in human society is the drastically increasing quantity of CO_2 in the atmosphere. CO_2 adsorption or capture has become one of the most important technologies. MOF composites have a wide range of applications. This is because graphite oxide and MOF composites have been extensively researched for their reactive adsorption of CO_2 and synergistic effects in accordance with the active sites observed near the junction between graphene sheets and the MOFs (*19*).

Another remarkable application of MOF composites is in hydrogen storage. When it burns, it does not produce CO_2, and it has a substantially higher energy density than gasoline. For describing the total H_2 uptakes, the excess and total amounts of adsorption are considered (*20*). The highest reported data for H_2 storage capacity by MOF was previously 99.5 mg g^{-1} at 56 bar and 77 K in NU-100. MOF-210's highest total H_2 storage capacity is 176 mg g^{-1} (excess 86 mg g^{-1}) at 77 K and 80 bar (*21, 22*). However, MOFs can absorb hydrogen even at low temperatures. They have no advantages in terms of hydrogen storage in high temperatures over other porous materials, such as clay, at room temperature (*23, 24*).

Liquid Adsorption

The use of metal clusters or ions or nanoparticles (NPs) inside MOFs has been considered to separate hazardous compounds from liquids. Using nanoparticles in MOFs may considerably improve their adsorption efficiency. Kampouraki et al. have extensively explored MOF composites with the shifting of sulfur-containing chemicals (such as thorium, *Bacillus thuringiensis*, and 4,6-Dimethyldibenzothiophene) and nitrogen-containing compounds (such as quinovose and investigational new drugs) in fuels. They discovered that the removal capacity of *B. thuringiensis* by $CuCl_2$-loaded MIL-47 (310 mg g^{-1} for $CuCl_2$ [0.05]/MIL-47) is significantly higher than that of pure MIL-47 (231 mg g^{-1}) and even CuI-Y, which was previously shown to have the highest capacity (254 mg g^{-1}) (*2*).

Molecule Separation

MOFs have become attractive materials for small molecular separation because of their large surface area, tunable pore diameter, and surface functions. The formation of high-quality MOF crystals is significantly difficult, however, because they are so brittle.

Catalysis

MOFs and their composites have been gaining attention as new catalytic materials. For pure MOFs, the catalytic capabilities can come from the MOFs themselves (*26, 27*), the functional species, or a mixture of both. In heterogeneous catalysis, Dhakshinamoorthy and Garcia have thoroughly examined the use of metal NP–MOF composites (*28*). There are two types of catalysis: catalysis by cluster–MOF composites and catalysis by semiconductor–MOF composites.

Biomedical Applications

Response to external stimuli such as heat, light, or magnetic force can be recorded after programing the MOFs. The structural evolutions that arise have implications for therapeutics, and the signal can be investigated for health diagnostics. Horcajada et al.'s review covers all aspects of pure MOFs as drug carriers, which include the cytotoxicity, interactions with loaded molecules, and regulated drug release via tailored MOF structures (*29*). This section focuses on use of MOF composites in the biological field. It includes MOFs with surface coatings to improve biocompatibility or MOFs with implanted nanostructures to diversify their applicability.

Electrical Applications

Since one of the key features of MOFs is porosity, their combined protonic and electronic behavior would allow them to have a wide range of applications, including in fuel cells and electrical sensors. In MOFs, as the metal nodes are not in a redox-active state, they are often poor conductors, and most organic ligands utilized in MOF production do not allow for electron transport (*29*). The usage of redox-active metal centers or organic linkers, or electron donor and acceptor building blocks, in the synthesis of MOFs with improved conductivity suffers from some major restrictions in terms of structural integrity and circumstances to activate their conductivities. One of the most crucial factors in this process is thought to be the temperature.

Sensing Applications

Metals are unique combinations of properties. They are appealing in different applications because of their organic frameworks. Vast surface areas paired with the right lighting can make a big difference. The pore structure and chemical functionality have been fine-tuned, implying that it is possible to attain great analyte sensitivity and selectivity (*9*).

QCM-Based Sensors

The use of quartz crystal microbalance (QCM) equipment is a simple technique for creating a sensor. Quartz plates covered with a substance such as silica, gold, or copper are common substrates for such devices (*30, 31*). Biemmi et al. have grown an HKUST-1 (HKUST = Hong Kong University of Science and Technology) film directly on SAM-modified gold QCM electrodes for water sorption study (*32*). SAM-modified QCM substrates were immersed in the HKUST-1 mother solution at room temperature to generate an opaque layer, as seen in Figure 3. After placing the substrate on the QCM setup, it was exposed to various water vapor concentrations. The water absorbed in the film was measured by the frequency shift, and it was determined in mass uptake linked to the mass of the empty film using the Sauerbrey equation. At $p/p^0 = 0.8$, the amount of water adsorbed was approximately 16 mmol g^1, which was the same as the quantity of water contained in a bulk HKUST-1 sample. The QCM-based method's water sorption isotherms are thus consistent with bulk data. Several other organizations have studied MOFs in water sorption since they are widely demanded in the domain of humidity sensing. Ameloot et al. (*17*), through they used electrochemical deposition technique, produced HKUST-1 coatings on top of QCM substrates and have investigated their reactions to increasing relative humidity gas flows. The film's water absorption was calculated in a way similar to previously mentioned studies. Furthermore, the signal was found to be highly reproducible by the authors. Dry and water-containing nitrogen fluxes are cycled in order.

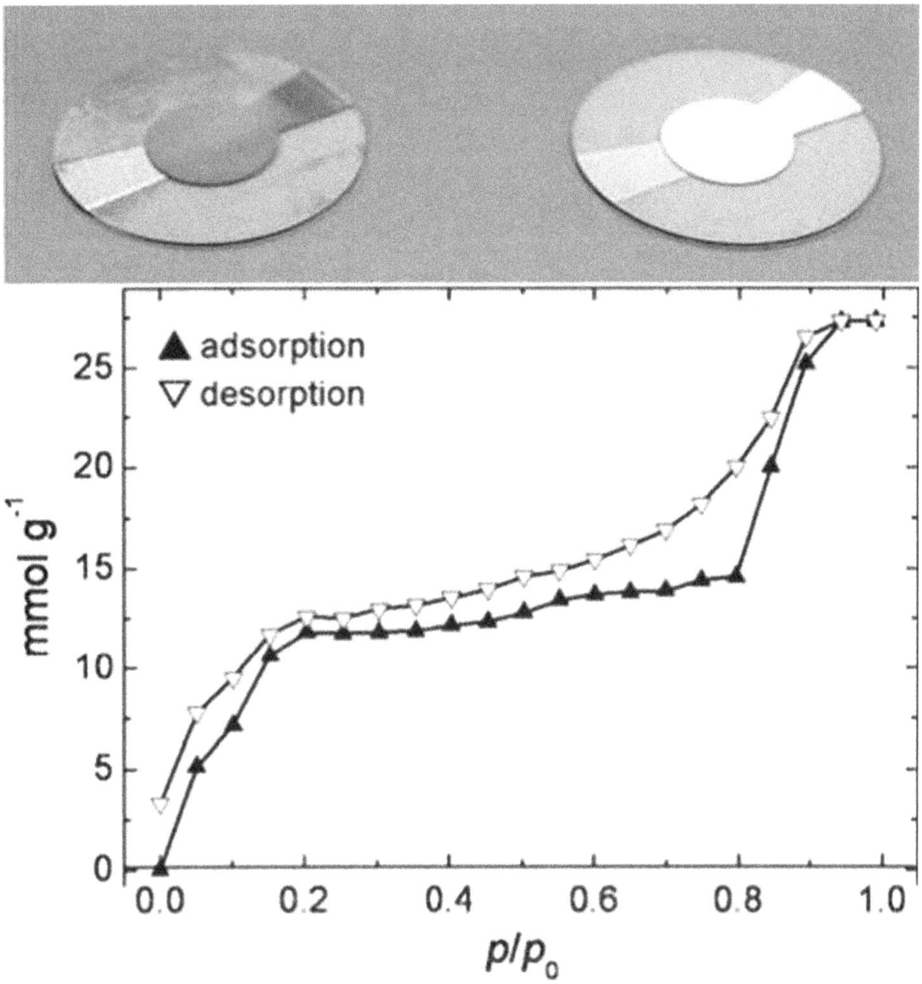

Figure 3. (Top) QCM devices after growth of [Cu_3 (BTC)$_2$(H_2O)$_3$ 3xH_2O]: opaque layer on the SAM-modified gold electrode (left); reference chip without SAM on the gold surface (right). (Bottom) Water sorption isotherm of a thin film of [Cu_3(BTC)$_2$] recorded at 294 K with the QCM setup. Reproduced with permission from reference (32). Copyright 2008 Elsevier.

Stress-Induced Chemical Detection

Allendorf et al. (33) have been interested in microcantilever-based sensing system by considering HKUST-1 as a test MOF. Indeed, because of the repulsive or attractive interactions between the guest framework and the host molecules, the lattice parameters of the MOF may enlarge or decrease when molecules are introduced into the pores. At the MOF film-to-substrate interface, a mechanical stress can be observed, which can cause the substrate to bend. The scientists were able to create a reactive, selective, reversible sensor for water (see Figure 4), methanol, and ethanol vapors using this method, but N_2 and O_2 elicited no response.

In terms of microcantilevers, by following the stepwise deposition technique in the liquid phase or straight growth in a base solution at room temperature (as Biemmi et al. (32) demonstrate),

HKUST-1 films were grown. Before they were used, the microcantilevers were layered with gold and functionalized with SAM.

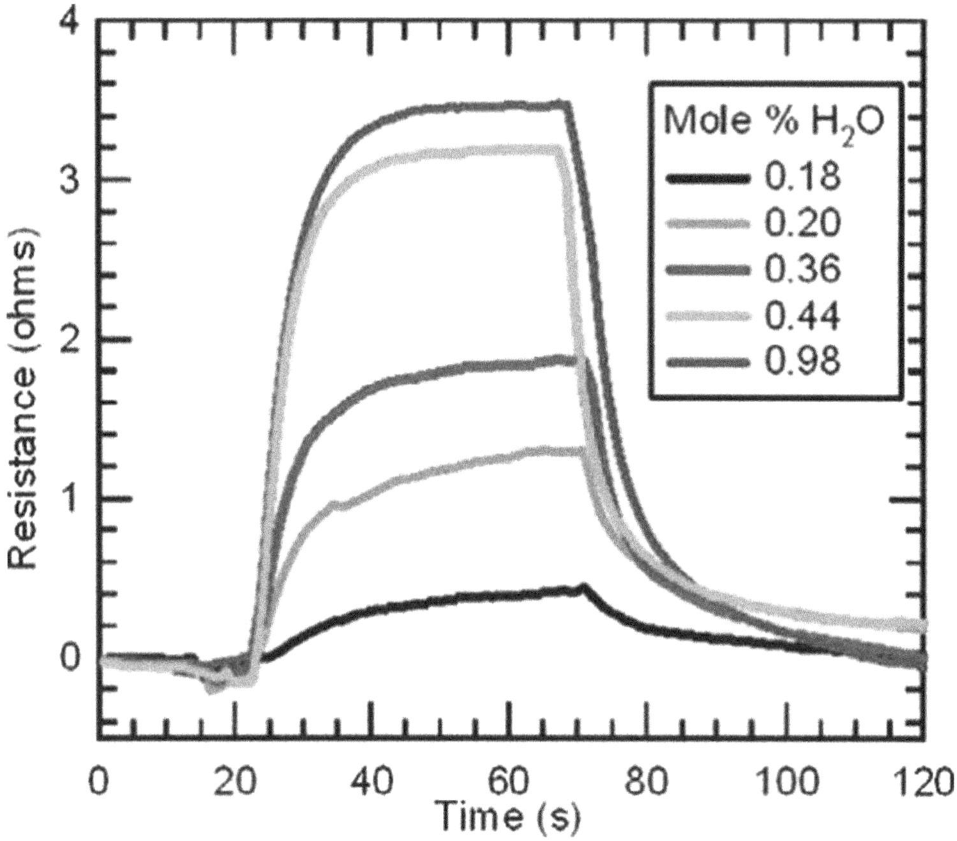

Figure 4. Temporal response of the cantilever piezoresistive sensor to water vapor diluted in N_2 (room temperature, 1 atm). Reproduced with permission from reference (33). Copyright 2008 American Chemical Society.

Devices Based on Optical Properties

In the sensing field, several optical qualities (such as color and refractive index) are often considered. Lu and Hupp et al. discovered that the hue pattern of a MOF film can be predetermined from its interference properties. The authors used a silicon substrate for growing micrometer-thick homogenous ZIF-8 films. The film was produced by recurrently dipping the substrate in a freshly produced mother solution. The colorless film showed a variety of colors because of the constructive and destructive interferences inside the film, which were dependent on the thickness.

When the films are formed on glass surfaces, ultraviolet–visible spectrophotometry can be used to precisely quantify this effect. The packing degree can influence the overall refractive index when constructing MOF films by assembling preformed colloids. The refractive index is also affected by the swelling and contracting of the film (in cases where a flexible MOF was utilized) and by filling pores with solvents. Environmental ellipsometry (*34*) was used to study MOF films; with this technique, the optical index and thickness of the film can be measured in situ in a controlled atmosphere containing gas or vapors at a specific concentration. The ability to extinguish fluorescence has also been demonstrated (*35*). When submerged in a solution of diethylamine, the MOF's fluorescence intensity drops, but it is not in a linear fashion when the concentration of analyte rises. In other analytes, no reaction is elicited by methanol, acetone, or aniline. However, strong fluorescence emitters are required for applications as thin films, which make this mechanism less intriguing than others.

Most studies have used the MOF as a sensor. The MOF simultaneously adsorbs the analyte and generates a detectable response. Silver nanoparticles supply the response in case of nonselective sensing device, and it increases the selectivity. By using localized surface plasmon resonance, minute variations in the refractive index of bulk silver can be measured. The gaseous molecule accumulations around the nanoparticles could trigger these alterations. The selective storage of one type of gas molecules is improved by the MOF coating surrounding the nanoparticle. By using the liquid phase epitaxy growth, the authors grew a covering of HKUST-1 on their array of silver nanoparticles (see Figure 5). A COOH-terminated, thiol-based SAM was used to initially functionalize the nanoparticles.

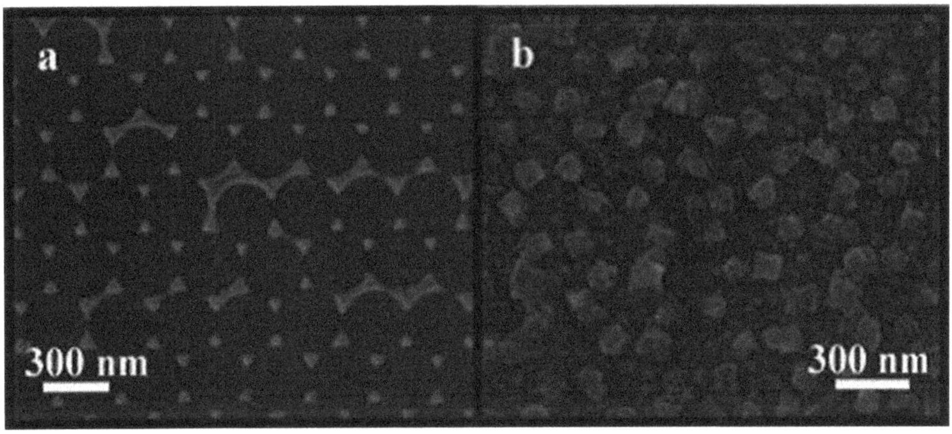

Figure 5. Scanning electron microscope (SEM) images of (a) a triangular Ag nanoparticle array fabricated by nanosphere lithography on a glass coverslip and (b) Ag nanoparticle array coated with 20 cycles of polycrystalline HKUST-1 film. Reproduced with permission from reference (35). Copyright 2010 American Chemical Society.

CO_2 was chosen as the analyte because HKUST-1 can adsorb it in considerable amounts. The effect of coating thickness was explored, and it was discovered that 37 cycles of deposition were ideal. The MOF thin film's crystallinity was validated by X-ray diffraction, and the sensing signal was increased 14-fold. The signal was successfully calibrated in relation to the CO_2 content in an N_2 flow. This approach has a substantial benefit in that it is not limited to a single MOF or analyte; various sensing devices for different gases might simply be created using other MOFs that are selective to different analytes.

Other Mechanisms

Andres et al. (*36*) have shown that MOF films are capable of being employed as capacitive humidity sensors. A copper net-supported HKUST-1 membrane was initially produced by the same group (Figure 6). A substrate of copper slice is used and through evaporation, the aluminum electrodes were deposited on the film, as observed from the optical study (Figure 7).

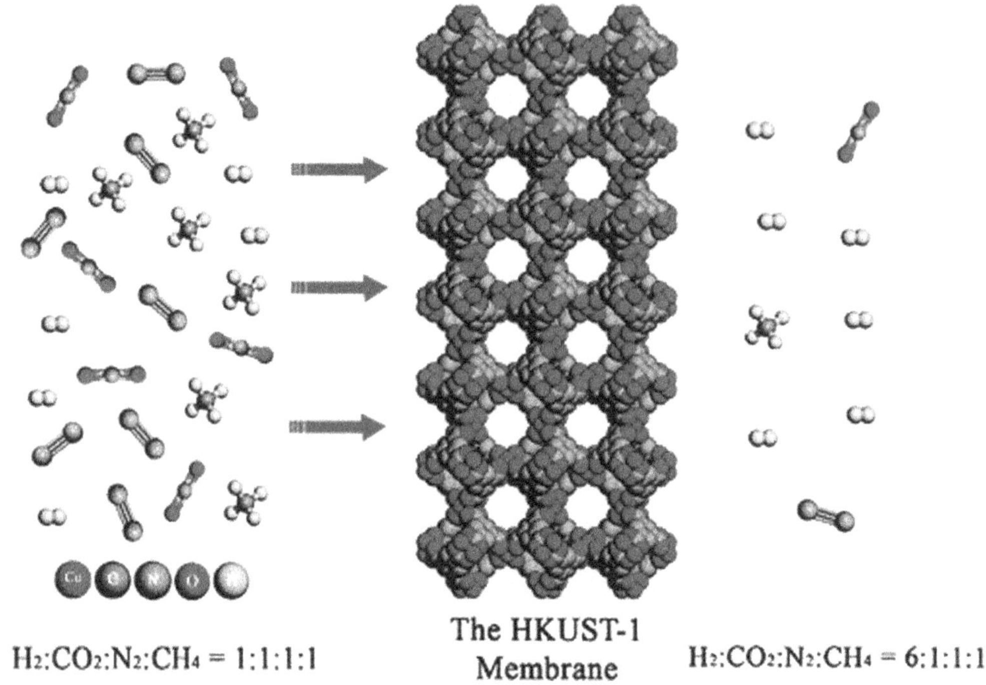

Figure 6. Copper net-supported $Cu_3(BTC)_2$ membranes have been successfully synthesized by means of a "twin copper source" technique. Reproduced with permission from reference (36). Copyright 2009 American Chemical Society.

Figure 7. Optic micrographs of the (a) copper net and (b) net-supported $Cu_3(BTC)_2$ membrane; SEM images of (c) the surface and (d) a cross section of the membrane. Reproduced with permission from reference (36). Copyright 2009 American Chemical Society.

The film's capacitance was tested at a variety of relative humidity settings, and a linear response of carbon was observed when the relative humidity changed. In addition, the measurements had good repeatability and stability over time. Transduction processes for sensing with MOFs have been considered, but they have not been applied to films. The development of MOF-based detecting devices would benefit greatly from manufacturing such MOFs into thin films.

Selectivity and Introduction of Artificial Antibodies

Although MOFs have great potential to be used in real-time sensing applications, their selectivity is still not at an acceptable limit. To increase the selectivity, researchers have started to use molecularly imprinted polymers (MIPs) as artificial antibodies (*37–39*). MIPs are gaining a lot of scholarly attention because of their predetermined selectivity toward the analytes, and they have a higher sensitivity and stability than biological antibodies (*40–43*). By incorporating nanosized materials, their performance can be greatly enhanced (*44–46*). Figure 8 shows how MIPs generally form.

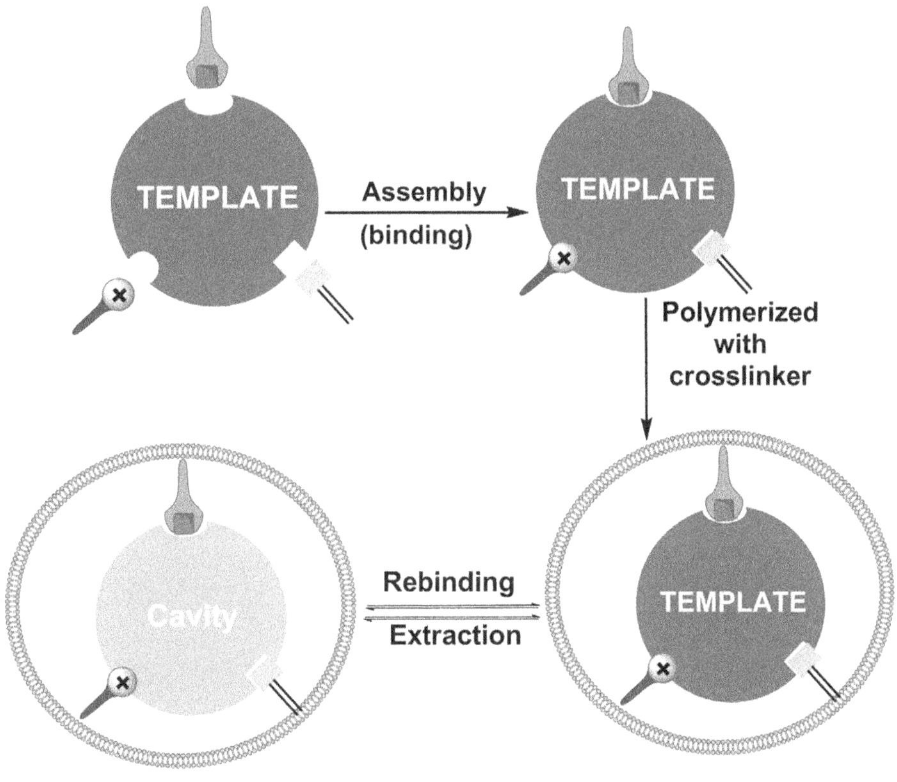

Figure 8. General protocol for fabrication of molecularly imprinted polymers (MIPs). Reproduced with permission from reference (45). Copyright 2016 Elsevier.

Typically, there are two key methods for fabricating MIPs, which are based on noncovalent–covalent interactions between the functional monomer and the template molecules. Covalent imprinting is known as the stoichiometric imprinting technique, and it only requires the functional monomer residues to be in the cavities. Table 1 represents the MOF-MIP composite used to detect several drugs and biomolecules.

Table 1. Application of MOF-MIPs as Sensors

Sl No.	*Template*		*Technique*	*LOD*	*Linear Range*	*Reference*
	Drug	Furosemide	Differential pulse voltammetry (DPV)	8.0 nM	0.08-100 μM	(*47*)
		Nitrofurazone	DPV	3×10^{-9} M	$5 \times 10^{-9} - \times 10^{-5}$ M	(*48*)
		Diclofenac	DPV	0.1 μM	6.5 μM to $1.5 \times 10^{+3}$ μM	(*49*)
		17β-estradiol	DPV	6.19×10^{-15} mol L^{-1}	10^{-14} to 10^{-9} mol L−1	(*50*)
	Biomolecules	Patulin	DPV	0.0007 ng mL^{-1}	0.001 to 70.0 ng mL^{-1}	(*51*)
		cysteine enantiomers	DPV		10^{-13} to 10^{-7} mol L^{-1}	(*52*)
		Sarcosine	CV	0.4 pM	1-100 pM	(*53*)

Detection of Drugs by MOF-MIPs

Although drugs can save human lives, they often have side effects. People might suffer cancer-like diseases with trace level variations of drugs in the body. Thus, it is highly urgent to create trace level determination drugs. Generally, furosemide (2-(2-furylmethyl) amino-5-(sulfonylamino)-4-chlorobenzoic acid) is a derivative of anthranilic acid. When the renal tubules become unable to adsorb the salts and fluids present in urine, a significant urine output can be observed. Furosemide drugs are sometimes used as diuretics, and such an effect can be enhanced with the increase in the dosage. Because of these properties, furosemide has been a major cause of heart failure, congestion, and cirrhosis. Furosemide has classified under the category of stimulants, which are used in sports. As such, it can be illegally used to mask other drugs by reducing the concentration of urine. Coming up with a method of detecting furosemide is highly urgent (47).

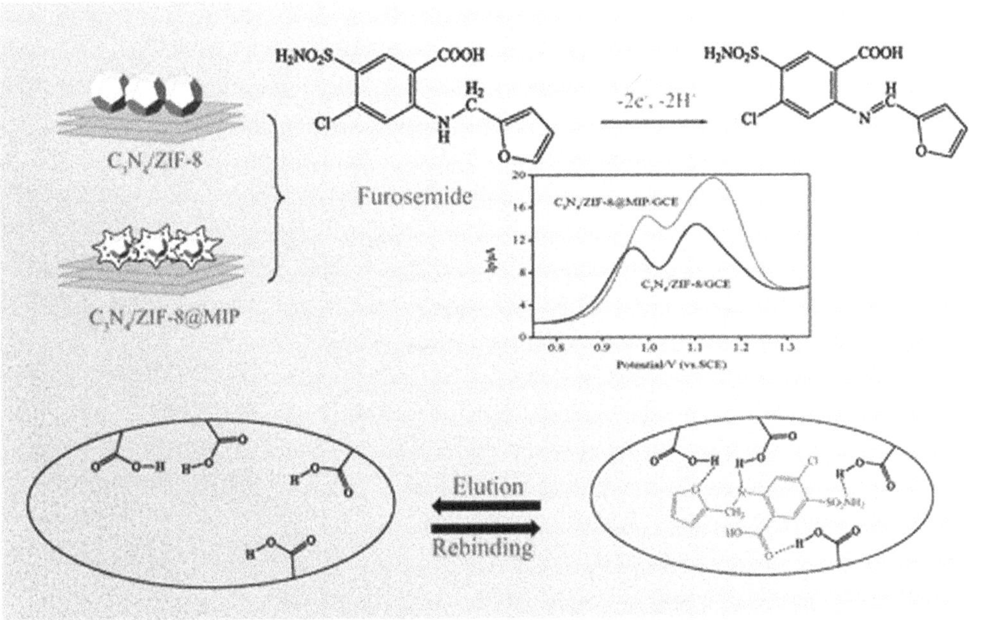

Figure 9. Schematic representation showing the preparation of C_3N_4/ZIF-8@MIP/glassy carbon electrode (GCE) and electrochemical detection of furosemide. Reproduced with permission from reference (48). Copyright 2018 American Chemical Society.

Scholarly research has been focusing on the use of electrochemical techniques in detecting biomolecules and other environmental pollutants. These techniques are useful because of their low cost, higher response rates, and easy technical operations (54–57). A C_3N_4 nanosheet MOF (ZIF-8) wrapped with an MIP sensor was developed by Wang et al. (48) to detect diuretic furosemide (Figure 9). By using precipitation polymerization, ZIF-8@MIP was synthesized in the presence of furosemide in the solution. The C_3N_4 nanosheets and ZIF-8@MIP were modified on the glassy carbon electrode (GCE). Prior to detection, furosemide was removed from the ZIF-8@MIP's membrane. By following the differential pulse voltammetry (DPV) technique, furosemide was detected. π-π stacking interactions and hydrogen bonding were also involved in this process. The surface and the conductivity of the electrode gets enhanced by the C_3N_4 nanosheet present within it. The cavities of MIP expand because of the ZIF-8 (Zn) framework structure, as broad improvement in the sensitivity for furosemide detection was observed.

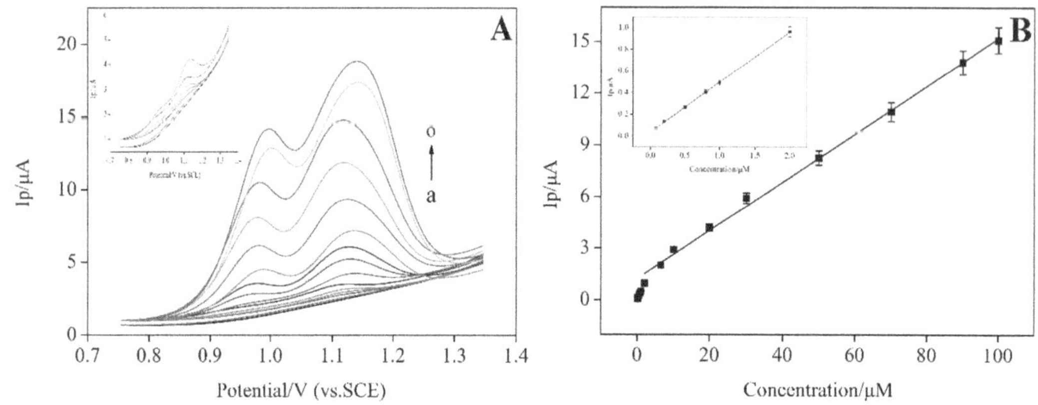

Figure 10. (A) DPV response of the C_3N_4/ZIF-8@MIP/GCE electrode towards different concentration of furosemide on in 0.04 M pH = 5.0 BR buffer; (B) Calibration curve representing the linear relationship between furosemide concentration and peak current of C_3N_4/ZIF-8@MIP/GCE (n = 3, 95% confidence level). Reproduced with permission from reference (47). Copyright 2015 Elsevier.

Figure 10 shows the DPV plot toward detecting furosemide at different increasing concentrations. The DPV response was found to increase with an increase in the concentration of furosemide. The current response has exhibited an excellent linear behavior in a range of 0.08–2 μM and 2–100 μM, (Figure 10[B] and inset). The limit of detection (LOD) was measured to be as low as 8.0 nM ($S/N = 3$).

Nitrofurazone (NFZ) (5-nitro-2-furaldehyde, NFZ) is a nitrofuran derivative. Generally, it acts as a broad-spectrum antibiotic. It is highly resilient to a variety of Gram-positive and negative bacteria, and it treats skin infections and trypanosomiasis. The NFZ antibiotic is potentially carcinogenic in nature, and it can have a harmful effect on humans in increased concentration.

Cheng et al. synthesized a molecularly imprinted electrochemical sensor for detecting NFZ based on biomass carbon (BC) and decorated with MOF-derived Cr_2O_3 and silver nanoparticles (Figure 11) (*48*).

The MIP was developed using precipitation polymerization, as shown in Figure 11(A). The silver nanoparticles were grown on the Cr-MIL-101 under its calcinations with BC, which has large active surface area. BC/Cr_2O_3/Ag nanocomposites were developed under the nitrogen atmosphere. The presence of Cr_2O_3/Ag in the MOF enhances the large specific surface area in the framework, as well as improves the electrical conductivity. The surface of the composite was exposed to MIP through its dropwise addition. NFZ was specifically recognized by hydrogen bonding (*48*).

Figure 12(A) shows the SEM and transmission electron microscopy (TEM) images of Cr-MIL-101, which have octahedral morphology and are uniform in nature (particle sizes of approximately 400 nm). In Figure 12(B), the silver nanoparticles, which were synthesized in situ on the exterior of Cr-MIL-101, were found to be uniformly distributed on the Cr-MIL-101's surface. The amorphous BC was synthesized by treating the walnut shells with nitric acid, as shown in Figure 12(C). The BC simultaneously increases the conductivity and enables electron transfer for improving the electrochemical response. After calcinations at 900 °C, BC/Cr-MIL-101/Ag as changed to BC/Cr_2O_3/Ag, which maintains the octahedral morphology of Cr-MIL-101, as shown in Figure 12(D). By following the precipitation polymerization method, a spherical MIP was obtained, as shown in Figure 12(E). Through hydrogen bonding, the MIP can specifically recognize NFZ, for which the sensitivity of the modified electrode increases. Figure 12(F) shows the SEM image of the electrode modified with $BC/Cr_2O_3/Ag/MIP$ shows that the rough surfaces have many voids, which facilitate the electrochemical response toward the detection of NFZ.

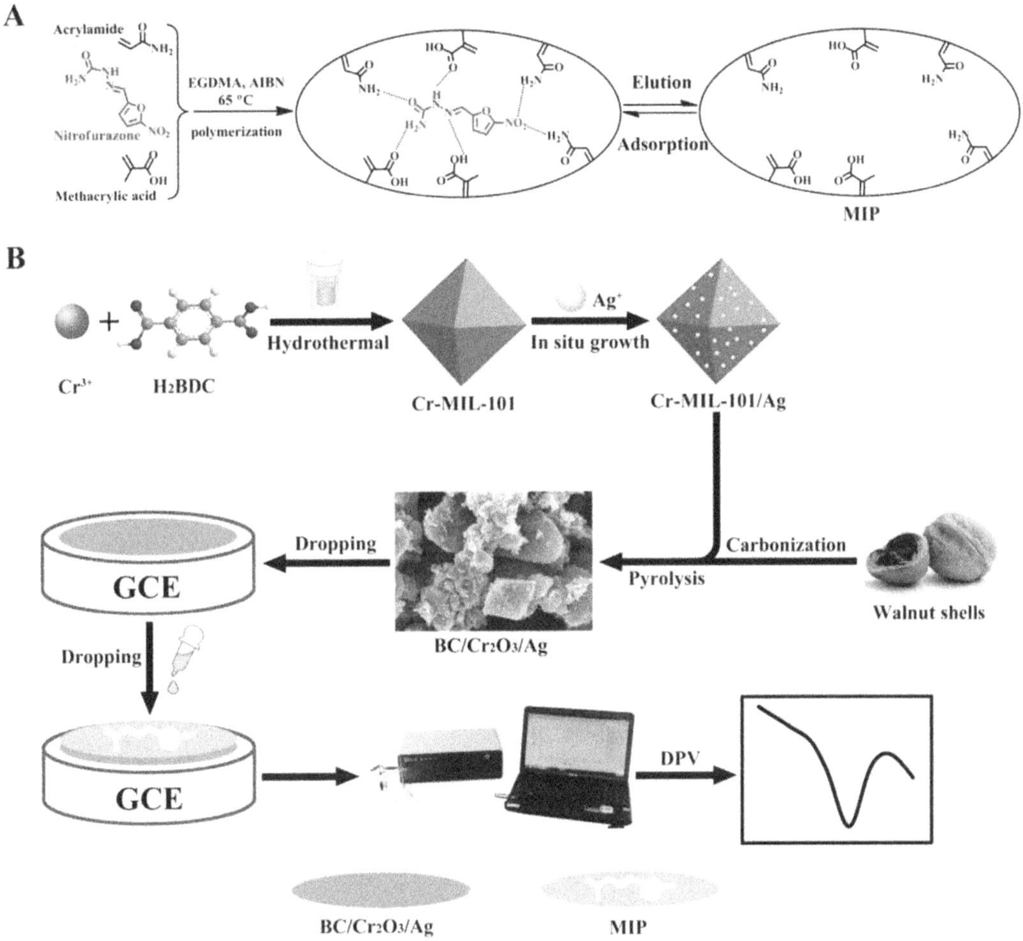

Figure 11. (A): Synthesis of MIP by precipitation polymerization. (B): The preparation process of BC/Cr_2O_3/Ag/MIP/GCE-modified electrode and the electroreduction. Reproduced with permission from reference (49). Copyright 2020 Elsevier.

Through the DPV technique, the $BC/Cr_2O_3/Ag/MIP/GCE$-modified electrode was used to detect NFZ under certain optimal experimental conditions. A series of concentrations of NFZ were taken to perform electrochemical detection. As shown in Figure 13(A) and (C), it was observed that with the increase of the concentrations of the NFZ, the corresponding current response gets enhanced. A linear range of 5×10^{-9} to 1×10^{-5} M was recorded in Figure 13(B) and (D). The limit of detection (LOD) was calculated to be is 3×10^{-9} M (S/N=3). Malekzadeh et al. detected diclofenac by developing an electrochemical sensor out of a Zr-MOF/MIP nanocomposite. The linear range and LOD were calculated to be 6.5 μM to $1.5\times10^{+3}$ μM and 0.1 μM, respectively (*50*). Duan et al. detected 17β-estradiol by developing a MOF/CNT and Prussian blue electrode. The linear range and LOD were calculated to be 10^{-14} to 10^{-9} mol L^{-1} and 6.19×10^{-15} mol L^{-1}, respectively (*51*).

Figure 12. SEM and transmission electron microscopy (TEM) spectra of each modified material prepared: Cr-MIL-101 (A), Cr-MIL-101/Ag (B), BC (C), BC/Cr_2O_3/Ag (D), MIP (E), BC/Cr_2O_3/Ag/MIP/GCE (F). Reproduced with permission from reference (49). Copyright 2020 Elsevier.

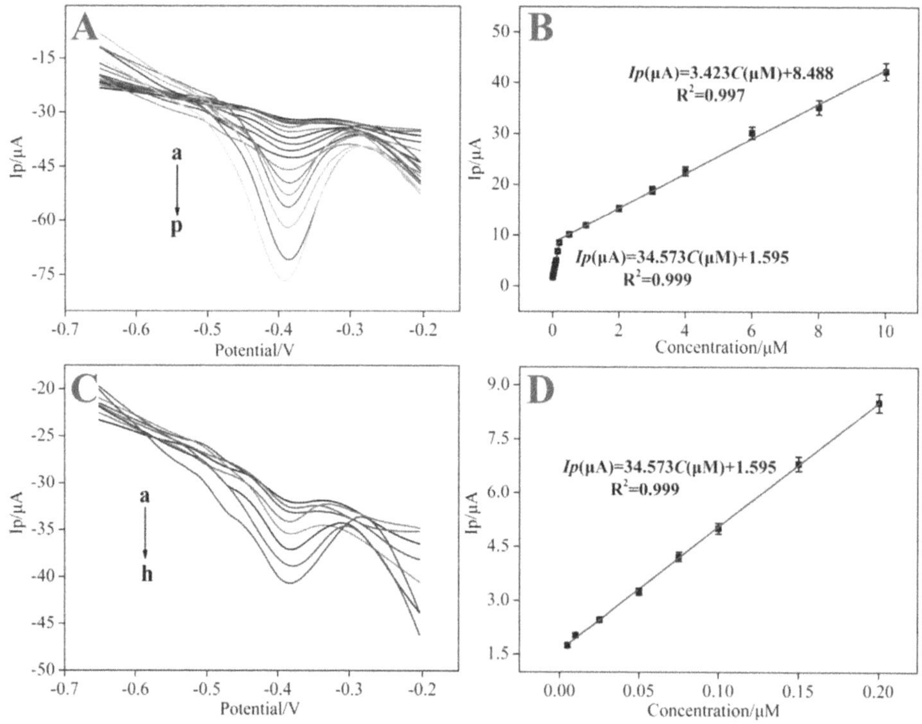

Figure 13. (A) and (C): DPV curve of BC/Cr2O3/Ag/MIP/GCE in phosphate buffer saline (PBS) (pH = 7) buffer containing a series of concentrations of NFZ under optimal experimental conditions (background deducted); (B) and (D): linear fitting between the current peak (μA) and the corresponding concentration (μM). Among them, the concentration (a–p) is 0.005, 0.01, 0.025, 0.05, 0.075, 0.1, 0.15, 0.2, 0.5, 1, 2, 3, 4, 6, 8, and 10 μM, respectively, and the concentration (a–h) is 0.005, 0.01, 0.025, 0.05, 0.075, 0.1, 0.15, and 0.2 μM, respectively. Reproduced with permission from reference (49). Copyright 2020 Elsevier.

Detection of Biomolecules by MOF-MIPs

Among different fruit toxins, patulin is one of the most important ones. It is also a type of toxic secondary metabolite, which is normally produced by certain species of fungus. Apples and several apple-processed products have the highest patulin content. Because it is very soluble in water, it easily contaminates aquatic environments. Patulin has acid-resistant and thermal stability properties. These properties make it difficult to completely remove it from food products after they have been manufactured. The mutagenic, neurotoxic, immunosuppressive, and teratogenic effects have made patulin a serious concern, as higher concentrations can damage the activity of some crucial enzymes, as well as DNA.

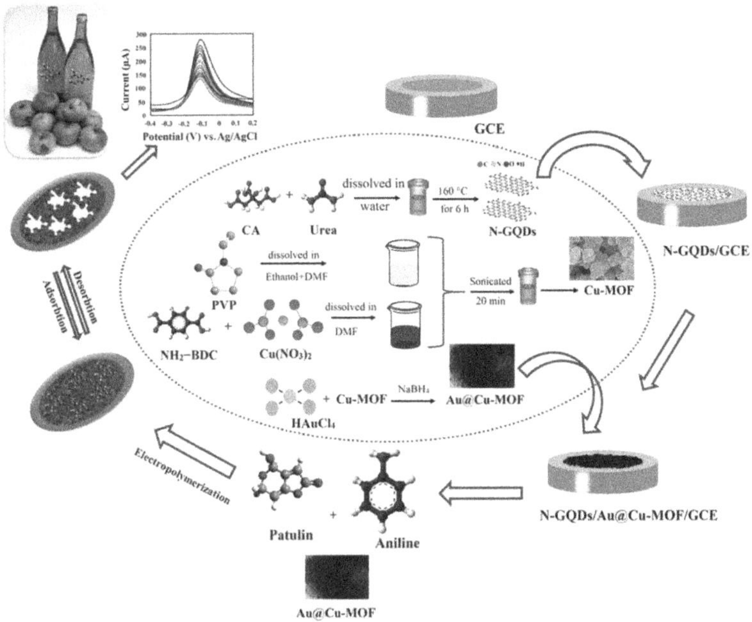

Figure 14. A schematic preparation of MIP/Au@Cu-MOF/N-DGQs/GCE and electrochemical detection of patulin. Reproduced with permission from reference (52). Copyright 2020 Elsevier.

Hatamluyi et al. have developed an ultra-sensitive molecularly imprinted electrochemical sensor based on Au@Cu-MOF/N-GQDs to detect patulin (Figure 14) (52). N-GQDs and Au@Cu-MOF are utilized in the GCE's surface, as they show greater potential for sensor applications. Au@Cu-MOF gives the MIP an enhanced surface area and can be used as an electroactive species in electrochemical analysis. The Au@Cu-MOF/N-GQDs/GCE was developed through the electropolymerization method, where the aniline and patulin were used as a functional monomer and template, respectively.

Through the DPV technique, a quantitative analysis of patulin was performed. Different concentrations of patulin were taken for the study. With the increased concentration of patulin, the MIP/Au@Cu-MOF/N-GQDs/GCE response decreased, as shown in Figure 15(A). This happens because the patulin molecules have occupied the imprinted cavities. Figure 15(B) shows a linear range of 0.001–70.0 ng mL^{-1} that was observed toward the detection of patulin. The LOD and limit of quantitation (LOQ) were calculated to be 0.0007 and 0.0023 ng mL^{-1}, respectively. With the MIP/Au@Cu- MOF/GCE electrode, the same procedure was followed to detect patulin. To study the influence of the N-GQDs, MIP/Au@Cu-MOF/GCE electrode was utilized to detect patulin by following the similar approach.

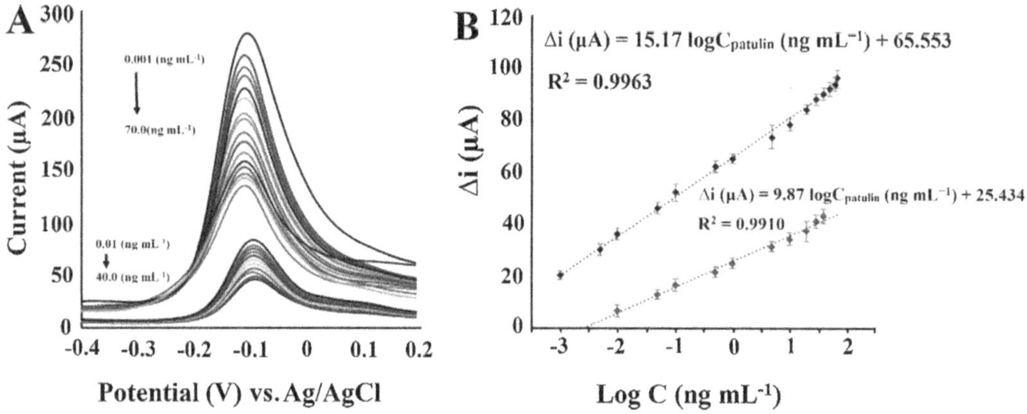

Figure 15. (A) DPV study of MIP/Au@Cu-MOF/N-GQDs/GCE and MIP/Au@Cu-MOF/GCE electrode in PBS solution (B) Linear relationships of the current difference (Δi) and logarithm concentration of patulin for the MIP/Au@Cu-MOF/N-GQDs/GCE (blue line) and MIP/Au@Cu-MOF/GCE (red line). Reproduced with permission from reference (52). Copyright 2020 Elsevier.

A comparison shown in Figure 15(A) and (B) shows that the current change for MIP/Au@Cu-MOF/GCE was very weak and exhibits a low linear range. Hence it is clear that the MIP/Au@Cu-MOF/N-GQDs/GCE benefits from the synergistic effects of N-GQDs and Au@Cu-MOF.

Duan et al. fabricated an MIP electrochemical sensor from Prussian blue-porous carbon-CNTs hybrids modified glassy carbon electrode (Figure 16) (53). The electrode was used for selective and sensitive recognition of cysteine enantiomers. Among different non-essential amino acids, L-Cystine has greater importance in metabolism, protein synthesis, and detoxification reactions. Different amino acid enantiomers show different pharmacological, biochemical, and physiological effects. As such, it is necessary to detect and separate amino acid enantiomers. The two examples of amino acid enantiomers present can have diverse effects (i.e., one might have positive effects, and another can have serious side effects to the organism) (*51*).

The porous 3D imprint enhances the electrical conductivity, electroactive area, and electron transfer facility. This study was conducted using polypropylene (PPy) as functional monomers and cystine as templates, and conductive permselective imprinted thin films were developed using the MIP. The process involved simple overoxidation and dedoping processes of PPy. This event performed under the medium. It led to the removal of template cysteine molecules. Positively charged enantiomers cysteine was recognized because of the imprinted cavities inside it.

MOF-Based MIP Film

MIP-based thin films have gained a lot of scholarly attention. After incorporating the MOF into the MIP, the film can provide additional advantages. This type of film will be particular interest for sensing applications. Iskierko et al. fabricated a sacrificial MOF that supported a surface-created, conducting polymer-based MIP film for the Lipocalin determination (*58*).

This method is very productive, simple, and it has the potential to be used in real time applications in near future.

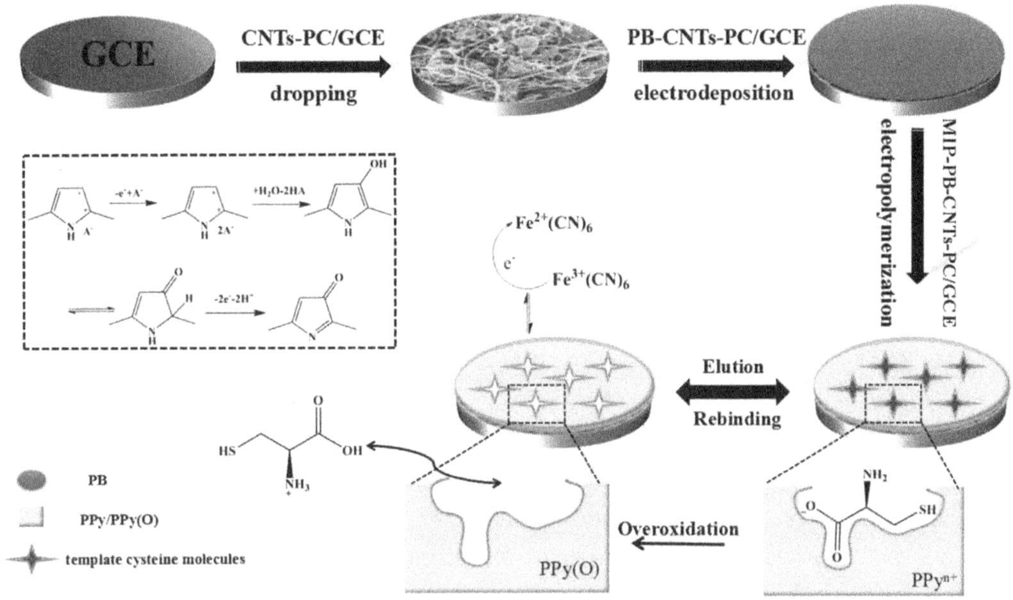

Figure 16. Schematic representation showing the process involved in MIP-PB-PC-CNTs/GCE preparation. Reproduced with permission from reference (53). Copyright 2019 Elsevier.

Conclusion and Future Scope

In summary, we can understand that MOF-based MIP sensors have gained scholarly attention. Although they have great potential, only a few MIP-MOF sensors have been reported so far. They have been employed to sense various analytes of environmental and biomedical activities. This innovative porous MIP-MOF hybrid composite with high sensitivity and selectivity has been combined with optical, electrochemical, or mass spectroscopic transducers in sensing. Additionally, the sensors have a high surface area, high transfer rate, and porosity, as well as provide an enhanced stability with its high adsorption capacity. Among all the reported sensors, electrochemical sensors are the most popular. Core-shell nanostructures using MOFs as the core and MIP layers as shells are of particular interest because of their enhanced sensing performances (i.e., broad detection range and low LOD values). Despite this, the sensing mechanism of MOF-MIP sensors has not been fully explored. The outcomes of the literature so far provide many opportunities for fabricating low-cost, highly sensitive, selective portable sensors with many practical applications. In future, we may combine this type of materials with wearable or flexible technology, which can be very useful for the onsite detection of any analyte of interest. With increasing cases because of recent COVID-19 and post-COVID side effects such as mucormycosis, there is an urgent need for people to be able to monitor their symptoms outside the hospital. Thus, this MIP-MOF sensor has proven useability as a highly selective, stable receptor and can be highly useful as a home monitor.

Authors Declaration

T.R.D. and P.B. contributed equally to writing this chapter.

References

1. Li, H.; Eddaoudi, M.; O'Keeffe, M.; Yaghi, O. M. *Nature* **1999**, *402* (6759), 276–279.
2. Kitagawa, S.; Kitaura, R.; Noro, S. *Angew. Chem. Int. Ed.* **2004**, *43* (18), 2334–2375.

3. Eddaoudi, M.; Kim, J.; Rosi, N. L.; Vodak, D.; Wachter, J.; O'Keeffe, M.; Yaghi, O. *Med. Sci.* **2002**, *295*, 469.
4. Wang, Z.; Cohen, S. M. *Chem. Soc. Rev.* **2009**, *38* (5), 1315–1329.
5. Yamada, T.; Kitagawa, H. *J. Am. Chem. Soc.* **2009**, *131* (18), 6312–6313.
6. Rasheed, T.; Rizwan, K.; Bilal, M.; Iqbal, H. M. N. *Molecules* **2020**, *25* (7), 1598.
7. Knobler, C. B.; Wang, B.; Yaghi, O. *Med. Sci.* **2010**, *327*, 846.
8. Furukawa, S.; Hirai, K.; Nakagawa, K.; Takashima, Y.; Matsuda, R.; Tsuruoka, T.; Kondo, M.; Haruki, R.; Tanaka, D.; Sakamoto, H.; Shimomura, S.; Sakata, O.; Kitagawa, S. *Angew. Chem. Int. Ed.* **2009**, *48* (10), 1766–1770.
9. Koh, K.; Wong-Foy, A. G.; Matzger, A. J. *Chem. Commun.* **2009**, *6162*.
10. Allendorf, M. D.; Betard, A.; Fischer, R. A. In *Metal-Organic Frameworks: Applications from Catalysis to Gas Storage*; Farrusseng, D., Ed.; Wiley-VCH Press, 2011.
11. Shekhah, O.; Chernikova, V.; Belmabkhout, Y.; Eddaoudi, M. *Crystals* **2018**, *8* (11), 412.
12. Lew, C. M.; Cai, R.; Yan, Y. *Acc. Chem. Res.* **2010**, *43* (2), 210–219.
13. Goujon, A.; Varret, F.; Escax, V.; Bleuzen, A.; Verdaguer, M. *Polyhedron* **2001**, *20* (11–14), 1339–1345.
14. Culp, J. T.; Park, J.-H.; Benitez, I. O.; Huh, Y.-D.; Meisel, M. W.; Talham, D. R. *Chem. Mater.* **2003**, *15* (18), 3431–3436.
15. Culp, J. T.; Park, J.-H.; Frye, F.; Huh, Y.-D.; Meisel, M. W.; Talham, D. R. *Coords. Chem. Rev.* **2005**, *2005*, 2642.
16. Al-Kutubi, H.; Gascon, J.; Sudhölter, E. J. R.; Rassaei, L. *ChemElectroChem* **2015**, *2* (4), 462–474.
17. Ameloot, R.; Stappers, L.; Fransaer, J.; Alaerts, L.; Sels, B. F.; De Vos, D. E. *Chem. Mater.* **2009**, *21* (13), 2580–2582.
18. Demessence, A.; Boissière, C.; Grosso, D.; Horcajada, P.; Serre, C.; Férey, G.; Soler-Illia, G. J. A. A.; Sanchez, C. *J. Mater. Chem.* **2010**, *20* (36), 7676.
19. Horcajada, P.; Serre, C.; Grosso, D.; Boissiere, C.; Perruchas, S.; Sanchez, C.; Férey, G. *Adv. Mater.* **1931**, *21*, 2009.
20. Petit, C.; Bandosz, T. J. *J. Colloid Interface Sci.* **2015**, *447*, 139–151.
21. Salem, M. M. K.; Braeuer, P.; Szombathely, Mv; Heuchel, M.; Harting, P.; Quitzsch, K.; Jaroniec, M. *Langmuir* **1998**, *14* (12), 3376–3389.
22. Farha, O. K.; Yazaydın, A. Ö; Eryazici, I.; Malliakas, C. D.; Hauser, B. G.; Kanatzidis, M. G.; Nguyen, S. T.; Snurr, R. Q.; Hupp, J. T. *Nat. Chem.* **2010**, *2* (11), 944–948.
23. Furukawa, H.; Go, N.; Ko, Y. B.; Aratani, N.; Choi, S. B.; Choi, E.; Yazaydin, A. O.; Snurr, R. Q.; O'Keeffe, M.; Kim, J.; Yaghi, O. *Med. Sci.* **2010**, *329*, 424.
24. Thomas, K. M. *Catal. Today* **2007**, *120* (3–4), 389–398.
25. Kampouraki, Z. C.; Giannakoudakis, D. A.; Nair, V.; Hosseini-Bandegharaei, A.; Colmenares, J. C.; Deliyanni, E. A. *Molecules* **2019**, *24* (24), 4525.
26. Khan, N. A.; Jhung, S. H. *Angew. Chem. Int. Ed.* **2012**, *51* (5), 1198–1201.
27. Yoon, M.; Srirambalaji, R.; Kim, K. *Chem. Rev.* **2012**, *112* (2), 1196–1231.
28. Dhakshinamoorthy, A.; Garcia, H. *Chem. Soc. Rev.* **2012**, *41* (15), 5262–5284.
29. Horcajada, P.; Gref, R.; Baati, T.; Allan, P. K.; Maurin, G.; Couvreur, P.; Férey, G.; Morris, R. E.; Serre, C. *Chem. Rev.* **2012**, *112* (2), 1232–1268.

30. D'Alessandro, D. M.; Kanga, J. R. R.; Caddy, J. S. *Aust. J. Chem.* **2011**, *64* (6), 718.
31. Ferreira, G. N. M.; Da-Silva, A. C.; Tomé, B. *Trends Biotechnol.* **2009**, *27* (12), 689–697.
32. Biemmi, E.; Darga, A.; Stock, N.; Bein, T. *Micropor. Mesopor. Mater.* **2008**, *114* (1–3), 380–386.
33. Allendorf, M. D.; Houk, R. J. T.; Andruszkiewicz, L.; Talin, A. A.; Pikarsky, J.; Choudhury, A.; Gall, K. A.; Hesketh, P. J. *J. Am. Chem. Soc.* **2008**, *130* (44), 14404–14405.
34. Biemmi, E.; Scherb, C.; Bein, T. *J. Am. Chem. Soc.* **2007**, *129* (26), 8054–8055.
35. Kreno, L. E.; Hupp, J. T.; Van Duyne, R. P. *Anal. Chem.* **2010**, *82* (19), 8042–8046.
36. Andrés, M. A.; Vijjapu, M. T.; Surya, S. G.; Shekhah, O.; Salama, K. N.; Serre, C.; Eddaoudi, M.; Roubeau, O.; Gascón, I. *ACS Appl. Mater. Interfaces* **2020**, *12* (3), 4155–4162.
37. Maspoch, D.; Ruiz-Molina, D.; Wurst, K.; Domingo, N.; Cavallini, M.; Biscarini, F.; Tejada, J.; Rovira, C.; Veciana, J. *Nat. Mater.* **2003**, *2* (3), 190–195.
38. Roy, E.; Patra, S.; Tiwari, A.; Madhuri, R.; Sharma, P. K. *Biosens. Bioelectron.* **2017**, *89* (1), 234–248.
39. Patra, S.; Roy, E.; Madhuri, R.; Sharma, P. K. *Environ. Sci. Technol.* **2015**, *49* (10), 6117–6126.
40. Patra, S.; Roy, E.; Saha, S.; Kumar, D.; Madhuri, R.; Sharma, P. K. *Chem. Eng. J.* **2017**, *321*, 195–206.
41. Patra, S.; Roy, E.; Madhuri, R.; Sharma, P. K. *Chem. Eng. J.* **2016**, *304*, 259–270.
42. Patra, S.; Roy, E.; Madhuri, R.; Sharma, P. K. *Biosens. Bioelectron.* **2015**, *66*, 1–10.
43. Patra, S.; Roy, E.; Madhuri, R.; Sharma, P. K. *R.S.C. Adv.* **2014**, *4*, 56690–56700.
44. Roy, E.; Patra, S.; Madhuri, R.; Sharma, P. K. *Talanta* **2014**, *120*, 198–207.
45. Patra, S.; Roy, E.; Madhuri, R.; Sharma, P. K. *Anal. Chim. Acta* **2016**, *918*, 77–88.
46. Kumar, S.; Karfa, P.; Patra, S.; Madhuri, R.; Sharma, P. K. *R.S.C. Adv.* **2016**, *6* (43), 36751–36760.
47. Patra, S.; Roy, E.; Madhuri, R.; Sharma, P. K. *Biosens. Bioelectron.* **2015**, *15*, 1–10.
48. Wang, Y.; Cheng, J.; Liu, X.; Ding, F.; Zou, P.; Wang, X.; Zhao, Q.; Rao, H. *ACS Sustainable Chem. Eng.* **2018**, *6* (12), 16848–16858.
49. Cheng, J.; Li, Y.; Zhong, J.; Lu, Z.; Wang, G.; Suna, M.; Jiang, Y.; Zou, P.; Wang, X.; Zhao, Q.; Wanga, Y.; Rao, H. *Chem. Eng. J.* **2020**, *398*, 125664.
50. Malekzadeh, M.; Mohadesi, A.; Karimi, A. M.; Ranjbar, M. *Anal. Bioanal. Electrochem.* **2020**, *12* (3), 402–414.
51. Duan, D.; Si, X.; Ding, Y.; Li, L.; Ma, G.; Zhang, L.; Jian, B. *Bioelectrochemistry* **2019**, *129*, 211–217.
52. Hatamluyia, B.; Rezayic, M.; Beheshtie, H. R.; Boroushakia, M. T. *Sens. Actuators B* **2020**, *318*, 128219.
53. Duan, D.; Yang, H.; Ding, Y.; Li, L.; Ma, G. *Electrochim. Acta* **2019**, *302*, 137–144.
54. Tang, P.; Wang, Y.; He, F. *J. Saudi Chem. Soc.* **2020**, *24* (8), 620–630.
55. Das, T. R.; Sharma, P. K. *Microchem. J.* **2019**, *147*, 1203–1214.
56. Das, T. R.; Jena, S. K.; Madhuri, R.; Sharma, P. K. *Appl. Surf. Sci.* **2018**, *449*, 304–313.
57. Das, T. R.; Sharma, P. K. *Microchem. J.* **2020**, *156*, 105850.
58. Iskierko, Z.; Sharma, P. S.; Prochowicz, D.; Fronc, K.; D'Souza, F.; Toczydłowska, D.; Stefaniak, F.; Noworyta, K. *ACS Appl. Mater. Interfaces* **2016**, *8* (31), 19860–19865.

Chapter 7

Zinc-Based Metal–Organic Framework for Heavy Metal Sensing

Afsana Khan,[1] Mamta Giri,[2] Kalpna,[2] Smita S. Kumar,[2] and Sonia Bansal[*,1]

[1]Department of Physics, J.C. Bose University of Science & Technology, YMCA, Faridabad, Haryana 121006, India

[2]Department of Environmental Sciences, J.C. Bose University of Science & Technology, YMCA, Faridabad, Haryana 121006, India

[*]Email: soniabansal248@gmail.com

Increasing toxicity due to organic and inorganic chemicals, heavy metals, and other factors in the environment is negatively affecting humans and the ecosystem. Among various heavy metal ion contaminants, arsenic, cadmium, lead, mercury, and chromium are considered very harmful to living organisms. This creates a demand for new sensing techniques for the selective determination of toxic heavy metal ions. A variety of effective detection techniques have been developed and investigated. Zinc-based metal–organic frameworks (MOFs) have gained worldwide attention for the same. They show outstanding sensing performance and are the best candidate for the detection of various organic and inorganic chemical species such as heavy metal ions, biomolecules, and gas molecules. The use of MOFs for heavy metal sensing has become an interesting and promising field for research and development. Because of their three-dimensional organic–inorganic complexes, large surface areas, highly porous structures, and high-density adsorptive sites, MOFs have attracted considerable attention among researchers developing sensors for the detection of heavy metal ions. This chapter is a systematic review of the synthesis and application of Zn-based MOFs for effective heavy metal detection in various mediums.

Introduction

In the last few decades, rapid population growth and industrial development have created unacceptable levels of environmental pollution, which is not only hazardous to human health but also to our environment and biological systems. Common pollutants, including heavy metal ions such as arsenic, cadmium, chromium, lead, mercury, and iron, are discharged from many industries in huge quantities daily; these pollutants later mix with our water and contaminate our aquatic systems. This is now becoming a significant global concern. Heavy metal ions are highly toxic and nonbiodegradable (*1–3*). Their presence in environmental and biological systems has several side effects on human health.

Nowadays, the heavy metal As(III) is significant because of its high toxicity toward human health and the environment (*4*, *5*). Many sources introduce As(III) into the environment, including natural, anthropogenic, industrial, and administered sources. It enters the human body via ingestion, inhalation, and skin absorption and affects our integumentary, nervous, respiratory, and cardiovascular systems. The World Health Organization stated that the permissible limit of As(III) in drinking water is 10 μg/L, and the daily intake of As per kilogram of body weight must not be more than 2 μg/L.

Industries discharge deadly poisonous Hg(II) ions as well, which are transformed into neurotoxic methylmercury via bacteria and micro-organisms present in the aquatic system. They enter the human body when we consume seafood. According to the International Agency for Research on Cancer, methylmercury is carcinogenic and causes disorders in nervous functions, delayed developmental skills, and many other negative outcomes. According to the U.S. Environmental Protection Agency, the permissible limit of Hg(II) in drinking water is 2 ppb.

A heavy toxic metal that is well known for its severe effect even at very low concentrations is cadmium. Usually, the main source of cadmium toxicity in the environment is industry. It enters living organisms and accumulates via the food chain, and it alters the function of various important organ systems (*6–8*). Its permissible limit in drinking water is 0.003 mg/L, as established by the U.S. Environmental Protection Agency (*9*).

Lead is normally found in our natural environment, but as we know, the presence of anything in excessive amounts always creates problems. In the case of toxic heavy metals, this is a serious concern. The amount of lead in the environment is increasing due to anthropogenic activities including fossil fuel burning in automobiles and domestic burning of solid fuels. Lead acts as a threatening poison, especially for children 6 years or younger. The limit of lead in drinking water should not exceed 5 μg/L (*10*, *11*).

Chromium is also one of the toxic heavy metals present in our environment due to various sources. These sources can be natural or anthropogenic, including soil and rock erosion and industrial activities (iron and steel production, wearing of brakes and tires, ore processing plants, etc.). Cr(III) and Cr(VI) are the most common forms of chromium in the environment, but Cr(VI) is more soluble. Chromium is a carcinogenic and mutagenic ion that has severe effects on the human body and can create allergies (*12*). The permissible limit of chromium in drinking water set by the World Health Organization is 2 μg/L.

Researchers have explored different methods for the detection of heavy metal ions. Recent studies revealed that zinc-based metal-organic frameworks (Zn-MOFs) are well suited for sensing toxic heavy metal ions (*13*, *14*). Metal-organic frameworks (MOFs) are crystalline materials, an attractive class of coordination polymers, that consist of a combination of metal ions and organic linker groups (*15*, *16*). In the last few years, the chemistry of MOFs has developed significantly. MOFs are primarily prepared via solvothermal methods under controlled conditions. MOFs have large surface areas, high tenability, highly ordered structures, excellent light absorption, good fluorescence-sensing capability, and many other features as shown in Figure 1. Moreover, because of the high porosity and high surface areas of MOFs, they can integrate an abundance of molecules inside the framework and modify them with guest interaction (*17–19*). MOFs also have good thermal stability. Because of their tunable structural properties, MOFs are considered useful in energy storage, purification technology, and catalysis. They can be widely used in electrical conduction, molecular magnets, nanocatalysis, and so on (*20–26*). MOFs play an important role in functional applications because of their large internal pore volume and well-defined pore sizes. They have the highest specific surface area among porous materials. All these features indicate that Zn-MOFs have great potential for practical applications related to the sensing and analysis of toxic chemical substances and heavy metal ions in biological systems and the environment as shown in Table 1.

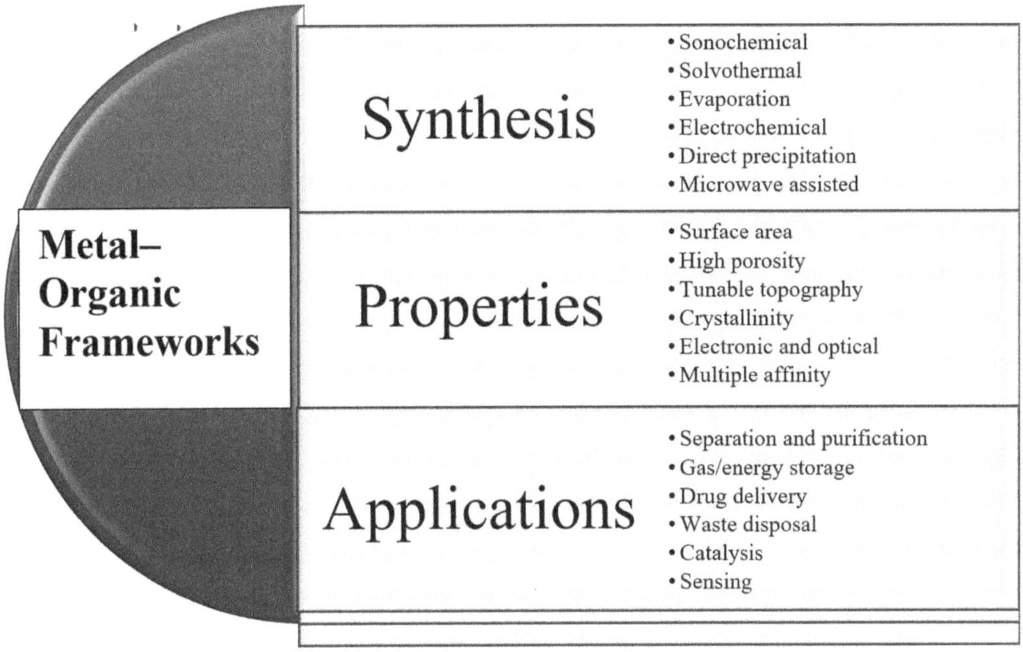

Figure 1. Representation of MOF.

Classification of MOFs

MOFs are classified into two groups based on their structures (27): rigid frameworks and flexible frameworks. Rigid frameworks are highly stable. During the adsorption and desorption of guest particles, they can retain their shape and are also used for molecular sieving (*28*). Flexible frameworks, on the other hand, are not stable. Their shape changes with changes in pressure and temperature. These frameworks also change with the insertion or removal of guest molecules, but they can retain their porous structure at high pressure (*29*) during the adsorption and insertion of gas molecules. There are some MOFs that exhibit a breathing effect during adsorption and desorption (*30–33*).

Synthesis of Zn-MOFs

Two major constituents of MOFs are metal ions and organic ligands. Different methods have been developed to synthesize MOFs, such as solvothermal/hydrothermal, unconventional, microwave-assisted, electrochemical, ultrasound, and layer-by-layer synthesis methods.

Table 1. List of Zn-MOFs for Heavy Metal Ion Detection

Serial no.	*MOF Used for Sensing*	*Ligands*	*Synthesis Method*	*Metal Ion Detected*	*Sensing Category*	*Detection Limit*	*Remarks*	*Ref.*
1	UMCM-4-NH_2	2-Amino-1,4- benzenedicarboxylic acid (NH_2-H_2BDC); 4,4′,4″-Tricarboxytriphenylamine (TPA)	Postsynthetic covalent modification	Fe^{3+}	Luminescent	10^{-6} mol L^{-1}	Fast response time (1 min)	(*106*)
2	ZnL·5H_2O·0.75DMF	H_2L = 9-(pyridin-4-yl)-9*H*-carbazole-3,6-dicarboxylic acid; DMF = *N,N*-dimethylformamide	Solvothermal method	Fe^{3+}	Luminescent	10,800 M^{-1}	Stable in the pH range of 4–9	(*107*)
3	$[Zn_2(OH)(sip)(bpe)(H_2O)]H_2O$	5-sulfoisophthalic acid; 1,2-bis(4-pyridyl)ethane	Hydrothermal method	Fe^{3+}	Luminescent	1.46 × 104 M^{-1}	Used at room temperature	(*108*)
4	$[H(H_2O)_8][DyZn_4(imdc)_4(im)_4]$	H_3imdc = 4,5-imidazole dicarboxylic acid; im = imidazole	Hydrothermal method	Fe^{3+}	Luminescent	9.29 × 10^5 M^{-1}	High sensitivity	(*109*)
5	$[Zn_5(hfipbb)_4(trz)_2(H_2O)_2]$ (NNU-1)	$H_2hfipbb$ = 4,4′-(hexafluoroisopropylidene)bis(benzoic acid); Htrz = 1H-1,2,3-tria-zole	Hydrothermal method	Fe^{3+}	Luminescent	0.20 mmol·L^{-1}	High selectivity	(*110*)
6	$[Zn(tpbpc)_2]$·solvent	Htpbpc=4′-[4,2′;6′,4″]-terpyridin-4′-yl-biphenyl-4-carboxylic acid	Hydrothermal method	Hg^{2+}	Luminescent	10^{-8} M	High selectivity	(*111*)
7	$\{[Zn(4,4'\text{-}AP)(5\text{-}AIA)]\cdot(DMF)_{0.5}\}_n$	4,4 AP = 4,4′-azopyridine; 5-AIA = 5-amino isophthalic acid; DMF = N,N′-dimethylformamide	Solvothermal method	Hg^{2+}	Luminescent	10^{-11} M	Detected Hg^{2+} ion in spiked environmental water samples	(*112*)
8	$\{[Zn_2(\mu_3\text{-}OH)(cpta)(4,4'\text{-}bipy)]\cdot H_2O\}_n$	4,4′-bipy = 4,4′-bipyridine	Solvothermal method	Cr^{3+} Cr^{6+}	Luminescent	5.55 × 10^{-6} M 6.91 × 10^{-6} M	Recycled at least four times	(*113*)

Table 1. (Continued). List of Zn-MOFs for Heavy Metal Ion Detection

Serial no.	*MOF Used for Sensing*	*Ligands*	*Synthesis Method*	*Metal Ion Detected*	*Sensing Category*	*Detection Limit*	*Remarks*	*Ref.*
9	$[Zn_2(4,4'\text{-nba})_2(1,4\text{-bib})_2]_n$	4,4′-H_2nba = 3-nitro-4,4′-biphenyldicarboxylic acid; 1,4-bib = 1,4-bis(imidazole-1-ylmethyl)benzene	Solvothermal method	Cr^{3+} Cr^{6+}	Luminescent	3.25 µM 3.8 µM	Recycled at least five times	(*114*)
10	$\{Zn_2(L)_2(DMF)_2H_2O\}_n$	L = 2,5-bis(phenylamino)-1,4-benzenedicarboxylic acid	Solvothermal method	Cd^{2+}	Luminescent	0.12 µM	High sensitivity	(*70*)

Solvothermal/Hydrothermal Method

The synthesis of MOFs mainly depends on the solvent, as the solvent affects the morphology of crystals. In the solvothermal synthesis method, a cluster of metal ions and organic ligands is heated with the solvent. The equipment used for heating is dependent on temperature. Below 400 K, glass vials are used; above this temperature, Teflon-lined autoclaves or bomb reactors are used (*34*). Hydrothermal reaction occurs when the temperature is higher than the boiling point of the solvent. On the other hand, a lower temperature leads to a nonisothermal reaction. It has been reported that some MOFs have been synthesized at ambient conditions (*35–37*).

Mechanochemical Method

MOFs are sometimes prepared using unconventional methods. Pichon et al. described the mechanochemical method, which does not use a solvent (*38*). This method has many advantages. It is an environmentally friendly process, and it has a high yield with a short reaction time. In this method, clusters of metal salt and an organic linker are ground without any kind of solvent. Some volatile compounds are formed in the reaction mixture. The final MOF is obtained with gentle heating of the byproducts (*39–41*).

Alternative Methods

MOFs are also prepared by some alternative methods, which are discussed in the following sections.

Microwave-Assisted Method

This method is generally used for quick synthesis of organic and nanoporous materials. The energy required for the reaction is obtained from microwave (MW) radiation. The advantages of this method include fast crystallization, narrow particle size distribution, and morphological controls (*30, 42–47*). The first MOF synthesized through the MW method was Cr-MIL-100, after which several other MOFs containing Al^{3+} and Cr^{3+} were also synthesized using this method. MOF formation is typically carried out at 100 °C with a reaction time exceeding 1 h. Compared to the conventional electrical heating process, MW radiation can form MOFs more quickly. IRMOF1 and HKUST-1 are the two widely studied MOFs that implemented MW synthesis (*48, 49*).

Electrochemical Method

Researchers at BASF were the first to synthesize MOFs through the electrochemical route (*50*). To prepare MOFs, they used cathode material and organic linkers. Instead of using metal salts, anodic dissolution was used, which acts as the metal source and helps supply metal ions continuously for the synthesis of the MOF. Hydrogen is released via this method. Protic solvents are used to avoid the gathering of metal ions on the cathode (*51*). MOFs such as $[Zn(MIm)_2]$ and $[Zn(BIm)_2]$, which contain ionic liquid as a linker, can also be synthesized through the electrochemical method (*52*).

Ultrasonic Method

In this method, MOFs are synthesized with the use of ultrasonic radiation (ranging from 20 kHz to 10 MHz). This method is very advantageous, as it is a rapid, environmentally friendly, and low-cost technique. This method can achieve reduced crystallization time and smaller particle size via homogeneous and accelerated nucleation (*53, 54*). When high-energy ultrasonic radiation interacts with liquid, it undergoes cavitation and provides energy (*55*). Qiu et al. reported the ultrasonic method to synthesize MOFs in ethanol solvent (*56*).

Layer-by-Layer Method

MOF thin films are prepared via this method, in which a solution of metal ion and organic linker is placed on the functionalized organic surface continuously. The orientation of thin film depends on the array of reactants added (*57*). The metal source used is the largest factor affecting growth of the MOF film (*58, 59*). Substrates that were functionalized with different functional groups showed high oriented growth (*60*).

In addition to these synthesis methods, other methods [such as chemical solution deposition (*61, 62*), postsynthesis modification (*63*), and ionothermal methods (*64, 65*)] have also been applied for MOF synthesis.

Application of Zn-MOFs as Heavy Metal Ion Sensors

Several Zn-MOFs have been extensively explored and reported on for their potential use as highly selective sensors for heavy metal ion detection (*66–101*).

Detection of Arsenic

Recently, a ratiometric fluorescent sensor with a multifunctional Zn-MOF loaded with acid phosphatase (ACP) and hemin (ACP/hemin@Zn-MOF) was reported for the detection of arsenate. It was created through one-step self-assembly of Zn^{2+} ions, terephthalic acid (BDC), 2-aminoterephthalic acid (NH_2-BDC), ACP, and hemin in one pot at room temperature, as shown in Figure 2 (*66*). This MOF is highly sensitive, providing a lower detection limit of 1.05 μgL^{-1} (*66*). In addition, Baghayeri et al. prepared a graphene oxide– and zinc-based MOF through a simple solvothermal method. This MOF exhibited good reproducibility with a detection limit of 0.06 ppb (*67*).

Detection of Cadmium

Among the heavy toxic metals that industries produce, cadmium is the most toxic because it influences our immune and nervous systems. It is extremely important to screen these contaminants using a simple analysis method because of their toxic and carcinogenic effects on human health. Zn-MOFs have been used to trace the concentration of cadmium by detecting the cadmium ion. Recently, Farahani and Safarifard reported a luminescent Zn(II)-based MOF with azine-decorated pores, $[Zn_2(H_2BDC)_2(4\text{-bpdh})]\cdot 3DMF$ (TMU-16). It can detect Fe^{3+} and Cd^{2+} with detection limits of 0.2 and 0.5 μM, respectively (*68*). This MOF has a remarkable fluorescence-quenching effect, as shown in Figure 3 (*68*). In addition, a three-dimensional Zn-MOF formulated as $\{Zn_2(L)_2(DMF)_2H_2O\}_n$ [where L is 2,5-bis(phenylamino)-1,4-benzenedicarboxylic acid] was prepared using the solvothermal method. The ion-exchange interaction between Cd^{2+} and Zn^{2+} is a possible sensing mechanism for the detection of Cd^{2+}. This MOF provided a cadmium detection limit as low as 0.12 μM (*69*).

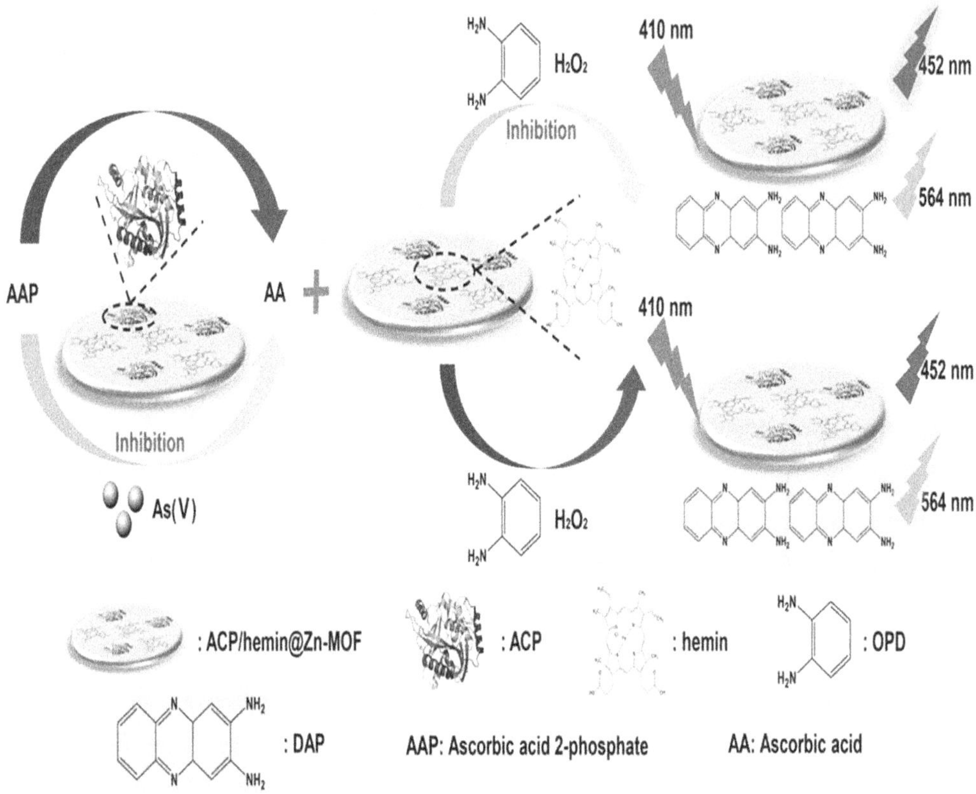

Figure 2. Illustration of sensing mechanism of ACP/hemin@Zn-MOF to detect arsenate. Reproduced with permission from reference (66). Copyright 2021 Elsevier.

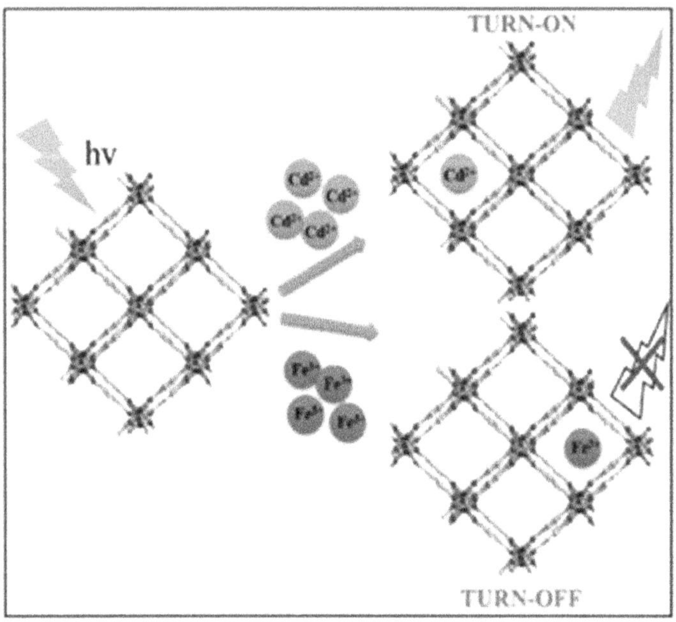

Figure 3. Schematic of fluorescence-quenching sensing mechanism of TMU-16 toward Fe^{3+} and Cd^{2+}. Reproduced with permission from reference (68). Copyright 2019 Elsevier.

A Zn-MOF with benzene-1,3-dicarboxylic acid as the organic linker was hydrothermally synthesized by Nazari et al. (*70*). This MOF is used as a selective nanoadsorbent for the extraction of cadmium ions. The detection limit for cadmium obtained by this MOF is 0.2 μg L^{-1} (*70*). A Zn(II)-based MOF, formulated as [Zn2(TPOM)(NDC)2]·3.5H2O [where TPOM is tetrakis(4-pyridyloxymethylene)methane and H2ndc is 2,6-naphthalenedicarboxylic acid] was successfully synthesized via hydrothermal reaction. The MOF can detect and differentiate Fe^{3+} and Cr^{6+} ions with high selectivity (*71*). Moreover, Roushani et al. synthesized an electrochemical sensor TMU-16-NH2 for the efficient detection of the Cd^{2+}ion using an MOF formulated as ([Zn2(NH2-BDC)2(4-bpdh)]·3DMF). This MOF exhibits a lower limit of detection of 0.2 μg L^{-1} (72).

Detection of Iron

Iron is a widespread heavy metal contaminant in our environment, and it may cause harm to living organisms. The recommended maximum permissible limit of iron in drinking water is as low as 0.3 ppm (*73*). Several Zn-MOFs have been synthesized for the detection of Fe^{3+} ions.

Moradi et al. synthesized the highly luminescent 3-D MOF $[Zn_2(oba)_2(bpy)]$ [where oba is 4,4′-oxybis(benzoicacid) and bpy is 4,4′-bipyridine] via the sonochemical method. This Zn-MOF is highly selective for Fe^{3+} ions in methanol, with a response time of less than 2 min and a limit of detection of 0.3 μM (*74*). In addition, Hou et al. proposed a Zn-MOF (ZnMOF-74) for highly selective and sensitive detection of Fe^{3+} in aqueous solution that is based on 2,5-dihydroxyterephthalic acid (DOBDC)-Zn2+ MOFs. The cation exchange between Fe^{3+} and Zn^{2+} formed a blue weak-fluorescent iron(III)-phenol salt complex that can be considered a possible fluorescence-quenching sensing mechanism, as shown in Figure 4 (*75*). This MOF can be used as a probe for the detection of Fe^{3+} with a good linear range of 0.1–100 μM and a lower detection limit of 40 nM, making ZnMOF-74 a suitable sensor (*75*).

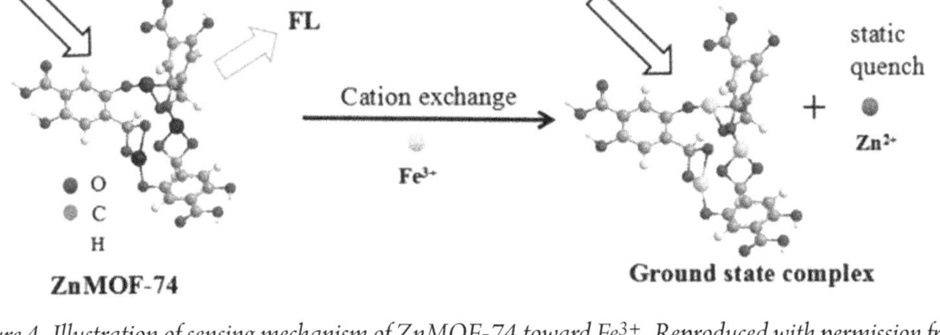

Figure 4. Illustration of sensing mechanism of ZnMOF-74 toward Fe^{3+}. Reproduced with permission from reference (75). Copyright 2019 Elsevier.

Zhang et al. prepared a chiral 3-D luminescent Zn(II)-MOF formulated as $[Zn(BPDPE)_2(dca)_2]_n$ [where BPDPE is 4,4′-bis (pyridy)diphenyl ether and H_2dca is dextro-camphoric acid] that provides fluorescence-sensing material for detecting metal ions. This MOF is highly selective and sensitive enough to detect Fe^{3+} ions through luminescence quenching, with a binding constant of 1.10×10^5 M^{-1} (*76*). An amine/azine-functionalized Zn-MOF, TMU-17-NH_2, was formulated as $[Zn(NH_2\text{-}BDC)(4\text{-}bpdb)]\cdot 2DMF$ [where NH_2-BDC is amino-1,4-benzenedicarboxylic acid and 4-bpdb is 1,4-bis(4-pyridyl)-2,3-diaza-2,3-butadiene] for highly selective and sensitive detection of Fe^{3+} in dimethylformamide solution. This MOF provides fast recognition of Fe^{3+} ions, with a response time of less than 1 min. The detection limit was estimated to be 0.7 μM (*77*).

Sun and Chen solvothermally synthesized a highly porous MOF, $\{[Zn_3(bpydb)_2(atz)_2(DMF)](DMF)_6\}_n$, using 4,4′-(4,4′-bipyridine-2,6-diyl) dibenzoic acid ($bpydbH_2$), 3-amino-1,2,4-triazole (Hatz), and $Zn(NO_3)_2\cdot 6H_2O$ as the starting materials. Because of its strong fluorescent emission and high porosity, this MOF behaves as a chemosensor for the selective detection of Fe^{3+} ions, with high selectivity and sensitivity in water (*78*). Du et al. prepared a luminescent Zn(II)-MOF, $[Zn_3(\mathbf{L})_2(bipy)(\mu_3\text{-}OH)_2]\cdot 3H_2O$ (**1**), where bipy is 4,4′-bipyridine and $\mathbf{H_2L}$ is 9H-carbazolyl-3,6-dicarboxylic acid, from the reaction of a rigid ligand $H_2\mathbf{L}$ and Zn(II) ions. Compound **1** created an interesting 3-D network with $[Zn_9(\mu_3\text{-}OH)_6(CO_2)_{12}]$ and exhibited high selectivity toward Fe^{3+} ions (*79*). Two new Zn-MOF formulated as $[Zn(H_2BPTC)](H_2O)_2$ and $[Zn_4(H_2BPTC)_2(SO_4)_2][(CH_3)_2NH_2]_4(H_2O)_4$ were synthesized using zinc salts and 3,3′,5,5′ biphenyltetracarboxylic acid (H_4BPTC) as building blocks via an anion modulate method. These two MOFs selectively detect Fe^{3+} ions (*80*).

Detection of Mercury

Invisible, toxic mercury ion pollution of water has an extremely adverse effect on the environment and living organisms, even at very low concentrations. Because of rapid industrialization, hazardous mercury (II) ions can accumulate in the water or soil. Thus, establishing quick and convenient sensing devices to detect mercury (II) ions in water is a priority to protect human health. Zn-MOFs have been extensively studied and tested for their use as sensors of Hg^{2+} ions (*81*–*89*).

Recently, Jiang et al. constructed the 3-D supramolecular network Zn-APA with uncoordinated carboxylic oxygen atoms, where APA is 5-aminoisophthalic acid, which exhibits high selectivity and fast fluorescence response to Hg^{2+} ions over other metal ions in aqueous solutions (*81*). Pankajakshan et al. solvothermally synthesized a water-stable zinc-based porous structure $[Zn(4,4'\text{-}AP)(5\text{-}AIA)]\cdot(DMF)_{0.5}]_n$, where 4,4′AP is 4,4′-azopyridine and 5-APA is 5-amino isophthalic acid. This MOF offers highly selective and sensitive detection of mercury ions in the presence of other interfering metal ions in water. The lower detection limit of this MOF for Hg^{2+} is 10^{-11} M, with a very high binding constant of 1.011×109 $M^{-1}s^{-1}$ (*82*). Khatun et al. prepared a new luminescent, pillared paddle-wheel Zn-MOF, $Zn_2(NDC)_2(DPTTZ)$, featuring naphthalene dicarboxylate (NDC) as antenna chromophores and an energy donor, while N,N′-di(4-pyridyl)thiazolo-[5,4-d]thiazole (DPTTZ) acts as an energy acceptor and light emitter. The photoluminescence of this MOF went through a significant red shift to 450 nm, followed by quenching toward Hg^{2+} in the presence of other transition metal ions (Mn^{2+}, Fe^{2+}, Co^{2+}, Ni^{2+}, Cu^{2+}, and Cd^{2+}) (*83*). Kang et al. hydrothermally synthesized a ternary porous zinc(II) coordination polymer $[Zn_2(bbmb)_2\ (tdc)_2]{\cdot}2H_2O$, where bbmb is 1,4-bis(benzimidazol-1-yl-methyl) benzene and H_2tdc is 2,5-thiophene dicarboxylic acid. This MOF possesses high sensitivity toward mercury (*84*). Wan et al. synthesized a novel anionic Zn-anionic MOF, Zn-TPTC, via the solvothermal method. This MOF possesses high selectivity and sensitivity for detecting low levels of Hg^{2+}, with detection limits as low as 3.67 nM (*85*). Razavi et al. synthesized a Zn-MOF, $[Zn(OBA)\text{-}(DPT)_{0.5}]{\cdot}DMF$ (TMU-34(-2H)), through the photoluminescence method, with 4,4′-oxybis(benzoic acid) used as the organic ligand and 3,6-di(pyridine-4-yl)-1,2,4,5-tetrazine as the spacer. TMU-34(-2H) shows different photoluminescence emissions in water and acetonitrile due to hydrogen bonding between tetrazine moieties and water molecules, as shown in Figure 5 (*86*). TMU-34(-2H) can selectively detect Hg^{2+} in water with a limit of detection of 1.8 μM and a short response time (15 s) (*86*). Rudd et al. prepared a series of facile Zn-based luminescent MOFs (LMOFs): $Zn_2(ofdc)_2(tppe)$ (LMOF-261), $Zn_2(hfdc)_2\ (tppe)$ (LMOF-262), and $Zn_2(dbtdcO_2)_2\ (tppe)$ (LMOF-263). The MOFs were prepared by incorporating a strongly emissive molecular fluorophore and functionally diverse colinkers into Zn-based structures. LMOF-263 exhibited impressive water stability and high sensitivity and selectivity toward Hg^{2+} at very low concentrations (3.3 ppb). This MOF is an excellent candidate for a fluorescent chemical sensor and adsorbent for aqueous contaminants (*87*). Wu et al. developed a novel fluorescent sensor using the combination of the MOF UiO-66-NH_2 and a T-rich fluorophore labeled single-stranded DNA. This hybrid system exhibited excellent sensing performance with a lower detection limit of 17.6 Nm (*88*).Wen et al. prepared the amino-decorated Zn-MOF, which is highly selective and sensitive for aqueous Hg^{2+} and Cr^{6+} metal ions (*89*).

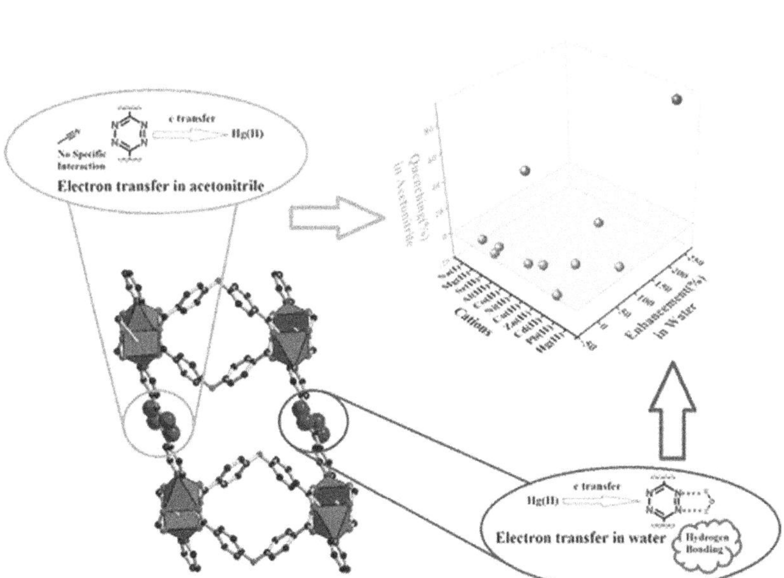

Figure 5. Illustration of sensing mechanism of TMU-34(-2H) toward Hg^{2+}. Reproduced with permission from reference (86). Copyright 2017 American Chemical Society.

Detection of Chromium

Recently, a highly luminescent Zn(II)-MOF, $[Zn_2(T_ZT_Z)_2(BDC)_2]\cdot 2DMF$ [where T_ZT_Z is 2,5-di(4-pyridyl)thiazolo[4,5-d]thiazole and BDC is terephthalic acid], was solvothermally prepared. This Zn(II)-MOF can act as a highly selective and recyclable luminescent probe for the quantitative detection of Cr^{6+} ions in aqueous solutions, with a detection limit as low as 4 μM (*90*). Another solvothermally synthesized luminescent Zn-MOF, $\{[Zn(L)(bpe)_{0.5}]\cdot DMF\}_n$ [where H_2L is 4,4′-((naphthalene-1,4-dicarbonyl)bis(azanediyl))dibenzoic acid and bpe is 1,2-bis(4-pyridyl)ethene], is highly sensitive and selective and can reversibly detect $Cr_2O_7^{2-}$, 2,4,6-trinitrophenol, and nitrofurazone in aqueous media, as shown in Figure 6 (*91*).

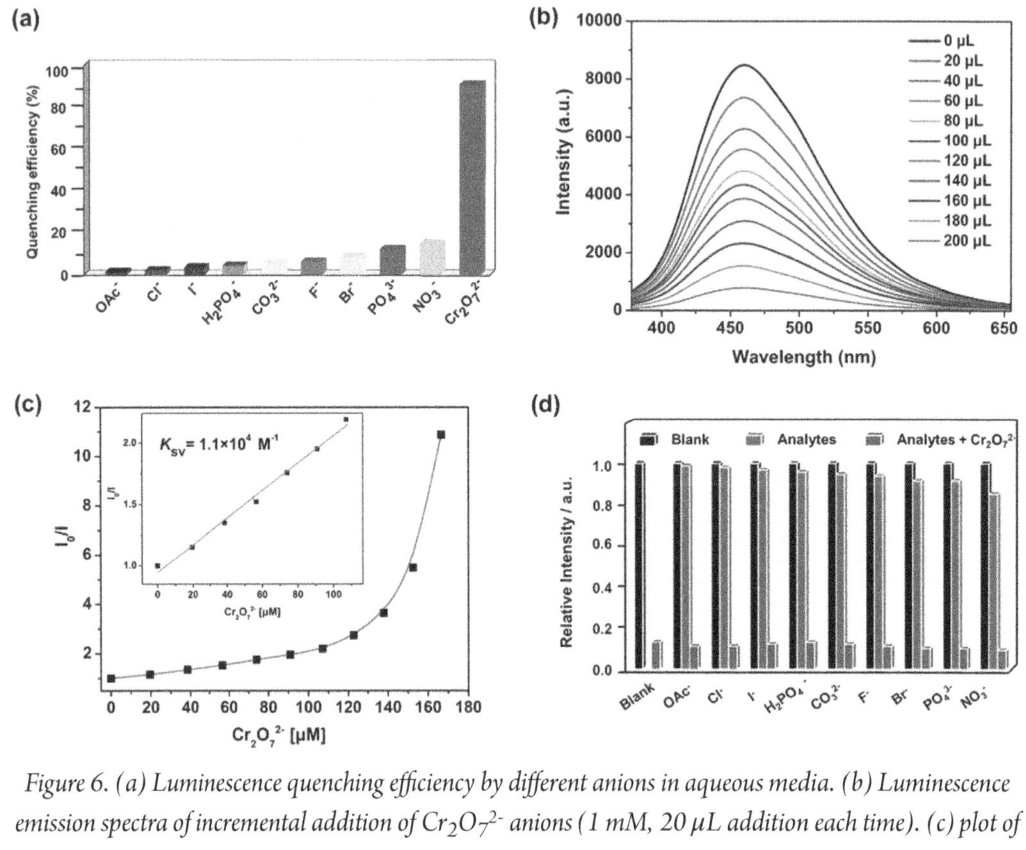

Figure 6. (a) Luminescence quenching efficiency by different anions in aqueous media. (b) Luminescence emission spectra of incremental addition of $Cr_2O_7^{2-}$ anions (1 mM, 20 μL addition each time). (c) plot of $Cr_2O_7^{2-}$ anions(d) Selective detection of $Cr_2O_7^{2-}$ anions in the presence of other anions in aqueous media. Reproduced with permission from reference (91). Copyright 2021 Royal Society Of Chemistry.

The two-dimensional Zn-MOF $\{[Zn(PA^{2-})(4,4'\text{-bpy})](H_2O)\}_n$, where PA is pamoic acid and 4,4′-bpy is 4,4′-bipyridine, was synthesized via a solvothermal method. The MOF detected Cr^{6+} with a detection limit of 4.12 μM and was also used as a photocatalyst to convert Cr^{6+} to Cr^{3+} (*92*). Li et al. prepared a hydrolytically stable Zn(II)-MOF, $\{[Zn(afsba)(bbtz)_{1.5}(H_2O)_2]\cdot 2H_2O\}_n$, based on the amino/sulfonato-functionalized ligand 2-amino-5-sulfobenzoic acid (H_2afsba) and the auxiliary flexible ligand 1,4-bis(triazol-1-ylmethyl)benzene (bbtz) via the hydrothermal method. This MOF provided a low detection limit of 0.22 ppm/0.26 ppm for $CrO_4^{2-}/Cr_2O_7^{2-}$ and 0.07 ppm for Fe^{3+} (*93*). In addition, He and coworkers utilized the hard/soft–acid/base and dual-ligand strategies to construct a pillar-layer LMOF (94).

The MOF $\{[Zn_2(TRZ)_2(DBTDC\text{-}O_2)]\cdot DMAc\}_n$, where $H_2DBTDC\text{-}O_2$ is S,S-dioxodibenzothiophen-3,7-dicarboxylic acid and HTRZ is 1,2,4-triazole, is a multiresponsive luminescent sensor for Fe^{3+}, Al^{3+}, SiF_6^{2-}, $Cr_2O_7^{2-}$, nitrofurantoin, and nitrofurazone in water (*94*). An ultrastable amino-decorated Zn(II)-MOF, $[Zn(NH_2\text{-bdc})(4,4'\text{-bpy})]$ (where $NH_2\text{-}H_2bdc$ is 2-amino-1,4 benzenedicarboxylic acid and 4,4′-bpy is 4,4′-bipyridine), was successfully synthesized and used as a multiresponsive fluorescent sensor for highly selective detection, with lower detection limits of 2.21 μM for CrO_4^{2-} and 1.30 μM for $Cr_2O_7^{2-}$ (*95*).

Recently, the highly fluorescent thiazolo[5,4-d]thiazole-extended viologen chromophore was incorporated for the first time to obtain a luminescent Zn(II)-MOF. The luminescent framework exhibits excellent stability in common solvents and can selectively sense CrO_4^{2-} and $Cr_2O_7^{2-}$ in water with high sensitivity (*96*). Six functional MOFs were constructed from semiflexible 4,4′-bis(imidazolyl)diphenyl ether (bidpe), via the solvothermal method, namely $[Zn(bidpe)(mpa)]_n$, $[Cu(bidpe)(mpa)(H_2O)]_n$, $[Cd(bidpe)(mpa)]_n$, $[Zn_{1.5}(bidpe)_2(smpa)(H_2O)]_n$, $[Cu_{1.5}(bidpe)_2(smpa)(H_2O)]_n$, and $[Cd(bidpe)(hmpa)]_n$ (where H_2mpa is m-Phthalic acid, H_2smpa is 5-sulfonatobenzene-1,3-dicarboxylate, and H_2hmpa is 5-Hydroxyisophthalic acid). These MOFs are used for sensing Cr^{6+} anions, Fe^{3+} ions, and nitrobenzene with high selectivity and sensitivity (*97*).

A Zn(II) coordination polymer formulated as $\{[Zn_3(mtrb)_3(btc)_2]\cdot 3H_2O\}_n$ [where mtrb is 1,3-bis(1,2,4-triazole-4-ylmethyl)benzene and btc is 1,3,5-benzenetricarboxylate] was synthesized. It can detect Fe^{3+}, $Cr_2O_7^{2-}$, and CrO_4^{2-} with high selectivity and sensitivity and has detection limits of 1.78 μM, 2.83 μM, and 4.52 μM, respectively (*98*). Guo et al. synthesized a recyclable multiresponsive Zn-MOF fluorescence sensor formulated as $[Zn_2(L)_2(TPA)]\cdot 2H_2O$ (where L is 4-(tetrazol-5-yl)phenyl-4,2′:6′,4″-terpyridine and TPA is terephthalic acid) via the solvothermal method. It can detect pesticide 2,6-dichloro-4-nitroaniline, Fe^{3+}, and Cr^{6+} with high sensitivity and selectivity and exhibits a low detection limit in methanol or water via fluorescence quenching (*99*).

A 3-D Zn-MOF, formulated as $[Zn(tpbpc)_2]$·solvent (where Htpbpc is 4′-4,2′;6′,4″]-terpyridin-4′-yl-biphenyl-4-carboxylic acid), which exhibits unique (2,8)-connected 8T20 topology with binuclear $[Zn_2(COO)_4N_4]_n$ secondary building units as 8-c nodes and tpbpc− ligands as 2-c nodes. This multiresponsive MOF's fluorescent sensor simultaneously detects Hg^{2+} and Cr^{6+} ions with high sensitivity and low detection limits in aqueous solution (*100*). Moreover, the Zn(II)-MOF $[Zn_2(TPOM)(NDC)_2]$·3.5H2O (where TPOM is tetrakis(4-pyridyloxymethylene)methane and

H_2ndc is 2,6-naphthalenedicarboxylic acid) was constructed hydrothermally and exhibited excellent selectivity toward Fe(III) and Cr(VI) (*101*).

Other Heavy Metal Ion Detection

Recently, a novel dual-functional composite (MOF_L–TpBD) was synthesized and characterized via solvothermal methods. The prepared MOF was used to extract and detect Pb^{2+} ions, with a detection limit of 0.32 μg L^{-1} (*102*). In addition, MOF-5 $Zn_4O(TPA)_3$ (where TPA is terephthalic acid) was used as a rapid, "turn-on" phosphorescent sensor for highly selective and sensitive detection of Pb^{2+} ions in mussels. This sensor provides a lower detection limit of 2 nmol/L for the determination of Pb^{2+} ions (*103*). A 3-D Zn-MOF formulated as $Zn_2(BDC)_2(DABCO)$ (where H_2BDC is 1,4-benzene dicarboxylic acid and DABCO is 1,4-diazabicyclo[2.2.2]octane) was prepared for the detection of Zn^{2+} ion. It provides a detection limit of 0.7 μM for Zn^{2+} ions within a concentration range of 50–200 μM (*104*). Moreover, a luminescent Zn(II)/Na(I)-MOF $\{[Zn_2Na(L)(HL)_2(H_2O)_2][OAc]\cdot 2H_2O\}_n$ (where 1, H_2L is 5-methyl-1,3-benzenedicarboxylic acid) was constructed for the fluorescence detection of Cu^{2+} ions with excellent stability. It provides a detection limit down to 0.65 μM for Cu^{2+} ions (*105*).

Other Metal Ion-Based MOFs Used for Heavy Metal Sensing

Other metal ions are also used in the construction of LMOFs for toxic heavy metal ion sensing, as shown in Table 2. Cadmium-based LMOFs show excellent selectivity and sensitivity toward heavy metal ions (Fe^{3+}, Hg^{2+}, Cu^{2+}, and Cr^{3+}) in water. Other lanthanide-based, LMOFs using terbium (Tb^{3+}) were reported to have impressive sensitivity toward Zn^{2+} and Fe^{3+} with extraordinary luminescence (*121*, *123*). Furthermore, MOFs based on copper and other metal ions were successfully developed for the detection of toxic heavy metal contaminants in aqueous solutions.

Future Scope

Knowledge of Zn-MOFs is rapidly growing among researchers. There is a vast range of uses for MOFs as sensors of organic and inorganic chemical species, but there are still major gaps in the understanding of their structural stability and properties and modes of action. Investigating these factors can lead to the destruction of the crystal structure of certain MOF sensors with time stability and decomposition mechanisms and thus cannot be accomplished systematically. Future work is still needed to improve the sensing performance of MOFs in terms of stability, sensitivity, selectivity, and reusability.

Table 2. Other Metal Ion MOFs for Heavy Metal Ion Detection

Serial no.	*MOFs Used for Sensing*	*Ligands*	*Synthesis Method*	*Element Detected*	*Sensing Category*	*Detection Limit*	*Remarks*	*Ref.*
1	$\{[Zn(L)(dcdps)]\}_n$, $\{Zn(L)(bdc)\}_n$, $\{[Cd(L)(oba)]\cdot 0.5DMF\}_n$, and $\{[Cd(L)(bdc)\cdot 2H_2O]\cdot 2DMF\}_n$	**L** = 4,4′-(2,5-bis(methylthio)-1,4-phenylene)dipyridine; H_2dcdps = 4,4′-sulfonyldibenzoic acid; H_2bdc = 1,4-dicarboxybenzene; H_2oba = 4,4′-oxybisbenzoic acid	Hydrothermal method	Fe^{3+}	Luminescent	6.21×10^{-5} M	Also detected organic molecules	(*115*)
2	$[Cd(L)(atpa)]_n$ and $[Cd(L)(tbta)(H_2O)]_n$	H_2atpa = 2-aminoterephthalic acid; H_2tbta = tetrabromoterephthalic acid; L = 1,4-bis(benzimidazol-1-yl)-2-butene	Hydrothermal method	Cu^{2+} Hg^{2+}	Luminescent	1.2 μM and 0.043 μM	Dual responsive	(*116*)
3	$[\{Cd_2(5\text{-}F\text{-}1,3\text{-}bpeb)_2(1,3\text{-}BDC)_2\}\cdot 0.5DMF\cdot 2H_2O]_n$	$[\{Cd_2(syn\text{-}dftpmcp)(1,3\text{-}BDC)_2\}\cdot 0.5DMF\cdot H_2O]_n$ (*syn*-dftpmcp = *syn*-3,4,12,13-tetrakis(4-pyridyl)-8,17-bisfluoro-1,2,9,10-diethano[2.2]metacyclophane)	Solvothermal method	Al^{3+}	Luminescent	183 ppb	High sensitivity and selectivity	(*117*)
4	$\{[Cd_2(L)(FA)(H_2O)_3]\cdot H_2O\}_n$	1,6-bis(benzimidazol-1-yl)hexane (L); 4,4-(hexafluoroisopropylidene)diphthalic acid (H_4FA)	Hydrothermal method	$Cr_2O_7^{2-}$	Luminescent	2.16 μM	High quenching efficiency	(*118*)
5	$[Cd(OBA)(L_1)\cdot 0.75DMA]_n$	H_2OBA = 4,4′-oxybis(benzoic acid); L_1 = 4-amino-3,5-bis(4-pyridyl)-1,2,4-triazole)	Hydrothermal method	Cr^{3+} $CrO_4{}^22-$ $Cr_2O_7^{2-}$	Luminescent	10.25, 0.30, and 9.18 μM	Excellent recyclability	(*119*)
6	$[Cd(L)(tbta)(H_2O)]$n	H_2tbta = tetrabromoterephthalic acid; L = 1,4-bis(benzimidazol-1-yl)-2-butene	Hydrothermal method	Hg^{2+} $Cr_2O_7^{2-}$	Luminescent	0.043 μM and 0.13 μM	Dual responsive	(*120*)

Table 2. (Continued). Other Metal Ion MOFs for Heavy Metal Ion Detection

Serial no.	*MOFs Used for Sensing*	*Ligands*	*Synthesis Method*	*Element Detected*	*Sensing Category*	*Detection Limit*	*Remarks*	*Ref.*
7	TbTATB	TATB = 4,4′-4″- striazine 2,4,6-triyl-tribenzoic acid	Solvothermal method	Zn^{2+}	Luminescent	10.5 nM	Excellent sensitivity	(*121*)
8	$[Co(DCTP)(L)(H_2O)_2]_n$	L = 1,4-bis(benzimidazol-1-ylmethyl)-benzene; H2DCTP = 2,5-dichloroterephthalic acid	Hydrothermal method	Hg^{2+}	Luminescent	1.10 μM	Fast response time of 3 s	(*122*)
9	$\{Tb_3(\mu_3HPyIDC)_4(H_2O)_8]\cdot NO_3\cdot 4H_2O\}_n$	μ3-HPyIDC = bis(carbamate) derivatives of 4,5-bis(hydroxymethyl)imidazole	Solvothermal method	Fe^{3+}	Luminescent	10^{-4} M	Unique fluorescence changing from green to blue\~for Fe3+	(*123*)
10	$[Cd_3\{Ir(ppy\text{-}COO)_3\}_2(DMF)_2(H_2O)_4]\cdot 6H_2O\cdot 2DMF$	ppy-COOH = methyl-3-(pyridin-2-yl)benzoic acid; DMF = N,N-dimethylformamide	Solvothermal method	Fe^{3+} $Cr_2O_7^{2-}$	Luminescent	67.8 ppb and 145.1 ppb	Good water stability and high selectivity	(*124*)

Conclusions

In this chapter, we have summarized advancements in newly synthesized Zn-MOF sensors and their applications for the detection of toxic heavy metal ions. MOFs have been used as sensing materials for toxic heavy metal ions, organic compounds, and gas molecules. Their features include low toxicity, a large surface area, antibacterial activity and biocompatibility, tunable porosity, reversible adsorption, high catalytic ability, and tunable chemical functionalization. Table 1 summarizes Zn-MOF sensors used for heavy metal sensing. In these sensors, the guest molecules incorporated within the MOFs can emit or induce luminescence. MOFs have high potential to monitor both large and trace concentrations of inorganic contaminants that enter our environmental and biological systems via anthropogenic activities.

References

1. Jian, M.; Liu, B.; Zhang, G.; Liu, R.; Zhang, X. Adsorptive removal of arsenic from aqueous solution by zeoliticimidazolate framework-8 (ZIF-8) nanoparticles. *Colloids Surf. A-Physicochem. Eng. Asp.* **2015**, *465*, 67–76.
2. Ling, L. L.; Liu, W. J.; Zhang, S.; Jiang, H. Magnesium oxide embedded nitrogen self-doped biochar composites: fast and high-efficiency adsorption of heavy metals in an aqueous solution. *Environ. Sci. Technol.* **2017**, *51*, 10081–10089.
3. Zou, Y.; Wang, X.; Khan, A.; Wang, P.; Liu, Y.; Alsaedi, A.; Hayat, T.; Wang, X. Environmental remediation and application of nanoscale zero-valent iron and its composites for the removal of heavy metal ions: a review. *Environ. Sci. Technol.* **2016**, *50*, 7290–7304.
4. Alizadeh, T.; Rashedi, M. Synthesis of nano-sized arsenic-imprinted polymer and its use as As3+ selective ionophore in a potentiometric membrane electrode: Part 1. *Analytica Chimica Acta* **2014**, *843*, 7–17.
5. Chen, H.-H.; Huang, J.-F. EDTA assisted highly selective detection of As3+ on Au nanoparticle modified glassy carbon electrodes: facile in situ electrochemical characterization of Au nanoparticles. *Analytical Chemistry* **2014**, *86*, 12406–12413.
6. Sun, S. L.; Yu, H. N.; Ling, C. J.; Shen, S. R.; Miao, A. Q.; Pan, S. S.; Zhao, C. Y. Inhibitory effects of epigallocatechin gallate (EGCG) and Cd2+ on the growth of prostate cancer PC-3 cell. *Chin. J. Agric. Biotechnol.* **2009**, *17*, 902–907.
7. Verougstraete, V.; Lison, D.; Hotz, P. Cadmium, lung and prostate cancer: a systematic review of recent epidemiological data. *J. Toxicol. Env. Heal. B* **2003**, *6*, 227–256.
8. Chow, W. H.; Dong, L. M.; Devesa, S. S. Epidemiology and risk factors for kidney cancer. *Nat. Rev. Urol.* **2010**, *7*, 245–257.
9. *World Health Organization, Aluminium in Drinking-Water: Background Document for Development of WHO Guidelines for Drinking-Water Quality*, 2003. https://apps.who.int/iris/handle/10665/75362.
10. WaldM. L. *EPA Process New Limits on Lead in the Air, The First Revision In 30 years*," 2008, http://www.nytimes:com/Washington/02 epa.html2008/05/02 (accessed May 22, 2021).
11. Underwood, E. J. Trace Elements and Health: An Overview. *Phil. Trans. R. Soc. Land.* **1979**, *288*, 5–14.
12. Lv, R.; Wang, J. Y.; Zhang, Y. P.; Li, H.; Yang, L. Y.; Liao, S. Y.; Gu, W.; Liu, X. An amino-decorated dual-functional metal–organic framework for highly selective sensing of Cr(III) and

Cr(VI) ions and detection of nitroaromatic explosives. *J. Mater. Chem. A* **2016**, *4*, 15494–15500.

13. Merin, M.; Anke, M.; and Ihnat. *Trace Elements and Compounds in Soil*. M. S. Wiley VCH Verlag GmbH & KGaA, Weinhein, 2004.
14. Wu, Z.-F.; Gong, L.-K.; Huang, X.-Y. A Mg-CP with in Situ Encapsulated Photochromic Guest as Sensitive Fluorescence Sensor for Fe 3+ /Cr 3+ Ions and Nitro-Explosives. *Inorg. Chem.* **2017**, *56*, 7397–7403.
15. Zhou, H. C.; Long, J. R.; Yaghi, O. M. Introduction to Metal–Organic Frameworks. *Chem. Rev.* **2012**, *112* (2), 673–674doi:10.1021/cr300014x.
16. Cai, F.; Wang, Q.; Chen, X.; Qiu, W.; Zhan, F.; Gao, F.; Wang, Q. Selective binding of Pb2+ with manganese-terephthalic acid MOF/SWCNTs: theoretical modeling, experimental study and electroanalytical application. *Biosensors Bioelectronics* **2017**, *98*, 310–316.
17. Gao, C.; Tian, H.; Ai, J.; Li, L.; Dang, S.; Lan, Y.; Sun, Z. A microporous Cu-MOF with optimized open metal sites and pore spaces for high gas storage and active chemical fixation of CO2. *Chem. Commun.* **2016**, *52*, 11147–11150.
18. Liu, J.; Chen, L.; Cui, H.; Zhang, J.; Zhang, L.; Su, C.-Y. Applications of metal–organic frameworks in heterogeneous supramolecular catalysis. *Chem. Soc. Rev.* **2014**, *43*, 6011–6061.
19. Zhang, Y.; Yuan, S.; Day, G.; Wang, X.; Yang, X.; Zhou, H.-C. Luminescent sensors based on metal-organic frameworks. *Coord. Chem. Rev.* **2018**, *354*, 28–45.
20. Saha, D.; Grappe, H. A.; Chakraborty, A.; Orkoulas, G. Postextraction Separation, On-Board Storage, and Catalytic Conversion of Methane in Natural Gas: A Review. *Chem. Rev.* **2016**, *116*, 11436–11499doi:10.1021/acs.chemrev.5b00745.
21. Lim, D.-W.; Kitagawa, H. Proton Transport in Metal-Organic Frameworks. *Chem. Rev.* **2020**, *120* (16), 8416–8467doi:10.1021/acs.chemrev.9b00842.5.
22. Li, D.; Xu, H.-Q.; Jiao, L.; Jiang, H.-L. Metal-organic frameworks for catalysis: State of the art, challenges, and opportunities. *Energy Chem.* **2019**, *1*, 100005.
23. Li, H.; Li, L.; Lin, R.-B.; Ramirez, G.; Zhou, W.; Krishna, R.; Zhang, Z.; Xiang, S.; Chen, B. Microporous Metal–Organic Framework with Dual Functionalities for Effcient Separation of Acetylene from Light Hydrocarbon Mixtures. *ACS Sustain. Chem. Eng.* **2019**, 7, 4897–4902.
24. Zhu, Q. L.; Xu, Q. Metal organic framework composites. *Chem. Soc. Rev.* **2014**, *43*, 5468–5512.
25. Ploetz, E.; Engelke, H.; Lächelt, U.; Wuttke, S. The Chemistry of Reticular Framework Nanoparticles: MOF, ZIF, and COF Materials. *Adv. Funct. Mater.* **2020**, *30* (41), 1909062.
26. Chidambaram, A.; Stylianou, K. C. Electronic metal–organic framework sensors. *Inorg. Chem. Front.* **2018**, 5, 979–998. https://doi.org/10.1039/c7qi00815e
27. Sabouni, R. (2013) *Carbon Dioxide Adsorption by Metal Organic Frameworks (Synthesis, Testing and Modeling)*. Ph.D. thesis, University of western Ontario, Electronic thesis and dissertation repository, 28-29.
28. Dincă, M.; Long, J. R. Strong H2 binding and selective gas adsorption within the microporous coordination solid Mg3(O2C-C10H6-CO2)3. *J. Am. Chem. Soc.* **2005**, *127*, 9376–9377.
29. Horike, S.; Shimomura, S.; Kitagawa, S. Soft porous crystals. *Nat. Chem.* **2009**, *1* (9), 695–704.

30. Hamon, L.; Llewellyn, P. L.; Devic, T.; Ghoufi, A.; Clet, G.; Guillerm, V.; Pimgruber, G. D.; Maurin, G.; Serre, C.; Driver, G.; Beek, W. V.; Jolimaitre, E.; Vimont, A.; Daturi, M.; Ferey, G. Co-adsorption and separation of CO2-CH4 mixtures in the highly flexible MIL-53(Cr) MOF. *J. Am. Chem. Soc.* **2009**, *131*, 17490–17499.
31. Llewellyn, P.; Bourrelly, S.; Serre, C.; Filinchuk, Y.; Férey, G. How hydration drastically improves adsorption selectivity for CO2 over CH4 in the flexible chromium terephthalate MIL-53. *Angew. Chem. Int. Ed.* **2006**, *45*, 7751–7754.
32. Serre, C.; Mellot, C.; Surblé, S.; Audebrand, N. Y.; Filinchuk, Y.; Férey, G. Role of solvent-host interactions that lead to very large swelling of hybrid frameworks. *Science* **2007**, *315*, 1828–1831.
33. Choi, H.; Suh, M. Highly selective CO2 capture in flexible 3d coordination polymer networks. *Angew Chem. Int. Ed.* **2009**, *48*, 6865–6869.
34. Wang, C.; Ying, J. Y. Sol-gel synthesis and hydrothermal processing of anatase and rutile titania nanocrystals. *Chem. Mater.* **1999**, *11*, 3113–3120Doi:10.1021/cm990180f.
35. Huang, L.; Wang, H.; Chen, J.; Wang, Z.; Sun, J.; Zhao, D.; Yan, Y. Synthesis, morphology control and properties of porous metal-organic coordination polymers. *Microporous Mesoporous Mater.* **2003**, *58*, 105–114.
36. Tranchemontagne, D. J.; Hunt, J. R.; Yaghi, O. M. Room temperature synthesis of metal-organic frameworks: MOF-5, MOF-74, MOF-177, MOF-199 and IRMOF-0. *Tetrahedron* **2008**, *64*, 8553–8557.
37. Cravillion, J.; Munzer, S.; Lohmeier, S. J.; Feldhoff, A.; Huber, K.; Wiebcke, M. Rapid room temperature synthesis and characterization of nanocrystals of a prototypical zeoliticimidazolate framework. *Chem. Mater.* **2009**, *21* (8), 1410–1412.
38. Pichon, A.; Lazuen-Garay, A.; James, S. L. Solvent free synthesis of a microporous metal-organic framework. *CrystEngComm* **2006**, *8*, 211–214.
39. Nouar, F.; Eckert, J.; Eubank, J. F.; Forster, P.; Eddaoudi, M. Zeolite like metal-organic framework (ZMOFs) as hydrogen storage platform: Lithium and magnesium ion exchange and H2-(rho-ZMOF) interaction studies. *J Am ChemSoc* **2009**, *131* (8), 2864–2870.
40. Braga, D.; Giaffreda, S. L.; Grepioni, F.; Chierotti, M. R.; Gobetto, R.; Palladino, G.; Polito, M. Solvent effect in a "solvent free" reaction. *CrystEngComm* **2007**, *9*, 879–881.
41. Pichon, A.; James, S. L. An array based study of reactivity under solvent free mechanochemical conditions - insights and trends. *CrystEngComm* **2008**, *10*, 1839–1847.
42. Jhung, S. H.; Chang, J. S.; Hwang, J. S.; Park, S. E. Selective formation of SAPO-5 and SAPO-34 molecular sieves with microwave irradiation and hydrothermal heating. *Microporous Mesoporous Mater.* **2003**, *64*, 33–39Doi:10.1016/S1387-1811(03)00501-8.
43. Jhung, S. H.; Lee, J. H.; Yoon, J. W.; Hwang, J. S.; Park, S. E.; Chang, J. S. Selective crystallization of CoAPO-34 and VAPO-5 molecular sieves under microwave irradiation in an alkaline or neutral condition. *Microporous Mesoporous Mater.* **2005**, *80*, 147–152Doi:10.1016/j.micromeso.2004.11.013.
44. Kang, K. K.; Park, C. H.; Ahn, W. S. Microwave preparation of titanium-substituted mesoporous molecular sieve. *Catal. Lett.* **1999**, *59* (1), 45–49.
45. Jhung, S. H.; Chang, J. S.; Hwang, Y. K.; Park, S. E. Crystal morphology control of AFI type molecular sieves with microwave irradiation. *J. Mater. Chem.* **2004**, *14*, 280–285.

46. Hwang, Y. K.; Chang, J. S.; Park, S. E.; Kim, D. S.; Kwon, Y. U.; Jhung, S. H.; Hwang, J. S.; Park, M. S. Microwave fabrication of MFI zeolite crystals with a fibrous morphology and their applications. *Angew. Chem. Int. Ed.* **2005**, *44*, 556–560.
47. Jhung, S. H.; Lee, J. H.; Chang, J. S. Microwave synthesis of nanoporous hybrid material, Chromium trimesate. *Bull. Korean Chem. Soc.* **2005**, *26*, 880–881.
48. Lu, C. M.; Liu, J.; Xiao, K.; Harris, A. T. Microwave enhanced synthesis of MOF-5 and its CO2 capture ability at moderate temperatures across multiple capture and release cycles. *Chem. Eng. J.* **2010**, *156*, 465–470.
49. Schlesinger, M.; Schulze, S.; Hiestchold, M.; Mehring, M. Evolution of synthetic methods for microporous metal-organic frameworks exemplified by the competitive formation of [Cu2(btc)3(H2O)3] and [Cu2(btc)(OH)(H2O)]. *Microporous Mesoporous Mater.* **2010**, *132*, 121–127.
50. Mueller, U.; Puetter, H.; Hesse, M.; Wessel, M. *Method for Electrochemical Production of a Crystalline Porous Metal Organic Skeleton Material*. Patent - WO2005/049892.
51. Mueller, U.; Schubert, M.; Teich, F.; Puetter, H.; Schierle-Arndt, K.; Pastre, J. Metal-organic frameworks – Prospective industrial applications. *J. Mater. Chem.* **2006**, *16*, 626–636.
52. RicherI.; SchubertM.; MüllerU. *Porous Metal Organic Framework Based on Pyrroles and Pyridinones*. Patent WO2007/131955.
53. Suslick, K. S.; Choe, S. B.; Cichowlas, A. A.; Grinstaff, M. W. Sonochemical synthesis of amorphous iron. *Nature* **1991**, *353*, 414–416.
54. Gedanken, A. Sonochemical synthesis of amorphous iron. *UltrasonSonochem* **2004**, *11*, 47–55.
55. SonoT.; Mingos, D. M. P.; Baghurst, D. R.; Lickiss, P. D. Novel energy source for reactions. *In the New Chemistry*: Hall, N. , Ed.; University of Cambridge, Cambridge, U.K., 2004; Chapter 4.
56. Qiu, L. G.; Li, Z. Q.; Xu, T.; Jiang, S. Fecile synthesis of nanocrystals of a microporous metal-organic framework by an ultrasonic method and selecting sensing of orgnoamines. *Chem. Commun.* **2008**, 3642–3644.
57. Li, Z. Q.; Qiu, L. G.; Su, T.; Wu, Y.; Wang, W.; Wu, Z. Y.; Jiang, X. Ultrasonic synthesis of the microporous metal-organic framework Cu3(BTC)2 at ambient temperature and pressure: An efficient and environmentally friendly method. *Mater. Lett.* **2009**, *63*, 78–80.
58. Zacher, D.; Yusenko, K.; Bétard, A.; Henke, S.; Molon, M.; Ladnorg, T.; Shekhah, O.; Schüpbach, B.; de los Arcos, T.; Krasnopolski, M.; Meilikhov, M.; Winter, J.; Terfort, A.; Wöll, C.; Fischer, R. A. Liquid – phase epitaxy of multicomponent layer based porous coordination polymer of thin films of [M(L)(P)0.5] type: importance of deposition sequence on the oriented growth. *Chem. Eur. J.* **2011**, *17*, 1448–1455.
59. Mei Bai, M.; Zhang, J. B.; Cao, L. H.; Lia, Y. P.; Wanga, D. Z. Zinc(II) and cadmium(II) metal complexes with bis(tetrazole) ligands: synthesis and crystal structure. *Journal of the Chinese Chemical Society* **2011**, *58*, 69–74.
60. Shekhah, O.; Wang, H.; Zacher, D.; Fischer, R. A.; Wöll, C. Growth mechanism of metal-organic framework: insights into the nucleation by employing a step by step route. *Angew. Chem. Int. Ed.* **2009**, *48*, 5038–5041.
61. Horcajada, P.; Serre, C.; Grosso, D.; Boissière, C.; Perruchas, S.; Sanchez, C.; Férey, G. Colloidal route for preparing optical thin films of nanoporous metal-organic frameworks. *Adv. Mater.* **2009**, *21*, 1931–1935.

62. Demessence, A.; Boissière, C.; Grosso, D.; Horcajada, P.; Serre, C.; Férey, G.; Soler-Illia, G. J. A. A.; Sanchez, C. Adsorption properties in high optical quality nanoZIF-8 thin folms with tunable thickness. *J. Mater. Chem.* **2010**, *20*, 7676–7681.
63. Stock, N.; Biswas, S. Synthesis of metal-organic frameworks (MOFs): Routes to various MOFs topologies, morphologies and composites. *Chem. Rev.* **2012**, *112*, 933–969.
64. Cooper, E. R.; Andrews, C. D.; Wheatley, P. S.; Webb, P. B.; Wormald, P.; Morris, R. E. Ionic liquids as eutectic mixtures as solvent and template in synthesis of zeolite analogues. *Nature* **2004**, *430*, 1012–1016.
65. Parnham, E. R.; Morris, R. E. Ionothermal synthesis of zeolites, metal-organic frameworks and inorganic-organic hybrids. *Acc. Chem. Res.* **2007**, *40*, 1005–1013.
66. Xu, X.; Luo, Z.; Ye, K.; Zou, X.; Jianming Pan, X. One-pot construction of acid phosphatase and hemin loaded multifunctional metal–organic framework nanosheets for ratiometric fluorescent arsenate sensing. *Journal of Hazardous Material* **2021**, *412*, 124407.
67. Baghayeria, M.; Ghanei-Motlaghb, M.; Tayebeea, R.; Fayazic, M.; Narenjia, F. Application of graphene/zinc-based metal-organic framework nanocomposite for electrochemical sensing of As(III) in water resources. *Anal. Chim. Acta* **2020**, *1099*, 60–67.
68. Farahani, Y. D.; Safarifard, V. Highly Selective Detection of Fe3+, Cd2+ and CH2Cl2 Based on a Fluorescent Zn-MOF with Azine-decorated Pores. *J. Solid State Chem.* **2019**, *275*, 131–140.
69. Pan, Y.; Wang, J.; Guo, X.; Liu, X.; Tang, X.; Zhang, H. A new three-dimensional zinc-based metal-organic framework as a fluorescent sensor for detection of cadmium ion and nitrobenzene. *Journal of Colloid and Interface Science* **2018**, *513*, 418–426.
70. Zahra, N.; Tahera, M. A.; Fazelirad, H. A Zn based metal organic framework nanocomposite: synthesis, characterization and application for preconcentration of cadmium prior to its determination by FAAS. *RSC Adv.* **2017**, *7*, 44890–44895.
71. Lv, R.; Li, H.; Su, J.; Fu, X.; Yang, B.; Gu, W.; Liu, X. Zinc Metal–Organic Framework for Selective Detection and Differentiation of Fe(III) and Cr(VI) Ions in Aqueous Solution. *Inorg. Chem.* **2017**, *56* (20), 12348–12356.
72. Roushani, M.; Valipour, A.; Saedi, Z. Electroanalytical sensing of Cd2+ based on metal–organic framework modified carbon paste electrode. *Sensors and Actuators B: Chemical* **2016**, *233*, 419–425.
73. Cotruvo, J. A. 2017 WHO Guidelines for Drinking Water Quality: First Addendum to the Fourth Edition. *J. Am. Water Works Assoc.* **2017**, *109*, 44–51.
74. Moradi, E.; Rahimi, R.; Safarifard, V. Sonochemically Synthesized Microporous Metal–organic Framework Representing Unique Selectivity for Detection of Fe3+ Ions. *Polyhedron* **2019**, *159*, 251–258.
75. Hou, Y.; Song, Y.; Xiao, Y.; Wu, R.; Wang, L. ZnMOF-74 Responsive Fluorescence Sensing Platform for Detection of Fe3+. *Microchem. J.* **2019**, *150*, 104154.
76. Zhang, X.; Hu, J.; Wang, B.; Li, Z.; Xu, S.; Chen, Y.; Ma, X. A Chiral zinc(II) Metal-organic Framework as High Selective Luminescent Sensor for Detecting Trace Nitro Explosives Picric Acid and Fe3+ Ion. *J. Solid State Chem.* **2019**, *269*, 459–464.

77. Farahani, Y. D.; Safarifard, V. A Luminescent Metal-organic Framework with Pre-designed Functionalized Ligands as an Efficient Fluorescence Sensing for Fe3+ Ions. *J. Solid State Chem.* **2019**, *270*, 428–435.
78. Sun, M.; Chen, D. A Porous Zn(II)-based Metal–organic Framework for Highly Selective and Sensitive Fe3+ Ion Detection in Water. *Polyhedron* **2018**, *147*, 80–85.
79. Du, J.; Zhang, X.; Li, C.; Gao, J.; Hou, J.; Jing, X.; Mu, Y.; Li, L. A Bi-functional Luminescent Zn(II)-MOF for Detection of Nitroaromatic Explosives and Fe3+ Ions. *Sensors Actuators B, Chem.* **2018**, *257*, 207–213.
80. Hao, H.; Wang, Y.; Yuan, S.; Chen, D.; Li, D.; Dou, J. Two Zn(II)-based Metal-organic Frameworks for Selective Detection of Nitroaromatic Explosives and Fe3+ Ion. *Inorg. Chem. Commun.* **2018**, *98*, 120–126.
81. Jiang, J.; Lu, Y.; Zhou, Y.; Zhao, D.; Li, C. An Acid-base Resistant Zn-based Metal-organic Framework as a Luminescent Sensor for mercury(II). *J. Solid State Chem.* **2020**, *283*, 121–153.
82. Pankajakshan, A.; Kuznetsov, D.; Mandal, S. Ultrasensitive Detection of Hg(II) Ions in Aqueous Medium Using Zinc-based Metal–organic Framework. *Inorg. Chem.* **2019**, *58*, 1377–1381.
83. Khatun, A.; Panda, D. K.; Sayresmith, N.; Walter, M. G.; Saha, S. Thiazolothiazole-based Luminescent Metal–organic Frameworks with Ligand-to-ligand Energy Transfer and Hg2+-sensing Capabilities. *Inorg. Chem.* **2019**, *58*, 12707–12715.
84. Kang, W.; Han, C.; Liu, D.; Cui, G. A Bifunctional Benzimidazole-based Luminescent Zn(II) Coordination Polymer for Detection of Hg2+ and Photocatalytic Degrading of Methylene Blue. *Inorg. Chem. Commun.* **2019**, *106*, 81–85.
85. Wan, Y.; Zou, D.; Cui, Y.; Yang, Y.; Qian, G. A Zn Based Anionic Metal-organic Framework for Trace Hg2+ Ion Detection. *J. Solid State Chem.* **2018**, *266*, 70–73.
86. Razavi, S. A.; Masoomi, M. Y.; Morsali, A. Double Solvent Sensing Method for Improving Sensitivity and Accuracy of Hg(II) Detection Based on Different Signal Transduction of a Tetrazine-functionalized Pillared Metal–organic Framework. *Inorg. Chem.* **2017**, *56*, 9646–9652.
87. Rudd, N. D.; Wang, H.; Fuentes-Fernandez, E. M.; Teat, S. J.; Chen, F.; Hall, G.; Chabal, Y. J.; Li, J. Highly Efficient Luminescent Metal–organic Framework for the Simultaneous Detection and Removal of Heavy Metals from Water. *ACS Appl. Mater. Interfaces* **2016**, *8*, 30294–30303.
88. Wu, L.-L.; Wang, Z.; Zhao, S. N.; Meng, X.; Song, X.-Z.; Feng, J.; Song, S.-Y.; Zhang, H.-J. A Metal–Organic Framework/DNA Hybrid System as a Novel Fluorescent Biosensor for Mercury(II) Ion Detection. *Chemistry* **2016**, *22*, 477–80.
89. Wen, L.; Zheng, X.; Lv, K.; Wang, C.; Xu, X. Two Amino-decorated Metal–organic Frameworks for Highly Selective and Quantitatively Sensing of HgII and CrVI in Aqueous Solution. *Inorg. Chem.* **2015**, *54*, 7133–7135.
90. Safaeia, S.; Wang, J.; Junk, P. Incorporation of Thiazolothiazole Fluorophores into a MOF Structure: A Highly Luminescent Zn(II)-based MOF as a selective and reversible sensor for Cr2O7 2− and MnO4 – anions. *Journal of Solid State Chemistry* **2021**, *294*, 121762.
91. Wang, X.; Lei, M.; Zhang, T.; Zhang, Q.; Zhang, R.; Yang, M. A water-stable multi-responsive luminescent Zn-MOF sensor for detecting TNP, NZF and Cr2O72− in aqueous media. *Dalton Trans.* **2021**, *50*, 3816–3824.

92. Kaur, H.; Sinha, S.; Krishnan, V.; Koner, R. R. Photocatalytic Reduction and Recognition of Cr(VI): New Zn(II)-Based Metal–Organic Framework as Catalytic Surface. *Ind. Eng. Chem. Res.* **2020**, *59* (18), 8538–8550.
93. Li, P.; Zhang, L.; Yang, M.; Zhang, K. A Novel Luminescent 1D→2D Polyrotaxane Zn(II)-organic Framework Showing Dual Res ponsive Fluorescence Sensing for Fe3+ Cation and Cr(VI) Anions in Aqueous Medium. *J. Lumin.* **2019**, *207*, 351–360.
94. He, H.; Zhu, Q.; Li, C.; Du, M. Design of a Highly-stable Pillar-layer Zinc(ii) Porous Framework for Rapid, Reversible, and Multi-responsive Luminescent Sensor in Water. *Cryst. Growth Des.* **2019**, *19*, 694–703.
95. Wiwasuku, T.; Boonmak, J.; Siriwong, K.; Ervithayasuporn, V.; Youngme, S. Highly sensitive and selective fluorescent sensor based on a multi-responsive ultrastable amino-functionalized Zn(II)-MOF for hazardous chemicals. *Sensors and Actuators B: Chemical* **2019**, *284*, 403–413.
96. Li, P.; Yin, X.; Gao, L.; Yang, S.; Sui, Q.; Gong, T.; Gao, E. Modulating Excitation Energy of Luminescent Metal–organic Frameworks for Detection of Cr(VI) in Water. *ACS. Appl. Nano Mater* **2019**, *2*, 4646–4654.
97. Zong, Z.; Fan, C.; Bi, C.; Zhang, X.; Luo, R.; Meng, X.; Jin, F.; Fan, Y. Highly Selective Functional Luminescent Sensor toward Cr(VI)/Fe(III) Ion and Nitrobenzene Based on Metal–organic Frameworks: Synthesis, Structures, and Properties. *J. Solid State Chem.* **2019**, *270*, 651–665.
98. Ya-Qian, Z.; Vladislav, B.; Tian-Rui, Z.; Chang-Hao, Y.; Lin-Lu, Q.; Ke, L.; Bao-Long, L.; Bing, W. A Luminescent Zinc(ii) Coordination Polymer with Unusual (3,4,4)-coordinated Self-catenated 3D Network for Selective Detection of Nitroaromatics and Ferric and Chromate Ions: A Versatile Luminescent Sensor. *Dalton Trans.* **2018**, *47*, 6189–6198.
99. Guo, X.-Y.; Dong, Z.-P.; Zhao, F.; Liu, Z.-L.; Wang, Y.-Q. Zinc(ii)–organic framework as a multi-responsive photoluminescence sensor for efficient and recyclable detection of pesticide 2,6-dichloro-4-nitroaniline, Fe(iii) and Cr(vi). *New J. Chem.* **2019**, *43*, 2353–2361.
100. Xiao, J.; Liu, J.; Gao, X.; Ji, G.; Wang, D.; Liu, Z. A Multi-chemosensor Based on Zn-MOF: Ratio-dependent Color Transition Detection of Hg (II) and Highly Sensitive Sensor of Cr (VI). *Sensors Actuators B, Chem.* **2018**, *269*, 164–172.
101. Lv, R.; Li, H.; Su, J.; Fu, X.; Yang, B.; Gu, W.; Liu, X. Zinc Metal–organic Framework for Selective Detection and Differentiation of Fe(III) and Cr(VI) Ions in Aqueous Solution. *Inorg. Chem.* **2017**, *56*, 12348–12356.
102. Lia, W. T.; Hua, Z.-J.; Meng, J.; Zhang, X.; Gao, W.; Chen, M. L.; Wang, J. H. Zn-based metal organic framework-covalent organic framework composites for trace lead extraction and fluorescence detection of TNP. *Journal of Hazardous Materials* **2021**, *411*, 125021.
103. Xu, S.; Zhan, L.; Hong, C.; Chen, X.; Chen, X.; Oyama, M. Metal–organic Framework-5 as a Novel Phosphorescent Probe for the Highly Selective and Sensitive Detection of Pb(II) in Mussels. *Sensors Actuators B, Chem.* **2020**, *308*, 127733.
104. Shayegan, H.; Farahani, Y. D.; Safarifard, V. A Pillar-layer Metal-organic Framework as A Turn-on Luminescent Sensor for Highly Selective and Sensitive Detection of Zn(II) Ion. *J. Solid State Chem.* **2019**, *279*, 120968.
105. Liu, L.-L.; Yu, Y.-Z.; Zhao, X.-J.; Wang, Y. R.; Cheng, F.-Y.; Zhang, M.-K.; Shua, J.-J.; Liu, L. A robust Zn(ii)/Na(i)-MOF decorated with [(OAc)2(H2O)2]n2n− anions for the

luminescence sensing of copper ions based on the inner filter effect. *Dalton Trans.* **2018**, *47*, 7787–7794.

106. Du, Y.; Song, N.; Lv, X.; Hu, B.; Zhou, W.; Jia, Q. Establishment of Highly Sensitive and Selective Fluorescent Sensor for Fe3+ Based on a Post-synthetic Modification of Metal-organic Framework. *Dyes Pigm.* **2017**, *138*, 15–22.
107. Zhu, Y.; Zhou, X.; Li, L.; Yu, Y.; Huang, W. A Water-stable Metal-organic Framework as A Luminescent Fe3+ Sensor under Weak Acidic and Weak Basic Conditions. *Sci. China Chem.* **2017**, *60*, 1581–1587.
108. Zhu, X.; Zhang, K.; Zhou, W.; Li, Q.; Fu, Y.; Liu, R.; Qian, G. A Luminescent Two-fold Interpenetrated Pillared-layer Metal-organic Framework for Highly Selective and Sensitive Sensing of Fe3+. *Inorg. Chem. Commun.* **2016**, 73, 90–93.
109. Li, Y.-F.; Wang, D.; Liao, Z.; Kang, Y.; Ding, W.-H.; Zheng, X.-J.; Jin, L.-P. Luminescence Tuning of the Dy–Zn Metal–organic Framework and Its Application in the Detection of Fe(III) Ions. *J. Mater. Chem. C.* **2016**, *4*, 4211–4217.
110. Hou, B.; Tian, D.; Liu, J.; Dong, L.; Li, S.; Li, D.; Lan, Y. A Water-stable Metal–organic Framework for Highly Sensitive and Selective Sensing of Fe3+ Ion. *Inorg. Chem.* **2016**, *55*, 10580–10586.
111. Xiao, J.; Liu, J.; Gao, X.; Ji, G.; Wang, D.; Liu, Z. A Multi-chemosensor Based on Zn-MOF: Ratio-dependent Color Transition Detection of Hg (II) and Highly Sensitive Sensor of Cr (VI). *Sensors Actuators B, Chem.* **2018**, *269*, 164–172.
112. Pankajakshan, A.; Kuznetsov, D.; Mandal, S. Ultrasensitive Detection of Hg(II) Ions in Aqueous Medium Using Zinc-based Metal–organic Framework. *Inorg. Chem.* **2019**, *58*, 1377–1381DOI:10.1021/acs.inorgchem.8b02898.
113. Jin, H.; Xu, J.; Zhang, L.; Ma, B.; Shi, X.; Fan, Y.; Wang, L. Multi-responsive luminescent sensor based on Zn (II) metal-organic framework for selective sensing of Cr(III), Cr(VI) ions and *p*-nitrotolune. *Journal of Solid State Chemistry* **2018**, *268*, 168–174.
114. Xu, T.-Y.; Li, J.-M.; Han, Y.-H.; Wang, A.-R.; Hea, K.-H.; Shi, Z.-F. A new 3D four-fold interpenetrated dia-like luminescent Zn(ii)-based metal–organic framework: the sensitive detection of Fe3+, Cr2O72−, and CrO42− in water, and nitrobenzene in ethanol. *New J. Chem.* **2020**, *44*, 4011–4022.
115. Ge, F.; Sun, G.; Meng, L.; Ren, S.; Zheng, H. Four New Luminescent Metal–organic Frameworks as Multifunctional Sensors for Detecting Fe3+, Cr2O7 2– And Nitromethane. *Cryst. Growth Des.* **2020**, *20*, 1898–1904.
116. Yang, Y.; Liu, D.; Li, Y.; Dong, G. Two New Luminescent Ternary Cd(II)-MOFs by Regulation of Aromatic Dicarboxylate Ligands Used as Efficient Dual-responsive Sensors for Toxic Metal Ions in Water. *Polyhedron* **2019**, *159*, 32–42.
117. Li, W.; Gu, J.; Li, H.; Dai, M.; Young, D.; Li, H.-Y.; Lang, J.-P. Post-synthetic Modification of a Two-dimensional Metal–organic Framework via Photodimerization Enables Highly Selective Luminescent Sensing of Aluminum (III). *Inorg. Chem.* **2018**, 57, 13453–13460.
118. Li, J.; Liu, D.; Hao, Z.; Cui, G. An Unusual (3,4,5)-connected 3D Cadmium Metal– organic Framework as a Luminescent Sensor for Detection of Cr2O7 2− in Water. *Inorg. Chem. Commun.* **2018**, 97, 79–82.

119. Qin, B.; Zhang, X.; Zhang, J. A Stable Multifunctional Cadmium-organic Framework Based on 2D Stacked Layers: Effective Gas Adsorption, and Excellent Detection of Cr3+, CrO4 2−, and Cr2O7 2−. *Dyes Pigm.* **2020**, *174*, 108011.
120. Yang, Y.; Liu, D.; Li, Y.; Dong, G. Two New Luminescent Ternary Cd(II)-MOFs by Regulation of Aromatic Dicarboxylate Ligands Used as Efficient Dual-responsive Sensors for Toxic Metal Ions in Water. *Polyhedron* **2019**, *159*, 32–42.
121. Fan, T.; Xia, T.; Zhang, Q.; Cui, Y.; Yang, Y.; Qian, G. A Porous and Luminescent Metal-organic Framework Containing Triazine Group for Sensing and Imaging of Zn2+. *Micropor. Mesopor. Mater.* **2018**, *266*, 1–6.
122. Xiao, Q.; Liu, D.; Wei, Y.; Cui, G. A New Multifunctional Two-dimensional cobalt(II) Metal–organic Framework for Electrochemical Detection of Hydrogen Peroxide, Luminescent Sensing of Metal Ions, and Photocatalysis. *Polyhedron* **2019**, *158*, 342–351.
123. Cai, S.; Zheng, S.; Fan, J.; Xiao, T.; Tan, J.; Zhang, W. A New Sensor Based on Luminescent Terbium–organic Framework for Detection of Fe3+ in Water. *Inorg. Chem. Commun.* **2011**, *14*, 937–939.
124. Fan, K.; Bao, S.; Nie, W.; Liao, C.; Zheng, L. Iridium(III)-based Metal–organic Frameworks as Multiresponsive Luminescent Sensors for Fe3+, Cr2O7 2–, and ATP2– In Aqueous Media. *Inorg. Chem.* **2018**, *57*, 1079–1089.

Chapter 8

Metal–Organic Frameworks for Fluorescent Detection of Biomolecules

Yi-Hui Wang,[1] Zhen Song,[1] Xin-Yuan Hu,[1] and Huai-Song Wang[*]

Department of Pharmaceutical Analysis, China Pharmaceutical University, Nanjing 210009, China

[*]Emails: wanghuaisong@cpu.edu.cn, wanghuai1234@gmail.com
[1]These authors contributed equally to this work.

Metal-organic frameworks (MOFs) are organic-inorganic hybrid materials with intramolecular pores formed by the self-assembly of organic ligands and metal ions or clusters through coordination bonds. In MOFs, the arrangement of organic ligands and metal ions or clusters has obvious directionality, which can form different frame pore structures, thus showing various adsorption properties, optical properties, electromagnetic properties, etc. On the basis of their properties, MOFs have shown great attractive and potential applications in imaging applications and biological sensing. Metal ions/clusters, organic ligands and loaded guest molecules of MOFs can be used as fluorescent emission centers, and can produce fluorescence response to specific analytes. In this chapter, strategies to fabricate MOF-based luminescent biomolecule sensing platforms will be summarized. Furthermore, MOFs for in vitro sensing and intracellular imaging will be introduced. The future perspectives and potential challenges of MOFs for fluorescence detection of biomolecules were discussed.

Introduction

Biomolecules, ranging from proteins, lipids, nucleic acids and metabolites, are actively involved in various biochemical reactions in the human body, which also play key roles in maintaining human health and resisting diseases. Besides, as biomarkers, biomolecules are commonly used targets for disease diagnosis. Therefore, rapid, sensitive, and accurate detection methods for biomolecules have become a research hotspot, which has an important significance in predicting biological state and condition of human health (*1*, *2*). Up to now, polymerase chain reaction (PCR), colonies forming units (CFU), and enzyme-linked immunosorbent assays (ELISA) are mainly applied for biomolecule detection. However, the detection signal is related to the number of target biomolecules, thus it cannot be measured when the number of target biomolecules is lower than the instrument resolution, which leads to the limitation of traditional detection (*2*, *3*). Compared with traditional techniques, fluorescent nanomaterials with high sensitivity is a good choice for detecting biomolecules.

Based on its high sensitivity, selectivity and simplicity, fluorescent nanomaterials can be widely used for analysis of biomolecules (*4*). The approach adopts an external light source, which can excite the fluorophores of analytes, while realizing fast, real-time detection of biomolecules and reducing costs. By measuring the concentration of these substances (such as, aromatic amino acids and enzyme cofactors), the desired information about the cell condition and metabolism can be obtained (*5*). Especially, fluorescence detection based on metal-organic frameworks (MOFs) for biomolecules presents advantages because of their special physical and chemical properties (*6*).

MOFs, crystals of coordination polymers (CPs) or coordinate networks, are an intriguing class of hybrid functional materials that are self-assembled from metal ions and organic ligands. The "MOFs" was first mentioned by Yaghi and co-works (*7*). In general, the size, structure, function and other properties of MOFs depend on various combinations between different metal ions and organic ligands (*8*). Therefore, MOFs as a kind of promoting materials are expected to apply in numerous areas, such as gas storage, separations, drug delivery, sensing and imaging (*7–10*). In particular, MOFs are potential to be used for fluorescence detection by making fluorescence or luminescent sensors, since fluorescence can be generated from the metal ions, organic ligands and other additions, also can be limitlessly tuned via the connection mode of the composition (*10*).

Furthermore, in order to make MOFs have greater application prospect and more extensively used in fluorescence detection, the classical MOFs are modified to formed MOFs with specific functions. For instance, nanoscale MOFs (NMOFs) which is the size of MOFs scaled down to nanoscale, biological MOFs (bio-MOFs) which is the MOFs coordinated with biological ligands, and lanthanide-based MOFs (MOFs-Ln) which is the MOFs modified by lanthanide ions (*6*, *11*, *12*). With the unique properties and multiple functions of MOFs, it has been adapted for the analysis of a variety of biomolecules, such as, DNA/RNA, enzymes and other small biomolecules.

In this chapter, we will introduce the preparation strategies of MOFs-based luminescent sensing platforms. (The ligands mentioned in this chapter are shown in Table 1) MOFs for in vitro sensing and intracellular imaging will also be discussed in detail. Ultimately, the opportunities and challenges for MOFs for fluorescence detection of biomolecules will be addressed.

Strategies to Fabricate MOF-Based Sensing Platforms

In recent decades, attentions have been paid to the development of MOF-based luminescent sensing platforms for analytical chemistry. So far, different MOFs have been used to detect targets, including small organic molecules, gases, ions, and biomolecules (*13–15*). Compared with inorganic nanomaterial platforms (e.g., quantum dots, graphene oxide, gold nanoparticles/nanoclusters, and MoS_2), MOFs have multiple characteristics including pore size, controllable structure, and acting sites for the construction of various luminescent sensors. In addition, due to its inherent biodegradability and the ability to use biocompatibility building blocks, MOFs have potential advantages for biological applications. The excellent properties of MOFs, including biodegradability, fluorescence, low toxicity, are main drivers for their development in the fields of fluorescent chemistry and biosensing.

The metal ions, guest molecules, or organic ligands could induce luminescence of MOFs. (Figure 1) Luminescent lanthanide ions and organic ligands are the most general choices for the fabrication of luminescent MOFs. Ligands with aromatic or π-conjugated backbones play key roles in the luminescence of MOFs. Most of the ligands have antenna effects on lanthanide MOFs–based luminescence. For constitutively non-luminescent MOFs, guest luminescent species can be accurately encapsulated in the pores of MOFs to fabricate luminescent MOFs. Using their ordered pore structure, MOFs could encapsulate a variety of luminescent guests, including metal complexes, organic dyes and quantum dots (QDs), which is an important strategy for building luminescent MOFs. In addition, some MOFs with enzymatic activity could catalyze specific non-fluorescent analytes to emit fluorescence to "turn-on" sensing, while other homologues remain unreacted without interfering with the analysis. The characteristic of the emission of MOFs makes them excellent candidates in biological sensing and luminescent chemical.

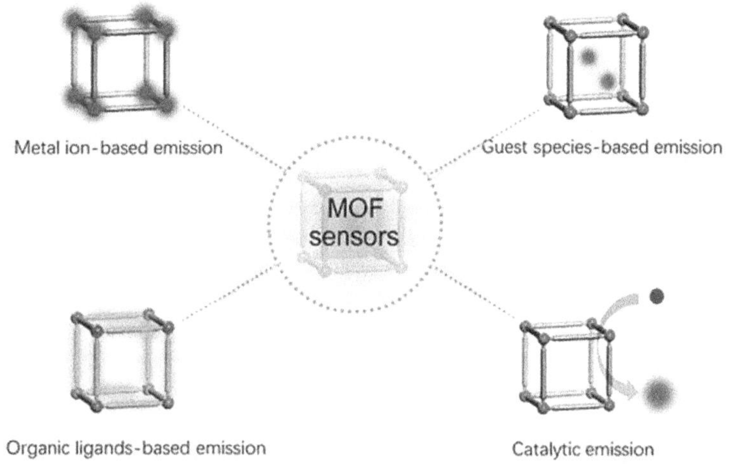

Figure 1. Common strategies for generating luminescence from fabricated MOFs. Reproduced with permission from reference (9). Copyright 2019 Royal Society of Chemistry.

Emissive Metal Ion-Based MOFs

Using emissive metal ions as coordination centers is a general strategy for fabricating luminescent MOFs. Lanthanide (Ln^{3+}) ions are most commonly used in the metal ions. Ln^{3+} ions have unique photophysical properties, including large Stokes shift, sharp emission peak, high photostability and fixed energy level (*16*, *17*). The pore properties and functional groups of MOFs provide a potential platform for molecular recognition through modulation of Ln^{3+} and organic bridged ligands in the frameworks. So Ln^{3+} ions are used to prepare luminescent MOFs for visible or near-infrared (NIR) sensing. In general, the luminescence intensity of Ln^{3+} ions is weak due to the fact that the lanthanide transition is prohibited by Laporte selection rules. There have been many proposals designed to enhance their luminscence. For example, Ln^{3+} ions can be close to the organic linker to provide an antenna effect resulting in a significant increase in the luminescence intensity. During the analysis, the analyte can selectively adjust energy transfer (EnT) efficiency from "antenna" to the Ln^{3+} ions. In addition, introducing different Ln^{3+} ions into the same framework to construct mixed lanthanide MOFs enables ratiometric sensing by influencing the EnT process between Ln^{3+} ions. In other words, the lanthanide MOFs structure can be precisely designed and controlled to meet specific requirements. So far, many lanthanide MOFs have been successfully designed for luminescence sensing of small molecules (e.g., nitroaromatics and picric acid) and inorganic ions (e.g., Fe^{3+}, Cu^{2+}, Ag^+ and $Cr_2O_7^{2-}$) (*18*, *19*).

Organic Ligands-Based Emission

A large number of organic linkers are used in MOF chemistry. Organic ligands composed of *p*-electron-rich aromatic groups can produce luminescent properties in MOFs. As luminescent sensors, the emission of MOFs is affected by the interaction of *p*-electron-rich aromatic ligands with the analytes. The most commonly used method for designing such kind of MOF is the luminescence quenching effect (signal off). There are three main mechanisms leading to fluorescence quenching: photo-induced electron transfer (PET), resonance energy transfer (RET) and inner filter effect (IFE) or the combination. Only some MOFs can utilize the luminescent enhancement effect (signal-on) as a sensor, which depends on bonding or reaction between the ligand and the analytes.

Guest Species-Based Emission

Many of the MOF structures of porous and high specific surface area allows molecular absorption for analyte sensing object, which is undoubtedly a quite simple and convenient method. There have been some strategies using fluorescent guest molecules well demonstrated for the design of luminescent MOFs, such as rhodamine-6G (R6G), fluorescin-isothiocyanate (FITC), and dye-labeled aptamers (*20*, *21*). In addition, encapsulated luminescent nanomaterials [e.g., QDs, carbon dots (CDS), silver nanoclusters (AgNCs), and gold nanoclusters (AuNCs)] can be introduced into MOFs to fabricate luminescent MOFs for sensing (*22–24*). The sensing process is mostly achieved by "signal off" or "ratiometric sensing".

Catalytic Emission

It is well known that MOFs have all kinds of applications, of which catalysis is an important area of research. The uniform and precise crystal structure of MOFs can ensure their unique pore size and large specific surface area, thus improving their catalytic selectivity (*25*, *26*). If MOFs have good catalytic performance for fluorogenic reactions, it will provide an opportunity to establish a new type of high sensitivity catalytic sensing system based on MOFs. A few MOFs have been used for catalytic sensing. For example, HKUST-1, MIL-88, and MIL-53(Fe) have peroxidase-like catalytic activity and have been used to simulate peroxidases for catalytic reactions of H_2O_2 with chromogenic reagents such as methylene blue (*27–29*).

MOFs for In Vitro Sensing

MOF-based methods can be mainly divided into the following two methods: (1) MOFs as fluorescent quenchers on the analyte fluorophore based on fluorescence resonance energy transfer (FRET), PET or charge transfer (*30–32*); (2) MOFs that combine fluorescence or luminescence properties are prepared to sensitively respond to the environment or guest species. MOFs for sensing biomolecules (DNA/RNA and enzyme), as well as small molecules (e.g., amino acids, glucose, reactive oxygen species, dopamine, etc.), were mainly discussed (*10*).

DNA and RNA Sensing

Traditionally, fluorescent nucleic acid detection can be achieved on the nanomaterial-based platforms, including single-walled carbon nanotubes, GO, carbon and gold nanoparticles (GNPs), as commonly used "quenching agents" (*33*, *34*). Recently, studies have shown that MOF shows good selective adsorption affinity and fluorescence quenching ability toward single-stranded DNA (*ss*DNA) and double-stranded DNA (*ds*DNA). In 2013, Chen and co-workers demonstrated that MOF, $Cu(H_2DTOA)$, is reliable and effective in detecting HIV-1 DNA sequences and thrombin. Since then, they have studied various MOFs to detect viral RNA, or DNA. Recently, they prepared a series of water-stable MOFs, $[Cu_3(Cmdcp)_2(dps)_4(H_2O)_4(SO_4)]_n$, $\{[Dy(Cmdcp)(H_2O)_3](NO_3)\cdot 2H_2O_n$, $\{[Zn(HCbdcp)_2]\cdot H_2O_n$ and $[Cu(dcbb)_2]_n$, using the ligands $H_3CbdcpBr$, $H_3CmdcpBr$, and $H_2dcbbBr$ (*30*, *35–37*). The fluorescence of Carboxyfluorescein (FAM) was quenched by MOF through PET process. When the target DNA existed, the *ds*DNA was removed away from the surface of the MOF, and the fluorescence intensity of the FAM was restored. The established system can be used as an effective fluorescent sensing platforms for detection of Sudan virus RNA (SUDV RNA) and human immunodeficiency virus 1 *ds*DNA (HIV *ds*DNA) sequences with detection limits of 73 pM and 196 pM, respectively.

In order to understand the relationship between fluorescence quenching and the surface properties of NMOFs, Xia and co-workers used self-assembly technology to adjust the surface charge properties of nano-MIL-101 (Cr) and lanthanide nanosheet NMOFs (*38*). They labeled two different electrically charged fluorophores FAM and carboxytetramethylrhodamine (TAMRA) on P1 (an aptamer of Homo sapiens tumor suppressor gene). For TAMRA-P1, the fluorescence intensity decreased rapidly when MIL-101 was added to the aptamer solution. After interacting with the target T1, the strong fluorescence emission is recovered. This suggests that MIL-101 has a weaker affinity for *ds*DNA than *ss*DNA. However, the fluorescence quenching behavior of FAM-P1 is opposite. Fluorescence was partially quenched when FAM-P1 interacted with MIL-101, and further quenched after hybridization with T1. They thought that FAM-*ds*DNA (FAM-P1+T1) did not leave MIL-101, but readsorbed to the surface of MIL-101. Since negatively charged carboxyl and phenolic hydroxyl groups on FAM can replace HPO_4^{2-} ions adsorbed/coordinated on the surface of MIL-101 previously, FAM can interact with Cr^{3+} ions on the surface, and FAM-dsDNA is resorbed on MIL-101, leading to further fluorescence quenching. In contrast, TAMRA-*ds*DNA (TAMRA-P1+T1) contains a positively charged group (quaternary ammonium groups) that cannot replace the previously adsorbed HPO_4^{2-}, so TAMRA-*ds*DNA leaves MIL-101 and its fluorescence recovered accordingly. The results show that the surface properties of NMOFs play an important role in the fluorescence change of charged dye molecules during sensing. When the target DNA is present, the FAM-labeled *ss*DNA adsorbed on NMOFs will be further quenched by van der Waals force (Figure 2). This fluorescence quenching behavior is unique to MOFs, which differs from traditional nanoscale quenching agents (GO, CNTs, AuNPs, and MoS_2 nanosheets). The unique fluorescence quenching characteristics of NMOFs provide a new idea for the construction of ratiometric fluorescent DNA biosensors.

Zeolitic imidazolate framework-8 (ZIF-8) is one of the most promising NMOFs due to its good biocompatibility and easy preparation. Nevertheless, ZIF-8 is rarely used as a DNA fluorescence sensing platform due to its poor fluorescence quenching properties. Recently, our group doped lanthanide ions into ZIF-8 to modulate its fluorescence quenching behavior (*39*). La^{3+} doped ZIF-8 enhanced the fluorescence quenching ability of ZIF-8, thus providing an ideal fluorescence sensing platform for DNA detection. The La^{3+} in ZIF-8-La can decrease the background fluorescence, resulting in a signal-to-noise ratio approximately three times higher than that of pure ZIF-8 (Figure 3). The signal-to-background ratio is ideally linear with the analyte (DNA or microRNA) concentration in the range of 1-30 nM. Core-shell CS-ZIF-8-La, where more La^{3+} ions were modified on the surface of ZIF-8, showed the same fluorescence sensing behavior for positive dye-labeled DNA (TAMRA-P1 and TexasRed-P1), but showed completely opposite fluorescence quenching behaviors for negative dye-labeled DNA (FAM-P1 and Cy5-P1). CS-ZIF-8-La serves as a ratiometric sensing quenching platform for fluorescence detection of DNA and microRNA. This work expands the application field of ZIF-8 in biomolecular sensing.

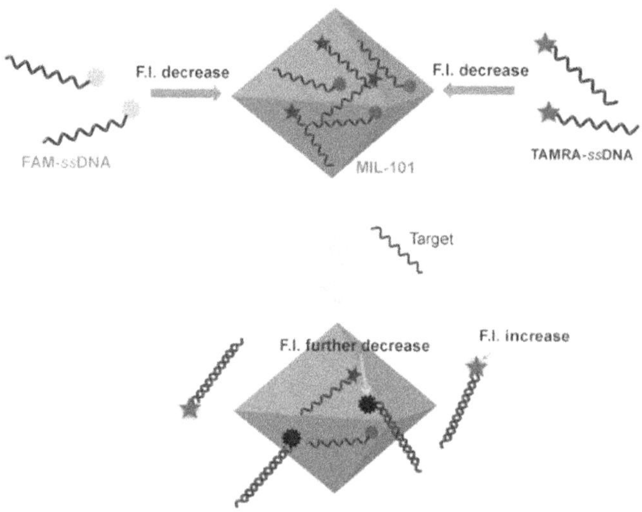

Figure 2. Schematic illustration of the fluorescent DNA assay with MIL-101(Cr) as a sensing platform. Reproduced with permission from reference (38). Copyright 2017 American Chemical Society.

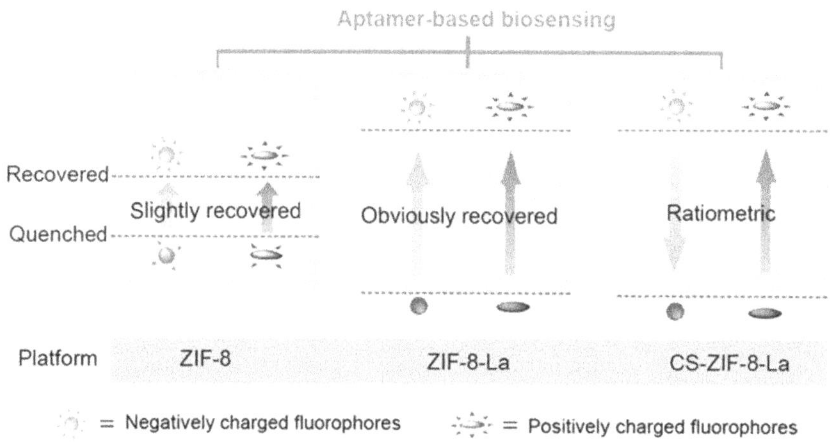

Figure 3. Schematic illustration of the fluorescent DNA/microRNA sensing on the platform of ZIF-8 or lanthanide ion (Ln^{3+}) doped ZIF-8. Reproduced with permission from reference (39). Copyright 2019 American Chemical Society.

Subsequently, Our group investigated the aptamer adsorption capacity and fluorescence quenching capacity of different MOFs, including ZIF-8, MIL-88, MIL-100, MIL-101, Eu-TDA and Tb-TDA [coordinated by 2,2′-thiodiacetic acid (TDA) and Eu^{3+} or Tb^{3+}] in terms of nucleic acid sensing and labled dyes (*6*). FAM-labeled miR-21 aptamer (FAM-Ap) and TexasRed-labeled miR-21 aptamer (TexasRed-Ap), were designed as two types of dye aptamer and their adsorption properties on NMOFs were compared. It was found that TexasRed-Ap can be used for nucleic acid (miR-21) extraction and sensing with pH-responsive nucleotide chain (TexasRed-Ap-pH) or additional random chain ssDNA-1′ (Texasred-Ap-a). After exposed to the target miR-21 in biological samples, Texasred-dsDNA + NMOFs composite materials can be collected to form TexasRed-dsDNA, which could be achieved by changing pH or adding ssDNA-1 matched with ssDNA-1′ (Figure 4). The linear relationship between miR-21 detection was from 0.1 to 200 pM. The results indicate that NMOFs are promising platforms for nucleic acid extraction and fluorescence detection.

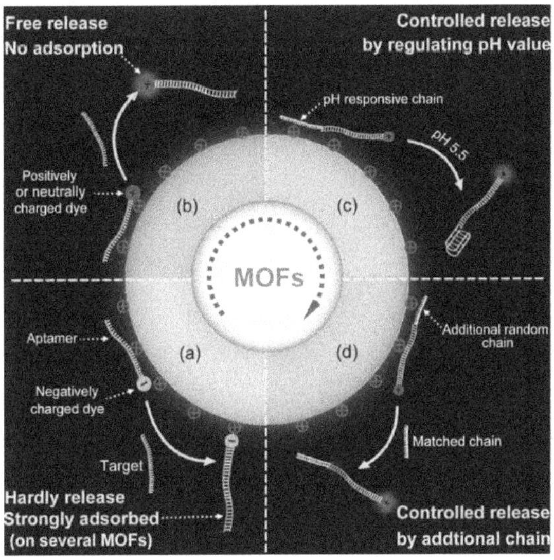

Figure 4. Adsorption behaviors of dye-labeled aptamers on NMOFs. Reproduced with permission from reference (6). Copyright 2021 Elsevier.

Enzyme-Activity Sensing

In order to improve the recyclability of enzymes, reduce the contamination of enzymes in products, and explore a new field for the application of enzymes in biomedicine, the in vitro application of enzymes in various processes especially through the immobilization of enzymes has been widely studied. Having remarkable flexibility in structural design and almost unparalleled surface controllability, MOFs are becoming more and more popular as a candidate material for enzyme immobilization platform (*40–42*). Many MOF-enzyme composites are used to enhance electrochemical signals and show great prospects in protein / enzyme sensing (*43*). Based on zirconium-based PCN-222, Zhang and co-workers studied a photoelectrochemical (PEC) sensor for phosphoprotein (*44*). As shown in Figure 5a, it is sensing mechanism. As a visible light collection unit, the TCPP ligand in PCN-222 can enhance the photocurrent response to dopamine (DA) in oxygen-saturated aqueous medium. The existence of phosphate groups in proteins can coordinate with Zr-O clusters, which can be applied to the unlabeled detection of phosphoprotein (α-casein) at PCN-222 modified electrodes (Figure 5b). The photocurrent signal decreases with the increase of α-casein content because the steric hindrance effect of α-casein prevents the interfacial electron transfer between dissolved O_2, PCN-222, and DA (Figure 5c).

In recent years, some MOFs composites for enzyme activity detection have been designed (*45*, *46*). Lin and co-workers developed a fluorescent biosensor based on gelatinase A, also known as an important member of matrix metalloproteinase-2 (MMP-2). The sensor is based on the FRET, between MOF $[Cu(H_2DTOA)]_n$ and poly(9,9-dioctylfluorenyl-2,7-diyl) dots (Pdots), which are connected by polypeptide chains (*45*). With the aid of a polypeptide chain linker, The fluorescence of Pdots was quenched by $[Cu(H_2DTOA)]_n$ via a FRET process. Polypeptide chains include specific MMP-2 substrates. Therefore, the participation of MMP-2 can release Pdots from MOF and restore the fluorescence of the system. MMP-2 not only can be quantitatively detected, but also has high sensitivity.

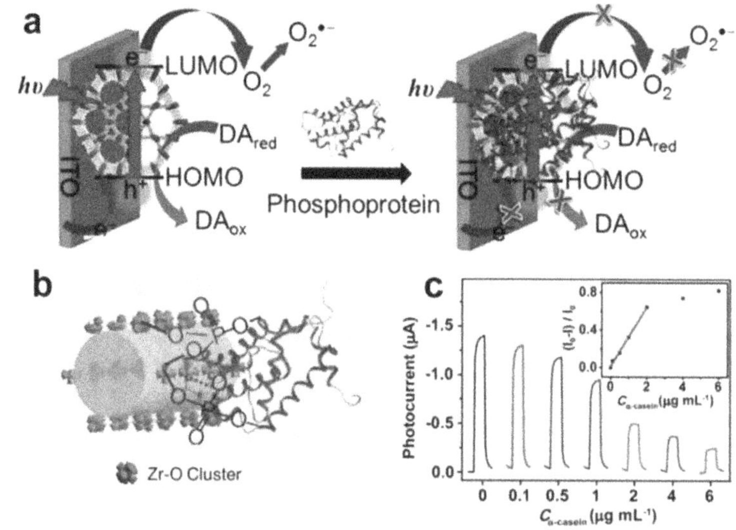

Figure 5. The PCN-222 modified electrode for label-free detection of phosphoprotein. Reproduced with permission from reference (44). Copyright 2016 American Chemical Society.

For dye-labeled *ss*DNA, many MOFs showed obvious fluorescence quenching ability. Based on this phenomenon, MOFs can be designed as a quenching platform for enzyme activity detection (*47*). A probe has recently been developed by Ruan and co-workers to detect deoxyribonuclease I. A type of 24-mer *ss*DNA probe (FAM-P) was labeled by FAM and adsorbed on MOF, $[Ba_2(L)(H_2O)]\cdot(DMF)_{0.5}(H_2O)_3(H_4L$=bis-(3,5-dicarboxy-phenyl)terephthalamide), which showed different affinities to FAM-P with different lengths. MOF has high quenching efficiency for long *ss*DNA probes. Deoxyribonuclease I can shorten the long *ss*DNA chain of the probe, thus preventing fluorescence quenching caused by MOF. This method is suitable for the determination of deoxyribonuclease I activity.

Small-Biomolecule Sensing

Due to its highly regular channel structure and controllable pore size, MOF is suitable for the encapsulation of guest molecules and has potential advantages in chemical sensing. In recent years, enzyme-coated polymetallic oxide optical fibers have achieved fruitful results in the fields of biocatalysts, proteome protection and biosensors (*40*, *48*, *49*). For example, the anti-metabolic drug 6-mercaptopurine (6-MP) is an anti-cancer drug that usually has serious side effects. It is very important to monitor the concentration of serum 6-MP during chemotherapy. Recently, fluorescence sensors based on MOFs have been reported for highly selective detection of 6-NP (*50, 51*). At the excitation wavelength of 335 nm, double emission nanosensors were prepared by combining NH_2-MIL-53(Al) (λ_{em}= 425 nm) with 3-mercaptopropionic acid modified CdTe QDs (MPA-CdTe QDs) (λ_{em} = 528 nm). After adding 6-MP, the fluorescence of NH_2-MIL-53(Al) is selectively quenched because of its strong internal filtration effect, while the fluorescence of MPA-CdTe QDs is enhanced. The nanosensor has been applied to the determination of 6-MP in human urine samples with satisfactory results.

For biological small molecule sensing (*52*), Ma and co-workers used a Tb^{3+}/Eu^{3+}co-doped bio-MOF-1 (expressed as Tb/Eu@bio-MOF-1) to rate dipicolinic acid (DPA) (*53*). DPA is a major and unique component of bacterial spores, including Bacillus anthracis (anthrax) (*54*). Eu^{3+} and Tb^{3+} ions were encapsulated into bio-MOF-1 to synthesize Tb/Eu@bio-MOF-1 luminescent materials, which showed an orange-red emission color (615 nm) due to the efficient energy transfer from Tb^{3+} to Eu^{3+} (Figure 6a). In the presence of DPA, the fluorescence color of Tb/Eu@bio-MOF-1 gradually changes to green in the presence of 545 nm, because DPA chelates with Tb^{3+} ions to form ligand-metal energy transfer. Using this method, DPA visual colorimetric sensing is carried out with the naked eye under ultraviolet light (Figure 6b).

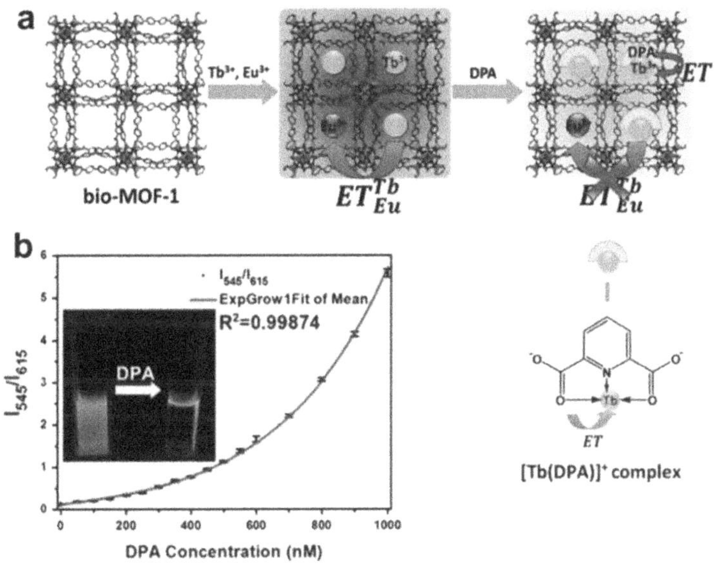

Figure 6. Ratiometric detection of DPA by regulating energy transfer process in bio-MOF-1. Reproduced with permission from reference (53). Copyright 2016 Elsevier.

The detection of biomarkers in biological samples is an important means of clinical diagnosis and follow-up effective treatment. Several MOF-based fluorescent probes with high selectivity and sensitivity have been developed for specific indicators. Yan and co-workers developed a fluorescent probe using MIL-121 to detect hippuric acid (HA), in urine, a biological indicator of toluene exposure (*55*). Eu^{3+} ions were introduced into the channel of MIL-121 to produce fluorescent Eu^{3+}@MIL-121, which is suitable for the analysis of HA in human urine. On the test strip, in the urine added by HA, the fluorescent color of the sensor changed from bright red to crimson. The color of different intensity can be distinguished by the naked eye, and the degree of toluene poisoning can be diagnosed (*40*).

Sensing Other Molecules Existing in Biosystem

Because of their good pore size and adjustable pore wall, MOF have been used as matrix candidates for biomolecules with various structures, such as reactive oxygen species (ROS), including H_2O_2, $^1O_2$, RO·, O_2^-, OH·, and so on. They can also be detected in biological samples on MOF platform. Due to the high oxidation activity of ROS, the sensing of ROS can be realized by quenching the fluorescence of MOF or electrochemical methods (*56–58*). Xia and co-workers investigated that the fluorescence of sulfur-labeled $\{[Eu_2(TDA)_3]\cdot 2H_2O\}_n$ can be selectively quenched by ROS. The mechanism of fluorescence quenching is that the oxidation of TDA sulfide to sulfoxide leads to a significant change in the polarity of the coordination environment of Eu^{3+}. The fluorescence of intramolecular charge transfer (ICT), $\{[Eu_2(TDA)_3]\cdot 2H_2O\}_n$ via ICT to Eu^{3+} via sulfoxide is quenched accordingly. The method has been successfully applied to the monitoring of ROS in biological samples and environmental.

In addition, in order to detect ROS sensitively. Our group discovered a novel fluorescent coordination polymer (NCP_{Cd}), which was synthesized by one-pot method based on dithiocarboxylic acid ligands (*59*). With the reaction of NCP_{Cd} and ROS, the morphology of NCP_{Cd} changed from nanorods to hexagonal particles and then broken into nano-fragments (Figure 7). At the same time, the fluorescence of NCP_{Cd} at 421nm was quenched accordingly. In order to design a highly sensitive ROS probe, rhodamine 6G (R6G) was doped into NCP_{Cd}. In the presence of ROS, the fluorescence of NCP_{Cd} at 421nm was quenched, while R6G was released from the broken nanorods and the fluorescence at 555 nm was restored. R6G-doped NCP_{Cd} (NCP_{Cd}-R) was used as a highly sensitive and selective probe for hydrogen peroxide (H_2O_2). The detection limit is 12.4 nM. In addition, NCP_{Cd}-R is further extended to glucose sensing combined with glucose oxidase (GOx) to oxidize glucose and produce H_2O_2, which shows the potential of practical application.

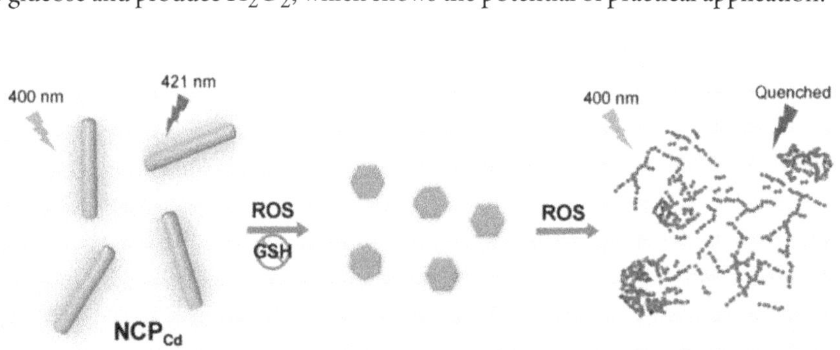

Figure 7. Schematic illustration of fluorescent coordination polymers based on disulfide carboxylate ligand for ROS sensing. Reproduced with permission from reference (59). Copyright 2018 Japan Society for Analytical Chemistry.

Yan has reported the first example based on metal ion fluorescence sensing, which is based on the cation exchange mechanism between target metal ions and frame metal ions in metal ion fluorescence sensing. They reported the exploration of fluorescent MIL-53 (Al) with high selectivity and sensitivity for the detection of Fe^{3+} in aqueous solution (*60*). The cation exchange between Fe^{3+} and Al^{3+} in MIL-53 (Al) results in the fluorescence quenching of MIL-53 (Al). This is due to the transition from strong fluorescence MIL-53 (Al) to weak fluorescence MIL-53 (Fe), which allows the detection of Fe^{3+} in aqueous solution with high selectivity and high sensitivity. The linear range is 3-200 μm, and the detection limit is 0.9 μ m. This method can be used to determine Fe^{3+} in human urine, which provides a new cation exchange mechanism for fluorescence MOFs detection of metal ions, and provides a new strategy for designing selective MOF sensor to detect other metal ions.

MOFs for Intracellular Imaging

For bioimaging, MOF crystals need to be reduced to the nanometer scale to facilitate the endocytosis process of cells. Recent studies have shown that NMOFs can be synthesized by a variety of different technologies, which are mainly used in inorganic and organic polymer nanoparticles (*61*). The NMOFs still retain the structural diversity and physicochemical properties as bulk MOFs, and exhibit particle sizes of tens to hundreds of nanometers. On the one hand, NMOF can show a higher level of structural adaptability (including pore size, morphology, and shape) by selecting suitable metal ions and bridging ligands. On the other hand, after completing the predetermined task, NMOF is essentially biodegradable. The characteristics of NMOFs make it a potential platform for bioimaging.

At present, it has been reported that several kinds of NMOF have been used as fluorescence quenching platforms for dye-labeled aptamers for intracellular RNA or DNA sensing. To monitor ATP activity in living cells, Xia and co-workers created an adenosine-5′-tripho-sphate (ATP)-aptamer/MOF-Ln nanosheet complex for DNA sensing. Multiple dye-aptamers/MOF-Ln nanosheets can be assembled by FAM and TAMRA labeled ATP aptamers (*31*). The ratio of FAM-aptamer to TAMRA-aptamer loaded on dye aptamer/MOF-Ln nanosheets was adjusted to be used as a two-color nanoprobe for the determination of ATP activity. Based on the characteristics of the aforementioned complexe, the authors used dye-P2+MOF-La biosensor to detect adenosine in MCF-7 human breast cancer cells by confocal microscope. After MCF-7 detection, dye-P2+MOF-La was incubated with cells, and intracellular adenosine molecules interacted with dye P2 (Figure 8a). Under the irradiation of 488 nm light, the fluorescence of FAM (green) at 500-550 nm and TAMRA (red) at 550-650 nm changed obviously. The adenosine activity of individual cells can be monitored by dye-P2+MOF-La biosensor. After incubating for 4 h through these steps, the fluorescent color of different cells changed from green to red. As shown in Figure 8c-e, differences in color emission were also monitored in individual cells. The activity of adenosine can be clearly observed in different regions of individual cells and different cells. The results show that the two-color nanoprobe designed by the authors has obvious changes in the color or fluorescence of ATP, and the sensor may be an ideal diagnostic method for measuring ATP in living cells.

Intracellular pH plays a significant role in control cell proliferation, apoptosis, and cell metabolism (*62*, *63*). Lin and colleagues reported a method that allows real-time monitoring of pH in living cells through NMOFs (*64*). Using amino-triphenyldicarboxylic acid (amino-TPDC) and $ZrCl_4$ as raw materials, a kind of UiO NMOF was synthesized by heating dimethylfor-mamide (DMF) solution. Then, UiO NMOF is covalently coupled with FITC (pH-dependent ratiometric sensor) to obtain F-UiO NMOFs with remarkable ratiometric pH-sensing performence (Figure 9a and b). Meanwhile, the F-UiO NMOFs exhibit efficient and rapid endocytosis. As shown in Figure 9c, the UiO NMOFs were spread in the nuclear endosome and maintained the structural integrity of the nuclear endosome in the TEM images. Also, in living cells, F-UiO NMOFs were used to detect the pH ranging from 5.0 to 7.0 (Figure 9d). Besides, after binding to rhodamine-B isothiocyanate (RITC), the UiO NMOF also showed good ratiometric O_2 sensing ability in live cells (*65*).

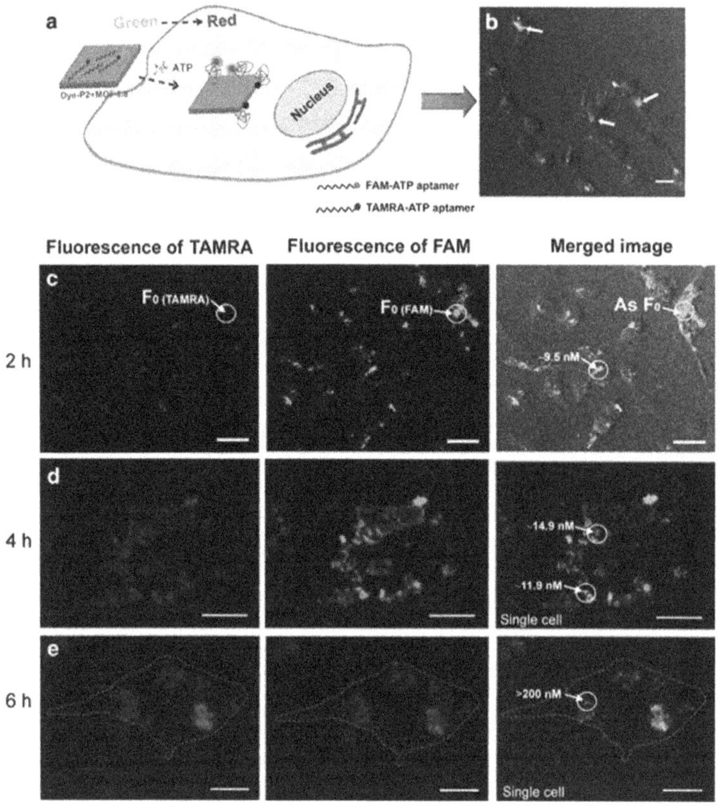

Figure 8. Intracellular fluorescent imaging of adenosines. Reproduced with permission from reference (31). Copyright 2017 Nature Publishing Group.

H_2S is a colorless flammable acid gas, which plays a significant role in human pathology and physiology. Abnormal levels of H_2S in cells are associated with cancer, Alzheimer's disease, and diabetes (*66*, *67*). These characteristics make H_2S a potential target for the diagnosis and treatment of these diseases. To detect H_2S, Ghosh and co-workers invented a strategy of reducing azide to amine based on H_2S for detection (*68*). The sensing platform is the luminous UiO-66-NH_2, which allows the functional synthesis of amine to be modified into azide by diazotization. Because H_2S can mediate the reduction of azides to amines, the prepared UiO-66-N_3 can successfully monitor H_2S by fluorescence opening. Sun and co-workers reported another kind of β-cyclodextrin (CD-MONT-2)-based NMOF for fluorescence detection of H_2S. In the CD-MONT-2, the uncoordinated cyclohexanol molecules will be completely removed to form guest-free CD-MONT-2 when heated at 120°C (*69*). Under the excitation of 330 nm, CD-MONT-2 showed weak fluorescence at 409 nm. The fluorescence intensity increased significantly in the presence of H_2S, and could be observed in living cells.

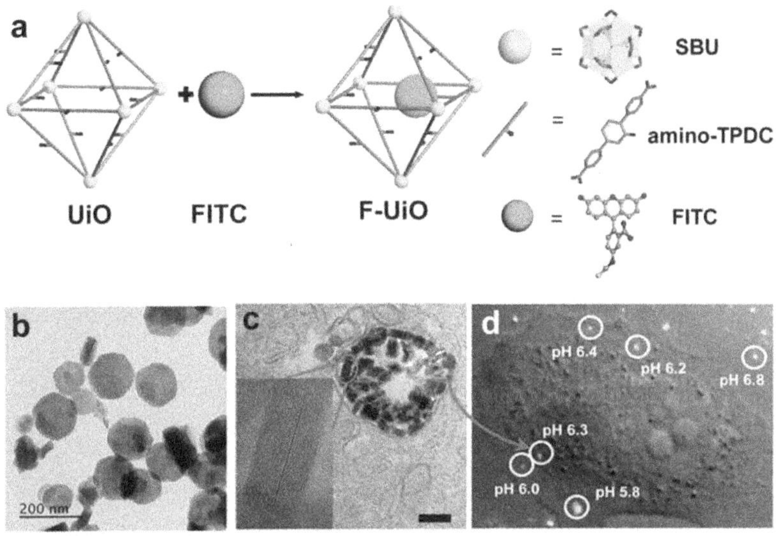

Figure 9. F-UiO NMOFs for real-time intracellular pH sensing. Reproduced with permission from reference (64). Copyright 2014 American Chemical Society.

Table 1. Ligands and Their Abbreviations of MOFs

MOFs	*Ligands*
MIL-101 (Cr)	terephthalic acid
ZIF-8	2-methyl imidazole
PCN-222	TCPP
Tb/Eu@bio-MOF-1	4,4′-biphenyl dicarboxylic acid(BPDC)
$\{[Eu_2(TDA)_3]\cdot 2H_2O\}_n$	2,2′-thiodiacetic acid (TDA)
NCP_{Cd}	dithiocarboxylic acid
ATP-aptamer/MOF-Ln	2,2′-thiodiacetic acid (TDA)
UiO NMOF	amino-triphenyldicarboxylic acida(mino-TPDC)
CD-MONT-2	Cyclodextrin (CD)

Conclusions and Future Perspectives

In this chapter, the fabrication methods of luminescent sensors based on MOF are reviewed. MOFs have the following potential advantages in sensor designing: (1) the emission characteristics of MOFs can be derived from metal ions, organic ligands, guest species or their catalytic activity, making them excellent candidates for fluorescent chemistry and biosensing; (2) simply adjusting organic ligands, metal ions or reaction conditions can achieve structural diversity; (3) Changing the reaction conditions can adjust the morphology of MOFs. At the same time, in the field of biosensing and biological imaging, MOFs as a potential application platform, has also received attention and discussion. For the design of biological probes, MOFs offer various compelling features, such as direct synthesis, porosity, and predictable structure. These allow MOFs to interact specifically with many biomolecules. In addition, post-synthetic modification of MOF surface can greatly improve its biocompatibility. Because of their diversity of composition and structure, MOF-based probes have emerged in vitro and in vivo for sensing and imaging.

MOFs can be used as a fluorescence quenching agent for dye-labeled aptamers. It selectively detects a variety of molecules, including RNA, DNA, and biological small molecules, through the fluorescence "turn-on" strategy. Compared with inorganic nanomaterials such as graphene oxide, graphene, MoS_2 and GNPs, MOFs have potential advantages in bioimaging and biosensing due to their ability to utilize biocompatible building blocks and inherent biodegradability. In addition, MOF, with self-fluorescence properties can be prepared. Because MOF has its own fluorescence properties, it is easily affected by biomolecules.

Despite of wide variety of MOF sensors, their sensing targets are still limited. Sensing targets usually include metal ions, inorganic anions, and explosives. In particular, more research is needed in the application of biomolecular sensing and in vivo sensing. In terms of biosensing, MOF-based sensors still have some problems to be solved. First of all, the interaction between metal ions and biological ligands need systematic research, can provide reasonable bio-MOFs design principle of the system. Second, the stability of MOFs need to be enhanced to avoid degradation before completing sensing, and they also need to consider and improve biocompatibility. Thermochemical stability and water stability of MOFs still need to be strengthened, otherwise it will limit its development in biological applications. Third, in biological environment, the morphology (such as shape and size) of MOF affects its interaction with biomolecules, especially its application in vivo. In order to obtain better imaging results in living organisms or living cells, the size of MOF must be reduced to the nanometer scale so that they can be absorbed by living cells and improve their systemic circulation in the blood. It is well known that the size of bulk MOF crystals is synthesized on the millimeter or micron scale. The major challenge facing researchers now is how to split large-sized MOF crystals into nanomaterials or directly synthesize NMOFs. It is necessary to develop a simple, controllable, cost-effective preparation method, and their influence on the synthesis of materials such as solvent, temperature, pH value, in order to ensure suitable size, uniform shape and processing performance of MOFs.

Therefore, in order to create probes with excellent molecular specificity and precise targeting in living systems, intelligent design of MOF structures will be needed in the future. In vivo efficacy needs to be systematically studied for potential clinical applications, including the accumulation and metabolism of residual substances. In addition, possible clinical application of MOFs has also aroused concerns about their toxic side effects or unknown biological activities. Toxic evaluation of MOFs is essential in the diagnosis and treatment of clinical research. Selecting metal ions and ligands with low toxicity is a possible solution to this problem. Finally, the mechanism of bio-MOFs overcoming barriers in vivo and a series of subsequent processes (such as biological absorption, distribution, transformation, metabolism and excretion) require in-depth study.

Until now, MOF-based ratio sensing with self-calibration ability has received extensive attention. Compared with conventional single peak measurement method, the ratio sensing method can reduce the impact on the environment, and the color and intensity changes are easy to distinguish. MOFs have potential advantages in the design of ratio sensors due to the variety of emission ligands and metal ions and guest molecules that can be selected. Particularly, lanthanide MOFs, can be projected as ratio sensors. In addition, diverse Ln^{3+} based emission in co-doped mixed lanthanide MOFs have encourged researchers to better design multicolor or color-tunable sensors. NMOFs based biocompatible molecules could be modified, which would lead to great advances in intracellular or in vivo imaging, in situ theranostic. That is said, broad application of functional luminescent NMOFs in chemical and biomolecular sensing or imaging should be pursued.

References

1. Zhang, H.-C.; Guo, Y.-M. Advances of Carbon Quantum Dots for Fluorescence Turn-on Detection of Reductive Small Biomolecules. *Chinese. J. Anal. Chem.* **2021**, *49*, 14–23.
2. Yanawut, M.; Takuya, K.; Koji, O. Recent Developments of Point-of-Care (POC) Testing Platform for Biomolecules. *Trac-Trend. Anal. Chem.* **2020**, 116160.
3. Liu, H.; Lei, Y. A Critical Review: Recent Advances in "Digital" Biomolecule Detection with Single Copy Sensitivity. *Biosens. Bioelectron.* **2021**, *177*, 112901.
4. Doan Thanh, N.; Nguyen Khoa, H.; Hoang Van, D.; Nguyen Thi Ai, N.; Nguyen Tien, T.; Dang Ung, V.; Shin, W. S.; Kim, J. S.; Duong Tuan, Q. A Hemicyanine Complex for the Detection of Thiol Biomolecules by Fluorescence. *Dyes Pigments.* **2016**, *131*, 301–306.
5. Lipovsek, B.; Joskowiak, A.; Krc, J.; Topic, M.; Prazeres, D. M. F.; Chu, V.; Conde, J. P. Characterisation of Hydrogenated Silicon-Carbon Alloy Filters with Different Carbon Composition for on-Chip Fluorescence Detection of Biomolecules. *Sensor Actuat. A-Phys.* **2010**, *163*, 96–100.
6. Xie, X.; Ke, R.; Cheng, C.; Wang, Y.-H.; Song, Z.; Zhang, C.-D.; Wang, H.-S. Multiple Adsorption Properties of Aptamers on Metal-Organic Frameworks for Nucleic Acid Assay. *Biosens. Bioelectron.* **2021**, *176*, 112896.
7. Li, H.; Eddaoudi, M.; O'Keeffe, M.; Yaghi, O. M. Design and Synthesis of an Exceptionally Stable and Highly Porous Metal-Organic Framework. *Nature.* **1999**, *402*, 276–279.
8. Della Rocca, J.; Liu, D.; Lin, W. Nanoscale Metal-Organic Frameworks for Biomedical Imaging and Drug Delivery. *Accounts Chem. Res.* **2011**, *44*, 957–968.
9. Liu, Y.; Xie, X.-Y.; Cheng, C.; Shao, Z.-S.; Wang, H.-S. Strategies to Fabricate Metal-Organic Framework (MOF)-based Luminescent Sensing Platforms. *J. Mater. Chem. C* **2019**, *7*, 10743–10763.
10. Wang, H.-S. Metal-Organic Frameworks for Biosensing and Bioimaging Applications. *Coordin. Chem. Rev.* **2017**, *349*, 139–155.
11. Wang, H.-S.; Wang, Y.-H.; Ding, Y. Development of Biological Metal-Organic Frameworks Designed for Biomedical Applications: From Bio-Sensing/Bio-Imaging to Disease Treatment. *Nanoscale Adv.* **2020**, *2*, 3788–3797.
12. Wang, H.-S.; Li, J.; Li, J.-Y.; Wang, K.; Ding, Y.; Xia, X.-H. Lanthanide-based Metal-Organic Framework Nanosheets with Unique Fluorescence Quenching Properties for Two-Color Intracellular Adenosine Imaging in Living Cells. *NPG Asia Mater.* **2017**, *9*.
13. Allendorf, M. D.; Bauer, C. A.; Bhakta, R. K.; Houk, R. J. T. Luminescent Metal–Organic Frameworks. *Chem. Soc. Rev.* **2009**, *38*, 1330–1352.
14. Cui, Y.; Zhang, J.; He, H.; Qian, G. Photonic Functional Metal–Organic Frameworks. *Chem. Soc. Rev.* **2018**, *47*, 5740–5785.
15. Zhou, J.; Tian, G.; Zeng, L.; Song, X.; Bian, X.-w. Nanoscaled Metal-Organic Frameworks for Biosensing, Imaging, and Cancer Therapy. *Adv. Healthc. Mater.* **2018**, *7*, 1800022.
16. Yan, B. Lanthanide-Functionalized Metal–Organic Framework Hybrid Systems to Create Multiple Luminescent Centers for Chemical Sensing. *Accounts Chem. Res.* **2017**, *50*, 2789–2798.
17. Rocha, J.; Brites, C. D. S.; Carlos, L. D. Lanthanide Organic Framework Luminescent Thermometers. *Chem. Eur. J.* **2016**, *22*, 14782–14795.

18. Meyer, L. V.; Schönfeld, F.; Müller-Buschbaum, K. Lanthanide Based Tuning of Luminescence in MOFs and Dense Frameworks from Mono-and Multimetal Systems to Sensors and Films. *Chem. Commun.* **2014**, *50*, 8093–8108.
19. Rocha, J.; Carlos, L. D.; Paz, F. A. A.; Ananias, D. Luminescent Multifunctional Lanthanides-Based Metal–Organic Frameworks. *Chem. Soc. Rev.* **2011**, *40*, 926–940.
20. Han, T.-T.; Yang, J.; Liu, Y.-Y.; Ma, J.-F. Rhodamine 6g Loaded Zeolitic Imidazolate Framework-8 (ZIF-8) Nanocomposites for Highly Selective Luminescent Sensing of Fe^{3+}, Cr^{6+} and Aniline. *Micropor. Mesopor. Mat.* **2016**, *228*, 275–288.
21. Schrimpf, W.; Jiang, J.; Ji, Z.; Hirschle, P.; Lamb, D. C.; Yaghi, O. M.; Wuttke, S. Chemical Diversity in a Metal–Organic Framework Revealed by Fluorescence Lifetime Imaging. *Nat. Commun.* **2018**, *9*, 1647.
22. Hill, S. A.; Benito-Alifonso, D.; Davis, S. A.; Morgan, D. J.; Berry, M.; Galan, M. C. Practical Three-Minute Synthesis of Acid-Coated Fluorescent Carbon Dots with Tuneable Core Structure. *Sci. Rep.* **2018**, *8*, 12234.
23. Huang, R.-W.; Wei, Y.-S.; Dong, X.-Y.; Wu, X.-H.; Du, C.-X.; Zang, S.-Q.; Mak, T. C. W. Hypersensitive Dual-Function Luminescence Switching of a Silver-Chalcogenolate Cluster-Based Metal–Organic Framework. *Nat. Chem.* **2017**, *9*, 689–697.
24. Peng, D.; Zhang, L.; Li, F.-F.; Cui, W.-R.; Liang, R.-P.; Qiu, J.-D. Facile and Green Approach to the Synthesis of Boron Nitride Quantum Dots for 2,4,6-Trinitrophenol Sensing. *ACS. Appl. Mater. Inter.* **2018**, *10*, 7315–7323.
25. Dhakshinamoorthy, A.; Li, Z.; Garcia, H. Catalysis and Photocatalysis by Metal- Organic Frameworks. *Chem. Soc. Rev.* **2018**, *47*, 8134–8172.
26. Yang, Q.; Xu, Q.; Jiang, H.-L. Metal–Organic Frameworks Meet Metal Nanoparticles: Synergistic Effect for Enhanced Catalysis. *Chem. Soc. Rev.* **2017**, *46*, 4774–4808.
27. Lin, Y.; Dai, Y.; Sun, Y.; Ding, C.; Sun, W.; Zhu, X.; Liu, H.; Luo, C. A Turn-on Chemiluminescence Biosensor for Selective and Sensitive Detection of Adenosine Based on HKUST-1 and QDs-Luminol-Aptamer Conjugates. *Talanta* **2018**, *182*, 116–124.
28. Lin, T.; Qin, Y.; Huang, Y.; Yang, R.; Hou, L.; Ye, F.; Zhao, S. A Label-Free Fluorescence Assay for Hydrogen Peroxide and Glucose Based on the Bifunctional MIL-53(Fe) Nanozyme. *Chem. Commun.* **2018**, *54*, 1762–1765.
29. Zhang, Y.; Zhang, W.; Chen, K.; Yang, Q.; Hu, N.; Suo, Y.; Wang, J. Highly Sensitive and Selective Colorimetric Detection of Glutathione via Enhanced Fenton-Like Reaction of Magnetic Metal-Organic Framework. *Sensor Actuat. B-Chem.* **2018**, *262*, 95–101.
30. Yang, S.-P.; Chen, S.-R.; Liu, S.-W.; Tang, X.-Y.; Qin, L.; Qiu, G.-H.; Chen, J.-X.; Chen, W.-H. Platforms Formed from a Three-Dimensional Cu-Based Zwitterionic Metal–Organic Framework and Probe ss-DNA: Selective Fluorescent Biosensors for Human Immunodeficiency Virus 1 ds-DNA and Sudan Virus RNA Sequences. *Anal. Chem.* **2015**, *87*, 12206–12214.
31. Wang, H.-S.; Li, J.; Li, J.-Y.; Wang, K.; Ding, Y.; Xia, X.-H. Lanthanide-Based Metal-Organic Framework Nanosheets with Unique Fluorescence Quenching Properties for Two-Color Intracellular Adenosine Imaging in Living Cells. *NPG Asia. Mater.* **2017**, *9*, e354.
32. Tan, H.; Tang, G.; Wang, Z.; Li, Q.; Gao, J.; Wu, S. Magnetic Porous Carbon Nanocomposites Derived from Metal-Organic Frameworks as a Sensing Platform for DNA Fluorescent Detection. *Anal. Chim. Acta.* **2016**, *940*, 136–142.

33. Xu, J.-J.; Zhao, W.-W.; Song, S.; Fan, C.; Chen, H.-Y. Functional Nanoprobes for Ultrasensitive Detection of Biomolecules: An Update. *Chem. Soc. Rev.* **2014**, *43*, 1601–1611.
34. Chung, C.; Kim, Y.-K.; Shin, D.; Ryoo, S.-R.; Hong, B. H.; Min, D.-H. Biomedical Applications of Graphene and Graphene Oxide. *Accounts Chem. Res.* **2013**, *46*, 2211–2224.
35. Qin, L.; Lin, L.-X.; Fang, Z.-P.; Yang, S.-P.; Qiu, G.-H.; Chen, J.-X.; Chen, W.-H. A Water-Stable Metal–Organic Framework of a Zwitterionic Carboxylate with Dysprosium: A Sensing Platform for Ebolavirus RNA Sequences. *Chem. Commun.* **2016**, *52*, 132–135.
36. Zhao, H.-Q.; Yang, S.-P.; Ding, N.-N.; Qin, L.; Qiu, G.-H.; Chen, J.-X.; Zhang, W.-H.; Chen, W.-H.; Hor, T. S. A. A Zwitterionic 1d/2d Polymer co-Crystal and its Polymorphic sub-Components: A Highly Selective Sensing Platform for HIV ds-DNA Sequences. *Dalton Trans.* **2016**, *45*, 5092–5100.
37. Zhao, H.-Q.; Qiu, G.-H.; Liang, Z.; Li, M.-M.; Sun, B.; Qin, L.; Yang, S.-P.; Chen, W.-H.; Chen, J.-X. A Zinc(II)-Based Two-Dimensional MOF for Sensitive and Selective Sensing of HIV-1 ds-DNA Sequences. *Anal. Chim. Acta* **2016**, *922*, 55–63.
38. Wang, H.-S.; Liu, H.-L.; Wang, K.; Ding, Y.; Xu, J.-J.; Xia, X.-H.; Chen, H.-Y. Insight into the Unique Fluorescence Quenching Property of Metal-Organic Frameworks Upon DNA Binding. *Anal. Chem.* **2017**, *89*, 11366–11371.
39. Hao, Y.-B.; Shao, Z.-S.; Cheng, C.; Xie, X.-Y.; Zhang, J.; Song, W.-J.; Wang, H.-S. Regulating Fluorescent Aptamer-Sensing Behavior of Zeolitic Imidazolate Framework (ZIF-8) Platform Via Lanthanide Ion Doping. *ACS. Appl. Mater. Inter.* **2019**, *11*, 31755–31762.
40. Lyu, F.; Zhang, Y.; Zare, R. N.; Ge, J.; Liu, Z. One-Pot Synthesis of Protein-Embedded Metal-Organic Frameworks with Enhanced Biological Activities. *Nano Lett.* **2014**, *14*, 5761–5765.
41. Bhardwaj, S. K.; Sharma, A. L.; Bhardwaj, N.; Kukkar, M.; Gill, A. A. S.; Kim, K.-H.; Deep, A. TCNQ-Doped Cu-Metal Organic Framework as a Novel Conductometric Immunosensing Platform for the Quantification of Prostate Cancer Antigen. *Sensor Actuat. B-Chem.* **2017**, *240*, 10–17.
42. Guo, T.; Deng, Q.; Fang, G.; Gu, D.; Yang, Y.; Wang, S. Upconversion Fluorescence Metal-Organic Frameworks Thermo-Sensitive Imprinted Polymer for Enrichment and Sensing Protein. *Biosens. Bioelectron.* **2016**, *79*, 341–346.
43. Ling, P.; Lei, J.; Jia, L.; Ju, H. Platinum Nanoparticles Encapsulated Metal-Organic Frameworks for the Electrochemical Detection of Telomerase Activity. *Chem. Commun.* **2016**, *52*, 1226–1229.
44. Zhang, G.-Y.; Zhuang, Y.-H.; Shan, D.; Su, G.-F.; Cornier, S.; Zhang, X.-J. Zirconium-Based Porphyrinic Metal-Organic Framework (PCN-222): Enhanced Photoelectrochemical Response and its Application for Label-Free Phosphoprotein Detection. *Anal. Chem.* **2016**, *88*, 11207–11212.
45. Yang, W.; Zhang, G.; Weng, W.; Qiu, B.; Guo, L.; Lin, Z.; Chen, G. Signal on Fluorescence Biosensor for MMP-2 Based on Fret between Semiconducting Polymer Dots and a Metal Organic Framework. *RSC. Adv.* **2014**, *4*, 58852–58857.
46. Wang, K.; Li, N.; Zhang, J.; Zhang, Z.; Dang, F. Size-Selective QD@MOF Core-Shell Nanocomposites for the Highly Sensitive Monitoring of Oxidase Activities. *Biosens. Bioelectron.* **2017**, *87*, 339–344.

47. Wang, Y.; Zhu, Y.; Binyam, A.; Liu, M.; Wu, Y.; Li, F. Discovering the Enzyme Mimetic Activity of Metal-Organic Framework (MOF) for Label-Free and Colorimetric Sensing of Biomolecules. *Biosens. Bioelectron.* **2016**, *86*, 432–438.
48. Li, Y.; Guo, A.; Chang, L.; Li, W.-J.; Ruan, W.-J. Luminescent Metal-Organic-Framework-Based Label-Free Assay of Polyphenol Oxidase with Fluorescent Scan. *Chem. Eur. J.* **2017**, *23*, 6562–6569.
49. Cao, X.; Ni, Y.; Zhang, A.; Xu, S.; Chen, K.; Ouyang, P. Encapsulation of Enzymes in Metal Ion-Surfactant Nanocomposites for Catalysis in Highly Polar Solvents. *Chem. Commun.* **2017**, *53*, 3134–3137.
50. Jin, M.; Mou, Z.-L.; Zhang, R.-L.; Liang, S.-S.; Zhang, Z.-Q. An Efficient Ratiometric Fluorescence Sensor Based on Metal-Organic Frameworks and Quantum Dots for Highly Selective Detection of 6-Mercaptopurine. *Biosens. Bioelectron.* **2017**, *91*, 162–168.
51. Sun, Z.; Liu, Y.; Li, Y. Selective Recognition of 6-Mercaptopurine Based on Luminescent Metal-Organic Frameworks Fe-MIL-88NH_2. *Spectrochim. Acta, Pt. A: Mol. Biomol. Spectrosc.* **2015**, *139*, 296–301.
52. He, C.; Liu, D.; Lin, W. Nanomedicine Applications of Hybrid Nanomaterials Built from Metal-Ligand Coordination Bonds: Nanoscale Metal-Organic Frameworks and Nanoscale Coordination Polymers. *Chem. Rev.* **2015**, *115*, 11079–11108.
53. Zhang, Y.; Li, B.; Ma, H.; Zhang, L.; Zheng, Y. Rapid and Facile Ratiometric Detection of an Anthrax Biomarker by Regulating Energy Transfer Process in Bio-Metal-Organic Framework. *Biosens. Bioelectron.* **2016**, *85*, 287–293.
54. Tan, H.; Ma, C.; Chen, L.; Xu, F.; Chen, S.; Wang, L. Nanoscaled Lanthanide/Nucleotide Coordination Polymer for Detection of an Anthrax Biomarker. *Sensor Actuat. B-Chem.* **2014**, *190*, 621–626.
55. Hao, J.-N.; Yan, B. Recyclable Lanthanide-Functionalized MOF Hybrids to Determine Hippuric Acid in Urine as a Biological Index of Toluene Exposure. *Chem. Commun.* **2015**, *51*, 14509–14512.
56. Wang, Q.; Yang, Y.; Gao, F.; Ni, J.; Zhang, Y.; Lin, Z. Graphene Oxide Directed One-Step Synthesis of Flowerlike Graphene@HKUST-1 for Enzyme-Free Detection of Hydrogen Peroxide in Biological Samples. *Acs. Appl. Mater. Inter.* **2016**, *8*, 32477–32487.
57. Wang, H.-S.; Bao, W.-J.; Ren, S.-B.; Chen, M.; Wang, K.; Xia, X.-H. Fluorescent Sulfur-Tagged Europium(III) Coordination Polymers for Monitoring Reactive Oxygen Species. *Anal. Chem.* **2015**, *87*, 6828–6833.
58. Tan, H.; Ma, C.; Li, Q.; Wang, L.; Xu, F.; Chen, S.; Song, Y. Functionalized Lanthanide Coordination Polymer Nanoparticles for Selective Sensing of Hydrogen Peroxide in Biological Fluids. *Analyst* **2014**, *139*, 5516–5522.
59. Hao, Y.-B.; Yuan, D.; Chang, Y.-X.; Song, W.-J.; Wang, H.-S. Fluorescent Disulfide-Functional Coordination Polymers for Sensitive Detection of Hydrogen Peroxide. *Anal. Sci.* **2018**, *34*, 1379–1384.
60. Yang, C. X.; Ren, H. B.; Yan, X. P. Fluorescent Metal-Organic Framework MIL-53(Al) for Highly Selective and Sensitive Detection of Fe^{3+} in Aqueous Solution. *Anal. Chem.* **2013**, *85*, 7441–7446.

61. Lin, W. B. Nanoscale Metal-Organic Frameworks for Biomedical Imaging and Drug Delivery. *Abstr. Pap. Am. Chem. Soc.* **2014**, *248*, 1.
62. Casey, J. R.; Grinstein, S.; Orlowski, J. Sensors and Regulators of Intracellular pH. *Nat. Rev. Mol. Cell Biol.* **2010**, *11*, 50–61.
63. Webb, B. A.; Chimenti, M.; Jacobson, M. P.; Barber, D. L. Dysregulated pH: A Perfect Storm for Cancer Progression. *Nat. Rev. Cancer.* **2011**, *11*, 671–677.
64. He, C.; Lu, K.; Lin, W. Nanoscale Metal-Organic Frameworks for Real-Time Intracellular pH Sensing in Live Cells. *J. Am. Chem. Soc.* **2014**, *136*, 12253–12256.
65. Xu, R.; Wang, Y.; Duan, X.; Lu, K.; Micheroni, D.; Hu, A.; Lin, W. Nanoscale Metal-Organic Frameworks for Ratiometric Oxygen Sensing in Live Cells. *J. Am. Chem. Soc.* **2016**, *138*, 2158–2161.
66. Eto, K.; Asada, T.; Arima, K.; Makifuchi, T.; Kimura, H. Brain Hydrogen Sulfide is Severely Decreased in Alzheimer's Disease. *Biochem. Biophys. Res. Commun.* **2002**, *293*, 1485–1488.
67. Kamoun, P.; Belardinelli, M. C.; Chabli, A.; Lallouchi, K.; Chadefaux-Vekemans, B. Endogenous Hydrogen Sulfide Overproduction in Down Syndrome. *Am. J. Med. Genet. A* **2003**, *116A*, 310–311.
68. Szabo, C.; Coletta, C.; Chao, C.; Modis, K.; Szczesny, B.; Papapetropoulos, A.; Hellmich, M. R. Tumor-Derived Hydrogen Sulfide, Produced by Cystathionine-Beta-Synthase, Stimulates Bioenergetics, Cell Proliferation, and Angiogenesis in Colon Cancer. *Proc. Natl. Acad. Sci. U. S. A* **2013**, *110*, 12474–12479.
69. Wei, Y.; Sun, D.; Yuan, D.; Liu, Y.; Zhao, Y.; Li, X.; Wang, S.; Dou, J.; Wang, X.; Hao, A.; Sun, D. Pb(II) Metal–Organic Nanotubes Based on Cyclodextrins: Biphasic Synthesis, Structures and Properties. *Chem. Sci.* **2012**, *3*, 2282–2287.

Chapter 9

Recent Advances and Challenges in Selective Environmental Applications of Metal–Organic Frameworks

Kajal Saini,[1] Shivbabu Yadav,[2] Monika Jain,[3] Arvind Gupta,[2,4] and Smita S. Kumar[*,1,2]

[1]Department of Environmental Sciences, J.C. Bose University of Science & Technology, YMCA, Faridabad, Haryana 121006, India

[2]Centre for Energy Studies, J.C. Bose University of Science & Technology, YMCA, Faridabad, Haryana 121006, India

[3]Department of Natural Resource Management, College of Forestry, Banda University of Agriculture & Technology, Banda, Uttar Pradesh 210001, India

[4]Department of Mechanical Engineering, Centre for Energy Studies, J.C. Bose University of Science & Technology, YMCA, Faridabad, Haryana 121006, India

[*]Email: smita@jcboseust.ac.in

Metal Organic Frameworks (MOFs) are an evolving class of crystalline porous materials that have emerged as a recent research hotspot in material science. MOFs possess varied structural as well as physical and chemical functional properties such as porosity, stability, surface area. Such potential and multifunctional properties of MOFs make it preferable over conventionally used technologies.Recent advanced developments in this field focus on computational screenings, biosensing, adsorption, thermal energy storage and photolysis of water to generate Hydrogen by using MOFs. Metal Organic Frameworks (MOFs) have been used for the removal of a vast range of substances from the environment ranging from heavy metal ions (Cr, Hg, Pb, As and Cd), persistent organic pollutants (dyes, phenols, herbicides and pesticides), radionuclides (Xe, Ur, Th, Sr, Cs and I_2), and pathogens to greenhouse gases (carbon dioxide) which pose a great threat to the ecosystem because of their harmful impacts on living beings and their surroundings. The ability of MOFs to adsorb heavy metals, to adsorb and photo degrade persistent organic pollutants, to sequester carbon dioxide and other substancessuch as radionuclides in order to remove them from surroundings has proven to be useful and has potential to be scaled up for industrial applications thus making them a class of material most useful in the environmental sector. Hereby in this manuscript, the recent environmental applications of MOF based materials have been discussed along with challenges and future prospects.

Introduction

Metal organic frameworks (MOFs) as the name suggests are organic-inorganic materials having metal ions bound to organic ions/molecules known as ligands. MOFs possess some unique characters like thermal stability, greater surface area, nanoscale cavities with uniform structure, tunable pore size, tunnels and accessible cages, controllable morphology and various particle dimensions, surface modification and in-pore functionality (*1*). These mentioned properties of MOFs have received great attention and make it promising material for removal of heavy metals, radio nuclides, persistent organic pollutants like dyes, phenols, pesticides, herbicides etc. and also for separation and sequestration of gases like Carbon dioxide. Many studies have demonstrated that MOFs perform efficiently in various applications in comparison to conventional porous materials, such as zeolites and porous carbon-based substances (2).

Initially, these MOFs exhibited low stability in aqueous media but now it has been modified to make it more water stable. Modification makes it to act as an effective adsorbent for pollutant removal because occurrence of these heavy metals in surface and ground water is a genuine worldwide concern. UiO-66, MIL-53(Al@100aBDC, ZIF-67, MOR-1-HA, AMOF-1, 3D Co (II) MOF are some of the modified metal organic frameworks. These can be utilised in uptake of heavy metals from water, mentioned along with their uptake mechanism and method of synthesis briefly (*3*, *4*). Nowadays, the use of such materials for radionuclide uptake has beengoing through speedy improvement. In this, the current improvement ofmetal organic frameworks in sequestering radionuclide from radioactive wastes, such asCaesium, Strontium, Uranium, Thorium, Iodine, Xenon etc. The synthesis method of MOF composites for capturing radionuclides as well as the removal mechanisms is briefly mentioned, some of MOFs utilised in this includeHKUST-1 MOF, MIL-101-Cr-HMTA etc. (*5*, *6*).

Also, effective methods to remove Persistent Organic Pollutants are urgently required as these synthetic substances have a combination of chemical and physical features that allows them to travel far distances for a safe and sound environment. So far, POPs have been removed by photocatalytic oxidation, advanced oxidation and biodegradation etc. Adsorptive removal of persistent organic pollutants with suitable MOFs as adsorbents like NH2-MIL-101-Al,MOF-235UiO-66(Zr), MIL-53-Cr, ZIF-8 etc, can be one of the most favourable methods as it is a simple process, low cost and employed in mild working conditions (*7*–*11*).

Recently, progress on MOFs framework has been remarkable because of its properties that allow it to use as a biosensor for detection of food contaminants, as a drug carrier, in adsorption thermal energy storage application, photocatalysis of water to generate hydrogen upon irradiation with ultraviolet, visible and NIR. Researchers have also consolidated antibacterial materials into cages of MOFs to provide them with antibacterial properties to kill and prevent growth of microbes mainlybacteria and has also been utilised as an active cathode material in batteries (*12*–*22*). Such extensive applications face some challenges and obstructions that needfurther improvement accordingly. This chapter provides a mini-review of applications of MOFs, recent advancements and challenges.

Environmental Applications of Metal–Organic Frameworks

Removal of Heavy Metals Using MOFs

The contamination of surface and groundwater with heavy metalslike Arsenic, Chromium, Cadmium, Copper, Mercury and Lead is a genuine worldwide concern, both naturally, as well as for human wellbeing. Excess of these components presents extreme health hazards for people, and furthermore for other life structures through bioaccumulation along natural ways of life (*23*). Consequently, steps ought to be taken to reduce the concentration of such components in aqueous media to favourable limits. MOFs have been employed for removal of these metals from water because of its high potential for adsorption. Nowadays, the utilization of MOFs in eliminating contaminants from water is a progressing area of research (*24*).

MOF is a crystalline material having inorganic metal clusters/ions, through metal linker bonds they are connected to organic occurring linkers (*25*). MOFs generally have large surface area because of presence of large number of internal pores (*26*). It was found that the binding energy of MOFs with metal linker is higher than water and metal centre, this property resist the inclusion of water during internal water diffusion (*1*). There is also an essential inadequacy of early MOFs as they show low stability in water, which fundamentally restricts their pragmatic use with water but now it has been modified to make it more structurally stable, for example- UIO-66 (*3, 4*). These structurally Water-stable MOFs have potential to be utilized as adsorbents in the effective expulsion of target adsorbates from fluid arrangements. In the subsequent section, uptake of heavy metals from water using various MOFs are discussed.

Chromium

Functionalised MOFs with three imine and azine groups were synthesized- TMU-4, 5 and 6. **TMU-5** showed 3D structure of linked pores which was highly ordered and having metal ion removing capacity. It was observed that with increasing pH adsorption capacity of TMU-5 to adsorb Cr(III) increases up to pH 10 (*2*). Another significant MOF was reported by Wang et al., a **chitosan MOF (UiO-66)** composite was synthesized by grinding and irradiation with microwaves. It was employed to uptake Cr(VI) at low pH, increasing the pH led to decrease in adsorption capacities as active sites are blocked by OH groups. The removal capacities of UiO-66 were discovered to be greater than that for chitosan itself (*27*).

Saleem et al. documented the uptake of Cr(III) from water using synthetically modified MOFs. The most efficient MOFs was **UiO-66-NHC(S)NHMe – an S-functionalized UiO-66-NH$_2$** which exhibited improved water stability. This novel modified MOFs was synthesized by reacting UiO-66 with isothiocynates or diphosgene/thiophosgene. This modified MOFs displayed 25 times more adsorption capacity than unmodified one (*28*). Another Cr(VI) adsorbing MOF was documented by Li et al. named **ZIF-67**. It was found that uptake mechanism was ion exchange between the hydroxyl groups of MOF and Chromium (VI) anions that led to increase in pH from 7.6 to 9 pH in 3 hours during process of adsorption. It was also found that in acidic medium, electrostatic attraction between Cr(VI) anions and MOFs promoted adsorption process (*29*).

A safe and fast synthesis of **MOR-1-HA** was reported by Raptiet al. It is an amino functionalised UiO-66 framework coated with alginic acid. It was produced by reflux synthesis. The mechanism for adsorption was ion exchange in which Cr(VI) ions in the solution exchanged with Chloride anions of MOFs. It worked efficiently in the pH range of 2-8 without altering the adsorption capacity of MOR-1-HA. removal capacity was reported to be highly efficient around 280 mg/g that is highly appreciable compared to unmodified MOR-1. It was concluded that coating with alginic acid increases the adsorption capacity (*30*).

Cadmium

Chakraborty et al. synthesized **AMOF-1** based. It was found to be an efficient framework that can remove cadmium(II) ions from water. The adsorption capacity was found to be 41 mg/g. It was observed that it is highly selective for Cd(II) ions, but not completely (*31*). A 3-Dimensional cobalt and **TATAB**based **MOF (3D Co(II) MOF)** for removal of Cadmium(II) ions was reported by Abbasi et al. This MOF was obtained by Ultrasound Irradiation or hydrothermal synthesis to give small sized particles. It revealed that nanostructures of 3D Co (II) MOF have better adsorption capacities than 3D Co (II) MOF. It did not work efficiently in low pH due to absence or less no. of active ANH groups. It can be reusable as it remained intact after desorption (*32*). Another type of Cadmium (II) removal by **MOF ($Cu_3(BTC)_2$-SO_3H framework)**was documented by Wang et al. It was synthesized by modifying $Cu_3(BTC)_2$ with sulfonic acid. Adsorption was efficient at pH 6, below this pH there was hindrance and active sites were filled by hydrogen ions which limits the adsorption. This MOF can be reused after washing and drying (*33*).

Fang et al. documented an **MOF-PCN-100** which is constructed from $Zn_4O(CO2)_6$ SBUs (secondary building units) and TATAB linkers (i.e., 4,40 ,400-s-triazine-1,3,5-triyl tri-p-aminobenzoate). In this, metal ions were encapsulated due to chelating coordination mode of MOF. It was found that MOF-PCN-100 efficiently remove Cadmium(II) ions (*34*). Zhang et al. evaluated the Cadmium(II) ion adsorption capabilities of **HS-mSi@MOF-5**. It is an advanced version of MOF-5 which is a Zinc based with the help of hydrothermal method. HS-mSi@MOF-5 (98 mg/g) showed better adsorption capacity than that of MOF-5 (43.6 mg/g) (*35*). Rahimi and Mohaghegh documented the adsorption efficiency of **magnetic Cu-terephthalate MOF** for cadmium ions. It is a copper-based material which was synthesized by solvothermal methods. Cu-terephthalate MOF has readily available carboxyl groups which greatly boost the chemical adsorption process. The efficiency of removal of cadmium ions increased with increasing pH of the solution This MOF has the highest capacity to remove 100mg/g of Cadmium (II) ions from water (*36*).

Arsenic

Jian et al. showed that **ZIF-8** removed As(III) and As(V) moderately at neutral pH. This MOF was prepared by slightly modified method at room temperature. The rate of As species adsorption was controlled by particle diffusion and not through the pores of MOF. It is because large sized Arsenic cannot fit through the pores of ZIF-8 (*37*). Vu et al. documented that **MIL-53 (Fe)** can be synthesized by free-solvothermal methods. MIL-53 (Fe) showed uptake capacity of 21mg/g for removal of As(V). Lewis acid base interaction between anions and MOF was the basis for adsorption of As(V) (*38*). Adsorption capacities of **Fe-BTC** was reported by Zhu et al. to remove As(V). It was produced by an autogenous pressure synthesis. It was observed that As(V) removal efficiency of Fe-BTC at pH between 2-10 is higher so there is a need to regulate the optimum pH using NaOH and HCl. Fe-BTC in comparison to iron oxide is 37 times more effective in As(V) removal (*39*). Li et al. suggested that **MOF-808**as another adsorbent for As(V). It was produced by microwave irradiation, there is another process for the synthesis of this MOF is solvothermal synthesis. The crystals obtained by former method were smaller than the crystals obtained by the latter one. The maximum removal capacity achieved by MOF-808 was found to be 25mg/g (*40*). **UiO-66** was reported by Wang et al. for adsorption of As(V) with capacity of about 303mg/g at pH 2 which is optimum for UiO-66. It also worked in broad pH range from 1-10 and there was no hindrance by other anions present in water (*41*).

Mercury

Saleem et al. synthesized a MOF named **UiO-66-NHC(S)NHMe** this framework showed 25 times more removal of Hg(II) by adsorption as compared to UiO-66-NH_2. UiO-66 alone showed very low affinity towards mercury ions (*42*). Luo et al. synthesised a new class of MOF named **Zn(hip)(L).(DMF)(H_2O)** by solvo(hydro)thermal method. This MOF has a single channel which is hexagonal and occupied with acylamide and hydroxyl sites along with pores in inner walls. The maximum adsorption capacity observed by Zn(hip)(L).(DMF)(H2O) was estimated to be 333mg/g, optimum pH value for this was 5 and noticed that adsorption capacity was decreased in both acidic and alkaline conditions (*43*).Liu et al. investigated about **Cr-MIL-101-AS**, a modified framework for effective uptake of Mercury(II) by adsorption from water. In 6 hours, it can adsorb 99% of 10 ppm and this MOF can be used for further two cycles and showed no loss in capacity of MOF (*44*).

Additionally, in 2016, Rudd et al. documented**LMOF-263**showing very encouraging mercury adsorption capacities. LMOF-263 having extreme adsorption limit was assessed to be 380 mg/g. The uptakeinstrument discovered to be based on the principle of HSAB (*45*)

Adsorption of Mercury(II) from water by **MOF-74-Zn**was reported by Xiong et al. He synthesised MOF-74-Zn by solvo(hydro)thermal method and investigated that mechanism for this adsorption was result of both physisorption and chemisorptions. In this, physisorption played greater role. Maximum removal capacity was found to be at pH 6 i.e., 63mg/g (*46*).

Lead

Hs-mSi@MOF-5 is documented by Zhang et al. for removal of Pb(II). It acts as an efficient adsorbent as it reached at equilibrium in just 30 minutes and maximum capacity of its adsorption was about 312 mg/g. Optimum pH for working of HS-mSi@MOF-5 was found to be 6 as above this it started precipitating lead and at lower pH, protons and lead ions started to compete with each other. So optimum pH must be maintained for effective removal of Pb(II) ions (*35*).

Jamali et al. investigated another rod like, **MOF–Dy(BTC)(H_2O)(DMF)** for removing Pb(II) from aqueous medium. This MOF showed medium uptake of about 5mg/g) of Pb(II) but having ability to reach at equilibrium within 10 minutes. maximum removal capacity was observed at neutral pH i.e. 7. MOF–Dy(BTC)(H_2O)(DMF) found to be reusable for 5 cycles and its adsorption efficiency remains 94-98% (*47*).

MIL-53(Al@100aBDC), was identified by Ricco et al. to remove Pb(II). This framework was synthesised by combining iron oxide with MIL-53. The maximum adsorption capacity was 492mg/g and taken 6 hours to reach the equilibrium. It was also found that this equilibrium can be achieved in 2 hours when heating was supplied (*48*). Zou et al. investigated about **HKUST-1 - MW@H3$PW_{12}O_{40}$**, MOFwith high selectivity and adsorption of Pb(II). Its highest adsorption was found to be 98mg/g and reached equilibrium in 10 minutes (see Table 1) (*49*).

Table 1. Showing MOFs, ions adsorbed by them, their adsorption capacity and toxicity caused by these metals

Metal	*MOF*	*Ions uptake*	*Capacity (mg/g)*	*Toxicity (50)*	*Ref.*
Cr	TMU-5	Cr(III)	123	Cr(III) not detrimental to mammals, Cr(VI) toxic and carcinogenic	(2)
	Chitosan-MOF	Cr(VI)	94		(27)
	UiO-66-NHC(S)NHMe	Cr(III)	117		(28)
	ZIF-67	Cr(VI)	15		(29)
	MOR-1-HA	Cr(VI)	280		(30)
Cd	AMOF-1	Cd(II)	41	Carcinogenic phytotoxic, itai-itai disease, bio-accumulative,	(31)
	3D Co(II) MOF	Cd(II)	70%		(32)
	Cu3(BTC)$_2$-SO$_3$H	Cd(II)	89		(33)
	PCN-100	Cd(II)	1.6		(34)
	HS-mSi@MOF-5	Cd(II)	98		(35)
	Cu-terephthalate	Cd(II)	100		(36)
As	ZIF-8	As(III), As (V)	50, 60	Keratosis, carcinogenic, phytotoxic arsenicosis	(37)
	MIL-53(Fe)	As(V)	21		(38)
	Fe-BTC	As(V)	12		(39)
	MOF-808	As(V)	25		(40)
	UiO-66	As(V)	303		(41)
Hg	UiO-66-NHC(S)NHMe	Hg(II)	769	Minamata disease, Biomagnification	(42)
	Zn(hip)(L).(DMF)(HZn(hip)(L).(DMF)(H$_2$O)	Hg(II)	333		(43)
	Cr-MIL-101-AS	Hg(II)	99%		(44)
	LMOF-263	Hg(II)	380		(45)
	MOF-74-Zn	Hg(II)	63		(46)

Table 1. (Continued). Showing MOFs, ions adsorbed by them, their adsorption capacity and toxicity caused by these metals

Metal	*MOF*	*Ions uptake*	*Capacity (mg/g)*	*Toxicity* (50)	*Ref.*
Pb	HS-mSi@MOF-5	Pb(II)	312	Dyslexia, Autism, carcinogenic	(35)
	Dy(BTC)(H_2O)(DMF)	Pb(II)	5		(47)
	MIL-53(Al@100aBDC)	Pb(II)	492		(48)
	HKUST-1-MW@H3PWHKUST-1-MW@$H3PW_{12}O_{40}$	Pb(II)	98		(49)

Removal of POPs Using MOFs

Persistent Organic Pollutants (POPs) are synthetic compounds that have drawn impressive worldwide consideration in the course of recent years. These synthetic compounds have a mixture of physical and chemical properties that permit them to travel significant distances, oppose decomposition, and can bio accumulate. Contact with these substances has been related with a wider impact on people and untamed life (*7*). So, there is a need to manage these pollutants effectively. There are various technologies that are energy saving and cost effective like photocatalytic degradation and adsorptive removal of POPs from aqueous media. A new class of porous, functional organic-inorganic hybrid materials, MOFshave been synthesized that have shown tremendous potential and splendid viewpoint in removing these pollutants through process of photocatalysis and adsorption (*8*). Need to develop favorable water treatment techniques to efficiently remove these persistent pollutants is an instant demand and MOFs can be proved to be the solution for this.

MOFs become more significant than other conventional adsorbents because of its high porosity, high surface area. Easily modifiable structures and pore geometry (*9*). To describe the adsorption behaviours of MOFs for POPs, Freundlich, Langmuir ansSips isotherm models were generally used. MOFs can also act as a photocatalyst when irradiated and then started to show semiconductor like behaviour. Upon irradiation, MOFs get photo excited and generate electrons and holes that lead to catalytic redox reactions (*10*). Some MOFs when triggered they release ROS to degrade the POPs (*11*). removal by adsorption and photocatalytic degradation of POPs in aqueous medium using frameworks are described here.

Removal of Persistent Toxic Substances (PTS)

Leng et al. reported two MOFs, **ZIF-8** and **MIL-101(Cr)** employed to remove Uranine (azo dye) from wastewater. MIL-101(Cr) showed greater capacity for removal of Uranine in comparison to that of ZIF-8, the difference in adsorption capacities of both was due to presence of different size of pore apertures. MIL-101(Cr) has larger pore aperture of about 12-16 Å and ZIF-8 has only about 3.4 Å to adjust uranine. Another advantage of using MIL-101(Cr) is that it can be regenerated and highly stable. This was found to be an endothermic adsorption reaction and involved electrostatic force of attraction between Uranine (negatively charged) and MOF (positively charged) (*51*, *52*). Li et al. documented the removal of Dimethyl Phthalate (DMP) using framwork**MIL-53(Al).** When this framework was synthesised with oxides of Aluminium it exhibited highest capacity for removal of DMP (*53*).

MIL-101(Cr) and **MIL-53(Cr)** are two Cr based MOFs documented by Haque et al., having high porosity and exhibited high adsorption for Methyl Orange (MO) from aqueous media as compared to Activated carbon (AC) which is conventionally used. It was reported that MIL-101(Cr) was highly efficient for removal of MO in comparison to other Cr based MOFs. Its adsorption capacity was further improved by reacting it with protonated ethylenediamine and ethylenediamine. Another MOF was reported by them named **MIL-101-Al** that has capacity to adsorb both Methylene Blue (MB) and Methyl Orange (MO) (*54*). Adsorption capacity of MIL-101-Al enhanced by the introduction of $-NH_2$ group that led to adsorption uptake of MB of about 762 mg/g at 30 °C which was previously found to be only 195 mg/g. **MOF-235** (iron terephthalate), another MOF was investigated which shows higher uptake capacity for both MB and MO from wastewater. The reason for this higher capacity uptake was found to be strong electrostatic force of attraction between the framework and pollutants (*55*).

Liang et al. investigated that **MIL-100(Fe)** when photo deposited with Pt, Au and Pd showed efficient decomposition of Methyl Orange. When irradiated with visible light, these noble metals transfer photo generated electrons and lengthened the life of carriers for efficient degradation. It was also found that an extra path was offered by MOF channels for migration of induced electrons and led to separation of charge-carrier. MIL-100(Fe) when deposited with Pt it shows high stability and can be reused for further cycles (*56*).

Photodegradation of Nitroaromatic compounds was documented by Wu et al. with **$Ag_4(NO_3)_4$(dpppda)**. Various nitroaromatic compounds whose degradation was tested include p-Nitrophenol, 2,4-dinitrophenol and nitrobenzene. It was found that $Ag_4(NO_3)_4$(dpppda) completely degraded the nitrobenzene to give CO_2 when irradiated with UV light and also suggested that MOF can be utilised for 5 generation cycles as it is highly stable (*57*). A Fenton-like heterojunction photocatalyst was reported by Li et al. named, **g-C_3N_4/NH_2-MIL-88B(Fe).** It was experimented that this MOF act as an efficient photocatalyst for degradation of Methylene Blue (*11*). **Zn(5-aminoisophthalic acid)·H_2O, $Co_3(BPT)_2$(DMF)(bpp)·DMF** and **[Cd(5-aminoisophthalic acid)·$H_2O]_n$·$2nH_2O$** are some of the novel MOFs that was experimented for degradation of RhB dye using Visible light. When irradiated, they released highly reactive hydroxyl radical that degraded the dye efficiently and noticed that these MOFs were highly efficient than degradation by conventional photo catalysts (*58, 59*).

Removal of Herbicides and Pesticides

Some Scientists reported that MOFs exhibited efficient photocatalytic degradation of herbicides and pesticides. Mohaghegh et al. produced graphene based novel MOFs photo catalysts named **Ag_3PO_4/$BiPO_4$**and **AB/Cu(tpa).GR** employed for Atrazine degradation. They worked both under UV and visible lights but the latter MOF showed higher catalytic activity due to presence of GR species that facilitated transfer of electrons and separation of charge carriers than the former MOF (*60*). 2,4-D (2,4-dichlorophenoxyacetic acid) is a herbicide that can be adsorbed on **MIL-53(Cr),** investigate by Jung et al. The efficiency for adsorption of 2,4-D by MIL-53(Cr) was found to be greater than conventional AC. The mechanism for adsorption includes electrostatic force of attraction between 2,4-D and framework (*61*).

Methyl-chlorophenoxypropionic acid (MCPP) is a herbicide that cannot be degraded easily so it can be removed by adsorption on **UiO-66(Zr)** framework, this was reported by Seo et al. This framework has higher adsorption capacity (at low concentration of MCPP) i.e. 7.5 times higher than activated carbon and also exhibited 30 times more kinetic constant. At low pH, it showed rapid and high uptake because of electrostatic attractions and at high pH it was by π–π interactions between framework and herbicide. This MOF can also be regenerated (*62*). Zhu et al. reported a significant MOF named **UiO-67(Zr)** that can be used to efficiently remove Glufosinate and Glyphosate from aqueous solutions by chemisorption and not by electrostatic adsorption and included formation of Zr-O-P bonds that led to sequestration of Organophosphorous pesticides (*63*).

Removal of Phenols

Bisphenol-A (BPA) can be adsorbed by using **MIL-53(Cr)** from the aqueous solution, reported by Park et al. BPA is considered to be highly toxic phenolic compound that cannot be degraded easily so there is a need to adsorb it. Compared with a stable zeolite and activated carbon, MIL-53(Cr) possessed higher kinetic rate and adsorption capacity. Since BPA and Hydroxyl group of MIL-53(Cr) interacted with π–π bond and hydrogen bonding (64).

MIL-100(Fe) and **MIL-101(Cr)** removed BPA by adsorption process, reported by Qin et al. 252.5 mg/g of BPA was adsorbed by using MIL-101(Cr) that was the maximum adsorbance and capacity is 1.84 times higher than activated carbon and 4.54 times higher than MIL-100(Fe). π–π interaction and partial Hydrogen bonding between BPA rings and framework was responsible for this kind of adsorption (65).

Phenol from aqueous solution could be degraded using some kind of mixed MOFs, **[Cd0.15Zn0.85(oba)(4-bpdh)0.5]n·1.5DMF (TMU-5(Cd 15%)), [Zn(oba)(4-bpdh)0.5]n·1.5DMF (TMU-5), etc.** in the absence of auxiliary oxidants like hydrogen peroxide under irradiance of visible/UV light, examined by Morsel et al. when compared the phenol degradation efficiency Zn based MOFs founded to have greater efficiency than Cd based MOF (66). Li et al. examined that 2-Chlorophenol can be photodegraded by **$[Cd_2(Fe\text{-}L)_2\text{-}(\mu2\text{-}O)(H_2O)_2]\cdot 2DMF\cdot H_2O$** and **$[Zn_2(Fe\text{-}L)_2(\mu2\text{-}O)(H_2O)_2]\cdot 4DMF\cdot 4H_2O$** when irradiated with Visible light (see Table 2) (67).

Carbon Dioxide Sequestration

Due to many anthropogenic and natural factors, earth's global climate has been facing gradual changes like global warming mainly due to increase in concentration of CO_2 in the atmosphere, affecting both flora and fauna in catastrophic way. So, capturing of CO_2 became a need of the hour (73). There are various technologies but sequestration by MOFs can be proved to be the best solution as it has high porosity, high surface area, can be designed and synthesized easily, also allows a facile optimization of pores and other properties. MOFs are characterised as dynamic hence, exhibit super sensitivity to external stimuli like change in pressure and temperature. It is also characterised as rigid because of its strong and stable porous framework which makes it good adsorbent than conventional ones (74, 75). MOFs can be adopted to form a strong bond with Carbon dioxide to make it favourable for sequestration (76).

CO_2 emission can be mitigated from fuel combustion power plants by using MOFs during 3 stages of combustion process i.e. pre combustion, post combustion and oxy-fuel combustion. Oxy-fuel combustion capture proved to be effective for newly built power plants, pre combustion capture can be employed in gasification plants and for retrofitting power plants, post combustion carbon capture could be effective (77).

Table 2. Summary of POPs removal by MOFs, their performance and mechanism involved

Adsorbates/ Pollutants	*MOFs*	*Mechanism*	*POPs Conc. (mg/ L)*	*Efficiency %*	*Ref.*
Methyl orange (MO)	MIL-101-Cr	Adsorption (Electrostatic interaction)	50	114	(68)
	MOF-235		40	477	(55)
	M@MIL-100(Fe)	Photocatalytic degradation (Visible Light)	20	100	(56)
Methylene blue (MB)	MOF-235	Adsorption (Electrostatic interaction)	40	187	(55)
	NH2-MIL-10 1-Al		40	762	(54)
Dimethyl phthalate (DMP)	MIL-53(Al)	Adsorption (π-π stacking interaction)	25	206	(69)
Uranine	MIL-101-Cr	Adsorption (Electrostatic interaction)	75.3	126.9	(70)
2,4-Dichlorop henoxyacetic acid (2,4-D)	MIL-53(Cr)	Adsorption	10-150	556	(62)
Methylchlorophenoxypropionic acid (MCPP)	UiO-66(Zr)	Adsorption	20-170	370	(71)
gasiGlyphosate (GP)	UiO-67(Zr)	Chemisorption	5.4–72	540	(72)
Glufosinate (GF)				538	
Bisphenol-A (BPA)	MIL-100(Fe)	Adsorption (π–π interactions)	20-300	56	(64, 65)
	MIL-53(Cr)		5-80	421	
	MIL-101(Cr)		20-300	253	
RhB	(Zn(5-aminoisoph thalic acid)•H_2O	Photodegradation (UV)	.0048	63.1	(58)
Atrazine	AB/Cu(tpa).GR nanocomposite	Photodegradation (UV)	-	100	(60)
		Photodegradation (Visible)		85	
Nitrobenzene (NB)	$Ag_4(NO_3)_4$(dpppda)	Photodegradation (UV)	69.56	100	(57)

Table 2. (Continued). Summary of POPs removal by MOFs, their performance and mechanism involved

Adsorbates/ Pollutants	*MOFs*	*Mechanism*	*POPs Conc. (mg/ L)*	*Efficiency %*	*Ref.*
Phenol	[Cd(oba)(4-bpdh)] n·1DMF (TMU-7)	Photodegradation (UV; Visible)	25-100 ;10-70	33.3; 15.5	(66)
	[[Cd0.15Zn0.85 (oba)(4-bpdh)0.5]n·1.5DMF (TMU-5)			82.5 ;73.5	
	[Cd0.3Zn0.7(oba)(4-b pdh)0.5]n·1.5DMF (TMU-5)			87.6; 78	

Sun et al. documented **MOF-801** into **PEBA** polymer to form a **matrix material**for the capture of CO_2. It was observed that this framework preferentially adsorbs CO_2 and seemed to be highly selective for CO_2 over N_2. This mixed-matrix composite membrane of MOF-801/PEBA has shown stabilised separation performance (*78*). **Mg-MOF-74** crystals utilised by Majumdar et al. for synthesis of Mg-MOF-74 mixed matrix substance for separation of CO_2 from carbon dioxide/ methane mixture. They formed a denser mixed matrix substances using Elmer's glue MMM. With increase in the amount of MOF, the selectivity and permeability of CO_2 also gets improved. It was also observed that plasticization has been reduced when Mg-MOF-74 crystals were incorporated. This exhibits strong adsorption selectivity for CO_2 (*79*). A continuous and defect less membrane of **Ni-MOF-74** on α-alumina was prepared by Lee et al. using layer seeding along with crystallization. Their study showed that this prepared MMM exhibited a greater affinity, selectivity and permeability for uptake of CO_2 in comparison to CH_4, H_2 and N_2 (*80*) .**6FDA**-durene polyimide membrane incorporated with CO_2-philic **KAUST**-nanocrystals was prepared by Chen et al. This has a promising application as it increased the selectivity as well as permeability of mixed matrix material membrane to sequester CO_2 from natural and biogas (*81*). Hopefully, MOFs' characteristics are continually manipulated which permits them to fill in as the cutting edge for CO_2uptake.

Sequestration of Radionuclide by MOFs

Nuclear power is significant for the production of electric energy with no generation of GHGs. MOFs have explicit surface region, porosity, and adaptable postmodification, supplying MOFs with extraordinary properties in the removal of radioactive pollutants. As of late, the usage of such materials for radionuclide separation has been going through quick turn of events (*5*, *6*). Radioactive Csand Sr are found to be profoundly radiotoxic and hot. Decay of theseradiate β particles that can cause long term hazards to living cells. Cs and Sr are generally isolated using ion exchange method (*82*). **MIL-101** which is hydrolytically stable was grafted with sulfonic acid group to prepare a cation exchange material which possessed good sorption properties and can be employed for removal of Cs^+ and Sr^{2+}, this was reported by Aquila et al. It was noticed that percentage removal could reach 98.92% for Sr^{2+} and 99.99% for Cs^+ (*83*). Naeimi et al. experimented that **HKUST-1 MOF** upon impregnation with potassium nickel hexacyanoferrate exhibit efficient sorption of Cs^+, having fast sorption kinetics, better selectivity and high uptake capacity (*84*).

Uranium mostly occurs in UO_2^{2+} form in waste that can be captured from water systems as it is highly toxic. MOF showed their UO_2^{2+} capture capabilities. Lin et al. investigated that covalently modification of **UiO-68** by phosphorylurea exhibited high removal capacity for UO_2^{2+} of about 217 mg/g. Another MOF, **MIL-101** was also modified with carboxyl group for selective uptake of UO_2^{2+} (*85*, *86*).

Shi et al. documented sequestration of Thorium (IV) from acidic solution by employing MOF. They found when carboxyl group attached to **UiO-68 MOF**, its sorption properties was increased but decreased the volumes of pore and surface areas. The uptake capacity was found to be 350 mg/g when **UiO-66-(COOH)$_2$** was used which was higher than UiO-66 (*87*).

Iodine is another radionuclide that can persist for longer time. Uptake of iodine gas using MOF, ZiF-8 investigated by Sava et al. It can capture as high as 125 wt%. The sequestration of iodine was larger in cage of ZiF-8 rather than on the surface (*88*). According to Li et al., there are some Cr based MIL series MOFs that can also capture iodine efficiently (*89*). Xiong et al. documented a framework named Zinc tetrazolate that showed high uptake capacity for Xe having excellent selectivity (*90*). Another MOF, **Ag@Ni-DoBDC** was reported by Thallapally et al. exhibited highest capacity for uptake of Xenon (*91*).

Recent Advances and Challenges

"Recent Advances in Metal Organic Frameworks Materials" include Synthesis (computational design, high-throughput synthesis), characterization (present techniques, new methods), and utilization of MOFs to overcome lifelong obstructions in processes and applications. Some of the recent advances are mentioned here.

Over recent twenty years, MOFs have acquired huge consideration as film substances for gas separation and adsorbent. Because of the numerous and variety of current MOFs, distinguishing the favourablemetal frameworkfor gas detachment is extremely difficult. Computational screening contemplates have assumed a significant part in precisely surveying uptake method and layer-orientedgas division exhibitions of frameworks in a productive way. Computational techniques, basically moleculesimulations will be proved priceless in future trials for MOFs (*12*).

From recent studies, it was reported that MOFs have likewise been used in adsorption thermal energy storage (ATES) applications; but very restricted data is available to access the performance of MOFs in adsorption thermal energy storage as compared to data available for conventional adsorbents (*13*). Wu et al. reviewed that contrasted with metal sources having two valency; MOFs made of aluminium are more encouraging as their structures show significant stability. Aluminium is additionally cost effective and easily available contrasted with different metals and in this way could be a feasible choice for MOFs. It was also investigated that Aluminium MOFs showed promising potential to be used as a material for membranes. There are two groups of aluminium MOFs that have been broadly examined this far: MIL and CAU. The two groups offer astounding structure stability and accordingly make themto use for different applications including gas separation and adsorption chillers but there are some gaps for development. From the viewpoint, a more principal exploration needs to be coordinated to clarify the effect of variousgroup in the MOF (*14*).

Shi et al. reviewed, upon irradiation of ultraviolet, visible and near infra-red on MOFs it acts as a catalyst for breakdown of water to generate HydrogenOn the basis of composition of photocatlysts these are divided into 3 categories- MOFs, its composites and inferred catalysts. Inferable from high strength, stability and semiconductor properties, MOFs-based materials can adequately convert sun-oriented energy to hydrogen energy, and the greater part of them can be reused a few times but some challenges need improvement for the photo catalytic proficiency of MOFs-based materials for this purpose (*15*).

In the course of recent years, shielding substances from microbial contamination has attracted a great deal of consideration in biomedical exploration to stay away from complexities during clinical therapy. Now, Scientists have consolidated antibacterial substances into the cages of permeable frameworks to provide antibacterial features. RFPand o-NBAwere stacked in ZIF-8 pores (RFP&o NBA@ZIF-8) for the efficient utilisation of thisin vitro and in vivo bacterial infections and wound contamination treatment under ideal pH and UV irradiation (*16*, *17*).

Increasing demand, risk factors in food has been increasing rapidly that results in deteriorating food quality such risks involve presence of pesticide residues, veterinary drugs, micotoxins, pathogens, illegal additives, heavy metals, food additives and other contaminants. The conventional technologies are proved to show only the average result when it comes to detection of such risk factors. Cheng et al. reviewed those recent advances in MOF as sensing material in fabricating sensors such as electrical luminescent, electro chemiluminiscent, colorimetric and surface-enhanced Raman scattering sensors has made it possible to detect them with full efficiency, though some more work is needed in this area (*18*).

MOFs are also chosen to be used as an active material due to its large surface area for development of cathode and also because of its diffused sites which are catalytically active. They also offer high electrical conductivity and better functionality along with better storage for Oxygen to be used in Li-O_2 battery. (*19*). Some of such advanced MOFs that has designed to show excellent efficiencies are MIL-101(Fe), MIL-88B(Fe) and MIL-47. In the near future, MOFs properties can be combined with more advanced technologies to facilitate more functional and active cathode material for Li-O_2 battery (*20–22*). Improvement in metal–organic frameworks (MOFs) has improved from crucial chemistry to most of the engineering processes and applications that brings modern opportunities.

Challenges

The intensive research work ought to be done to conquer the accompanying difficulties and obstructions some of them are as follows:

- Mostly MOFs which are photo-active are still at the stage of lab-scale that cannot be made-up by one-time high-throughput synthesis. Besides, some preparation strategies are difficult and troublesome to regulate. The enlargement of additional artificial routes, particularly a single step approach under gentle conditions, is extremely fascinating for future mass-scale applications.
- In expansion to the discovery of dynamic species (like HO•, O_2 • and h^+) during photo catalysis, more subtleties related with the photo catalytic instrument need inside and out examination. For instance, the adsorption destinations of toxins on a superficial level/in the channel of MOFs, the interfacial electron move instrument just as the rate restricting advance ought to be distinguished. The plan of proper adjustment systems will be more logical and viable.
- The metal nanoparticles which used in loading to form MOF composites are mostly very expensive such as Au, Ag, Pd and Pt. There is a requirement for more exploration towards non precious metals like Cu & Bi to make it cost effective.
- To date, the multifunctional use of MOFs has been restricted. There are only sporadic reports of concurrent optical oxidation of dyes and reduction of Cr (VI) or H^+ (H_2 generation). More efficient and multifunctional organic frameworks are needed for the simultaneous and synergistic removal of contaminants from the surroundings and the generation of energy or desirable products in a single system.
- For better guideline of MOF photo catalysis, more endeavours are expected to seriously investigate the degradation processes. For instance, the harmfulness of degraded intermediates ought to be assessed because of fragmented mineralization of natural contaminations under most conditions.
- Various researches showed that there are many MOFs that are water resistant like ZIF series, MIL-101, MIL-125, UiO-66, but their resistance under extreme conditions like highly acidic or basic still needs to be upgrade. Besides, some Scientists suggested that

choosing high valence metals as central metal in synthesis of MOFs will have strong metal bond strength and will be proved beneficial and highly stable against extreme conditions.

- In addition to fixation on an inert conveyor and magnetic component insertion, fabricating MOF films is another attractive method for recycling MOFs. However, there was very narrow scope for use of MOF films in environmental photo catalysis. In particular, the fabrication of MOF films on a conductive substrate needs to be further expanded for the photoelectric decomposition of organic pollutants and heavy metal ion reduction (92–97).

Conclusions

The development of metal organic frameworks has unfolded solutions of many challenging and emerging problems like adsorption of heavy metals, sequestration of carbon dioxide, biosensing and uptake of radionuclide along with energy storage. The use of MOFs has given many ways to handle the contaminants by adsorbing it by electrostatic force of attraction, π-π stacking interaction and H-bonding. It photo degrades the POPs upon irradiation with visible, UV and NIR, conventionally it was done only by UV irradiation. Radionuclide uptake is mostly accomplished by selective insertion of groups onto MOF by post modification methods. For MOF itself, reusability and stability in aqueous media is a major challenge. But some MOFs are found to be highly stable like MIL-101, ZIF and UiO. Stability of MOFs can be enhanced by introducing high valence metal to the MOFs and it will also increase the degradability efficiency of MOFs. At long last, there is yet a dire need to foster novel vigorous MOFs with magnificent light reaping properties, high security and simple recyclability through the designing of metallic hubs and natural linkers. In synopsis, in spite of the fact that there are as yet numerous difficulties to be settled, in the previous years of improvement, scientists from everywhere in the world have gained extraordinary headway in MOF materials from foundational layout, controllable change to practical applications. It was accepted that with the joint endeavours of analysts in numerous fields, the possibilities of MOFs will be unquestionably splendid.

References

1. Bai, Y.; Dou, Y.; Xie, L.-H.; Rutledge, W.; Li, J.-R.; Zhou, H.-C. Zr-based metal–organic frameworks: design, synthesis, structure, and applications. *Chemical Society Reviews* **2016**, *45* (8), 2327–2367.
2. Tahmasebi, E.; Masoomi, M. Y.; Yamini, Y.; Morsali, A. Application of Mechanosynthesized Azine-Decorated Zinc(II) Metal–Organic Frameworks for Highly Efficient Removal and Extraction of Some Heavy-Metal Ions from Aqueous Samples: A Comparative Study. *Inorganic Chemistry* **2014**, *54* (2), 425–433.
3. Schoenecker, P. M.; Carson, C. G.; Jasuja, H.; Flemming, C. J.; Walton, K. S. Effect of water adsorption on retention of structure and surface area of metal–organic frameworks. *Industrial & Engineering Chemistry Research* **2012**, *51* (18), 6513–6519.
4. Cavka, J. H.; Jakobsen, S.; Olsbye, U.; Guillou, N.; Lamberti, C.; Bordiga, S.; Lillerud, K. P. A new zirconium inorganic building brick forming metal organic frameworks with exceptional stability. *Journal of the American Chemical Society* **2008**, *130* (42), 13850–13851.
5. Furukawa, H.; Cordova, K. E.; O'Keeffe, M.; Yaghi, O. M. The Chemistry and Applications of Metal-Organic Frameworks. *Science* **2013**, *341* (6149), 1230444–1230444.

6. Mon, M.; Bruno, R.; Ferrando-Soria, J.; Armentano, D.; Pardo, E. Metal–organic framework technologies for water remediation: towards a sustainable ecosystem. *Journal of Materials Chemistry A* **2018**, *6* (12), 4912–4947.
7. Fitzgerald, L.; Wikoff, D. S. Persistent Organic Pollutants. *Encyclopedia of Toxicology* **2014**, 820–825.
8. Hamon, L.; Serre, C.; Devic, T.; Loiseau, T.; Millange, F.; Férey, G.; Weireld, G. D. Comparative Study of Hydrogen Sulfide Adsorption in the MIL-53(Al, Cr, Fe), MIL-47(V), MIL-100(Cr), and MIL-101(Cr) Metal—Organic Frameworks at Room Temperature. *Journal of the American Chemical Society* **2009**, *131* (25), 8775–8777.
9. Rafique, M. M. Evaluation of Metal–Organic Frameworks as Potential Adsorbents for Solar Cooling Applications. *Applied System Innovation* **2020**, 3 (2), 26.
10. Zhang, Z.-M.; Zhang, T.; Wang, C.; Lin, Z.; Long, L.-S.; Lin, W. Photosensitizing Metal–Organic Framework Enabling Visible-Light-Driven Proton Reduction by a Wells–Dawson-Type Polyoxometalate. *Journal of the American Chemical Society* **2015**, *137* (9), 3197–3200.
11. Li, X.; Pi, Y.; Wu, L. Q.; Xia, Q.; Wu, J. L.; Li, Z.; Xiao, J. Facilitation of the visible light-induced Fenton-like excitation of H2O2 via heterojunction of g-C3N4/NH2-Iron terephthalate metal-organic framework for MB degradation. *Appl. Catal. B: Environ.* **2017**, *202*, 653–663.
12. Daglar, H.; Keskin, S. Recent advances, opportunities, and challenges in high-throughput computational screening of MOFs for gas separations. *Coordination Chemistry Reviews* **2020**, *422*, 213470.
13. Makhanya, N.; Oboirien, B.; Ren, J.; Musyoka, N.; Sciacovelli, A. Recent advances on thermal energy storage using metal-organic frameworks (MOFs). *Journal of Energy Storage.* **2021**Elsevier Ltd..
14. Wu, T.; Prasetya, N.; Li, K. Recent advances in aluminium-based metal-organic frameworks (MOF) and its membrane applications. *Journal of Membrane Science* **2020**, 118493.
15. Shi, Y.; Yang, A.-F.; Cao, C.-S.; Zhao, B. Applications of MOFs: Recent advances in photocatalytic hydrogen production from water. *Coordination Chemistry Reviews* **2019**, *390*, 50–75.
16. Song, Z.; Wu, Y.; Cao, Q. i.; Wang, H.; Wang, X.; Han, H. pH-Responsive, Light Triggered on-Demand Antibiotic Release from Functional Metal-Organic Framework for Bacterial Infection Combination Therapy. *Adv. Funct. Mater.* **2018**, *28* (23), 1800011.
17. Rubin, H. N.; Neufeld, B. H.; Reynolds, M. M. Surface-Anchored Metal–Organic Framework–Cotton Material for Tunable Antibacterial Copper Delivery. *ACS Applied Materials & Interfaces* **2018**, *10* (17), 15189–15199.
18. Cheng, W.; Tang, X.; Zhang, Y.; Wu, D.; Yang, W. Applications of metal-organic framework (MOF)-based sensors for food safety: Enhancing mechanisms and recent advances. *Trends in Food Science and Technology* **2021**, *112*, 268.
19. Jiang, Y.; Zhao, H.; Yue, L.; Liang, J.; Li, T.; Liu, Q.; Sun, X. Recent advances in lithium-based batteries using metal organic frameworks as electrode materials. *Electrochemistry Communications* **2020**, *122*, 106881.

20. Shen, L.; Song, H.; Wang, C. Metal-Organic Frameworks Triggered High-Efficiency Li storage in Fe-Based Polyhedral Nanorods for Lithium-ion Batteries. *Electrochimica Acta* **2017**, *235*, 595–603.
21. Shin, J.; Kim, M.; Cirera, J.; Chen, S.; Halder, G. J.; Yersak, T. A.; Meng, Y. S. MIL-101 (Fe) as a lithium-ion battery electrode material: a relaxation and intercalation mechanism during lithium insertion. *Journal of Materials Chemistry A* **2015**, 3 (8), 4738–4744.
22. Kaveevivitchai, W.; Jacobson, A. J. Exploration of vanadium benzenedicarboxylate as a cathode for rechargeable lithium batteries. *Journal of Power Sources* **2015**, *278*, 265–273.
23. Fu, F.; Wang, Q. Removal of heavy metal ions from wastewaters: A review. *Journal of Environmental Management* **2011**, *92* (3), 407–418.
24. Li, S.; Chen, Y.; Pei, X.; Zhang, S.; Feng, X.; Zhou, J.; Wang, B. Water Purification: Adsorption over Metal-Organic Frameworks. *Chinese Journal of Chemistry* **2016**, *34* (2), 175–185.
25. Valizadeh, B.; Nguyen, T. N.; Stylianou, K. C. Shape engineering of metal–organic frameworks. *Polyhedron* **2018**, *145*, 1–15.
26. Kumar, P.; Pournara, A.; Kim, K. H.; Bansal, V.; Rapti, S.; Manos, M. J. Metal-organic frameworks: challenges and opportunities for ion-exchange/sorption applications. *Progress in Materials Science* **2017**, *86*, 25–74.
27. Wang, K.; Tao, X.; Xu, J.; Yin, N. Novel Chitosan–MOF Composite Adsorbent for the Removal of Heavy Metal Ions. *Chemistry Letters* **2016**, *45* (12), 1365–1368.
28. Saleem, H.; Rafique, U.; Davies, R. P. Investigations on post-synthetically modified UiO-66-NH 2 for the adsorptive removal of heavy metal ions from aqueous solution. *Microporous and Mesoporous Materials* **2016**, *221*, 238244.
29. Li, X.; Gao, X.; Ai, L.; Jiang, J. Mechanistic insight into the interaction and adsorption of Cr(VI) with zeolitic imidazolate framework-67 microcrystals from aqueous solution. *Chemical Engineering Journal* **2015**, *274*, 238–246.
30. Rapti, S.; Pournara, A.; Sarma, D.; Papadas, I. T.; Armatas, G. S.; Hassan, Y. S.; Manos, M. J. Rapid, green and inexpensive synthesis of high quality UiO-66 amino-functionalized materials with exceptional capability for removal of hexavalent chromium from industrial waste. *Inorganic Chemistry Frontiers* **2016**, 3 (5), 635–644.
31. Chakraborty, A.; Bhattacharyya, S.; Hazra, A.; Ghosh, A. C.; Maji, T. K. Post-synthetic metalation in an anionic MOF for efficient catalytic activity and removal of heavy metal ions from aqueous solution. *Chemical Communications* **2016**, 52 (13), 2831–2834.
32. Abbasi, A.; Moradpour, T.; Van Hecke, K. A new 3D cobalt (II) metal–organic framework nanostructure for heavy metal adsorption. *InorganicaChimica Acta* **2015**, *430*, 261–267.
33. Wang, Y.; Ye, G.; Chen, H.; Hu, X.; Niu, Z.; Ma, S. Functionalized metal–organic framework as a new platform for efficient and selective removal of cadmium(ii) from aqueous solution. *Journal of Materials Chemistry A* **2015**, 3 (29), 15292–15298.
34. Fang, Q.-R.; Yuan, D.-Q.; Sculley, J.; Li, J.-R.; Han, Z.-B.; Zhou, H.-C. Functional Mesoporous Metal—Organic Frameworks for the Capture of Heavy Metal Ions and Size-Selective Catalysis. *Inorganic Chemistry* **2010**, *49* (24), 11637–11642.
35. Zhang, J.; Xiong, Z.; Li, C.; Wu, C. Exploring a thiol-functionalized MOF for elimination of lead and cadmium from aqueous solution. *Journal of Molecular Liquids* **2016**, *221*, 43–50.

36. Ghouri, Y. A.; Richards, D. M.; Rahimi, E. F.; Krill, J. T.; Jelinek, K. A.; DuPont, A. W. Systematic review of randomized controlled trials of probiotics, prebiotics, and synbiotics in inflammatory bowel disease. *Clinical and experimental gastroenterology* **2014**, 7, 473.
37. Jian, M.; Liu, B.; Zhang, G.; Liu, R.; Zhang, X. Adsorptive removal of arsenic from aqueous solution by zeolitic imidazolate framework-8 (ZIF-8) nanoparticles. *Colloids and Surfaces A: Physicochemical and Engineering Aspects* **2015**, *465*, 67–76.
38. Vu, T. A.; Le, G. H.; Dao, C. D.; Dang, L. Q.; Nguyen, K. T.; Nguyen, Q. K.; Lee, G. D. Arsenic removal from aqueous solutions by adsorption using novel MIL-53(Fe) as a highly efficient adsorbent. *RSC Advances* **2015**, *5* (7), 5261–5268.
39. Zhu, B.-J.; Yu, X.-Y.; Jia, Y.; Peng, F.-M.; Sun, B.; Zhang, M.-Y.; Huang, X.-J. Iron and 1,3,5-Benzenetricarboxylic Metal–Organic Coordination Polymers Prepared by Solvothermal Method and Their Application in Efficient As(V) Removal from Aqueous Solutions. *The Journal of Physical Chemistry C* **2012**, *116* (15), 8601–8607.
40. Li, Z.-Q.; Yang, J.-C.; Sui, K.-W.; Yin, N. Facile synthesis of metal-organic framework MOF-808 for arsenic removal. *Materials Letters* **2015**, *160*, 412–414.
41. Wang, C.; Xiao, S.; Chen, Y.; Liu, X. Effect of single Si1−xCx coating and compound coatings on the thermal conductivity and corrosion resistance of Mg–3Sn alloy. *Journal of Magnesium and Alloys* **2015**, *3* (1), 10–15.
42. Saleem, H.; Rafique, U.; Davies, R. P. Investigations on post-synthetically modified UiO-66-NH 2 for the adsorptive removal of heavy metal ions from aqueous solution. *Microporous and Mesoporous Materials* **2016**, *221*, 238–244.
43. Luo, F.; Chen, J. L.; Dang, L. L.; Zhou, W. N.; Lin, H. L.; Li, J. Q.; Luo, M. B. High-performance Hg2+ removal from ultra-low-concentration aqueous solution using both acylamide- and hydroxyl-functionalized metal–organic framework. *Journal of Materials Chemistry A* **2015**, *3* (18), 9616–9620.
44. Liu, T.; Che, J.-X.; Hu, Y.-Z.; Dong, X.-W.; Liu, X.-Y.; Che, C.-M. Alkenyl/Thiol-Derived Metal-Organic Frameworks (MOFs) by Means of Postsynthetic Modification for Effective Mercury Adsorption. *Chemistry - A European Journal* **2014**, *20* (43), 14090–14095.
45. Rudd, N. D.; Wang, H.; Fuentes-Fernandez, E. M. A.; Teat, S. J.; Chen, F.; Hall, G.; Li, J. Highly Efficient Luminescent Metal–Organic Framework for the Simultaneous Detection and Removal of Heavy Metals from Water. *ACS Applied Materials & Interfaces* **2016**, *8* (44), 30294–30303.
46. Xiong, Y. Y.; Li, J. Q.; Gong, L. L.; Feng, X. F.; Meng, L. N.; Zhang, L.; Luo, F. Using MOF-74 for Hg 2+ removal from ultra-low concentration aqueous solution. *Journal of Solid State Chemistry* **2017**, *246*, 16–22.
47. Jamali, A.; Tehrani, A. A.; Shemirani, F.; Morsali, A. Lanthanide metal–organic frameworks as selective microporous materials for adsorption of heavy metal ions. *Dalton Transactions* **2016**, *45* (22), 9193–9200.
48. Ricco, R.; Konstas, K.; Styles, M. J.; Richardson, J. J.; Babarao, R.; Suzuki, K.; Falcaro, P. Lead(ii) uptake by aluminium based magnetic framework composites (MFCs) in water. *Journal of Materials Chemistry A* **2015**, *3* (39), 19822–19831.
49. Wang, K.; Tao, X.; Xu, J.; Yin, N. Novel Chitosan–MOF Composite Adsorbent for the Removal of Heavy Metal Ions. *Chemistry Letters* **2016**, *45* (12), 1365–1368.

50. Adriano, D. C. *Trace Elements in the Terrestrial Environment*; Springer Science & Business Media, 2013, pp 1−517.
51. Ferey, G. A Chromium Terephthalate-Based Solid with Unusually Large Pore Volumes and Surface Area. *Science* **2005**, *309* (5743), 2040–2042.
52. Park, K. S.; Ni, Z.; Cote, A. P.; Choi, J. Y.; Huang, R.; Uribe-Romo, F. J.; Yaghi, O. M. Exceptional chemical and thermal stability of zeolitic imidazolate frameworks. *Proceedings of the National Academy of Sciences* **2006**, *103* (27), 10186–10191.
53. Li, Z.; Wu, Y.; Li, J.; Zhang, Y.; Zou, X.; Li, F. The Metal-Organic Framework MIL-53(Al) Constructed from Multiple Metal Sources: Alumina, Aluminum Hydroxide, and Boehmite. *Chemistry - A European Journal* **2015**, *21* (18), 6913–6920.
54. Haque, E.; Lo, V.; Minett, A. I.; Harris, A. T.; Church, T. L. Dichotomous adsorption behaviour of dyes on an amino-functionalised metal–organic framework, amino-MIL-101(Al). *J. Mater. Chem. A* **2014**, *2* (1), 193–203.
55. Haque, E.; Jun, J. W.; Jhung, S. H. Adsorptive removal of methyl orange and methylene blue from aqueous solution with a metal-organic framework material, iron terephthalate (MOF-235). *Journal of Hazardous Materials* **2011**, *185* (1), 507–511.
56. Liang, R.; Jing, F.; Shen, L.; et al. M@MIL-100(Fe) (M = Au, Pd, Pt) nanocomposites fabricated by a facile photodeposition process: Efficient visible-light photocatalysts for redox reactions in water. *Nano Res.* **2015**, *8*, 3237–3249.
57. Wu, X. Y.; Qi, H. X.; Ning, J. J.; Wang, J. F.; Ren, Z. G.; Lang, J. P. One silver(I)/tetraphosphine coordination polymer showing good catalytic performance in the photodegradation of nitroaromatics in aqueous solution. *Appl. Catal. B: Environ.* **2015** (168-169), 98–104.
58. Zhao, J.; Dong, W.-W.; Wu, Y.-P.; Wang, Y.-N.; Wang, C.; Li, D.-S.; Zhang, Q.-C. Two (3,6)-connected porous metal–organic frameworks based on linear trinuclear [Co3(COO)6] and paddlewheel dinuclear [Cu2(COO)4] SBUs: gas adsorption, photocatalytic behaviour, and magnetic properties. *Journal of Materials Chemistry A* **2015**, *3* (13), 6962–6969.
59. Wang, F.; Dong, C.; Wang, C.; Yu, Z.; Guo, S.; Wang, Z.; Li, G. Fluorescence detection of aromatic amines and photocatalytic degradation of rhodamine B under UV light irradiation by luminescent metal–organic frameworks. *New Journal of Chemistry* **2015**, *39* (6), 4437–4444.
60. Mohaghegh, N.; Tasviri, M.; Rahimi, E.; Gholami, M. R. Comparative studies on Ag3PO4/BiPO4–metal-organic framework–graphene-based nanocomposites for photocatalysis application. *Applied Surface Science* **2015**, *351*, 216–224.
61. Ding, L.; Lu, X.; Deng, H.; Zhang, X. Adsorptive Removal of 2,4-Dichlorophenoxyacetic Acid (2,4-D) from Aqueous Solutions Using MIEX Resin. *Industrial & Engineering Chemistry Research* **2012**, *51* (34), 11226–11235.
62. Seo, Y. S.; Khan, N. A.; Jhung, S. H. Adsorptive removal of methylchlorophenoxypropionic acid from water with a metal-organic framework. *Chemical Engineering Journal* **2015**, *270*, 22–27.
63. Zhu, X.; Li, B.; Yang, J.; Li, Y.; Zhao, W.; Shi, J.; Gu, J. Effective Adsorption and Enhanced Removal of Organophosphorus Pesticides from Aqueous Solution by Zr-Based MOFs of UiO-67. *ACS Applied Materials & Interfaces* **2014**, *7* (1), 223–231.

64. Park, E. Y.; Hasan, Z.; Khan, N. A.; Jhung, S. H. Adsorptive Removal of Bisphenol-A from Water with a Metal-Organic Framework, a Porous Chromium-Benzenedicarboxylate. *Journal of Nanoscience and Nanotechnology* **2013**, *13* (4), 2789–2794.
65. Qin, F. X.; Jia, S. Y.; Liu, Y.; Li, H. Y.; Wu, S. H. Adsorptive removal of bisphenol A from aqueous solution using metal-organic frameworks, Desalin. *Water Treat.* **2014**, *54*, 93–102.
66. Masoomi, M. Y.; Bagheri, M.; Morsali, A.; Junk, P. C. High photodegradation efficiency of phenol by mixed-metal–organic frameworks. *Inorganic Chemistry Frontiers* **2016**, *3* (7), 944–951.
67. Li, J.; Yang, J.; Liu, Y.-Y.; Ma, J.-F. Two Heterometallic-Organic Frameworks Composed of Iron(III)-Salen-Based Ligands and d10Metals: Gas Sorption and Visible-Light Photocatalytic Degradation of 2-Chlorophenol. *Chemistry - A European Journal* **2015**, *21* (11), 4413–4421.
68. Haque, E.; Lee, J. E.; Jang, I. T.; Hwang, Y. K.; Chang, J. S.; Jegal, J.; Jhung, S. H. Adsorptive removal of methyl orange from aqueous solution with metal-organic frameworks, porous chromium-benzenedicarboxylates. *Journal of hazardous materials* **2010**Advance online publication.
69. Fan, X.; Wang, W.; Li, W.; Zhou, J.; Wang, B.; Zheng, J.; Li, X. Highly Porous ZIF-8 Nanocrystals Prepared by a Surfactant Mediated Method in Aqueous Solution with Enhanced Adsorption Kinetics. *ACS Applied Materials & Interfaces* **2014**, *6* (17), 14994–14999.
70. Zhang, H.; Xue, J.; Hu, N.; Sun, J.; Ding, D.; Wang, Y.; Li, L. Selective removal of U(VI) from low concentration wastewater by functionalized HKUST-1@H3PW12O40. *Journal of Radioanalytical and Nuclear Chemistry* **2015**, *308* (3), 865–875.
71. Barbero, N.; Vione, D. Why Dyes Should Not Be Used to Test the Photocatalytic Activity of Semiconductor Oxides. *Environmental Science & Technology* **2016**, *50* (5), 2130–2131.
72. De Smedt, C.; Spanoghe, P.; Biswas, S.; Leus, K.; Van Der Voort, P. Comparison of different solid adsorbents for the removal of mobile pesticides from aqueous solutions. *Adsorption* **2015**, *21* (3), 243–254.
73. Smithson, P. A. IPCC, 2001: climate change 2001: the scientific basis. Contribution of Working Group 1 to the Third Assessment Report of the Intergovernmental Panel on Climate Change, edited by J. T. Houghton, Y. Ding, D. J. Griggs, M. Noguer, P. J. van der Linden, X. Da. *International Journal of Climatology* **2002**, *22* (9), 1144–1144.
74. Qin, J.-S.; Yuan, S.; Alsalme, A.; Zhou, H.-C. Flexible Zirconium MOF as the Crystalline Sponge for Coordinative Alignment of Dicarboxylates. *ACS Applied Materials & Interfaces* **2017**, *9* (39), 33408–33412.
75. Li, J.-R.; Kuppler, R. J.; Zhou, H.-C. Selective gas adsorption and separation in metal–organic frameworks. *Chemical Society Reviews* **2009**, *38* (5), 1477.
76. Hu, Z.; Wang, Y.; Shah, B. B.; Zhao, D. CO2 Capture in Metal-Organic Framework Adsorbents: An Engineering Perspective. *Advanced Sustainable Systems* **2018**, 1800080.
77. Wang, Y.; Zhao, L.; Otto, A.; Robinius, M.; Stolten, D. A Review of Post-combustion CO_2 Capture Technologies from Coal-fired Power Plants. *Energy Procedia* **2017**, *114*, 650–665.
78. Sun, J.; Li, Q.; Chen, G.; Duan, J.; Liu, G.; Jin, W. MOF-801 incorporated PEBA mixed-matrix composite membranes for CO2 capture. *Separation and Purification Technology*. **2019**.

79. Majumdar, S.; Tokay, B.; Martin-Gil, V.; Campbell, J.; Castro-Muñoz, R.; Zamidi Ahmad, M.; Fila, V. Mg–MOF–74/Polyvinyl acetate (PVAc) mixed matrix membranes for CO2 separation. *Separation and Purification Technology* **2019**, 116411.
80. Lee, D.-J.; Li, Q.; Kim, H.; Lee, K. Preparation of Ni-MOF-74 membrane for CO2 separation by layer-by-layer seeding technique. *Microporous and Mesoporous Materials* **2012**, *163*, 169–177.
81. Chen, K.; Xu, K.; Xiang, L.; Dong, X.; Han, Y.; Wang, C.; Pan, Y. Enhanced CO_2/CH_4 separation performance of mixed-matrix membranes through dispersion of sorption-selective MOF nanocrystals. *Journal of Membrane Science* **2018**, *563*, 360–370.
82. Olatunji, M. A.; Khandaker, M. U.; Mahmud, H. N. M. E.; Amin, Y. M. Influence of adsorption parameters on cesium uptake from aqueous solutions- a brief review. *RSC Adv.* **2015**, *5* (88), 71658–71683.
83. Aguila, B.; Banerjee, D.; Nie, Z.; Shin, Y.; Ma, S.; Thallapally, P. K. Selective removal of cesium and strontium using porous frameworks from high level nuclear waste. *Chemical Communications* **2016**, *52* (35), 5940–5942.
84. Naeimi, S.; Faghihian, H. Performance of novel adsorbent prepared by magnetic metal-organic framework (MOF) modified by potassium nickel hexacyanoferrate for removal of Cs + from aqueous solution. *Separation and Purification Technology* **2017**, *175*, 255–265.
85. Carboni, M.; Abney, C. W.; Liu, S.; Lin, W. Highly porous and stable metal–organic frameworks for uranium extraction. *Chemical Science* **2013**, *4* (6), 2396.
86. De Decker, J.; Folens, K.; De Clercq, J.; Meledina, M.; Van Tendeloo, G.; Du Laing, G.; Van Der Voort, P. Ship-in-a-bottle CMPO in MIL-101(Cr) for selective uranium recovery from aqueous streams through adsorption. *Journal of Hazardous Materials* **2017**, *335*, 1–9.
87. Zhang, N.; Yuan, L.-Y.; Guo, W.-L.; Luo, S.-Z.; Chai, Z.-F.; Shi, W.-Q. Extending the Use of Highly Porous and Functionalized MOFs to Th(IV) Capture. *ACS Applied Materials & Interfaces* **2017**, *9* (30), 25216–25224.
88. Sava, D. F.; Rodriguez, M. A.; Chapman, K. W.; Chupas, P. J.; Greathouse, J. A.; Crozier, P. S.; Nenoff, T. M. Capture of Volatile Iodine, a Gaseous Fission Product, by Zeolitic Imidazolate Framework-8. *Journal of the American Chemical Society* **2011**, *133* (32), 12398–12401.
89. Li, B.; Peng, L.; Wei, D.; Lei, M.; Liu, B.; Lin, Y.; Gu, J. Enhanced flocculation and sedimentation of trace cadmium from irrigation water using phosphoric fertilizer. *Science of The Total Environment* **2017**, *601-602*, 485–492.
90. Xiong, S.; Liu, Q.; Wang, Q.; Li, W.; Tang, Y.; Wang, X.; Chen, B. A flexible zinc tetrazolate framework exhibiting breathing behaviour on xenon adsorption and selective adsorption of xenon over other noble gases. *Journal of Materials Chemistry A* **2015**, *3* (20), 10747–10752.
91. Banerjee, D.; Cairns, A. J.; Liu, J.; Motkuri, R. K.; Nune, S. K.; Fernandez, C. A.; Thallapally, P. K. Potential of Metal–Organic Frameworks for Separation of Xenon and Krypton. *Accounts of Chemical Research* **2014**, *48* (2), 211–219.
92. Hendon, C. H.; Rieth, A. J.; Korzyński, M. D.; Dincă, M. Grand Challenges and Future Opportunities for Metal–Organic Frameworks. *ACS Central Science* **2017**, *3* (6), 554–563.
93. Younas, M.; Rezakazemi, M.; Daud, M.; Wazir, M. B.; Ahmad, S.; Ullah, N.; Ramakrishna, S. Recent progress and remaining challenges in post-combustion CO2 capture using metal-organic frameworks (MOFs). *Progress in Energy and Combustion Science* **2020**, *80*, 100849.

94. Wu, H. B.; Lou, X. W. (David) Metal-organic frameworks and their derived materials for electrochemical energy storage and conversion: Promises and challenges. *Science Advances* **2017**, *3* (12), eaap9252.
95. Xu, X.; Liu, W.; Kim, Y.; Cho, J. Nanostructured transition metal sulfides for lithium ion batteries: progress and challenges. *Nano Today* **2014**, *9* (5), 604–630.
96. Hendon, C. H.; Tiana, D.; Fontecave, M.; Sanchez, C.; D'arras, L.; Sassoye, C.; Walsh, A. Engineering the Optical Response of the Titanium-MIL-125 Metal–Organic Framework through Ligand Functionalization. *Journal of the American Chemical Society* **2013**, *135* (30), 10942–10945.
97. Wang, H.; Zhu, Q.-L.; Zou, R.; Xu, Q. Metal-Organic Frameworks for Energy Applications. *Chem* **2017**, *2* (1), 52–80.

Chapter 10

Environmental Applications of Metal–Organic Frameworks

Atul Kumar,[1] Suresh Kumar,[2] Surender Duhan,[*,1] Ashok Kumar,[1] Sunita Devi,[3] and Parmod Kumar[4]

[1]Department of Physics, DCRUST, Murthal Sonipat, India 131039
[2]Department of Electronic Science, Kurukshetra University, Kurukshetra, India 136119
[3]Department of Chemistry, MKJK Mahavidhyalaya, Rohtak, Haryana, India 124001
[4]Department of Physics, J.C. Bose University of Science and Technology, YMCA, Faridabad, India 121006

[*]Email: surender6561@gmail.com

The field of metal-organic frameworks has experienced remarkable attention in recent years driven by the vast range of applications of nanomaterials. Due to critical global concerns the cleanup of environmental contaminants is essential. This chapter provides an overview of the many components involved in the material creation and assessment of specific applications in real time. Thefuture scope and significance of MOFs are also described in this chapter and a quick look at the coverage.

Introduction

Metal-organic frameworks (MOFs) are characterized by IUPAC as a, "coordination network with an open framework and possible voids (*1*)." MOFs were developed as a large group of ultra-high-porosity crystalline materials with large interior surfaces extending over 6000 m^2/g (2). Because of their large pore structures, porous materials have been widely used in energy and the environment. MOFs are new types of porous periodical materials, combining the features of organic and inorganic substances through the self-assembly of metal ions, and organic ligands. MOFs represent the remarkable chemical structures and the potential to combine organic and inorganic chemistry (3).

This field of chemical sciences has experienced unprecedented growth over the last 30 years and this is reflected in the continually increasing breadth of literature. One of the distinctive features of MOFs is their topologically varied and unique architectures, many of them originating from minerals. Materials researchers are constantly aiming to design structures with specific characteristics and functions. The initial step is to grasp the basic geometrical concepts to reverse engineer the magnificent buildings from nature (*4*).

MOFs' strong characteristics are broad surface area, adjustable pores, and high porosity compared to other conventional porous materials (e.g., activated carbon, zeolites, and silica). In addition, MOFs may be changed by altering their metal ion centers or organic ligands, or adding various groups to increase the porosity, thermal stability, and catalytical capabilities (5). To meet multiple demands, the adaptability of MOFs has more flexibility as compared to metals. In addition to the exceptional variability in the structures and components of organic and inorganic, these properties are of concern to MOF researchersfor potential clean energy applications, mainly as storage media for gasses (i.e., hydrogen and methane high-capacity adsorbents that meet various separating needs). Further applications are becoming more and more relevant in membranes, thin-film electronics, catalysis, and biological imaging (*6*).

MOFs may be achieved with high yields and high thermal stability as a form of multifunctional material through affordable starting ingredients and varied preparation techniques, making MOFs an excellent contender for energy and environmental applications (7). Because of its vast surface area, high porosity, and easy construction of molecular functional gaps, porous materials have emerged as appropriate receptors to capture environmental contaminants or detoxify them into safer compounds (*8*, *9*). MOFs are improvements over traditional porous adsorbents for various reasons. Recently, MOFs have begun to show significant potential for recognizing, capturing, and detoxifying environmental pollutants present in various media, including air, water, and soil (*10*). In this chapter, we cover MOFs for potential environmental applications.

Environmental Application of Metal–Organic Frameworks

Wastewater Treatment

In biological treatment, phase change, and sophisticated processes, oxidation can divide water treatment approaches. Aerobic or anaerobic sludge digesters are the most common technologies for natural wastewater treatment; phase-change methods involve the sorption of the contaminant from the aquic phase onto the surface of a pore solid material; advanced oxidation processes are based on nonselective, highly oxidating con-degradation methods (*11*). To secure the use of safe water and decrease environmental harm, the investigation and development of water treatment technologies that are successful in reducing this pollutant class are essential. Researchers are quickly developing novel nanomaterials that may be used as nano sorbents and improve performances (*12*).

MOFs have high prospects for application in water treatment due to the changeable pore width of the MOF membrane. Applications for wastewater treatment are also wider than MOF composite membranes, bare MOF membranes, or modified MOF membranes (*13*).

Microfiltration

MOFs are frequently utilized for membrane microfiltration because MOFs enhance membrane characteristics. To get the ZIF-8/PPTEE composite membrane used to remove micropollutants from water Ragab et al. modified the polytetrafluoroethylene micro-filtrated membrane using ZIF-8 (*14*), and the results indicated a growth in the adsorption capacity of over 40% and virtually a doubling in water permeability. One cause was the hydrophobicity of ZIF-8, which reduced friction between water and the surface of the membrane and thereby enhanced its membrane permeability. The MOF particles also create hydrogen connections to water for hydrophilicity enhancement and permeability (*15*).

Ultrafiltration

Ma et al. employed UIO-66 as a porous modification and developed a variety of novel polyethersulfone (PES) composite membranes (e.g., UiO-66@GO-PES membranes) to anchor it on a graphic oxide layer to create UiO-66@GO). They observed that the flow, reject rate, hydrophilicity, and antifouling were enhanced by 3.0 wt% of UiO-66. In addition, Sun et al. also combined a new hollow zeolite ester skeleton-8/ultrafiltration polysulfone (hZIF-8), which is functionalized with polysulfone (PSF) tannic acid to generate an advanced hollow zeolite imidazolate structure-8 (hzIF-8/PSF). The results demonstrated that, although good rejection and antifouling capabilities are preserved, the hZIF-8/PSF UF membrane demonstrated enhanced water permeability around 2.8 times that of the PSF membrane. Surprisingly, hZIF-8 was superior to ZIF-8 in (1) the presence of tannic acid at hZIF-8 in hubs the agglomeration of ZIF-8 in an effective manner resulted in a uniform-structured composite cone, (2) the hydrophobe surface changed from hydrophilic to ZIF-8, and (3) the internal vacuum improved hydrophobe porous and structure improved the flow rate of hZIF-8 and its vacuums (*16*, *17*).

Nanofiltration

Kinetic hydrate inhibitor (KHI) was created by Golpour et al. to treat the new type of nanofiltration MOF membrane. Results demonstrated that the MOF composite membrane improves the hydrophilicity of the composite membrane. Still, rejection remains near 96%, with the MOF composite membrane at its highest performance, 50% higher than the pure polymeric membrane. Initially, Yuan et al. created a novel ZIF-300 MOF-pure membrane to remove heavy metal ions from wastewater. The ZIF-300 membrane obtained showed excellent energy stability, and the water permit was 39.2 L/m2h bar, while the CuSO4 refusal rate was 99.21% (*18*).

Reverse Osmosis

Separation technology has evolved in recent years. In 2015, five different aqueous sage (KCl, NaCl, CaCl2, MgCl2, or AlCl3) solutions at a similar concentration (0.20 wt% were applied to Li group's in situ solvothermal synthesis to manufacture pure-phased Zr–MOF (i.e., UiO-66) polycrystalline membranes in alumina hollow fibers. The findings showed a great multivalent ion repulsion (e.g., 86.3% of Ca2 +, 98.0% of the Mg2 + and 99.3% of the Al3 +) and high permeability (0.28 L m2 h1 bar1 lm). In addition, Hee et al. created a new MOF composite reverse-osmose membrane using the combination of MOF sulfuric acid-treated HKUST-1[Cu3(BCT2)] for water stability with PSF membrane. Research indicated that the water flow rose by 33%. The refusion of salt rose slightly to 96% due to the hydrophilicity, pore shape, and antifouling ability of the MOF membrane. Xu et al. produced the UiO-66-NH2 ultra-thin film on the alumina surface, which successfully modeled the selectivity of the cation. They obtained the greatest possible cation separation (Na+ 2 /Mg 2+> 200 and Li+ /Mg 2+>60) recorded in the literature (*19*).

Carbon Capture

MOFs and associated materials (e.g., Hybrid Ultra-Micropores (HUMs)) have been developed in several C-capture applications as a possible absorber. The essential sorption qualities of MOFs and HUMs (e.g., common selectivity and working capacity), are more significant—probably far higher—than conventional C-capture sorbents for their use. Intrinsic modularity of MOFs allows "platforms" of linked sorbents with similar structures but with varying chemical compositions to explore structure-function links. To date, we have learnt that pores and pores are both significant in terms of their effects on selectiveness and capacity. Selectivity is essential for trace separations like Direct Air Capture (DAC), whereas the working capacity (e.g., NG sweetening and flue gas remedial processes), is particularly significant for concentration C-capture separation. However, to achieve the full potential of MOFs and HUMs, the range of issues must be addressed before they can be developed for industrial applications at a higher Technology Readiness Level (TRL) (*20*, *21*).

To Capture Toxic Chemicals

MOFs were regarded as next-generation adsorbents and catalytic materials due to air pollution produced by anthropogenic volatile organic compounds (VOCs) and hazardous molecule emissions. For the effectiveness of adsorption separations, especially for the capture of VOCs under wet circumstances, the competitive binding of various potential molecules must be addressed. MOFs might thus serve as an appropriate platform due to their high adsorption capacity and strong structural and chemical diversity. MOFs may be altered by modifying organic lines, metal nodes, pores functional groups, shape and pores width, regulating their hydrophilic/hydrophobic equilibrium, adsorption sites density, as well as type or molecular sieving impact (*22*).

These characteristics have proven important for absorbing many target compounds with high absorption capacities and selectivity. The useful strategies for designing MOF-based adsorbents to selectively adsorb toxic gasses (e.g., NOx and CO) have been examined in addition to a high degree of control over the amount of CUS or the exposed redox-active metals (Fe21, Cu2) and in the fine-tuning of the electron density available at these sites of adsorption through organic linker modification. More significantly, MOF form and design improvements can lead to novel functionalities and enhance adsorbent efficiency. For example, the modular steering development of hetero-structuring hierarchical MOF-on-MOF opens, the way for the building of modular and multifunctional devices combining catalytic and separate systems (*23*).

Removal of Organic Pollutants

There have been significant breakthroughs in the synthesis and development of MOF-based materials since 2000. Due to its varying chemical composition, high porosity, specific pore structure, and various functional features, MOFs have previously been proved to have considerable attention and promise in removing and processing organic contaminants. We have demonstrated in this chapter that MOFs are capable of functioning as catalysts of sensors and adsorbents. Nonetheless, most studies preliminary and the literature shows a significant gap between MOF chemistry research and environmental application. This gap cab be bridged with effort by both chemists and environmental engineers, the actual use of MOFs in organic pollutants removal jointly should be improved (*23*).

MOFs will be beneficial and environmentally friendly to produce MOFs using inexpensive metals. Green solvents, or solventless technologies, are essential to reduce the formation of residues during synthesis. Theoretical approaches to enhance the performance should also be applied to elucidate the interactions between MOF and contaminants as well as promote the growth of efficient molecular MOF structures (*24*).

To Capture VOCs

In the residential complexes and offices, frequent indoor air through numerous anthropogenic actions and products release of typical VOCs, aromatic mono- and mono-cyclical hydrocarbons, ketones and methyl isobutyl ketone, esters and alcohols, etc. Indoor air in sub-ppb levels is often present in most VOCs. Though the deliberations of ppm in industrial installations can be significantly greater where they are usually used as a solvent or as a precursor in numerous consumer items and chemical production/synthesis processes. Besides standard chemical industries, different VOCs from diverse sources (e.g., wastewater treatment, feeding of animals, oil refinement, and automotive paint), is also generated in considerable quantities. The expansion of appropriate technologies for their competent elimination was the critical job of various stakeholders, primarily environmental protection authorities, because of the severe well-being and ecological impacts of VOCs emissions (25).

Because of their high permeability, positive surface to volume ratio, and easy functionalization MOFs have been identified as innovative sorbents for removing different environmental contaminants. Over the years, MOFs have been examined in adsorptive ability to remove a vast diversity of VOCs. At times, the evaluation of this data is limited to useful utility since instead of being mixed in real-life situations, the database is generally restricted to only a few types of individual VOC components. Furthermore, the provided data have minimal practical importance as central portions of fictitious experimental facilities. Therefore, the study of MOFs against a mixture of VOCs is open research, mainly when various VOCs are present as mixes with comparatively low partial pressure in practical settings (*26*).

Detection of Explosives

MOFs emerge as valuable platform in the areas of efficient sensing and desensitization of energetic materials. The porosity of MOFs makes it beneficial to detecting energy materials since the targeted analyte may be incorporated into pores and can produce a noticeable interaction with the targeted analyst to create strong signals. The sensing and detection of several environmental toxins, explosive materials etc., is the subject of intensive exploration for MOFs. MOFs for the turn-off detection of nitroaromatic explosive materials may be used via the cargo transfer effect in the field of High-Energy Diffraction Microscopy (HEDMs) sensing, and a competitive coordination technique has been adopted for aggregation-induced emission phenomena and N-heterocyclic rich HEDMs to be targeted. These phenomena are highly beneficial when designing sensory materials to differentiate analytes through the transduction process by their minor structural changes in the molecules (*27*).

Astonishingly, high-energy ligands are produced using metal ions via self-assembly, theory, measuring techniques, and processing which are improved; also, that HEDM preparation and characterization is developed. For HEDMs, selecting metal nodes and changing the connector play crucial roles in generating the material's energy characteristics. The support of high-energy chelating linkers also results in diverse frame dimensions from 1D to 3D while the desired qualities are provided. Researchers use various energetic ligands (mainly N rich) to desensitize energy-efficient materials to produce insensitive energetic MOFs. Furthermore, due to their increased structural strengthening compared with 1D and 2D frameworks, 3D HEDMs also discovered huge promises in this arena. Stabilization and energy characteristics are provided by including inorganic anions into 3D HEDMs, including nitrate (NO3 2) and perchlorate (ClO4 2), into the pores for HEDMs. The other two ways to develop desensitized HEDMs using MOFs are the encapsulation of energetic material into MOF pores of MOF and MOF-reduced carbon to make powerfully but insensitive explosives (*28*).

Conversion of Biomass into Fuel

Recent scientific developments have been achieved by creating MOFs as catalysts for the conersion of fatty acids into liquid fuels. For instance, certain catalytic MOFs show improved decarboxylation/deoxygenation, which result in liquid fatty acid performance compared to the support systems for benchmark catalysts. Important aspects requiring scientific attention to be considered as effective catalysts for the conversion of liquid biomass to valuable liquid fuels are as follows:

1. Exploring Various Compositions of MOFs

The inclusion of a noble metal (pt) requires the existence of strong catalytic activity in most MOFs to convert the fatty acids into liquid fuel. However, a MOF that can be highly active without an expensive noble metal would be highly desired. Other MOFs with appropriate catalytic characteristics for decarboxylation and deoxygenation should be examined in this respect, apart from Ni-MOF (former parts). These MOFs should be thermally stable, chemically stable, contain active metal agents for deoxygenation/decarboxylation processes, and .Modeling studies can therefore be beneficial if these novel catalytical systems are developed and optimally developed (*29*).

2. Real Biomass Deoxygenation

Future studies should concentrate on accurate fats compositions as feeds, rather than just molecules with biomass. The lipid composition is far more complex than just one reactant model. For example, the saturated and unsaturated fatty acids with carbon chains spanning from 12 to 18 are common in heavy animals (e.g., cattle grease and butter). In comparison to one molecular model reactant, this should lead towards distinct and more complicated reaction processes. For example, the liquid product would yield more hydrocarbons. Furthermore, the more unsaturated reactant may need more hydrogen for saturated carbon-double bonds (*30*).

3. Experimental and Computer Efforts Integrated

Integration of fatty acid decarboxylation catalysts with simulation experiments should help establish fundamental structural performances and better understand the fatty acid conversion reaction processes into liquid fuels (*31*).

Radioactive Waste Sequestration

Recently, the cationic MOF studies have increased dramatically to remediate environmental contaminants. A significant amount of cationic-organic frameworks (i.e., AsO32 4, CrO22 4, and ClO2 4) for removal of nonradioactive anions are valuable points for creating comparable materials for the anionic radionuclide appropriation. An efficient technique for preparing cationic MOFs for anionic radionuclide separation remains neutral molecules that comprise several atoms with transition metals that are self-assembled. Three design rules may be helpful: (1) The presence in open-metal plants of Cationic MOFs is favorable to selectivity and, for separating TcO2 4 from SO22 4 and NO22; (2) The hydrophobicity of MOFs is another significant component; (3) A non/weakly coordinated MOF anion can serve as a template for the formation of appropriate pores during processing and easy to replace anion. Up to present MOF applications in I2 have been carried out mainly in a carbon-based solvent instead of in a gas state and some general guidelines for s electively trapping gasses are challenging to describe (*32*).

Produced pores of molybdenum must be somewhat more extensive than Xe/Kr. There are also different MOF groups, metal sites, and nanoparticles. MOFs may initially be determined with high-performance computer simulations via the structural feature relation by increasing computation power and improving simulations. There is also a developing tendency to use such large-scale calculations to discover strong MOF candidate radionuclide sequestration (*33*).

Finally, the preparation for the chromatography of commercial columns of MOFs with high-grain sizes is highly recommended. All of them are excellent substrates to mix with milled-organic frameworks to produce the packaging constituents except producing large crystals, chitosan, and ion exchange resins. The practical application for the cleanup of MOF/Magnet combination and membrane materials based on MOF will also be created in the future and radionuclides will be produced (*34*).

Conclusion

MOFs have a wide range of prospecting applications in several environmental areas, including treatment of wastewater, the movement of VOCs, radioactive sequencing, the production of biological fuel, explosive detection, and more. MOFs, in particular gases and wastes, are crucial functions in adsorption. However, there are also some conflicts worth attention, such as high production costs, poor stability, ecologically unfavorable synthesis, and obstructed industrial output. Although the above problems have been overcome with the detailed examination of MOFs, they can still not be eradicated entirely. Eliminating these deficiencies makes MOFs an outstanding multifunctional material to answer other application challenges.

References

1. Fonseca, J.; Gong, T.; Jiao, L.; Jiang, H.-L. Metal-Organic Frameworks (M.O.F.s) Beyond Crystallinity: Amorphous M.O.F.s, MOF Liquids and MOF Glasses. *J. Mater. Chem. A* **2021**, *9*, 10562–10611, DOI: 10.1039/D1TA01043C.
2. Xuan, W.; Zhu, C.; Liu, Y.; Cui, Y. Mesoporous Metal-Organic Framework Materials. *Chem. Soc. Rev.* **2012**, *41*, 1677–1695, DOI: 10.1039.
3. Duan, C.; Yu, Y.; Xiao, J.; Li, Y.; Yang, P.; Hu, F.; Xi, H. Recent Advancements in Metal-Organic Frameworks for Green Applications. *Green Energy Environ.* **2021**, *6*, 33–49, DOI: 10.1016/j.gee.2020.04.006.
4. Ding, M.; Cai, X.; Jiang, H. L. Improving MOF Stability: Approaches and Applications. *Chem. Sci.* **2019**, *10*, 10209–10230, DOI: 10.1039/C9SC03916C.
5. Liu, X.; Zhang, L.; Wang, J. Design Strategies for MOF-Derived Porous Functional Materials: Preserving Surfaces and Nurturing Pores. *J. Materiomics* **2021**, *7*, 440–459, DOI: 10.1016/j.jmat.2020.10.008.
6. Baumann, A. E.; Burns, D. A.; Liu, B.; Thoi, V. S. Metal-Organic Framework Functionalization and Design Strategies for Advanced Electrochemical Energy Storage Devices. *Commun. Chem.* **2019**, *2*, 1–14, DOI: 10.1038/s42004-019-0184-6.
7. Mallakpour, S.; Sirous, F.; Hussain, C. M. Metal-Organic Frameworks/Biopolymer Nanocomposites: From Fundamentals Toward Recent Applications in Modern Technologies. *New J. Chem.* **2021**, *45*, 8409–8426, DOI: 10.1039/D1NJ01302E.

8. Modak, A.; Bhanja, P.; Selvaraj, M.; Bhaumik, A. Functionalized Porous Organic Materials as Efficient Media for the Adsorptive Removal of Hg(II) Ions. *Environ. Sci. Nano* **2020**, 7, 2887–2923, DOI: 10.1039/D0EN00714E.
9. Zhang, W.; Aguila, B.; Ma, S. Retracted Article: Potential Applications of Functional Porous Organic Polymer Materials. *J. Mater. Chem. A* **2017**, 5, 8795–8824, DOI: 10.1039/C6TA11168H.
10. Brandt, P.; Nuhnen, A.; Öztürk, S.; Kurt, G.; Liang, J.; Janiak, C. Comparative Evaluation of Different MOF and Non-MOF Porous Materials for SO2 Adsorption and Separation Showing the Importance of Small Pore Diameters for Low-Pressure Uptake. *Adv. Sustain. Syst.* **2021**, 5, 2000285, DOI: 10.1002/adsu.202000285.
11. Roy, M.; Saha, R. 6 - Dyes and Their Removal Technologies from Waste-water: A Critical Review. In, Bhattacharyya, S., Mondal, N. K., Platos, J., Snášel, V., Krömer, P., Eds.; *Intelligent Environmental Data Monitoring for Pollution Management*; Academic Press, 2021; pp 127—160, DOI: 10.1016/B978-0-12-819671-7.00006-3.
12. Amin, M. T.; Alazba, A. A.; Manzoor, U. A Review of Removal of Pollutants from Water/Wastewater Using Different Types of Nanomaterials. *Adv. Mater. Sci. Eng.* **2014**, 1–24, DOI: 10.1155/2014/825910.
13. Wang, C.; Liu, X.; Keser Demer, N. K.; Chen, J. P.; Li, K. Applications of Water Stable Metal-Organic Frameworks. *Chem. Soc. Rev.* **2016**, *45*, 5107–5134, DOI: 10.1039/C6CS00362A.
14. Ragab, D.; Gomaa, H. G.; Sabouni, R.; Salem, M.; Ren, M.; Zhu, J. Micropollutants Removal from Water Using Microfiltration Membrane Modified with ZIF-8 Metal-Organic Frameworks (M.O.F.s). *Chem. Eng. J.* **2016**, *300*, 273–279, DOI: 10.1016/j.cej.2016.04.033.
15. Zhang, H.; James, J.; Zhao, M.; Yao, Y.; Zhang, Y.; Zhang, B.; Lin, Y. S. Improving Hydrostability of ZIF-8 Membranes via Surface Ligand Exchange. *J. Member. Sci.* **2017**, *532*, 1–8, DOI: 10.1016/j.memsci.2017.01.065.
16. Sutrisna, P. D.; Hou, J.; Zulkifli, M. Y.; Li, H.; Zhang, Y.; Liang, W.; D'Alessandro, D.; Chen, V. Surface Functionalized UiO-66/Pebax-based Ultrathin Composite Hollow Fiber Gas Separation Membranes. *J. Mater. Chem. A* **2018**, *6*, 918–931, DOI: 10.1039/C7TA07512J.
17. Li, J.; Wang, H.; Yuan, X.; Zhang, J.; Chew, J. W. Metal-Organic Framework Membranes for Wastewater Treatment and Water Regeneration. *Coord. Chem. Rev.* **2020**, *404*, 213116, DOI: 10.1016/j.ccr.2019.213116.
18. Chandrasekaran, P.; Ravindran, S.; Boovarahan, S. R.; Kurian, G. A. Hydrogen Sulfide-Mediated Cardioprotection Against Ischemia-Reperfusion Is Linked to KATP Channel for Mitochondrial Preservation but Not for Its Distinct Preference on Interfibrillar Mitochondria. *Bangladesh J. Pharmacol.* **2019**, *14*, 2.
19. Park, H. M.; Jee, K. Y.; Lee, Y. T. Preparation and Characterization of a Thin-Film Composite Reverse Osmosis Membrane Using a Polysulfone Membrane Including Metal-Organic Frameworks. *J. Member. Sci.* **2017**, *541*, 510–518.
20. Li, J. R.; Kuppler, R. J.; Zhou, H. C. Selective Gas Adsorption and Separation in Metal-Organic Frameworks. *Chem. Soc. Rev.* **2009**, *38*, 1477–1504, DOI: 10.1039/B802426J.
21. M. Mohamedali, D. Nath, and HI and A. Henni *Review of Recent Developments in CO2 Capture Using Solid Materials: Metal-Organic Frameworks (M.O.F.s)*; IntechOpen, 2016, DOI: 10.5772/62275.

22. Wang, B.; Xie, L.-H.; Wang, X.; Liu, X.-M.; Li, J.; Li, J.-R. Applications of Metal-Organic Frameworks for Green Energy and Environment: New Advances in Adsorptive Gas Separation, Storage and Removal. *Green Energy Environ.* **2018**, *3*, 191–228, DOI: 10.1016/j.gee.2018.03.001.
23. Bobbitt, N. S.; Mendonca, M. L.; Howarth, A. J.; Islamoglu, T.; Hupp, J. T.; Farha, O. K.; Snurr, R. Q. Metal-Organic Frameworks for the Removal of Toxic Industrial Chemicals and Chemical Warfare Agents. *Chem. Soc. Rev.* **2017**, *46*, 3357–3385, DOI: 10.1039/C7CS00108H.
24. Gandara-Loe, J.; Pastor-Perez, L.; Bobadilla, L. F.; Odriozola, J. A.; Reina, T. R. Understanding the Opportunities of Metal-Organic Frameworks (M.O.F.s) for CO2 Capture and Gas-Phase CO2 Conversion Processes: A Comprehensive Overview. *React. Chem. Eng.* **2021**, *6*, 787–814, DOI: 10.1039/D1RE00034A.
25. Ulker, O. C.; Ulker, O.; Hiziroglu, S. Volatile Organic Compounds (V.O.C.s) Emitted from Coated Furniture Units. *Coatings* **2021**, *11*Art. no. 7, DOI: 10.3390/coatings11070806.
26. Muir, B.; Sobczyk, M.; Bajda, T. Fundamental Features of Mesoporous Functional Materials Influencing the Efficiency of Removal of VOCs from Aqueous Systems: A Review. Sci. Total Environ. **2021**, *784*, 147121, DOI: 10.1016/j.scitotenv.2021.147121.
27. Lustig, W. P.; Mukherjee, S.; Rudd, N. D.; Desai, A. V.; Li, J.; Ghosh, S. K. Metal-Organic Frameworks: Functional Luminescent and Photonic Materials for Sensing Applications. *Chem. Soc. Rev.* **2017**, *46*, 3242–3285, DOI: 10.1039/C6CS00930A.
28. He, Z.; Li, M.; Que, W.; Stang, P. J. Self-Assembly of Metal-Ion-Responsive Supramolecular Coordination Complexes and Their Photophysical Properties. *Dalton Trans.* **2017**, *46*, 3120–3124, DOI: 10.1039/C7DT00174F.
29. Liao, Y.-T.; Matsagar, B. M.; Wu, K. C. -W. Metal-Organic Framework (M.O.F.)-Derived Effective Solid Catalysts for Valorization of Lignocellulosic Biomass. *ACS. Sustainable Chem. Eng.* **2018**, *6*, 13628–13643, DOI: 10.1021/acssuschemeng.8b03683.
30. *R. v. B and AVB Overview About Lipid Structure*; IntechOpen, 2013, DOI: 10.5772/52306.
31. Hossain, M. Z.; Chowdhury, M. B. I.; Jhawar, A. K.; Xu, W. Z.; Biesinger, M. C.; Charpentier, P. A. Continuous Hydrothermal Decarboxylation of Fatty Acids and Their Derivatives into Liquid Hydrocarbons Using Mo/Al2O3 Catalyst. *ACS Omega* **2018**, *3*, 7046–7060, DOI: 10.1021/acsomega.8b00562.
32. Wu, G.; Ma, J.; Li, S.; Wang, S.; Jiang, B.; Luo, S.; Li, J.; Wang, X.; Guan, Y.; Chen, L. Cationic Metal-Organic Frameworks as an Efficient Adsorbent for the Removal of 2,4-Dichlorophenoxyacetic Acid from Aqueous Solutions. *Environ. Res.* **2020**, *3*, 109542, DOI: 10.1016/j.envres.2020.109542.
33. Altintas, C.; Altundal, O. F.; Keskin, S.; Yildirim, R. Machine Learning Meets with Metal-Organic Frameworks for Gas Storage and Separation. *J. Chem. Inf. Model.* **2021**, *61*, 2131–2146, DOI: 10.1021/acs.jcim.1c00191.
34. Gawande, M. B.; Goswami, A.; Felpin, F. X.; Asefa, T.; Huang, X.; Silva, R.; Zou, X.; Zboril, R.; Varma, R. S. Cu and Cu-Based Nanoparticles: Synthesis and Applications in Catalysis. *Chem. Rev.* **2016**, *116*, 3722–3811, DOI: 10.1021/acs.chemrev.5b00482.

Chapter 11

Environmental Applications of Metal–Organic Frameworks and Derivatives: Recent Advances and Challenges

Adetola Christianah Oladipo,[1] Temitope Olabisi Abodunrin,[1] Deborah Temitope Bankole,[1] Oluwole Solomon Oladeji,[1] Godshelp Osas Egharevba,[1] and Olugbenga Solomon Bello[*,1,2]

[1]Department of Physical Sciences, Industrial Chemistry Programme, Landmark University, Omu-Aran, Kwara State 251101, Nigeria

[2]Department of Pure and Applied Chemistry, Ladoke Akintola University of Technology, P.M.B. 4000, Ogbomoso, Oyo State, Nigeria

[*]Emails: osbello06@gmail.com, osbello@lautech.edu.ng

The consequence of environmental pollution has raised the dire need for the discovery of efficient and potent methods for detection and removal of pollutants released into air and water bodies. Metal-organic frameworks (MOFs) are porous coordination polymers having intriguing features such as large surface areas, tailorable pore size and highly dense active sites reported for various environmental applications. Recent developments have focused on the modification of MOF structures, development of MOF-based materials including functionalized MOFs, MOF composites/hybrids and MOF derivatives. These modifications confer new and desirable properties over pristine MOFs and consequently lead to enhanced efficiency for pollutant sensing and adsorption applications. This chapter focuses on the recent developments and challenges in the use of MOF-based materials for sensing and adsorption of pollutants from air and water in the past seven years. Some challenges and future prospects are also discussed. In spite of the challenges encountered with the use of MOF-based materials for detection and removal of gaseous and water pollutants, they remain valuable materials for environmental applications.

1. Introduction

In recent times, environmental pollution has posed threats to the public and ecosystem. The criticality of this can be seen in the United Nations Sustainable Development Goals (SDG) which has five of her goals centered on the environment for improved standards of living (*1*). Increase in population and industrialization are key players responsible for environmental pollution and this results in the increase in toxic wastes and hazardous compounds in the atmosphere and water bodies (*2–5*).

Air pollution has recently become a global subject as a result of rapid industrialization and urbanization (*6, 7*). Forms of air pollutants include particulate matter, dust, toxic gases, liquid droplets and bio-aerosols which are major threats on human well-being (*4, 8*). Major air pollutants are oxides of sulfur (SO_x), nitrogen (NO_x), and carbon, chlorofluorocarbons, etc. These molecules have contributed to the formation of acid rain, smog, climate change and damage to ecological systems (*4, 5, 9*). Health effects of air pollution are exacerbation of asthma, increased respiratory mucosal symptoms, chronic obstructive pulmonary disease, cardiovascular disease and mortality (*10*).

Sulfur dioxide is a byproduct of fossil fuel emissions and is very toxic. Its presence as a pollutant is significant because trace quantities affect human health and form hazardous $PM_{2.5}$ products having deleterious effects on the environment (*4, 5*). Exposure to high concentrations results in difficulty in breathing, respiratory illness and aggravation of cardiovascular disease. In the environment, it causes foliar and crop damage, acidification of lakes and streams, acid deposition that accelerates the decay of buildings and monuments also contributes to visibility degradation (*11*). The two major oxides of carbon are carbon monoxide and carbon dioxide. Carbon monoxide is an important feedstock for the synthesis of various chemical compounds. Although a colorless and odorless gas, it is highly toxic to humans and the environment. Exposure to CO leads to unconsciousness, disorientation and death (*5*). A major greenhouse gas is carbon dioxide which is generated from the combustion of fossil fuels (*12*). The major sources of nitrogen oxides are anthropogenic sources such as agricultural activities, automobile exhausts, natural gas, coal and oil burning. Nitrogen oxides are highly toxic and hazardous to humans and are causes of severe environmental challenges. The most common ones present in the atmosphere are nitric oxide and nitrogen dioxide. The effects of exposure to humans are skin irritation, respiratory, nervous and immune damage. While on the environment, it generates acid rain, forms photochemical smog, depletes the ozone layer and increases global warming (*5, 10, 13*). Pesticides released through surface runoffs into water bodies damage the flora and fauna, cause diabetes, cancer, asthma, leukaemia and Parkinson's disease (*14*).

Water pollutants either occur naturally or through anthropogenic sources. Some water pollutants found in wastewater are dyes, pharmaceuticals and personal care products (PPCPs), polyaromatic hydrocarbons (PAHs), pesticides, heavy metals and radioactive substances (*15*). These pollutants are from industrial effluents discharge from manufacturing industries (such as food, leather, cosmetics, textile, petrochemical, paper and paint industries), oil spillage during exploration of crude oil, agricultural practices, etc. (*16*) The frequent and extensive use of these products, coupled with their poor removal by conventional wastewater treatment methods have made these pollutants ubiquitous in the environment.

Metals such as Cr(III) and Cr(VI) are mutagenic and carcinogenic due to their acute toxicity (*17*). The presence of Hg(II) leads to loss of vision and hearing, neurological deteriorations and damages to endocrine system (*18, 19*). Radioactive metals accumulate in living organisms and are highly toxic and carcinogenic (*20*). Dyes are known to cause human health disorders and disturbance in the ecological balance due to their stability to heat, light and oxidizing agents (*21*). Antibiotics which are examples of PPCPs threaten the environment and damage the human central nervous system (*22*). Dibutyl phthalate, an example of persistent organic chemicals (POP) in water effluents, causes disruption of the endocrine, reproductive and nervous systems in humans (*23*).

2. Metal–Organic Frameworks (MOFs)

MOFs constructed from metal ions/clusters and organic linkers were first reported in 1995 (*24*). MOFs are organic-inorganic crystalline hybrid materials which consist of strong bonds between organic linkers and metal ions/clusters. The metal clusters in this case are referred to as secondary building units (SBUs). They form connecting points that are linked together by the organic linkers, resulting in framework formation. The SBUs direct the MOF topology with their intrinsic geometries. The metal ions are locked into their positions by the linkers. This results in high structural stability of MOFs constructed by SBUs (*25*). MOFs are well known for their fascinating properties which include well-ordered structure, tunability, high porosity, diverse functionalities, large pore volume and surface area, good optoelectronic properties, high stability and the possibility to incorporate chemical functionalities without disrupting the framework. These properties distinguish them from other conventional porous solids like zeolites. One of the attractive features of MOF lies in the tendency of its structure to be tailored through modification of chemical nature, activity and surface properties for specialized applications in adsorption, separation, chemical sensing, catalysis, light harvesting, drug delivery, photocatalysis, bio-catalysis, gas sorption/separation, energy storage and conversion, etc. (*26–36*)

2.1. MOF-Based Materials

MOF-based materials have arisen due to their intriguing functionalities, tailorable and diversified structures, adjustable morphology, uniform heteroatom doping among others. MOF-based materials sometimes retain and extend the inherent features of pristine MOFs with greater advantages, applications and potentials for better performance. These materials are easily fabricated, increase flux mass transfer and provide easily accessible active sites. They also expose highly dense active sites, optimize size and morphology, and favor structure-performance relationship. The relationship between the morphology, physical and chemical properties of the organic and inorganic building blocks and the species encapsulated inside the pores is responsible for the variety of functionalities exhibited by MOFs (*37–44*). The applications of MOFs can be improved by tuning their chemical and physical properties through active groups grafting, change of organic linkers, impregnation of suitable active materials, post-synthetic modification and MOF composite/hybrid formation (*45*).

2.2. Classes of MOF-Based Materials

MOF-based materials are obtained with tailored structures and properties of MOF in pre-design (pre-synthesis) or post-synthetic modification where species, geometry, size and functionality are changed or varied (*38, 46*). MOFs either retain or sacrifice their backbone to yield various derivatives. Therefore, they can be broadly classified into nanoscale MOFs, functionalized MOFs, MOF composite/hybrid and MOF derivatives.

2.2.1. Nanoscale MOFs

MOFs with at least a dimension in the range of tens to hundreds of nanometers are referred to as nano-MOFs (n-MOFs). n-MOFs have larger surface area, improved dispersibility in aqueous media or other solvents than bulk MOF. MOF nanoparticles possess diverse morphologies such as sphere, cubic and octahedron (*47*). They can be prepared through solvothermal/hydrothermal, microwave assisted, ultrasound assisted, mechanochemistry, microemulsion, continuous flow production methods. Nanoscale materials are classified into particles, fibers, tubes, rods, thin films and membranes. Synthesis is carried out by controlled precipitation of self-assembled metal-organic polymers and confinement of supramolecular assembly at nanoscopic locations using templates (*48*).

2.2.2. Functionalized MOFs

Incorporation of multiple functionalities into a single network of MOF is of interest in recent times. Functionalization presents an effective strategy for tuning chemical and physical properties (*49*). Several MOF linkers which possess various functional groups such as amine, halogen, hydroxyl, nitro, carboxylic acid, sulfonic acid, among others, have been employed for synthesis to obtain desired functionality (*50*). The methods for functionalization are: Post-synthetic modification and Solvent-Assisted Linker Exchange (SALE) (*1*). Functionalization can be achieved through pre-design/post-synthetic modification of organic ligands or metal ion/clusters, use of functional ligands or metals for construction of frameworks and encapsulation of functional molecules into the pores (*51*).

The synthetic strategies employed for MOF synthesis for optimum design platform for functionalized MOFs involve the tailoring of either organic ligands or metal clusters with multiple functional sites or the introduction of multiple guest components as functional sites or construction of framework with functional sites and guest molecules (*50, 52*). Their modification can be accomplished before or after the process of synthesis (*50*). In these MOFs, the physicochemical features of each component are superimposed to accomplish an overall performance better than those of the individual components or their adducts (*53*).

2.2.3. MOF Composites/Hybrids

A multi-component material with multiple phases containing at least a continuous phase is known as composite. MOF composites are hybrid materials which comprise of MOF and other suitable constituent materials with noticeably different properties from those of the individual constituent. MOF composites usually combine the advantages of both constituents to get rid of their drawbacks(*54, 55*). MOF composites originate from the addition of MOF with functional substrates to improve stability and cause increase in adsorption sites (*49*). The integration of MOFs with functional materials such as carbon nanotubes, metal nanoparticles and nanorods, graphene, metal oxides, complexes, enzymes, etc. results in the formation of composites with multifunctionalities and generation of new physical and chemical properties (*38, 56, 57*). MOFs can also integrate their large surface area/ high porosity, crystallinity and flexibility along with optical, catalytic or electrical properties of other materials to form composites (*58*). This synergy gives the resulting composite a wider range of applicability than the individual component (*59–61*).

The synthetic method often employed for the synthesis of MOF composites are in situ synthesis and post-synthetic methods. In the in situ method, the composite is obtained by addition of pre-synthesized materials to the solution of the MOF (*57*). The in situ methods are either by bottle-around-ship method or ship-in-bottle method. In the bottle-around-ship method, large particles of composing materials are immobilized inside the cages of MOF via in situ building of MOF from precursors around the composing materials. While in the ship-in-bottle method, the composite is stabilized inside the pores via thermal or chemical methods where the composite materials form and is diffused by a solvent system inside the cages of MOF (*45*). In post-synthetic method, the composites are obtained by absorption of target precursors into the MOFs by immersion and vapor deposition methods followed by further processing for conversion (57). It involves simple loading or impregnation of composing materials in the MOF where the solubility of composite materials are of less concern (45).

MOF composites improve adsorption capacity, surface area and ease of separation. The suitability of MOF composites for environmental applications stem from their synthetic kinetics, morphology, physicochemical properties and stability (49).

2.2.4. MOF Derivatives

MOF derivatives are obtained by using MOFs as sacrificial templates or precursors. Usually the MOF template/structure backbone is destroyed (37, 52). Generally, the applications of MOFs are limited by their instability and poor conductivity (62, 63). A method designed to address this limitation is by the conversion of pristine MOFs to much more stable materials that can serve as precursors to other porous materials with large surface area (37, 64–66). MOF derivatives are obtained by thermal conversion (62, 67). Their properties include increase in pores, with high surface area and narrow pore-size distribution (38, 68). These derivatives contain any of porous carbon, metal/carbon, metal oxide/carbon, metal/metal-oxide/carbon-based materials, etc. The properties inherited from the parent MOFs are large surface area, uniform heteroatom doping, highly dispersed active sites and controllable compositions (67). Features of MOF derivatives are enhanced stability, large surface area, ordered porous structure and ease of pore size adjustment (39, 62, 69, 70).

MOF derivatives are commonly prepared using thermolysis, pyrolysis, carbonization and calcination. In thermolysis strategy, the metal nodes are transformed into metal nanoparticles while the organic bridging ligands form highly porous carbon materials (37, 39). Direct pyrolysis is a simple but effective method where metal-free carbons are prepared from MOF subjected to carbonization under an inert atmosphere of nitrogen or argon gas with in situ evaporation and/or subsequent leaching of metal species. In this method, partially preserved morphology is obtained where pyrolysis takes place under air to remove the carbons (43). The formation of metal/metal oxide embedded carbon can be obtained by simple pyrolysis of MOF (68). Pyrolysis usually enhances the chemical activity and durability of MOFs (38). Carbonization at higher temperature yields a higher degree of graphitization (66). Carbonized MOFs consist of both metal and carbon sources. Hybrid structures such as carbon-supported metal/metal oxides and hybrids from different metal-based compounds containing more than one component are prepared under inert atmosphere by direct carbonization of MOFs (38, 43). This process can take place with or without a secondary carbon. The nature of the carbonized material depends on the rate of heating, temperature of calcination, crystal size and component of MOF (38). Calcination yields porous carbon when the MOF is subjected to heat in air or an inert atmosphere at elevated temperatures over time (57, 69). This chapter explains the use of MOF-based materials for environmental applications. Specifically, the sensing and adsorption of several gaseous and water pollutants using MOF-based materials are discussed. The efficiency, mode of action of the materials for the various applications and chemical interactions between the sensors/adsorbents and the pollutants are reviewed.

3. Environmental Applications of MOF-Based Materials for Sensing of Pollutants

The hazardous effects that toxic pollutants have on the environment and human health make their detection necessary. The detection of pollutants is imperative for air monitoring and subsequent removal. Sensing materials, capable of rapid and selective detection of toxic species are required for this purpose. Several technologies have been employed for sensing of organic and inorganic pollutants (*71*). Some of these conventional methods are configured with high sensitivity and accuracy. However, they require expensive equipment, and the needed pre-treatment of water samples is highly complicated. Compared to conventional detection methods, the use of materials as chemical sensors is simple, economical and useful in real-time monitor (*72*). The operation of chemical gas sensors is based on the changes of some properties as a result of their interaction with analytes. MOFs are a relatively new burgeoning class of materials for gaseous and aqueous phase sensing. Measurable changes in intrinsic properties, when exposed to target analytes, qualify MOFs as sensors.

The fascinating and desirable properties possessed by MOFs make them suitable candidates for sensing applications: the porosity of MOFs enables the adsorption/desorption of guest molecules, in this case, the analyte to be sensed. Selective sensing capability for target analyte is made possible by the tunable pore size of MOFs. The pore dimensions of MOFs are modulated to allow the target analyte molecule to be included. Coordinatively unsaturated sites (CUS) created by the removal of occluded solvent molecules could be used to capture guests and enhance selectivity by reversibly binding with the target analyte (*73*).

The earliest MOFs synthesized were in form of bulk crystalline materials and are restricted by poor dispersibility in solution. These MOFs have limited applications in sensing pollutants, especially in aqueous solution. In recent times, nanoscale MOFs (n-MOFs) with larger surface areas than conventional MOFs have successfully overcome the barrier faced by conventional MOFs. The larger surface areas and more accessible active sites in n-MOFs, result in better sensing capability (*48*).

The water stability of MOFs is an essential property in selecting MOFs for sensing applications. It was acknowledged that MOF frameworks are destructible at the preliminary stage of synthesis since the ligand–metal bonds in the MOFs are easily susceptible to attack by water molecules. This barrier has been crossed by the introduction of water stable MOFs. Two of the categories according to their mechanism of formation, are stated as follows: First is the metal carboxylate frameworks consisting of high-valence metal ions, like Cr^{3+}, Fe^{3+} and Zr^{4+}. These ions possess high coordination number and charge density, which enhance the stability coordination bond of MOFs between metal ions and ligands. The second type is metal azolate frameworks comprising nitrogen-donor ligands, like tetrazolates, imidazolates, pyrazolates, and triazolates. The nitrogen-containing ligands are typically soft ligands, which can react with soft divalent metal ions to form MOFs with strong structural framework (*74*).

Incorporating additional functionality on the ligand, or nanoparticles into the accessible volume of the pores could also be a strategy to achieve selectivity (*75*). Metal oxide semiconductors derived from the pyrolysis of MOFs have also been reported to have large surface areas and surface active sites which offer advantages like high response, low working temperature, and low limit of detection in sensing applications (*72*).

3.1. Transduction Schemes for MOF-Based Sensors

The broad classification of MOF-based sensing techniques is based on the employed transducers. A transduction unit translates the sensed information from the sensing unit into a signal. The mode of action of the sensor is based on the transduction mechanism (*76*). Sensing in MOFs could be through optical, mechanical and electrical methods as shown in Figure 1.

The optical methods involve changes in color, luminescence features and refractive index. The emission of photon when the excited electron returns to the ground state is called photoluminescence. This emission could either be quenched or enhanced on exposure to an analyte. This depends on the extent of the host-guest interaction. Also, there could be shifts in the emission frequency. This is useful in the detection of analytes, in addition to the emission quenching or enhancement. Analytes with good electron donors or acceptors are easily detected in this case (*77*). Luminescent lanthanide-based MOFs have aroused great interest for use in sensing applications owing to their outstanding luminescent properties. Some of the properties include huge stokes shift, extensive fluorescence lifespan and sharp line emission (*78*). The chemosensing potentials of Ln-MOFs have been widely explored on organic small molecules and nitro explosives (*79*, *80*).

In chromism, there is a reversible change in color upon adsorption/desorption of analytes either in solution or gaseous state, these are termed solvatochromism and vapochromism respectively. The chromophore electronic transition from ground to excited state causes a large shift of the absorption spectrum which accompanies the change in color of the sensing material. In addition, there could be color change, resulting from the change in the coordination environment of the metal centers in the MOF.

Interferometry involves the measurement of the change in the refractive index on exposure to target analytes. This is done by monitoring the shifts of the interference peaks in the transmission spectrum. In this case, the MOF material has to be in form of thin film and the reflective surface on both sides of the film (*81*).

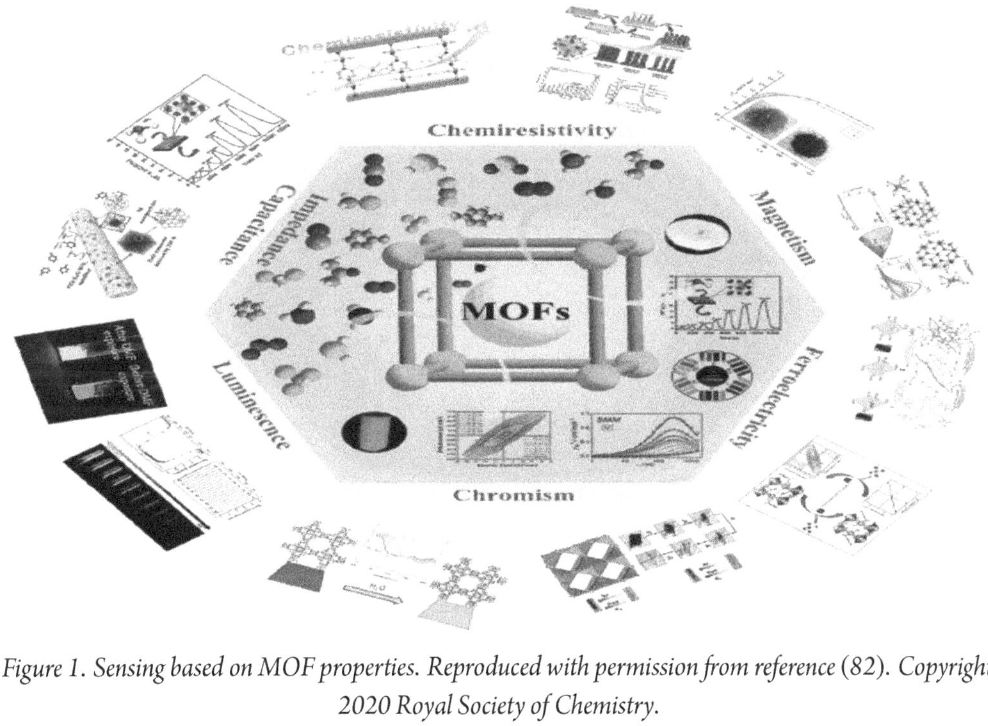

Figure 1. Sensing based on MOF properties. Reproduced with permission from reference (82). Copyright 2020 Royal Society of Chemistry.

Mechanical methods involve the absorption of analytes by MOF grown on electromechanical devices such as surface acoustic wave sensors, microcantilevers (MCL) and quartz crystal microbalance (QCM). These devices can convert gravimetric changes into vibrational signals. The gravimetric changes influence the frequency of acoustic waves, cantilever and resonant vibrations of the quartz crystal respectively, resulting into signal generation and detection (*82*).

The electrical method involves the reaction of target analytes with oxygen in the environment and detecting the changes in the electrical properties such as impedance, capacitance, work function and resistance (*83*). Here, the reaction of the target analyte with oxygen adsorbed on the surface of the MOF material at high temperature is required. The high porosity of MOFs is beneficial in the conductivity-based sensing applications (*84*).

Some recent exciting examples in the application of MOF-based materials in the detection of both gaseous and water pollutants are discussed in the subsequent section.

3.2. MOF-Based Materials for Sensing Gaseous Pollutants

The detection of volatile organic compounds (VOCs) from emission sources at working and residential places is important. The fabrication of capacitive sensor based on Cu-1,3,5-benzenetricarboxylic acid (BTC)MOF-199nanoporous film and coated on a copper substrate, was reported for the detection of ethanol, methanol, acetone and isopropanol vapors. The sensitivity was in the range of 250-1500 ppm and the sensor offered a reasonable response time and reversible detection of the analytes. Factors that affected the sensing ability for the different analytes were the dielectric constant, concentration of the analytes and the extent of adsorption into the MOF film (*85*). The same research group also reported that low limit of detection of 130.0 and 39.1 ppm for ethanol and methanol was obtained with the use of the same nanosensors. The reusability of the sensor for several sensing cycles was possible due to its reversible affinity toward the vapors of the analytes. There was selective sensing of the polar analytes over non-polar n-hexane (*86*). A zinc-based MOF was pyrolyzed to yield hierarchical hollow ZnO nanocages possessing hollow interiors covered by interpenetrated ZnO nanoparticles. It provided mesopore and macropore channels which aid the diffusion and subsequent reaction of gas molecules. The large surface area and exposed active sites offer a highly sensitive sensing toward acetone and benzene. The high sensitivity toward the VOCs, in sub-ppm and ppb levels, was attributed to the surface-adsorbed oxygen for the oxidation of the reducing analytes, thereby enhancing the sensing of the analytes (*87*). In an attempt to improve the sensing activity of the MOF 5-derived ZnO, the same research group synthesized Au@ZnO heterostructure by the pyrolysis of Au@MOF-5. The core-shell Au@ZnO nanoparticles showed a response value of 11 times higher than what was observed for just ZnO nanoparticles in the sensing of acetone. It was documented that Schottky junction formed by the Au-ZnO interface led to the enhanced acetone-sensing property of the Au-ZnO heterostructure (*87*). In another report, affinity layers were derived by the dispersion of NH_2-MIL-53(Al) nanoparticles in a Matrimid polymer matrix. The composite was deposited on aluminum electrodes and applied for the gas phase sensing of some alcohols. There was a fivefold increase in methanol sensitivity of the MOF composite device as compared to only the polymer-coated device. Increase in MOF concentration also increased the response time (*88*). Ghanbarian et al. developed a ternary nanocomposite of a bimetallic MOF, MIL-53(Cr-Fe), silver, and carbon nanotube (CNT) for the fabrication of a resistive gas sensor to detect isopropanol, ethanol and methanol at 25°C and 10% relative humidity. The analyte for which it had the highest response was methanol, with limit of detection (LOD) of 30.5 ppm. The sensing mechanism could be explained by the flexibility and reversible expansion of MIL-53 with the presence of analytes which led to the increase of the sensor resistance. The CNT helped with electron transport and tunable reactivity while silver spread electron through the composite and increased the conductivity. The conductivity decreased on exposure to the analytes (*89*). In another report on the sensing of methanol vapor, Zn-MOF of benzenedicarboxylate ligand having $Zn_xO_yC_z$ SBU, was synthesized and made a composite with manganese oxide. Although the crystallinity of the Zn-

MOF was sacrificed and the electrical resistivity lowered on addition of MnO_2, the composite could detect methanol vapor (*90*). In another report, double strategy was used for the modification of a Zr-MOF. It was functionalized with a sulfone group and also embedded with a noble metal, palladium to give Pd@Zr-BPDC-SO_2 (dibenzo[b,d]-thiophene-3,7-dicarboxylate5,5-dioxide). The composite was effective in ethanol sensing as a result of hydrogen bonding between the –OH group of ethanol, sulfonyl and Zr_6 clusters (*91*). Cao et al. reported on the preparation of Fe-doped ZnO lattice and the subsequent heterojunction formation with $ZnFeO_4$ nanoparticles. The acetone sensing of the heterostructure surpassed that of pristine ZnO in terms of response time, selectivity, sensitivity and stability. This was attributed to the large surface area, heterojunction formed and abundant adsorption sites for oxygen and analyte molecules (*92*). In a report that involved the study of the effect of carbonization conditions of zeolitic imidazolate frameworks (ZIFs) on surface area and the consequences on selective sensing of VOCs, it was observed that CNT grew on the surface of the nanoporous carbons after carbonization under hydrogen gas atmosphere. The CNT acted as π-rich antennae for the selective sensing of benzene and toluene vapors over n-hexane despite the fact that the high surface area of the CNT-free NPCs was sacrificed with the growth of CNTs. The enhanced and selective sensing of the aromatic hydrocarbons was as a result of the resulting π- π interaction between the aromatic rings of the hydrocarbons and the π-rich CNT (*93*).

The detection of CO_2 gas is paramount in air quality control. Therefore, highly sensitive sensors are required. Gassensmith et al. reported a γ-cyclodextrin-derived material, CDMOF-2, with high proton conductivity and free primary –OH group capable of reacting with CO_2 gas to form alkyl carbonate. This reaction brought about approximately 550-fold reduction in its conductivity. This positioned the material for selective CO_2 sensing by electrochemical impedance spectroscopy (*94*). Nanoporous Cu-BTC was used to develop an ultra-short near-IR fiber optic CO_2 sensor at wavelength of 1.57 μm. It had a rapid response time of 40 seconds and its limit of detection was 500 ppm. It was found suitable for the detection of the greenhouse gas (*95*). MOF analogs of M-MOF-74 (M- Mg, Ni, Co and Zn) and ethylenediamine functionalized Mg-MOF-74 were synthesized and employed for the selective sensing of CO_2 gas by monitoring the change in the work function of the sensing layer. The MOFs could interact with CO_2 gas through the coordinatively unsaturated metal centers. In addition, the NH_2 functional group also interacted with the gas, in the case of the functionalized Mg-MOF-74. Its sensitivity, stability and reversibility positioned it as the best of all the MOF materials used (*96*). In another report, ZIF-8 nanoparticles were integrated onto bimodal optical waveguides for the fabrication of an optical sensor for CO_2 detection. LOD of 3130 ppm at room temperature with a broad linear response was exhibited by the ZIF-8 nanoparticle sensor. There was no loss in performance of the sensor even under humid condition. Furthermore, its robustness, reusability and target analyte selectivity over water vapor and methane made it ideal for CO_2 sensing (*97*). Ye et al. employed impedance measurement for the detection of CO_2 using Zn-MOF-74 and NdMo-MOF. Increase in impedance was observed upon exposure to an increased concentration of CO_2 gas. With the use of Zn-MOF-74, there was a significant rate of CO_2 adsorption but a slow desorption. The rapid response to CO_2 was evidenced by the increase in the real part and decrease in the imaginary part of the impedance (*98*).

Over the years, metal oxide semiconductors like ZnO, TiO_2, SnO_2, etc. have found application in the detection of CO gas with high response. Many metal oxides derived from MOFs have also been used. In a report, n-type semiconductor SnO_2 nanoparticles was derived from a Sn-MOF and composited with molybdenum diselenide nanoflowers. Selective sensing of CO over other gases like SO_2, CH_4, H_2S, H_2 and even CO_2 was documented. As both SnO_2 and $MoSe_2$ are semiconductors, n-n heterojunction could be formed at the SnO_2 nanoparticle - $MoSe_2$ nanoflowers interface as electrons from $MoSe_2$ with high Fermi level were transferred to SnO_2 with low Fermi level until equilibrium was reached (*99*).

A lot of sensors developed for SO_2 sensing could only be used for its liquid phase detection (*100, 101*). Its gaseous phase detection is paramount as it is a class of air pollutants. Most reported luminescent Ln-MOF were obtained in powder form, which when applied in solution could be hard to recycle, unsuitable for use with gas flux blast and the dispersion could be time-consuming (*102*). As an improvement to overcome these drawbacks for the sensing of SO_2 gas, MOF-based sensors for SO_2 were developed. Chernikova et al. reported on SO_2 detection using an Indium-MOF (MFM-300). This was coated on a functionalized capacitive interdigitated electrode for the fabrication of a chemical capacitive sensor. This was the first MOF-based SO_2 sensor and coincidentally gave the lowest sensitivity at room temperature, compared to other sensors. The LOD was estimated to be 5 ppb. Furthermore, the sensor was exceptionally stable and reused for several sensing cycles. The selectivity over other gases like NO_2, CH_4 and H_2 was shown by the signal intensity for the target gaseous analyte compared to those of the other gases (*103*). Also, Eu(III)-organic framework film fabricated on UiO-66-NH_2 (UiO- University of Oslo) was developed. Upon the exposure of SO_2, the strong Eu^{3+} emission was rapidly quenched in just six seconds with LOD of 0.65 ppm due to the charge transfer from the occupied π-orbital of the ligand and the empty SO_2 orbital (*104*). Ingle et al. reported a MOF composite made up of Ni_3BTC_2 and OH-functionalized single walled carbon nanotubes (OH-SWNTs). The composite was drop-casted on gold micro electrode tip on Si/SiO_2 substrate. The composite gave a better performance in sensing SO_2 gas than pure Ni_3BTC_2 as measured by its increased sensitivity, fast response time and repeatability at room temperature (*105*). In another report, $ZnFe_2O_4$, derived from a Zn/Fe MOF, was functionalized with rGO. The formation of a p-n heterostructure between the two materials and the good electrical properties of rGO led to a better transient response, sensitivity and selectivity for the detection of SO_2 gas by the $ZnFe_2O_4$/rGO composite compared to single $ZnFe_2O_4$ and rGO (*106*).

For the detection of H_2S gas, malonitrile-functionalized ZIF-90 (MN-ZIF-90) which is a polymeric fluorescence probe, was synthesized. The fluorescence intensity was enhanced with the incorporation of malonitrile when exposed to H_2S gas. This aided the sensing ability of the material (*107*). Also, four MOF materials, including MIL-101(Cr), MIL-100(Fe), ZIF-8 and $Zn_3(BTC)_2.12H_2O$, were used for the development of cataluminescence sensors. After some experiments, the last two MOF materials were outstanding for good sensitivity and stability for H_2S sensing (*108*). A MOF based on rare-earth metal and fumarate ligand was fabricated into thin films on a capacitive interdigitated electrode. LOD of 5 ppb for H_2S detection was obtained. More interestingly, the sensor was selective for the target analyte over other gases like CH_4, NO_2, H_2 and C_7H_8 (*109*). In a report by Zhang et al., the poor selectivity and high working temperature of γ-Fe_2O_3 was addressed by making its composite with reduced graphene (rGO). γ-Fe_2O_3 was derived from MIL-88 and dispersed on reduced graphene oxide. Better sensing performance of the composite than γ-Fe_2O_3, even at room temperature, was as a result of the bulk resistance effect of γ-

Fe_2O_3, good conductivity and much active sites of rGO. The composite offered H_2S selectivity over $CHCl_3$, NH_3, SO_2 and HCHO (*110*). Ali et al. reported the fabrication of MOF-polymer mixed matrix for H_2S gas detection at room temperature. The introduction of MOF-5 led to the sensitivity of the composite to H_2S gas and was attributed to the increase in the surface area of MOF-5 with a special cage-bridge structure which enhanced energy transport. The fast response and high sensitivity (<8 s and 1 ppm respectively), coupled with the room temperature application, made the composite ideal for industrial applications (*111*).

The electrical method of sensing of NH_3 based on changes in normalized resistance was carried out using a MOF hybrid composed of Cu-BTC and either of graphite oxide or aminated graphite oxide. The chemical reaction of the analyte and the MOF component, evidenced by the irreversible increase in resistance upon exposure to NH_3, resulted in the structural collapse of the MOF component. Nevertheless, the electrical signal recorded as the amorphous phase adsorbed the NH_3 gas (*112*). The conductivity measurement of four MOF analogs synthesized from trimesic acid and either of Ba(II), Cd(II), Pb(II) and Zn(II) was carried out before and after exposure to NH_3 vapor. The Zn(II) analogs exhibited about 50 times increase in conductivity after exposure to NH_3, making it an ideal candidate for NH_3 sensing (*113*). Also, Zhang et al. documented the introduction of Eu^{3+} ions to the free carboxylic sites of MIL-124 to obtain MIL-124@Eu^{3+} as a luminescent MOF film. The MOF film was prepared on porous α-Al_2O_3 ceramic plate. In this case, NH_3 gas reacted chemically with the carboxylic group and formed the $-COONH_4$ moiety. The characteristic emission of MIL-124@Eu^{3+} was quenched by the transfer of energy between the ligand and the Eu^{3+} ion as a result of the effect of the resulting $-COONH_4$ moiety. The sensor was chemically stable, as confirmed by XRD analysis (*114*). Another vapoluminescent Eu-MOF was reported for the naked-eye sensing of NH_3 vapor. The solid sample, upon exposure to NH_3 and HCl vapor, exhibited a recyclable on-off-on sensing of the analytes. When viewed under UV light, it was noticed that the red luminescence of the MOF material which turned blue upon exposure to HCl vapor, was regained after being exposed to NH_3 vapor. To gain insight into this response, control experiments were performed and it was believed that the dissociation of Eu^{3+} from the Eu-MOF with exposure to the acidic HCL vapor, was reformed after exposure to the basic NH_3 vapor (*115*).

Two series of isoreticular Mg-MOFs with high thermal stability and large surface area were employed in the detection of NO_2 gas. The resistance of the gas sensor was monitored and found to decrease in the presence of NO_2 gas. The Mg-MOF with the larger surface area had more adsorption sites and adsorbed target gases, which in turn led to more change in resistance and sensor response. It was also documented that the pore size of the material had a great effect on the kinetics of the gas sensor (*116*). MOF analogs of lanthanide ions Eu^{3+} and Tb^{3+} were also explored for the luminescent sensing of NO_2 gas. The inclusion of the gas into the framework led to the increase and decrease in the luminescence intensity of Eu^{3+} and Tb^{3+} MOFs respectively. This was attributed to changes in the relative position of the ligand triplet and the Ln energy levels when NO_2 was adsorbed. Although the interaction between NO_2 gas and the MOF was weak, there was a high sensitivity of the MOF for the target analyte due to the affinity of the amino functional group to NO_2 gas (*117*).

For the chemiresistive sensing of acidic gases SO_2, NO_2, CO_2, pore surface functionalization of Zr-based UiO-66 MOF was explored by the modulation of the terephthalate ligand with the –OH and NH_2 groups (Figure 2). Pristine UiO-66, NH_2-UiO-66, OH-UiO-66 and NH_2-OH-UiO-66 were synthesized and the changes in their electrical resistance on exposure to the acidic gases were monitored. The basic NH_2-UiO-66 offered strong active sites for the acidic gases. Also, the interaction between the basic MOF and the acidic analytes led to a change in the electronic property of the sensing material (*118*).

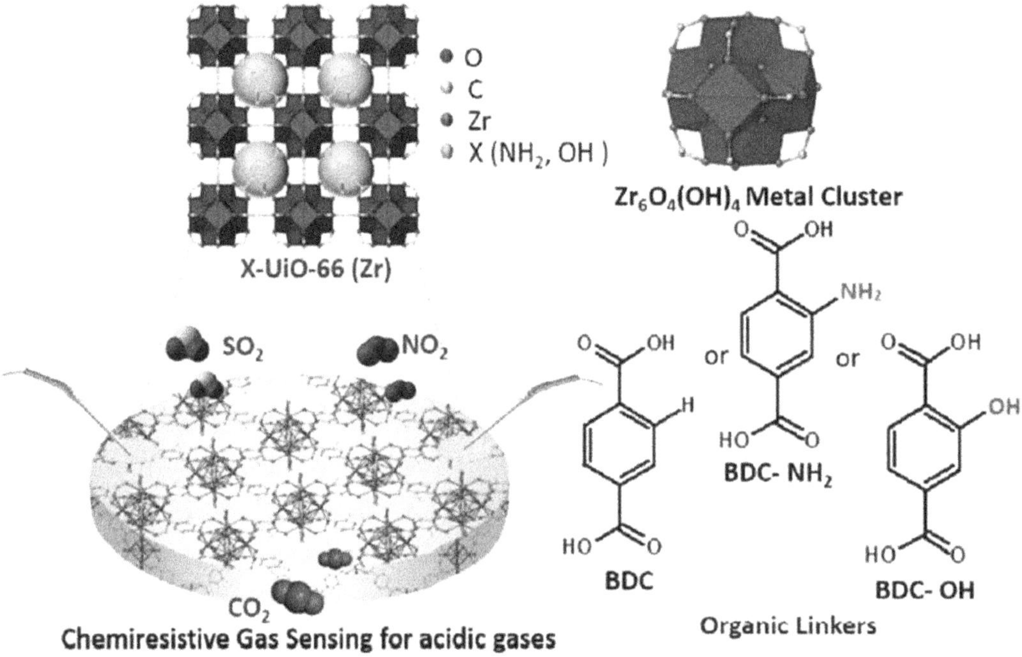

Figure 2. Illustration of chemiresistive sensing of acidic gases by the functionalization of organic linker in UiO-66 MOF with -OH and -NH_2 groups. Reproduced with permission from reference (118). Copyright 2019 Royal Society of Chemistry.

3.3. MOF-Based Materials for Sensing Water Pollutants

Among all the sensing techniques for MOFs, the luminescent technique is the dominant for cations detection. Luminescent MOFs have found a vast usage for the detection of toxic metal cations in water. Bhattacharyya et al. synthesized an anionic Mg-MOF with very interesting property. It contains cationic dimethyl ammonium (DMA) which could be replaced by Cu^{2+} only, of all the transition metal cations and resulted into a turn-off sensing of the copper cation. Also, the selective sensitization of Eu^{3+} resulted in a bright red luminescence (*119*). A turn-on fluorescent sensing mechanism was reported for the sensing of Hg^{2+} in water using lanthanide MOF (Eu-isophthalate) nanoparticles. An acid-base resistant Zn-based MOF constructed from 5-aminoisophthalic acid was utilized for the luminescent sensing of Hg^{2+}. The Zn-MOF contains uncoordinated carboxylic oxygen atoms and exhibited a pronounced fluorescent and selective response to Hg^{2+} over other metals present in the solution (*120*). A 3D cadmium MOF constructed from 5-(4-pyridyl)-tetrazolate and 1,4-naphthalenedicarboxylic acid was reported for its stability under heat, acidic and basic conditions. It was reported to be the first MOF that could selectively sense Cr^{3+} in aqueous solution by turn-on fluoresecence emission. The enhancement was as a result of the coordination of the analyte ion with oxygen and nitrogen atoms of the ligand (*84*). In another report, MIL-82 which has a carboxyl-rich group was designed as a parent MOF. In conjunction, a stable Ln-MOF was successfully developed by doping the Eu^{3+} cation on the coordination sites of MIL-82. The strong luminescence of Eu^{3+}- incorporated MIL-82 signified that the uncoordinated carboxyl group is a proficient scaffold for accommodating and sensitizing Eu^{3+} cations. This type of Lu-MOFs is a highly selective (relatively low detection limit of 0.09 mM) and effective chemical sensor for Ag^{+} in aqueous medium. This Lu-MOF has an outstanding water and thermal stability, which provided intermittent technique for detecting Ag^{+} in water samples (*71*). A Zinc-MOF of 2,6-naphthalenedicarboxylic acid was incorporated with Eu^{3+} ions via solvothermal and post-synthetic modification (PSM) techniques. Eu^{3+}@Zn-NDC exhibited three sensing channels for Cr^{3+} and three good linearities ranging from 1×10^{-3} to 7.5×10^{-5} mol/L. These parameters enhanced the sensitivity, selectivity and quantitative response of the Zn-MOF to Cr^{3+} in aqueous media (*121*). An electrochemical MOF-based sensor was reported for the detection of Pb^{2+} ions where NH_2-MIL-53(Cr) was incorporated on the surface of a glass carbon electrode (GCE). A good electronic response for Pb^{2+} was obtained as the oxidation current increased on increasing the concentration in the range 0.4-80 μm and an LOD value of 30.5 nM was obtained (*122*). Tran et al. also reported on the immobilization MIL-53(Fe), on GCE and its use for the electrochemical sensing of Cd^{2+} in water. The resulting electrode exhibited a stable electrochemical behavior and the response signal was enhanced. Therefore, it was used for the detection of Cd^{2+} using the differential pulse voltammetry technique and the LOD was reported to be 16 nM (*123*). The first recyclable MOF for the detection of phosphate anion was constructed from Eu and 1,3,5-benzenetribenzoate.The anion was selectively sensed in the presence of other anions. The luminescence intensity remained intact after five sensing cycles. This indicates a weak interaction between the MOF and phosphate anion (*124*). NH_2-functionalized terephthalic acid linker was employed in the synthesis of NH_2-MIL-53(Al) used for the detection of free ClO^- ions. The fluorescence of the MOF material was quenched on exposure to the free chlorine. The LOD and detection range were 0.04 μm and 0.05 – 15 μm respectively. The quenching was attributed to energy transfer through the hydrogen bond interaction between the ClO^- ions and the amino group of the MOF framework (*125*). In another report, CN^- anion was selectively sensed by post-synthetic modified ZIF-90 which incorporated specific active sites with the cyanide anion over some other anions. The LOD obtained met the WHO (World Health Organization) standard for the permissible limits of CN^- concentration in drinking water (*126*). A MOF hybrid, Cu-MOF/rGO was fabricated and electrode-modified for the electrochemical detection of nitrite anion. This sensor gave an LOD value of 33 nm and distinguishable selectivity for nitrite anion over interfering ions (*127*). Xia et al. reported the fluorescence quenching of a Cd(II)-MOF for the sensing of $Cr_2O_7^{2-}$ anion in water. The quenching was more on increasing the concentration of the anion, showing that the extent of quenching depended on the concentration of the analyte. The detection was as a result of the competition absorption and energy transfer mechanism (*128*). Also for the selective detection of $Cr_2O_7^{2-}$, a Zn-MOF was encapsulated with proflavin dye. The dye-encapsulated MOF exhibited green emission which was selectively quenched by the anion even in the presence of other competing anions. Fe^{3+} cation was also selectively detected by the same MOF material. The fluorescence quenching was attributed to light absorption by the analyte and the quenching mechanism was found to be static (*129*).

There are also reports on the detection of toxic organic pollutants. Yang et al. reported the electrochemical sensing of three co-existing dihydroxybenzene (DBI) isomers, namely catechol (CT), resorcinol (RS) and hydroquinone (HQ). Chitosan and rGO were coated on the surface of an electrode, resulting into a high electrical conductivity. In this composite was grafted an electroactive MOF, $Cu_3(BTC)_2$ (Figure 3). There were intense voltametric signals of the analytes at the electrode as a result of the porous framework of $Cu_3(BTC)_2$ and the electronic conductive composite. Quantitative analysis showed the LOD for the analytes were 0.44, 0.41 and 0.33 µm for HQ, CT and RS respectively. Satisfactory results were obtained in the use of the developed sensor for real water sample (*130*). $Cu_3(BTC)_2$-modified on carbon paste electrode was also used for the electrochemical detection studies of 2,4-dichlorophenol using cyclic voltammetry and differential pulse voltammetry. The sensor was also used for the analyte sensing in reservoir raw water samples. The sensor exhibited good selectivity toward the analyte, LOD of 9 nM and wide linear range detection within 0.04 – 1.0 µm (*131*). Further on DBI detection, Ni@graphene composites were obtained from a thermally-annealed Ni-BTC MOF. A sensor for the detection of HQ and CT was fabricated using the composite and glassy carbon electrode. Both DBIs could not be differentiated using the bare electrode, as only one oxidation peak was found from cyclic voltammetry measurements. Two distinct peaks were seen with the use of the composite and GCE, this indicated that both DBIs could be differentiated. In addition, the sensor was binder-free and easy to prepare (*132*). Two La(III)-based MOFs of 5-aminoisophthalic and oxalic acids were synthesized and found to display high sensitivity in the selective detection of acetone over several other analytes like acetonitrile, dimethylformamide, dimethylsulfoxide, ethyl acetate, etc. This was achieved by fluorescence quenching as a result of the interaction between the framework and acetone. Initially, there was reduction in the fluorescent emission intensity and eventually, there was full quenching (*133*). A Zn-MOF synthesized using 1,3,5-tris(3-carboxy-triazolemethylene)-2,4,6-tritoluene (H_3L) as organic ligand, produced a potential fluorescent material that could selectively and sensitively detectnitrobenzene in aqueous solution (*134*). In another report, a dual-functional Co(II) MOF was reported for its low detection limit of 14 and 4 µm for acetylacetone and Hg^{2+} by fluorescence emission quenching. The quenching was attributed to the resonance energy transfer from the excited Co-MOF to the analytes. The good recyclability of the MOF material was also reported (*135*). Li et al. reported on a yttrium-MOF of 2,5-furandicarboxylate for the selective recognition and luminescence sensing of Cu^{2+}, nitrobenzene and o-chlorophenol with LOD values of 0.1 µg/ml, 5 and 10 mg/L respectively. The fluorescence intensity was found to decrease on increasing the analytes concentration. The study presented a highly selective MOF for the detection of metal ions and small organic molecules (*136*).

Sensing of PPCPs is necessary due to their ubiquitous and hazardous nature when present in waterbodies in large quantities. The sensing of tetracycline (an antibiotic) using a stable luminescent Zr-MOF was documented. Results of the theoretical and experimental studies were in close consonance, the mechanism being fluorescence quenching attributed to the transfer of photo-induced electron from the organic ligand to tetracycline. Furthermore, the efficiency of the analyte sensing was enhanced by pre-concentrating the analyte in the pores of the MOF framework (*137*). A Zn-based MOF, NUC-6 (NUC-North University of China), was discovered to be luminescent and multi-responsive as it could detect both nitrofuran antibiotics (nitrofurazone and nitrofurantoin) and dichloronitroaniline pesticide in water by the quenching of its luminescence intensity. It exhibited detection limits of 116, 16 and 12 ppb for dichloronitroaniline, nitrofurantoin and nitrofurazone respectively. Optical spectroscopy and quantum chemical calculation inferred that the mechanism of sensing by luminescent quenching was dominated by internal filter effect and photo-induced electron transfer (*138*).

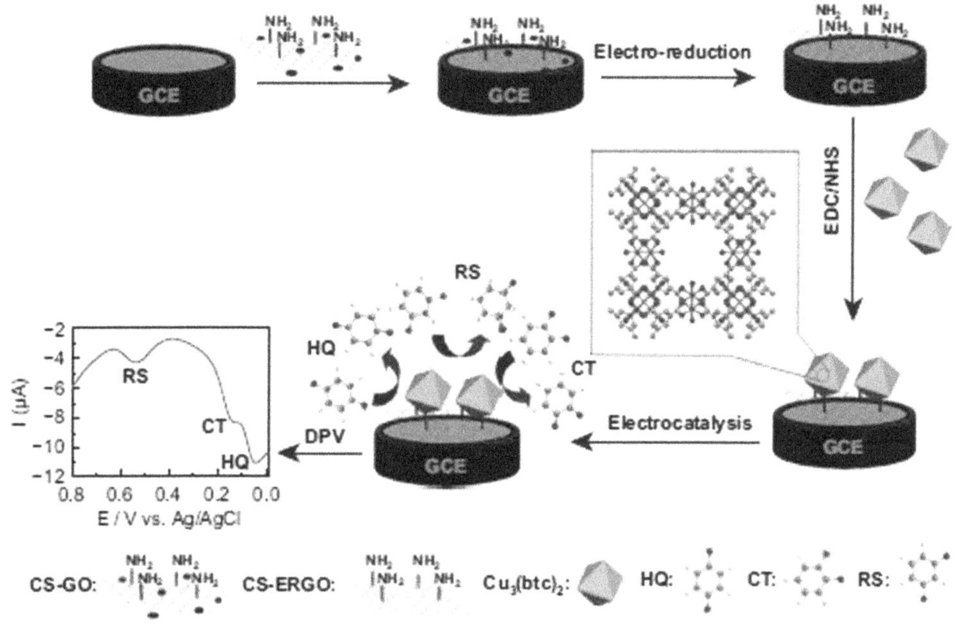

Figure 3. Detection of catechol (CT), resorcinol (RS), and hydroquinone (HQ) with MOF $Cu_3(BTC)_2$ and CS/rGO composite. Reproduced with permission from reference (130). Copyright 2016 American Chemical Society.

4. Environmental Applications of MOF-Based Materials for Adsorption of Pollutants

Adsorptive removal of toxic pollutants, either in gaseous state or from aqueous solution, is preferred to other techniques for remediation. This is because it is less energy intensive, relatively cheap, simple to operate, produces low harmful secondary products and could be carried out at low temperature. Furthermore, the adsorbent materials could be recycled (*45*). The selection of an ideal adsorbent is paramount to ensure the effectiveness of an adsorption operation. The adsorbent must possess high porosity and large surface area (*139*). MOFs are well known for their high surface area and high porosity. More interestingly, they are capable of selective adsorption of target species due to the presence of functionalities in their structure as shown in Figure 4. These functionalities could confer a profound impact on the adsorption efficiency through increased interaction between the functionalities and the adsorbates (*1*). These interactions are acid-base, electrostatic, coordination, hydrogen bonding, π-π, etc. (*140*) Electrostatic interactions are the commonest of these interactions. These are based on the surface charges of the material depending on the level of acidity or basicity of the medium. It could also be achieved when some particular species are present (*141*).

Consequently, electrostatic interaction could occur between the charged compound and oppositely charged adsorbates. π-π stacking interactions occur when aromatic rings are present in the structures of both the MOF and adsorbate (*142*). Hydrogen bonding is another important interaction for adsorption process. It is formed between two electrovalent atoms (N, O, F) where one withdraws electrons from hydrogen atom involved in a covalent bond and the other possesses a lone pair of electrons (*143*). Acid-base interaction also contributes to adsorption mechanisms in which Lewis bases could coordinate to CUS of MOFs (*144*).

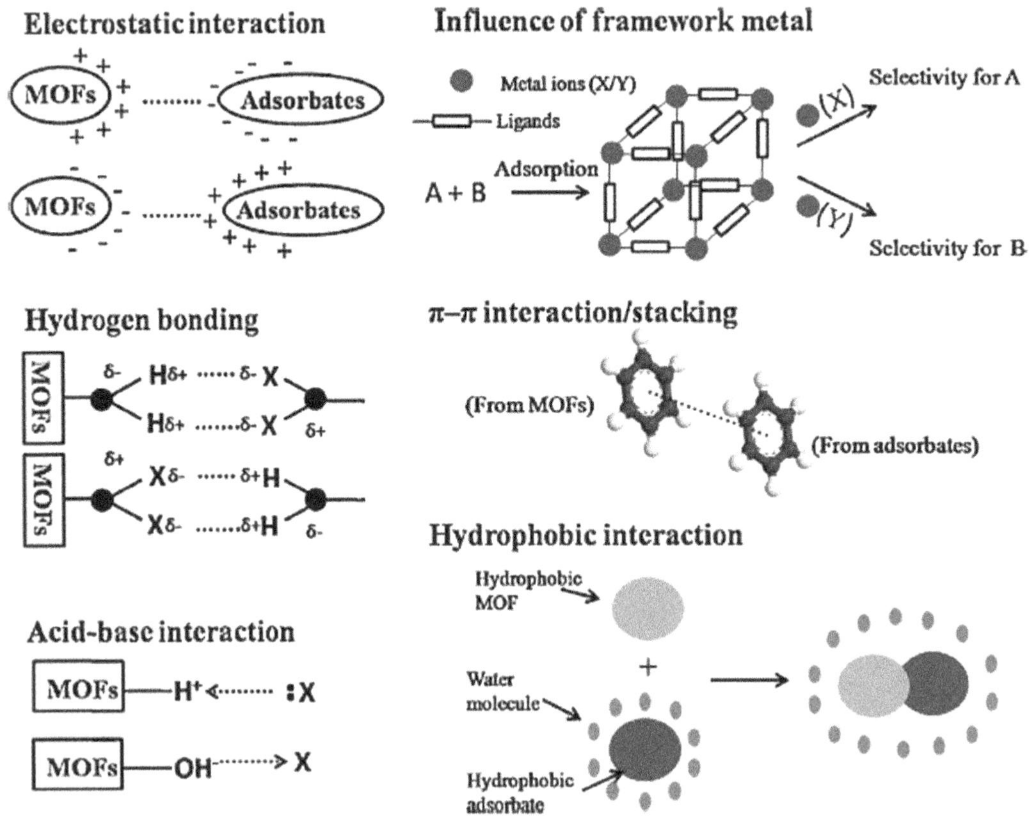

Figure 4. Possible adsorptive removal mechanisms of hazardous materials over MOFs. Reproduced with permission from reference (140). Copyright 2015 Elsevier.

However, as discussed earlier, there are new advances in the modification of MOFs to make them more effective in their use as adsorbents. For the functionalized MOFs, the introduced groups on the pristine MOFs could further interact with target adsorbate through different mechanisms (e.g., electrostatic, acid-base, hydrogen bonding, etc.), thus leading to better adsorption efficiency than the pristine form. MOF materials could also be composited with suitable materials for the enhancement of their adsorptive capability based on the synergy of the adsorptive efficiency of the compositing materials. The porosity of the MOFs is enhanced, this invariably gives rise to better adsorption performance. The functionalities of the MOF materials are tuned by the incorporation of the compositing materials, and in most cases, new properties that are desirable for adsorption processes are conferred on the MOF structure (*45*). Such materials include metal oxides, metal nanoparticles, graphene oxide, etc. (*145–147*) Another class of MOF-based materials are the MOF-derived carbons, also known as MOF derivatives. These are derived when MOF materials are pyrolyzed. They offer benefits like ordered structure, higher surface area, porosity, chemical stability and therefore higher adsorptive efficiency than the pristine MOFs (*148*).

In the subsequent section, the application of MOF-based materials in the adsorption of both gaseous and water pollutants would be discussed. Figure 5 illustrates some contaminants adsorbed using MOF-based materials.

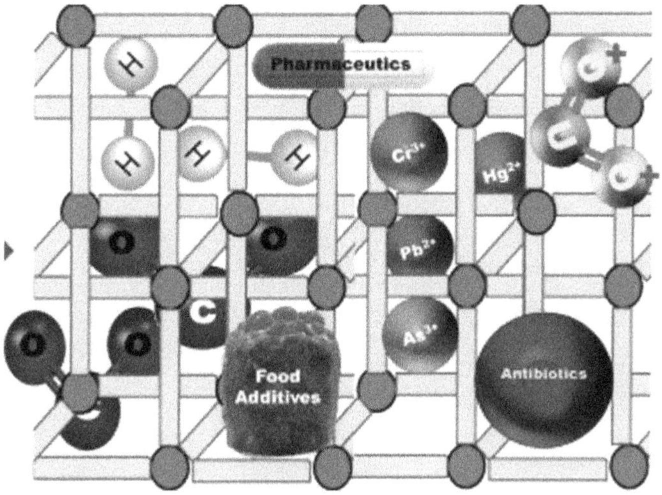

Figure 5. Application of MOF-based derivatives for adsorption of contaminants. Reproduced with permission from reference (149). Copyright 2020 Elsevier.

4.1. MOF-Based Materials for Adsorption of Gaseous Pollutants

MOFs have received a substantial application, in the removal of toxic air pollutants from flue gases due to their large surface areas, high porosity, pore tunability and distinct interactions (*150*). For the adsorption of CO_2, MOF-177 and MOF-5 are among the most studied MOFs and have been applied in real life scenarios. Particularly, MOF-177 with the brand name Basolite® Z377, have been commercialized (*151*). MOF-177 and MOF-5 were investigated for CO_2 adsorption and compared to 5A zeolite (traditional adsorbent). The adsorption capacity was found to be greater than that of 5A zeolite with high pressures in consideration. However, 5A zeolite has high affinity for CO_2 at sub-atmospheric pressure. MOF-177 displayed a fair CO_2/NO_2 selectivity and good regeneration capacity. Also, the potential of Cu-BTC MOF for CO_2 adsorption was investigated. The MOF was impregnated with 1-butyl-3-methylimidazolium acetate which doubled the adsorption capacity of CO_2 in comparison to pristine MOF at 15 kPa. The functionalized MOF was regenerated after many CO_2 desorption cycles (*152*). Cu-BTC was also impregnated with lithium nitrate, and was heat-treated under vacuum. The adsorption capacity of the impregnated MOF doubled that of the pristine MOF. Also, the adsorption process was completely reversible. This was attributed to the overlap of the CO_2 equilibrium adsorption and desorption curve of the lithium doped Cu-BTC MOF (*153*). A magnesium variant of MOF-74 (Mg-CPO-27) was also investigated for the adsorption of CO_2 and it exhibited an outstanding adsorption capacity under ambient temperature. Also, the regenerating capacity was studied and the CO_2 adsorption and desorption branches of isotherms were found to overlap (*154*). Recently, Edubilli et al. investigated a litmus test on UiO-66 MOF. This was carried out by packing a column with MOF constituting a combined pressure vacuum swing adsorption. The purity of CO_2 obtained and adsorption capacity were close to values obtained with other MOFs (*155*).

Gate opening phenomena is one of the attracting functions of flexible MOFs, during an adsorption process. This process is applicable to MOFs with no pores or even small pores, such that when introduced to certain gas molecules, the pores would expand. This expansion is directly dependent on the interaction of the MOF with the surface of the adsorbate (Figure 6). ZIF-20 has a 3D structure, which includes large cages connected through small windows. The selective adsorption of CO_2 over methane, at 760 torr and 273 K was five times greater. This was attributed to stronger connection between the pore surface and CO_2 molecules. The gas molecules were able to pass through the framework due to the process of window widening. Moreover, the flexibility of ZIF-20 framework was impacted by the exertion of quadrupole moment of CO_2 (*156*).

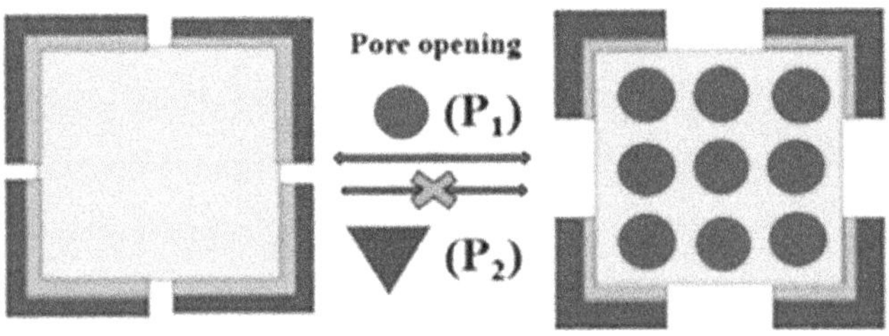

Figure 6. Schematic representation of selective gas adsorption in a flexible MOF centered on gate opening phenomena and specific pressure guests. Red circles and purple triangle represent sorbable and non-sorbable guest molecules. Reproduced with permission from reference (156). Copyright 2020 Elsevier.

Carbon monoxide is mainly produced from partial oxidation of compounds containing carbon, in a mixture constituting CO_2, N_2, hydrocarbons and H_2. The removal of carbon mononoxide (CO) is mainly carried by coordinating unsaturated metals with CO (*83*). Considering the capacity, MOF-74 is a very effective MOF, in which the adsorption strength could be attuned by changing the metals. CO-metal binding in Cu(aip)(H_2O) (aip = 5-azidoisophthalate), could lead to a deformation in the structure, which further improves the adsorption of CO. Also, a combination of spin-state transition and CO-metal binding in Fe-BTTri (H_3BTTri= 1,3,5-tris (1H-1,2,3-triazol-5-yl) benzene) have also been employed for CO removal. ZIF-8 was doped onto Au/TiO_2 for the oxidation of CO (*157*). A series of M-MOF-74 (where M= Mn^{2+}, Zn^{2+} Mg^{2+}, Fe^{2+}, Ni^{2+}, Co^{2+}), were investigated on CO adsorption. The CO-, Ni- and Fe-MOF-74 displayed an outstanding adsorption capacity at 1.2 bar and 298 K (*158*).

Table 1 shows the adsorption capacities of some MOF-based materials for CO_2 and other gases.

Table 1. Different Gas Adsorption Capacities of Selected MOFs

MOFs	*Adsorbate*	*Removal (mol/kg)*	*Operating conditions (Temperature/Pressure)*	*Reference*
ZIF-8	CO_2	9.10	303 K/4500 kPa	*(159)*
PCN-250(Fe_3)	CO_2	5.24	303 K	*(160)*
		3.02	298 K/100 kPa	*(161)*
UiO-66	CO_2	4.34	298 K/2000 kPa	*(162)*
Li-doped UiO-66	CO_2	2.80	298 K/100 kPa	*(163)*
MOF-177	CO_2	1.00	298 K/100 kPa	*(164)*
Cu-BTC	CO_2	11.70	297 K/1500 kPa	*(165)*
Ionic liquid-Functionalized MOF-177	CO_2	1.70	303 K/100 kPa	*(166)*
Tetraethylenepentamine-Funtionalized Mg-MOF-74	CO_2	6.11	298 K/100 kPa	*(167)*
Ionic liquid-Functionalized Cu-BTC	CO_2	1.70	303 K/15 kPa	*(152)*
MOF-177	SO_2	25.70	293 K/100 kPa	*(168)*
UiO-66	NO_2	1.59	298 K	*(169)*
Ba-doped Cu-BTC	SO_2	2.71	773 K	*(170)*
Cu-BTC		0.71		
Al-fumarate	DCM	3.40	298 K/44.70 kPa	*(171)*
	TCM	2.52	298 K/21.44 kPa	
UiO-66	Toluene	1.64	298 K	*(172)*
MOF-177	ethylbenzene	2.13	298 K/0.39 kPa	
	Ethenylbenzene Benzene	1.61	298 K/0.23 kPa	
	Toluene	8.82	298 K/4.88 kPa	*(173)*
	acetone	3.77	298 K/1.44 kPa	
		8.30	298 K/10.83 kPa	

Notes: DCM – dichloromethane; TCM – trichloromethane

Sulfur dioxide (SO_2) is a dominant oxide of sulfur, emitted in the environment via combustion of sulfur-containing fuels. SO_2 reacts with water to form acid rain which can have lethal effect on environmental health (*174*) HKUST-1 has been investigated for the removal of tetrahydrothiophene and thiophene. 78 wt% was achieved for sulfur removal from model oils containing thiophene. Conjugated polycarbazole porous MOFs ortho-Cz-POF, para-Cz-POF and meta-Cz-POF have been utilized in the removal of 3-methylthiophene (*175*). During adsorption process, MOFs interact with sulfur compounds available in the fuel through π-π coordination sites which have selected metal ions such as Cu^+, Ni^{2+}, Zn^{2+}, Cu^{2+} and Co^{2+}, hence, the effectiveness of the following MOFs: UMCM-152, CuCl/MIL-47(V), MIL-101(Cr), MIL-100(Fe), MOF-505, PWA/HKUST-1 and HKUST-1 (*174*). Functionalized and pristine UiO-66(Zr) have been tested for the adsorption of benzothiophene and thiophene. The modification was carried out by introducing an amino group on the linker and the introduction of carboxylic group at the defective site. HKUST-1, composited with graphene oxide (GO/HKUST-1) has also been reported as an effective adsorbent. The adsorption capacity was found to be 60.67 mg S/g, and was attributed to the improvement in the porosity with composite formation (*176*). Three different MOFs [MIL-160, MOF-177 and amine-functionalized MIL-125(Ti))] were investigated for SO_2 adsorption. MOF-177 shows a potential for industrial applications in a short-medium period with an adsorption capacity of 25.70 mol/kg (293 K). The adsorption isotherm was Type V, which indicated low affinity between the adsorbent and the adsorbate (*168*).

Ammonia (NH_3) gas can strongly bind to unsaturated metal sites and prevent recycling of materials (*177*, *178*). NH_3 removal has been investigated by utilizing robust MOFs with unsaturated metal sites such as $M_2Cl_2(BTDD)\text{-}(H_2O)_2$ (M = Mn, Co, Ni, H_2BTDD = 1*H*-1,2,3-triazolo[4,5-b],[4′,5′-i])dibenzo-[1,4]dioxin). The MOF showed high uptake (15.47 mmol/g) at 1 bar and 298 K as reported (*179*). Some of the other routes involve interaction between MOF functional groups (such as hydroxyl, ammonium and sulfonic) and NH_3. The importance of functional groups on the removal of NH_3 has been investigated by using Grand Canonical Monte Carlo simulations applied to a database of 137,953 hypothetical MOFs (*180*). Zr-based MOFs such as UiO-66(OH)$_2$, UiO-66OH, UiO-66, UiO-66-(COOH)$_2$ and UiO-66-NO_2 have also been investigated for the removal of NH_3. The presence of different functional groups such as -COOH, -NO_2, -SO_3H and NH_2 has greatly improved the adsorption of NH_3 under humid and dry conditions (*181*). Ammonia is a Bronsted base with high affinity for sulfonic and carboxylic acids. Hence, presence of such functional groups on MOFs could act as strong capture sites for ammonia. UiO-66-NH_2 was modified with anhydrous HCl. In comparison to pristine MOF, the UiO-66-NH_3Cl produced showed high uptake for ammonia in low pressure regions and at ambient temperature (*182*).

Hydrogen sulfide (H_2S) is an invisible gas that possesses an offensive odor. H_2S converts to a more stable product (sulfuric acid), in the atmosphere, thereby posing a threat to aquatic life and vegetation. The removal of H_2S was investigated using MIL-101(Cr)-SO_3Ag. The adsorption capacity was 96.75 mg/g with a very good cyclic stability. The H_2S adsorption was due to the interaction between Ag and S (*183*). The use of different functionalized UiO-66(Zr)-XN materials for H_2S removal by molecular investigation have been investigated. UiO-66-(Zr)-(COOH)$_2$ showed more activity (*184*). MIL-68(Al) was investigated using both theoretical and experimental approaches (high pressure of about 12 bar) for H_2S removal. A conclusion was made based on the obtained result, that the triangular pores of the MOF were partially filled by some solvent molecules, due to incomplete material activation (*185*). HKUST-1/GO and HKUST-1 have also been studied for H_2S removal. As a result of unsaturated copper sites in the structure of the MOF, both reactive and physical adsorption took place. Also, the stability of both materials, in the presence of water led to gradual degradation, indicating chemisorption (*186*). Pristine and functionalized UiO-66(Zr) were investigated in removing H_2S from binary gas mixtures. UiO-66-COOH and UiO-66$(COOH)_2$ showed the highest capacity, in comparison to other tested solids. This may be due to high isosteric heat of adsorption. This suggests an interaction between the polar H_2S molecules and the hydrophilic groups, which favors the adsorption process (*187, 188*). From a stability standpoint, ZIF-8 has also shown a promising activity for H_2S removal (*189, 190*).

A major class of air pollutants is VOCs with a potential for mutagenesis, carcinogenesis and tetratogenesis. These characteristics are capable of endangering human health and ecological environment (*191*). VOCs emission of anthropogenic nature arises from different industrial operations which include petroleum refining, natural gas and crude oil exploration (*192*). The potential of Cu-BTC MOF was investigated on fourteen VOCs and semi-VOCs. The adsorption capacity was higher than displayed by three non-MOF adsorbents (*193*). Al-fumarate MOF has also been tested for the adsorption of dichloromethane and trichloromethane (DCM and TCM respectively). The evaluated adsorption capacity was high for DCM and DCM/TCM selectivity. The phenomenon was attributed to differences in adsorbent molecule size. The small DCM molecules could easily have contact with the pores of the Al-fumarate and take advantage of the active site than that of DCM (*171*). The adsorption of toluene on UiO-66 was also investigated. The presence of moisture contributed negatively to the toluene adsorption capacity. This was an indication of competition in the adsorption process between water vapor and toluene. The reusability study was favorable under thermal treatment, after three cycles (*9*).

Fluorinated compounds such as chlorofluorocarbons (CFCs) belong to a class of gaseous pollutants, considered harmful to the atmosphere. In the last decades, there has been a gradual substitution with other service fluid types in place of CFCs, in refrigeration processes. This practice has helped in significant recovery of the functionality of the planetary ultraviolet shield (*191*). MOF-177 was investigated for the adsorption of sevoflurane (SF), the adsorbent showed high affinity and the adsorption efficiency was impressive. However, the presence of moisture observed in the vent line of the breathing system, challenged its implementation as an anesthetic scavenger (*194*). MIL-101 displayed a higher SF/H_2O selectivity with the packed column experiment in comparison to the conventional adsorbent, which suffers the effects of "roll-up" when saturation was attained by the test column. MIL-101 showed high stability after several desorption cycles (*195*).

4.2. MOF-Based Materials for Adsorption of Water Pollutants

MOF materials have been studied for water purification. These studies have proved the efficiency of MOFs for this intent. Their effectiveness is due of their high surface area, tunable porosity, controlled pore size, possibility of the modification of their structure, among others. Table 2 shows the adsorption capacities of some examples of MOF-based materials for the adsorption of water pollutants

For the adsorptive removal of heavy metals, a stable anionic MOF, AMOF-1 of 5,5'-(1,4-phenylenebis(methylene))bis(oxy) diisophthalic acid was developed and subsequently used to capture heavy metal ions Pb(II), Cd(II) and Hg(II) at ppm level. 98.7% removal of Hg(II) was attained after 24 hours. The adsorption capacities were 71, 41 and 78 mg/g respectively. The adsorption capacity was attributed to metal inclusion into the MOF structure by cation exchange (*196*). As a result of the affinity of sulfur atom to some metals, sulfur-containing functionalities are incorporated onto MOF structures. According to the hard/soft acid/base (HSAB) theory by Pearson (*197*), sulfur atom could interact with metal atoms and form chelates with them, thereby aiding their adsorption from aqueous solution. Example is the use of 2,2′-dithiodibenzoic acid for the synthesis of a zirconium MOF. This MOF was capable of selective removal of Au ions from water (*198*). NENU-400 (Northeast Normal University-400) contains free-standing thioether groups. It was prepared and employed for the adsorptive removal of Hg^{2+} and was found effective due to the interaction between the thioether group and the Hg^{2+} (*199*). The already modified NH_2-UiO-66 could still be further modified with other functionalities using several reagents which are linked by the exposed amino group (*200*). NH_2-UiO-66 was modified with DMTD (dimercapto-1,3,4-thiadiazole) for the introduction of hydrogen sulfide which enhanced the removal of Hg^{2+} (*201*). Other reagents like resorcyl aldehyde (RSA)and thioctic acid which were meant to increase the numbers of –OH groups, were used to modify UiO-66 for the removal of Pb(II) and Au(III) respectively (*201*, *202*). Carboxylic (-COOH), sulfonic (–SO_3H), nitro (-NO_2), etc. are other functional groups that have been reported to be incorporated into MOF structures, and gave better performance than pristine MOFs (*203*, *204*). The magnetic property of iron-based materials, useful for the separation of adsorbents from the adsorption medium after use, has motivated their use as composite materials (*205*). A magnetic MOF functionalized with CTAB and a MOF (Fe_3O_4@UiO-66@UiO-67) were used to remove Cr(VI) from water. UiO-66 served as the core and UiO-67 as the shell. Outstanding removal capacity was obtained and Fe_3O_4 was useful in separating the MOF composite from the adsorption medium after the removal process. Very high adsorption capacity of 932.1 mg/g obtained was due to tunable pores, surface active sites and electrostatic interactions. The composite showed a remarkable adsorption capacity in a wide pH range of 1.0–11.0 (*206*). Graphene oxide (GO), when incorporated into MOFs, confers enhanced adsorption capacity and water stability to the resulting MOF composite. Lu et al. reported the hydrothermal synthesis of MIL-101(Fe)/GO in which a sandwich structure was formed with the MOF distributed uniformly on the graphene oxide sheet. Its use for the adsorption of Pb^{2+} was reported (*146*). Chitosan was incorporated into UiO-66 and was found to have an adsorption capacity of 94 mg/g for Cr(VI). Its higher capacity than chitosan could be attributed to the electrostatic interaction between the amino group or the oxygen atom of the linker and the metal ions (*207*).

For the removal of pharmaceuticals and personal care products(PPCPs), adsorptive uptake of phenol and p-nitrophenol (PNP) from water on MIL-100 (Fe and Cr) and NH_2-MIL-101-Al) was studied (*208*). The latter showed a better adsorption capacity, as a result of H-bonding interaction between nitrophenol and the –NH_2 groups on the compound. A study reported the adsorption of amodiaquine, an antimalarial, from water. The zinc CPs used were [$Zn_2(fum)_2(bpy)$] (fum - fumarate; bpy - 4,4-bipyridine) and [Zn_4O $(bdc)_3$] (bdc-1,4-benzenedicarboxylate). The highest adsorption capacity was achieved within three hours with adsorption capacity of 0.478 mg/g on [$Zn_2(fum)_2(bpy)$] and 47.62 mg/g on [$Zn_4O(bdc)_3$] at pH of 4.3 (*209*). A 3D MOF {[$Cu_2(L).(H_2O)_2$](L=2,5-bis(3′,5′-dicarboxylphenyl)-benzoic acid) was used for the adsorption of chlorpromazine hydrochloride and diclofenac sodium. The adsorption of the latter was faster than the former. There was a weak interaction between the drugs and the –COO^- groups of the MOF.

The result presented an effective adsorbent for both pharmaceuticals (*204*). Several functionalities, capable of producing these interactions, have been introduced onto the MOF structure either by PSM or direct synthesis using already functionalized ligands (*210*). Hydroxyl group (-OH) introduced onto MOF structures could act as hydrogen bond donor or acceptor, and therefore bring about hydrogen-bonding interaction with the adsorbate. MIL-101(Cr)–OH and MIL-101(Cr)–$(OH)_3$ were prepared using ethanolamine or triethanolamine as reagents to supply the –OH groups, for the adsorption of some PPCPs namely, bisphenol A, triclosan, naproxen, etc. These –OH groups were noted to interact with oxygen atoms in the PPCPs, in a hydrogen donor-acceptor interaction (*211*). The adsorption efficiency was found to increase with an increase in the number of –OH groups in the MOFs. A decrease in the porosity of MIL-101 after functionalization with –OH group was observed. Despite this, there was an increased adsorptive uptake of the PPCPs adsorbed compared to pure MIL-101. This was also as a result of the hydrogen bond between the adsorbent and oxygen atoms of the adsorbate (*212*). In addition to the H-bonding interaction observed for the –OH functional group, amino (-NH_2) functional group could undergo electrostatic interaction with the target adsorbate. Lv et al. reported MIL-68(Ln)–NH_2 for p-Arsanilic acid adsorption. The high adsorptive removal was attributed to hydrogen bonding, electrostatic and π-π interaction between the functionalities of the adsorbate and the organic ligand (*213*). Although, this functionalization processes reduced the porosity of the pristine MOF, higher adsorptive removal was obtained due to the greater interaction between the adsorbate and the introduced functionalities on the MOF structure. The removal of diclofenac was achieved on CTAB-ZIF-67. Enhanced result (compared to CTAB-free ZIF-67) was obtained due to the electrostatic interaction caused by the positive surface charge of the functionalized MOF (*214*). The first ever synthesized MOF-5 which has a surface area 151 m^2/g, was carbonized and gave a porous carbon of surface area 1731 m^2/g. Higher adsorption capacities for bisphenol A, sulfamethoxazole and methyl orange dye than commercial activated carbon were reported. The excellent performance was attributed to the much higher surface area, preserved morphology, unique meso/macropore structure, cooperative interaction of a pore-filling mechanism, hydrogen bonding and electrostatic interaction (*215*). PPCPs, ibuprofen and diclofenac were adsorbed using pyrolyzed ZIF-8. The pyrolysis was done at different temperature ranging from 800-1200°C. It was noted that several functional groups like phenolic, lactonic and carboxylic groups served as hydrogen bond donors for interaction with the acceptors of the PPCPs (*216*). Another investigation on the use of MOF-derived carbon for the adsorption of PPCPs, atenolol and clofibric acid was carried out. An adsorption capacity of 552 and 540 mg/g respectively, was displayed and electrostatic interaction contributed to this. A simple solvent washing method regenerated the spent adsorbent (*217*).

For the adsorption of methylene blue and malachite green dyes, Fe_3O_4@AMCA-MIL53(Al) nanocomposite was prepared. Several chemical interactions namely hydrogen, pi-pi, electrostatic aided the adsorption process (*218*). Activated carbon is a widely used adsorbent, but its incorporation into MOFs has been explored. For instance, Hasanzadeh et al. reported its incorporation into MIL-101 (Cr) for the removal of direct red and acid blue. The composite was found to have higher surface area and pore volume than the individual materials. A large percentage of the dyes were removed within five minutes (*219*). Torad et al. pyrolyzed ZIF-67, made cobalt nanoparticle of the resulting nanoporous carbon and used same for the removal of methylene blue dye. An interesting observation was that the pristine MOF retained its original morphology even after pyrolysis. An outstanding adsorption capacity of 500 mg/g was obtained (*220*).

Table 2. Adsorption Capacities of Some Examples of MOF-Based Materials for the Adsorption of Water Pollutants

MOF	*Added functionality/ compositing material*	*Adsorbate*	*Adsorption capacity (mg/g)*	*Mode of interaction with adsorbate*	*Reference*
		Functionalized MOFs			
MIL-101	-OH	Naproxen	156	Hydrogen bonding	(*211*)
		Ketoprofen	80		
		Triclosan	112		
		bisphenol A	97		
		p-Chloro-m-xylenol	79		
MIL-68 (Ln)	$-NH_2$	p-Arsanilic acid	401	Electrostatic, hydrogen bonding, pi-pi	(*213*)
ZIF-67	CTAB	Diclofena (*211*)	61	Electrostatic interaction	(*80*)
UiO-66-NH_2	-SH	Hg^{2+}	670.5	Complexation reaction	(*19*)
		MOF composites			
MIL53(Al)	Fe_3O_4@AMCA	Methylene blue, malachite green	1.02 mmol/g 0.9	Hydrogen bonding, pi-pi and electrostatic interaction	(*218*)
CuMOF	Fe_3O_4	Malachite green	219	High surface area	(*221*)
		Pb^{2+}	114		
MIL-101 (Cr)	Activated carbon	Direct red acid blue	High surface area	(*219*)	
UiO-66	Chitosan	Cr(VI)	94	Electrostatic interaction	(*207*)
		MOF-derived Porous carbon			
Pyrolyzed ZIF-67	Co nanoparticles	Methylene blue	500	High surface area	(*220*)
MOF-5		Sulfamethoxazole, bisphenol A, methyl orange	625, 757 872	High surface area, preserved morphology	(*215*)
ZIF-8		Ibuprofen	320	Hydrogen bonding	(*216*)
		diclofenac	400		

5. Conclusions and Future Prospects

This chapter has revealed the potentiality of MOFs for environmental applications. Specifically, the usage of MOF-based materials for the detection and adsorption of gaseous and water pollutants have been discussed. Recent advances for the enhancement of the efficiency of MOFs include the incorporation of additional functionalities, development of MOF composites and pyrolysis of MOFs to obtain carbon materials. For sensing applications, conductive materials have been introduced into MOFs for better electrochemical sensing of analytes. These strategies have proven effective in achieving better results with MOFs for environmental applications.

Despite the fascinating results that have been achieved, there are still some challenges to be addressed as far as the environmental application of MOFs is concerned.

Firstly, the scalability of the materials is imperative. It is worthwhile to scale-up the synthesis of the materials for industrial use. Likewise, it is required that the scaled-up samples retain the properties possessed when prepared in laboratory scale. Secondly, studies on the application of these materials for real life scenarios are still scarce. Similarly, the performances of MOF-based adsorbents in real solutions containing mixed adsorbates, presence of salts or humic acids and in low concentration of contaminants should be investigated for commercial applications. Thirdly, the water stability of MOFs must be considered irrespective of the sensing/adsorption performance. In view of this, additional research is needed to develop MOF materials with better water stability. Fourthly, theoretical studies on process optimization and the interactions between the analytes and MOF materials should be embarked on for the design of MOF-based sensors and adsorbents. In addition, there is an urgent need to explore MOF-based adsorbents for purification applications not only for water and air, but for fuel-based contaminants. Lastly, studies focusing on the detection and removal of new contaminants are also essential to determine the potential of MOFs application in removing emerging contaminants / pollutants. Overall, this chapter has revealed that MOFs, with or without modifications, have great potentials in detecting and removing air and water pollutants.

Acknowledgments

Authors appreciate Landmark University, Nigeria, for the use of facilities to carry out this project.

References

1. Yoo, D. K.; Bhadra, B. N.; Jhung, S. H. Adsorptive Removal of Hazardous Organics from Water and Fuel with Functionalized Metal-Organic Frameworks: Contribution of Functional Groups. *J. Hazard. Mater* **2021**, *403* (August 2020), 123655. https://doi.org/10.1016/j.jhazmat.2020.123655.
2. Jang, S.; Song, S.; Lim, J. H.; Kim, H. S.; Phan, B. T.; Ha, K. T.; Park, S.; Park, K. H. Application of Various Metal-Organic Frameworks (Mofs) as Catalysts for Air and Water Pollution Environmental Remediation. *Catalysts* **2020**, *10* (2), 1–32. https://doi.org/10.3390/catal10020195.
3. Fang, X.; Zong, B.; Mao, S. Metal–Organic Framework-Based Sensors for Environmental Contaminant Sensing. *Nano-Micro Lett* **2018**, *10* (4), 1–19. https://doi.org/10.1007/s40820-018-0218-0.

4. Zhang, Y.; Yuan, S.; Feng, X.; Li, H.; Zhou, J.; Wang, B. Preparation of Nanofibrous Metal-Organic Framework Filters for Efficient Air Pollution Control. *J. Am. Chem. Soc* **2016**, *138* (18), 5785–5788. https://doi.org/10.1021/jacs.6b02553.
5. Islamoglu, T.; Chen, Z.; Wasson, M. C.; Buru, C. T.; Kirlikovali, K. O.; Afrin, U.; Mian, M. R.; Farha, O. K. Metal-Organic Frameworks against Toxic Chemicals. *Chem. Rev.* **2020**, *120* (16), 8130–8160. https://doi.org/10.1021/acs.chemrev.9b00828.
6. Li, P.; Li, J.; Feng, X.; Li, J.; Hao, Y.; Zhang, J.; Wang, H.; Yin, A.; Zhou, J.; Ma, X.; Wang, B. Metal-Organic Frameworks with Photocatalytic Bactericidal Activity for Integrated Air Cleaning. *Nat. Commun.* **2019**, *10* (1), 1–10. https://doi.org/10.1038/s41467-019-10218-9.
7. Nie, J.; Xie, H.; Zhang, M.; Liang, J.; Nie, S.; Han, W. Effective and Facile Fabrication of MOFs/Cellulose Composite Paper for Air Hazards Removal by Virtue of in Situ Synthesis of MOFs/Chitosan Hydrogel. *Carbohydr. Polym.* **2020**, *250* (February), 12125–12128. https://doi.org/10.1016/j.carbpol.2020.116955.
8. Liang, C. S.; Duan, F. K.; He, K. Bin; Ma, Y. L. Review on Recent Progress in Observations, Source Identifications and Countermeasures of PM2.5. *Environ. Int.* **2016**, *86*, 150–170. https://doi.org/10.1016/j.envint.2015.10.016.
9. Zhang, Q.; Li, Q.; Young, T. M.; Harper, D. P.; Wang, S. A Novel Method for Fabricating an Electrospun Poly(Vinyl Alcohol)/Cellulose Nanocrystals Composite Nanofibrous Filter with Low Air Resistance for High-Efficiency Filtration of Particulate Matter. *ACS Sustain. Chem. Eng.* **2019**, *7* (9), 8706–8714. https://doi.org/10.1021/acssuschemeng.9b00605.
10. Zhang, P.; Huang, Y.; Rao, Y.; Chen, M.; Li, X.; Ho, W.; Lee, S.; Cao, J. Chemical Etching Fabrication of Uniform Mesoporous Bi@Bi2O3 Nanospheres with Enhanced Visible Light-Induced Photocatalytic Oxidation Performance for NOx. *Chem. Eng. J.* **2021**, *406* (August 2020), 126910. https://doi.org/10.1016/j.cej.2020.126910.
11. Srivastava, R. K.; Jozewicz, W.; Singer, C. SO2 Scrubbing Technologies: A Review. *Environ. Prog.* **2001**, *20* (4), 219–228. https://doi.org/10.1002/ep.670200410.
12. Ryu, H. J.; Grace, J. R.; Lim, C. J. Simultaneous CO2/SO2 Capture Characteristics of Three Limestones in a Fluidized-Bed Reactor. *Energy and Fuels* **2006**, *20* (4), 1621–1628. https://doi.org/10.1021/ef050277q.
13. Li, J.; Han, X.; Zhang, X.; Sheveleva, A. M.; Cheng, Y.; Tuna, F.; McInnes, E. J. L.; McCormick McPherson, L. J.; Teat, S. J.; Daemen, L. L.; Ramirez-Cuesta, A. J.; Schröder, M.; Yang, S. Capture of Nitrogen Dioxide and Conversion to Nitric Acid in a Porous Metal–Organic Framework. *Nat. Chem.* **2019**, *11* (12), 1085–1090. https://doi.org/10.1038/s41557-019-0356-0.
14. Vikrant, K.; Tsang, D. C. W.; Raza, N.; Giri, B. S.; Kukkar, D.; Kim, K. H. Potential Utility of Metal-Organic Framework-Based Platform for Sensing Pesticides. *ACS Appl. Mater. Interfaces* **2018**, *10* (10), 8797–8817. https://doi.org/10.1021/acsami.8b00664.
15. Iranpour, R.; Stenstrom, M.; Tchobanoglous, G.; Miller, D.; Wright, J.; Vossoughi, M. Environmental Engineering: Energy Value of Replacing Waste Disposal with Resource Recovery. *Science (80-.)* **1999**, *285* (5428), 706–711. https://doi.org/10.1126/science.285.5428.706.
16. Saeed, T.; Naeem, A.; Ud Din, I.; Alotaibi, M. A.; Alharthi, A. I.; Wali Khan, I.; Huma Khan, N.; Malik, T. Structure, Nomenclature and Viable Synthesis of Micro/Nanoscale Metal

Organic Frameworks and Their Remarkable Applications in Adsorption of Organic Pollutants. *Microchem. J.* **2020**, *159*, 105579. https://doi.org/10.1016/j.microc.2020.105579.

17. Jain, S. N.; Shaikh, Z.; Mane, V. S.; Vishnoi, S.; Mawal, V. N.; Patel, O. R.; Bhandari, P. S.; Gaikwad, M. S. Nonlinear Regression Approach for Acid Dye Remediation Using Activated Adsorbent: Kinetic, Isotherm, Thermodynamic and Reusability Studies. *Microchem. J.* **2019**, *148* (May), 605–615. https://doi.org/10.1016/j.microc.2019.05.024.
18. Li, Y.; Hu, T.; Chen, R.; Xiang, R.; Wang, Q.; Zeng, Y.; He, C. Novel Thiol-Functionalized Covalent Organic Framework as Adsorbent for Simultaneous Removal of BTEX and Mercury (II) from Water. *Chem. Eng. J.* **2020**, *398* (April), 125566. https://doi.org/10.1016/j.cej.2020.125566.
19. Fu, L.; Wang, S.; Lin, G.; Zhang, L.; Liu, Q.; Fang, J.; Wei, C.; Liu, G. Post-Functionalization of UiO-66-NH2 by 2,5-Dimercapto-1,3,4-Thiadiazole for the High Efficient Removal of Hg(II) in Water. *J. Hazard. Mater.* **2019**, *368* (January), 42–51. https://doi.org/10.1016/j.jhazmat.2019.01.025.
20. Li, J.; Wang, X. X.; Zhao, G.; Chen, C.; Chai, Z.; Alsaedi, A.; Hayat, T.; Wang, X. X. Metal-Organic Framework-Based Materials: Superior Adsorbents for the Capture of Toxic and Radioactive Metal Ions. *Chem. Soc. Rev.* **2018**, *47* (7), 2322–2356. https://doi.org/10.1039/c7cs00543a.
21. Yi, X. H.; Wang, F. X.; Du, X. D.; Fu, H.; Wang, C. C. Highly Efficient Photocatalytic Cr(VI) Reduction and Organic Pollutants Degradation of Two New Bifunctional 2D Cd/Co-Based MOFs. *Polyhedron* **2018**, *152*, 216–224. https://doi.org/10.1016/j.poly.2018.06.041.
22. Abazari, R.; Reza Mahjoub, A.; Slawin, A. M. Z.; Carpenter-Warren, C. L. Morphology- and Size-Controlled Synthesis of a Metal-Organic Framework under Ultrasound Irradiation: An Efficient Carrier for PH Responsive Release of Anti-Cancer Drugs and Their Applicability for Adsorption of Amoxicillin from Aqueous Solution. *Ultrason. Sonochem.* **2018**, *42*, 594–608. https://doi.org/10.1016/j.ultsonch.2017.12.032.
23. Chi, H.; Wan, J.; Ma, Y.; Wang, Y.; Ding, S.; Li, X. Ferrous Metal-Organic Frameworks with Stronger Coordinatively Unsaturated Metal Sites for Persulfate Activation to Effectively Degrade Dibutyl Phthalate in Wastewater. *J. Hazard. Mater.* **2019**, *377* (May), 163–171. https://doi.org/10.1016/j.jhazmat.2019.05.081.
24. Jiang, J.; Yaghi, O. M. Brønsted Acidity in Metal – Organic Frameworks. *Chem. Rev.* **2015**, *115*, 6966–6997. https://doi.org/10.1021/acs.chemrev.5b00221.
25. Tranchemontagne, D. J.; Mendoza-Cortes, J. L.; O'Keeffe, M.; Yaghi, O. M. Secondary Building Units, Nets and Bonding in the Chemistry of Metal-Organic Frameworks. *Chem Soc Rev* **2009**, *38*, 1257.
26. Yaghi, O. M.; Li, G.; Li, H. Selective Binding and Removal of Guests in a Microporous Metal–Organic Framework. *Nature* **1995**, *378* (6558), 703–706. https://doi.org/10.1038/378703a0.
27. Lu, W.; Wei, Z.; Gu, Z. Y.; Liu, T.-F.; Park, J.; Park, J.; Tian, J. M.; Zhang, Q. Z.; Thomas, G.; Mathieu, B.; Hong-Cai, Z. Tuning the Structure and Function of Metal–Organic Frameworks via Linker Design. *Chem. Soc. Rev.* **2014**, *43*, 5561–5593.
28. Nagarkar, S. S.; Joarder, B.; Chaudhari, A. K.; Mukherjee, S.; Ghosh, S. K. Highly Selective Detection of Nitro Explosives by a Luminescent Metal-Organic Framework. *Angew. Chemie - Int. Ed.* **2013**, *52* (10), 2881–2885. https://doi.org/10.1002/anie.201208885.

29. Tran, T. Van; Cao, V. D.; Nguyen, V. H.; Hoang, B. N.; Vo, D. V. N.; Nguyen, T. D.; Bach, L. G. MIL-53 (Fe) Derived Magnetic Porous Carbon as a Robust Adsorbent for the Removal of Phenolic Compounds under the Optimized Conditions. *J. Environ. Chem. Eng.* **2020**, *8* (1), 102902. https://doi.org/10.1016/j.jece.2019.102902.
30. Yap, M. H.; Fow, K. L.; Chen, G. Z. Synthesis and Applications of MOF-Derived Porous Nanostructures. *Green Energy Environ.* **2017**, *2* (3), 218–245. https://doi.org/10.1016/j.gee.2017.05.003.
31. Yaghi, O. M.; O'Keeffe, M.; Ockwig, N. W.; Chae, H. K.; Eddaoudi, M.; Kim, J. Reticular Synthesis and the Design of New Materials. *Nature* **2003**, *423* (6941), 705–714. https://doi.org/10.1038/nature01650.
32. Evans, J. D.; Sumby, C. J.; Doonan, C. J. Post-Synthetic Metalation of Metal-Organic Frameworks. *Chem. Soc. Rev.* **2014**, *43*, 5933–5951.
33. Juan-Alcañiz, J.; Gascon, J.; Kapteijn, F. Metal-Organic Frameworks as Scaffolds for the Encapsulation of Active Species: State of the Art and Future Perspectives. *J. Mater. Chem.* **2012**, *22* (20), 10102–10119. https://doi.org/10.1039/c2jm15563j.
34. Kitagawa, S.; Kitaura, R.; Noro, S. I. Functional Porous Coordination Polymers. *Angew. Chemie - Int. Ed.* **2004**, *43* (18), 2334–2375. https://doi.org/10.1002/anie.200300610.
35. Murray, L. J.; Dincˇa, M.; Long, J. R. Hydrogen Storage in Metal-Organic Frameworks. *Chem. Soc. Rev.* **2009**, *38*, 1294–1314.
36. Van De Voorde, B.; Bueken, B.; Denayer, J.; De Vos, D. Adsorptive Separation on Metal-Organic Frameworks in the Liquid Phase. *Chem. Soc. Rev.* **2014**, *43* (16), 5766–5788. https://doi.org/10.1039/c4cs00006d.
37. Tan, H.; Ma, C.; Gao, L.; Li, Q.; Song, Y.; Xu, F.; Wang, T.; Wang, L. Metal-Organic Framework-Derived Copper Nanoparticle@carbon Nanocomposites as Peroxidase Mimics for Colorimetric Sensing of Ascorbic Acid. *Chem. - A Eur. J.* **2014**, *20* (49), 16377–16383. https://doi.org/10.1002/chem.201404960.
38. Wang, H.; Zhu, Q. L.; Zou, R.; Xu, Q. Metal-Organic Frameworks for Energy Applications. *Chem* **2017**, *2* (1), 52–80. https://doi.org/10.1016/j.chempr.2016.12.002.
39. Zhang, X.; Li, G.; Wu, D.; Li, X.; Hu, N.; Chen, J.; Chen, G.; Wu, Y. Recent Progress in the Design Fabrication of Metal-Organic Frameworks-Based Nanozymes and Their Applications to Sensing and Cancer Therapy. *Biosens. Bioelectron.* **2019**, *137* (May), 178–198. https://doi.org/10.1016/j.bios.2019.04.061.
40. Jiao, L.; Seow, J. Y. R.; Skinner, W. S.; Wang, Z. U.; Jiang, H. L. Metal–Organic Frameworks: Structures and Functional Applications. *Mater. Today* **2019**, *27* (August), 43–68. https://doi.org/10.1016/j.mattod.2018.10.038.
41. Hu, Y. H.; Zhang, L. Hydrogen Storage in Metal-Organic Frameworks. *Adv. Mater.* **2010**, *22* (20), 117–130. https://doi.org/10.1002/adma.200902096.
42. Li, Y.; Gai, T.; Lin, Y.; Zhang, W.; Li, K.; Liu, Y.; Duan, Y.; Li, B.; Ding, J.; Li, J. Eight Cd(II) Coordination Polymers with Persistent Room-Temperature Phosphorescence: Intriguing Dual Emission and Time-Resolved Afterglow Modulation. *Inorg. Chem. Front.* **2020**, *7* (3), 777–785. https://doi.org/10.1039/c9qi01273g.
43. Chen, Y. Z.; Zhang, R.; Jiao, L.; Jiang, H. L. Metal–Organic Framework-Derived Porous Materials for Catalysis. *Coord. Chem. Rev.* **2018**, *362*, 1–23. https://doi.org/10.1016/j.ccr.2018.02.008.

44. Cui, Y.; Li, B.; He, H.; Zhou, W.; Chen, B.; Qian, G. Metal-Organic Frameworks as Platforms for Functional Materials. *Acc. Chem. Res.* **2016**, *49* (3), 483–493. https://doi.org/10.1021/acs.accounts.5b00530.
45. Ahmed, I.; Jhung, S. H. Composites of Metal-Organic Frameworks: Preparation and Application in Adsorption. *Mater. Today* **2014**, *17* (3), 136–146. https://doi.org/10.1016/j.mattod.2014.03.002.
46. Liu, L.; Tai, X.; Zhou, X. Au3+/Au0 Supported on Chromium(III) Terephthalate Metal Organic Framework (MIL-101) as an Efficient Heterogeneous Catalystfor Three-Component Coupling Synthesis of Propargylamines. *Materials (Basel)* **2017**, *10* (2) https://doi.org/10.3390/ma10020099.
47. Qiu, Q.; Chen, H.; Wang, Y.; Ying, Y. Recent Advances in the Rational Synthesis and Sensing Applications of Metal-Organic Framework Biocomposites. *Coord. Chem. Rev.* **2019**, *387*, 60–78. https://doi.org/10.1016/j.ccr.2019.02.009.
48. Carne, A.; Carbonell, C.; Imaz, I.; Maspoch, D. Nanoscale Metal–Organic Materials. *Chem Soc Rev* **2011**, *40*, 291–305.
49. Jiang, D.; Chen, M.; Wang, H.; Zeng, G.; Huang, D.; Cheng, M.; Liu, Y.; Xue, W.; Wang, Z. W. The Application of Different Typological and Structural MOFs-Based Materials for the Dyes Adsorption. *Coord. Chem. Rev.* **2019**, *380*, 471–483. https://doi.org/10.1016/j.ccr.2018.11.002.
50. Rehman, A.; Farrukh, S.; Hussain, A.; Pervaiz, E. Synthesis and Effect of Metal–Organic Frame Works on CO2 Adsorption Capacity at Various Pressures: A Contemplating Review. *Energy Environ.* **2020**, *31* (3), 367–388. https://doi.org/10.1177/0958305X19865352.
51. Li, B.; Chrzanowski, M.; Zhang, Y.; Ma, S. Applications of Metal-Organic Frameworks Featuring Multi-Functional Sites. *Coord. Chem. Rev.* **2016**, *307*, 106–129. https://doi.org/10.1016/j.ccr.2015.05.005.
52. Mu, F.; Dai, B.; Zhao, W.; Zhang, L.; Xu, J.; Guo, X. A Review on Metal-Organic Frameworks for Photoelectrocatalytic Applications. *Chinese Chem. Lett.* **2020**, *31* (7), 1773–1781. https://doi.org/10.1016/j.cclet.2019.12.015.
53. Zheng, X.; Wang, L.; Liu, M.; Lei, P.; Liu, F.; Xie, Z. Nanoscale Mixed-Component Metal-Organic Frameworks with Photosensitizer Spatial-Arrangement-Dependent Photochemistry for Multimodal-Imaging-Guided Photothermal Therapy. *Chem. Mater.* **2018**, *30* (19), 6867–6876. https://doi.org/10.1021/acs.chemmater.8b03043.
54. Falcaro, P.; Hill, A. J.; Nairn, K. M.; Jasieniak, J.; Mardel, J. I.; Bastow, T. J.; Mayo, S. C.; Gimona, M.; Gomez, D.; Whitfield, H. J.; Riccò, R.; Patelli, A.; Marmiroli, B.; Amenitsch, H.; Colson, T.; Villanova, L.; Buso, D. A New Method to Position and Functionalize Metal-Organic Framework Crystals. *Nat. Commun.* **2011**, *2* (1), 237. https://doi.org/10.1038/ncomms1234.
55. Ricco, R.; Malfatti, L.; Takahashi, M.; Hill, A. J.; Falcaro, P. Applications of Magnetic Metal-Organic Framework Composites. *J. Mater. Chem. A* **2013**, *1* (42), 13033–13045. https://doi.org/10.1039/c3ta13140h.
56. Huang, D.; Wang, G.; Cheng, M.; Zhang, G.; Chen, S.; Liu, Y.; Li, Z.; Xue, W.; Lei, L.; Xiao, R. Optimal Preparation of Catalytic Metal-Organic Framework Derivatives and Their Efficient Application in Advanced Oxidation Processes. *Chem. Eng. J.* **2021**, *421* (November), 127817. https://doi.org/10.1016/j.cej.2020.127817.

57. Chen, X.; Peng, X.; Jiang, L.; Yuan, X.; Yu, H.; Wang, H.; Zhang, J.; Xia, Q. Recent Advances in Titanium Metal–Organic Frameworks and Their Derived Materials: Features, Fabrication, and Photocatalytic Applications. *Chem. Eng. J.* **2020**, *395* (April), 125080. https://doi.org/10.1016/j.cej.2020.125080.
58. Lykourinou, V.; Chen, Y.; Wang, X. Sen; Meng, L.; Hoang, T.; Ming, L. J.; Musselman, R. L.; Ma, S. Immobilization of MP-11 into a Mesoporous Metal-Organic Framework, MP-11@mesoMOF: A New Platform for Enzymatic Catalysis. *J. Am. Chem. Soc.* **2011**, *133* (27), 10382–10385. https://doi.org/10.1021/ja2038003.
59. Li, Z.; Zeng, H. C. Surface and Bulk Integrations of Single-Layered Au or Ag Nanoparticles onto Designated Crystal Planes {110} or {100} of ZIF-8. *Chem. Mater.* **2013**, *25* (9), 1761–1768. https://doi.org/10.1021/cm400260g.
60. Bradshaw, D.; Garai, A.; Huo, J. Metal–Organic Framework Growth at Functional Interfaces: Thin Films and Composites for Diverse Applications. *Chem. Soc. Rev.* **2012**, *41* (6), 2344–2381. https://doi.org/10.1039/c1cs15276a.
61. Uemura, T.; Yanai, N.; Kitagawa, S. Polymerization Reactions in Porous Coordination Polymers. *Chem. Soc. Rev.* **2009**, *38* (5), 1228–1236. https://doi.org/10.1039/b802583p.
62. Li, X.; Zhu, Q.-L. L. MOF-Based Materials for Photo- and Electrocatalytic CO2 Reduction. *EnergyChem* **2020**, *2* (3), 100033. https://doi.org/10.1016/j.enchem.2020.100033.
63. Zhang, W.; Hu, Y.; Ge, J.; Jiang, H. L.; Yu, S. H. A Facile and General Coating Approach to Moisture/Water-Resistant Metal-Organic Frameworks with Intact Porosity. *J. Am. Chem. Soc.* **2014**, *136* (49), 16978–16981. https://doi.org/10.1021/ja509960n.
64. Lee, K. J.; Lee, J. H.; Jeoung, S.; Moon, H. R. Transformation of Metal-Organic Frameworks/Coordination Polymers into Functional Nanostructured Materials: Experimental Approaches Based on Mechanistic Insights. *Acc. Chem. Res.* **2017**, *50* (11), 2684–2692. https://doi.org/10.1021/acs.accounts.7b00259.
65. Cao, X.; Tan, C.; Sindoro, M.; Zhang, H. Hybrid Micro-/Nano-Structures Derived from Metal-Organic Frameworks: Preparation and Applications in Energy Storage and Conversion. *Chem. Soc. Rev.* **2017**, *46* (10), 2660–2677. https://doi.org/10.1039/c6cs00426a.
66. Zhai, M.; Wang, F.; Du, H. Transition-Metal Phosphide-Carbon Nanosheet Composites Derived from Two-Dimensional Metal-Organic Frameworks for Highly Efficient Electrocatalytic Water-Splitting. *ACS Appl. Mater. Interfaces* **2017**, *9* (46), 40171–40179. https://doi.org/10.1021/acsami.7b10680.
67. Liu, H.; Ma, Y.; Chen, J.; Wen, M.; Li, G.; An, T. Highly Efficient Visible-Light-Driven Photocatalytic Degradation of VOCs by CO2-Assisted Synthesized Mesoporous Carbon Confined Mixed-Phase TiO2 Nanocomposites Derived from MOFs. *Appl. Catal. B Environ.* **2019**, *250* (February), 337–346. https://doi.org/10.1016/j.apcatb.2019.03.054.
68. Song, Y.; Cho, D.; Venkateswarlu, S.; Yoon, M. Systematic Study on Preparation of Copper Nanoparticle Embedded Porous Carbon by Carbonization of Metal-Organic Framework for Enzymatic Glucose Sensor. *RSC Adv.* **2017**, *7* (17), 10592–10600. https://doi.org/10.1039/c7ra00115k.
69. Dong, W.; Zhuang, Y.; Li, S.; Zhang, X.; Chai, H.; Huang, Y. High Peroxidase-like Activity of Metallic Cobalt Nanoparticles Encapsulated in Metal–Organic Frameworks Derived Carbon for Biosensing. *Sensors Actuators, B Chem.* **2018**, *255*, 2050–2057. https://doi.org/10.1016/j.snb.2017.09.013.

70. Chen, Y.; Zhang, R.; Jiao, L.; Jiang, H. Metal – Organic Framework-Derived Porous Materials for Catalysis. *Coord. Chem. Rev.* **2018**, *362*, 1–23. https://doi.org/10.1016/j.ccr.2018.02.008.
71. Ge, K. M.; Wang, D.; Xu, Z. J.; Chu, R. Q. A Luminescent Eu(III)-MOF for Selective Sensing of Ag+ in Aqueous Solution. *J. Mol. Struct.* **2020**, *1208*, 127862. https://doi.org/10.1016/j.molstruc.2020.127862.
72. Wang, X. F.; Song, X. Z.; Sun, K. M.; Cheng, L.; Ma, W. MOFs-Derived Porous Nanomaterials for Gas Sensing. *Polyhedron* **2018**, *152*, 155–163. https://doi.org/10.1016/j.poly.2018.06.037.
73. Lin, L. C.; Kim, J.; Kong, X.; Scott, E.; McDonald, T. M.; Long, J. R.; Al., E. Understanding CO2 Dynamics in Metal– Organic Frameworks with Open Metal Sites. *Angew Chem Int Ed* **2013**, *125* (16), 4506–4509.
74. Zhang, J. P.; Zhang, Y. B.; Lin, J. Bin; Chen, X. M. Metal Azolate Frameworks: From Crystal Engineering to Functional Materials. *Chem. Rev.* **2012,** *112* (2), 1001–1033. https://doi.org/10.1021/cr200139g.
75. Lei, J.; Qian, R.; Ling, P.; Cui, L.; Ju, H. Design and Sensing Applications of Metal – Organic Framework Composites. *Trends Anal. Chem.* **2014**, *58*, 71–73. https://doi.org/10.1016/j.trac.2014.02.012.
76. Kreno, L. E.; Leong, K.; Farha, O. K.; Allendorf, M.; Van Duyne, R. P.; Hupp, J. T. Metal–Organic Framework Materials as Chemical Sensors. *Chem. Rev.* **2012,** *112* (2), 1105–1125. https://doi.org/10.1021/cr200324t.
77. Li, W.; Wu, X.; Han, N.; Chen, J.; Tang, W.; Chen, Y. Core-Shell Au@ZnO Nanoparticles Derived from Au@MOF and Their Sub-Ppm Level Acetone Gas-Sensing Performance. *Powder Technol.* **2016**, *304*, 241–247.
78. Cui, Y.; Zhang, J.; He, H.; Qian, G. Photonic Functional Metal-Organic Frameworks. *Chem. Soc. Rev.* **2018**, *47* (15), 5740–5785. https://doi.org/10.1039/c7cs00879a.
79. Liu, F.; Gao, W.; Li, P.; Zhang, X. M.; Liu, J. P. Lanthanide Metal-Organic Frameworks as Multifunctional Luminescent Sensor for Detecting Cations, Anions and Organic Solvent Molecules in Aqueous Solution. *J. Solid State Chem.* **2017**, *253*, 202–210. https://doi.org/10.1016/j.jssc.2017.05.040.
80. Lin, X. M.; Niu, J. L.; Lin, J.; Hu, L.; Zhang, G.; Cai, Y. P. A Luminescent Tb(III)-MOF Based on Pyridine-3, 5-Dicarboxylic Acid for Detection of Nitroaromatic Explosives. *Inorg. Chem. Commun.* **2016**, 72, 69–72. https://doi.org/10.1016/j.inoche.2016.08.009.
81. Gładysiak, A.; Nguyen, T. N.; Navarro, J. A. R.; Rosseinsky, M. J.; Stylianou, K. C. A Recyclable Metal–Organic Framework as a Dual Detector and Adsorbent for Ammonia. *Chem. - A Eur. J.* **2017**, *23* (55), 13602–13606. https://doi.org/10.1002/chem.201703510.
82. Li, H. Y.; Zhao, S. N.; Zang, S. Q.; Li, J. Functional Metal-Organic Frameworks as Effective Sensors of Gases and Volatile Compounds. *Chem. Soc. Rev.* **2020**, *49* (17), 6364–6401. https://doi.org/10.1039/c9cs00778d.
83. Li, H.; Wang, K.; Sun, Y.; Lollar, C. T.; Li, J.; Zhou, H. C. Recent Advances in Gas Storage and Separation Using Metal–Organic Frameworks. *Mater. Today* **2018**, *21* (2), 108–121. https://doi.org/10.1016/j.mattod.2017.07.006.

84. Li, H.; Li, D.; Qin, B.; Li, W.; Zheng, H.; Zhang, X. Turn-on Fluorescence in a Stable Cd (II) Metal-Organic Framework for Highly Sensitive Detection of Cr3+ in Water. *Dye. Pigment.* **2020**, *178*, 108359. https://doi.org/10.1016/j.dyepig.2020.108359.
85. Homayoonnia, S.; Zeinali, S. Design and Fabrication of Capacitive Nanosensor Based on MOF Nanoparticles as Sensing Layer for VOCs Detection. *Sensors Actuators, B Chem.* **2016**, *237*, 776–786. https://doi.org/10.1016/j.snb.2016.06.152.
86. Hosseini, M. S.; Zeinali, S.; Sheikhi, M. H. Fabrication of Capacitive Sensor Based on Cu-BTC (MOF-199) Nanoporous Film for Detection of Ethanol and Methanol Vapors. *Sensors Actuators, B Chem.* **2016**, *230*, 9–16. https://doi.org/10.1016/j.snb.2016.02.008.
87. Li, W.; Wu, X.; Han, N.; Chen, J.; Qian, X.; Deng, Y.; Tang, W.; Chen, Y. MOF-Derived Hierarchical Hollow ZnO Nanocages with Enhanced Low-Concentration VOCs Gas-Sensing Performance. *Sensors Actuators, B Chem.* **2016**, *225*, 158–166. https://doi.org/10.1016/j.snb.2015.11.034.
88. Sachdeva, S.; Koper, S. J. H.; Sabetghadam, A.; Soccol, D.; Gravesteijn, D. J.; Kapteijn, F.; Sudhölter, E. J. R.; Gascon, J.; De Smet, L. C. P. M. Gas Phase Sensing of Alcohols by Metal Organic Framework-Polymer Composite Materials. *ACS Appl. Mater. Interfaces* **2017**, *9* (29), 24926–24935. https://doi.org/10.1021/acsami.7b02630.
89. Ghanbarian, M.; Zeinali, S.; Mostafavi, A. A Novel MIL-53(Cr-Fe)/Ag/CNT Nanocomposite Based Resistive Sensor for Sensing of Volatile Organic Compounds. *Sensors Actuators, B Chem.* **2018**, *267*, 381–391. https://doi.org/10.1016/j.snb.2018.02.138.
90. Patah, A.; Kasim, W.; Yuliarto, B. Potential Application Zn-MOF / MnO2 Composite as Methanol Gas Sensor. *Key Eng. Mater.* **2019**, *811*, 113–119. https://doi.org/10.4028/www.scientific.net/KEM.811.113.
91. Lee, J.-H.; Nguyen, T. T. T.; Nguyen, L. H. T.; Phan, T. B.; Kim, S. S.; Doan, T. L. H. Functionalization of Zirconium-Based Metal–Organic Frameworks for Gas Sensing Applications. *J. Hazard. Mater* **2020**, *403*, 124104.
92. Cao, E.; Guo, Z.; Song, G.; Zhang, Y.; Hao, W.; Sun, L.; Nie, Z. MOF-Derived ZnFe2O4 /(Fe-ZnO) Nanocomposites with Enhanced Acetone Sensing Performance. *Sensors Actuators B . Chem.* **2020**, *325*, 128783.
93. Torad, N. L.; Kim, J.; Kim, M.; Lim, H.; Na, J.; Alshehri, S. M.; Ahamad, T.; Yamauchi, Y.; Eguchi, M.; Ding, B.; Zhang, X. Nanoarchitectured Porous Carbons Derived from ZIFs toward Highly Sensitive and Selective QCM Sensor for Hazardous Aromatic Vapors. *J. Hazard. Mater.* **2021**, *405* (October 2020), 124248. https://doi.org/10.1016/j.jhazmat.2020.124248.
94. Gassensmith, J. J.; Kim, J. Y.; Holcroft, J. M.; Farha, O. K.; Stoddart, J. F.; Hupp, J. T.; Jeong, N. C. A Metal–Organic Framework-Based Material for Electrochemical Sensing of Carbon Dioxide. *J Am Chem Soc* **2014**, *136* (23), 8277–8282.
95. Chong, X.; Kim, K.; Ohodnicki, P. R.; Li, E.; Chang, C.; Wang, A. X. Ultrashort Near-Infrared Fiber-Optic Sensors for Carbon Dioxide Detection. *IEEE Sens. J.* **2015**, *15*, 5327–5332.
96. Pentyala, V.; Davydovskaya, P.; Ade, M.; Pohle, R.; Urban, G. Carbon Dioxide Gas Detection by Open Metal Site Metal Organic Frameworks and Surface Functionalized Metal Organic Frameworks. *Sensors Actuators B Chem* **2015**, *225*, 363–368.
97. Chocarro-Ruiz, B.; P´erez-Carvajal, J.; Avci, C.; Calvo-lozano, O.; Alonso, M. I.; Maspoch, D.; Lechuga, L. M. ACO2 Optical Sensor Based on Self-Assembled Metal – Organic

Framework Nanoparticles. *J. Mater. Chem. A* **2018**, *6*, 13171–13177. https://doi.org/10.1039/C8TA02767F.

98. Ye, B.; Gheorghe, A.; Hal, R. Van; Zevenbergen, M.; Tanase, S. CO2 Sensing under Ambient Conditions Using Metal–Organic Frameworks. **2020**, 1071–1076. https://doi.org/10.1039/d0me00004c.
99. Yang, Z.; Zhang, D.; Wang, D. Carbon Monoxide Gas Sensing Properties of Metal-Organic Frameworks-Derived Tin Dioxide Nanoparticles/Molybdenum Diselenide Nanoflowers. *Sensors Actuators, B Chem.* **2020**, *304* (August 2019), 127369. https://doi.org/10.1016/j.snb.2019.127369.
100. Li, H.; Zhu, H.; Sun, M.; Yan, Y.; Zhang, K.; Huang, D.; Wang, S. Manipulating the Surface Chemistry of Quantum Dots for Sensitive Ratiometric Fluorescence Detection of Sulfur Dioxide. *Langmuir* **2015**, *31* (31), 8667–8671. https://doi.org/10.1021/acs.langmuir.5b02340.
101. Che, S.; Dao, R.; Zhang, W.; Lv, X.; Li, H.; Wang, C. Designing an Anion-Functionalized Fluorescent Ionic Liquid as an Efficient and Reversible Turn-off Sensor for Detecting SO2. *Chem. Commun.* **2017**, *53* (27), 3862–3865. https://doi.org/10.1039/c7cc00676d.
102. Perea-Cachero, A.; Calvo, P.; Romero, E.; Téllez, C.; Coronas, J. Enhancement of Growth of MOF MIL-68(Al) Thin Films on Porous Alumina Tubes Using Different Linking Agents. *Eur. J. Inorg. Chem.* **2017**, *2017* (19), 2532–2540. https://doi.org/10.1002/ejic.201700302.
103. Chernikova, V.; Yassine, O.; Shekhah, O.; Salama, K. N. Highly Sensitive and Selective SO2 MOF Sensor: The Integration of MFM-300 MOF as a Sensitive Layer on a Capacitive Interdigitated Electrode. *J. Mater. Chem. A Mater. energy Sustain* **2018**, *6*, 5550–5554. https://doi.org/10.1039/C7TA10538J.
104. Zhang, J.; Xia, T.; Zhao, D.; Cui, Y.; Yang, Y.; Qian, G. In Situ Secondary Growth of Eu(III)-Organic Framework Film for Fluorescence Sensing of Sulfur Dioxide. *Sensors Actuators, B Chem.* **2018**, *260*, 63–69. https://doi.org/10.1016/j.snb.2017.12.187.
105. Ingle, N.; Mane, S.; Sayyad, P.; Bodkhe, G.; AL-Gahouari, T.; Mahadik, M.; Shirsat, S.; Shirsat, M. D. Sulfur Dioxide (SO2) Detection Using Composite of Nickel Benzene Carboxylic (Ni3BTC2) and OH-Functionalized Single Walled Carbon Nanotubes (OH-SWNTs). *Front. Mater.* **2020**, 7 (May), 1–7. https://doi.org/10.3389/fmats.2020.00093.
106. Zhou, L.-J.; Zhang, X.-X.; Zhang, W.-Y. Sulfur Dioxide Sensing Properties of MOF-Derived ZnFe2O4 Functionalized with Reduced Graphene Oxide at Room Temperature. *Rare Met.* **2020**, *40*, 1604–1613.
107. Li, H.; Feng, X.; Guo, Y.; Chen, D.; Li, R.; Ren, X.; Jiang, X.; Dong, Y.; Wang, B. A Malonitrile-Functionalized Metal-Organic Framework for Hydrogen Sulfide Detection and Selective Amino Acid Molecular Recognition. *Sci. Rep.* **2014**, *4* (4366), 1–5. https://doi.org/10.1038/srep04366.
108. Wan, X.; Wu, L.; Zhang, L.; Song, H.; Lv, Y. Novel Metal-Organic Frameworks-Based Hydrogen Sulfide Cataluminescence Sensors. *Sensors Actuators, B Chem.* **2015**, *220*, 614–621. https://doi.org/10.1016/j.snb.2015.05.125.
109. Yassine, O.; Shekhah, O.; Assen, A. H.; Belmabkhout, Y.; Salama, K. N.; Eddaoudi, M. H2S Sensors: Fumarate-Based Fcu-MOF Thin Film Grown on a Capacitive Interdigitated Electrode. *Angew. Chemie - Int. Ed.* **2016**, *55* (51), 15879–15883. https://doi.org/10.1002/anie.201608780.

110. Zhang, C.; Zhang, S.; Yang, Y.; Yu, H.; Dong, X. Highly Sensitive H2S Sensors Based on Metal-Organic Framework Driven γ-Fe2O3 on Reduced Graphene Oxide Composites at Room Temperature. *Sensors Actuators, B Chem.* **2020**, *325*, 128804. https://doi.org/10.1016/j.snb.2020.128804.
111. Ali, A.; Alzamly, A.; Greish, Y. E.; Bakiro, M.; Nguyen, H. L.; Mahmoud, S. T. A Highly Sensitive and Flexible Metal – Organic Framework Polymer- Based H 2 S Gas Sensor. **2021**. https://doi.org/10.1021/acsomega.1c02295.
112. Travlou, N. A.; Singh, K.; Rodríguez-Castellón, E.; Bandosz, T. J. Cu-BTC MOF-Graphene-Based Hybrid Materials as Low Concentration Ammonia Sensors. *J. Mater. Chem. A* **2015**, *3* (21), 11417–11429. https://doi.org/10.1039/c5ta01738f.
113. Sel, K.; Demirci, S.; Ozturk, O. F.; Aktas, N.; Sahiner, N. NH3 Gas Sensing Applications of Metal Organic Frameworks. *Microelectron. Eng.* **2015**, *136*, 71–76. https://doi.org/10.1016/j.mee.2015.04.035.
114. Zhang, J.; Yue, D.; Xia, T.; Cui, Y.; Yang, Y.; Qian, G. A Luminescent Metal-Organic Framework Fi Lm Fabricated on Porous Al2O3 Substrate for Sensitive Detecting Ammonia. *Microporous Mesoporous Mater* **2017**, *253*, 146–150. https://doi.org/10.1016/j.micromeso.2017.06.053.
115. Wang, D.; Liu, J.; Liu, Z. A Chemically Stable Europium Metal-Organic Framework for Bifunctional Chemical Sensor and Recyclable on – o Ff – on Vapor Response. *J. soli* **2017**, *251*, 243–247. https://doi.org/10.1016/j.jssc.2017.04.032.
116. Lee, J.; Nguyen, T.; Nguyen, D.; Kim, J. Gas Sensing Properties of Mg-Incorporated Metal – Organic Frameworks. **2019**, *74*, 1–11.
117. Gamonal, A.; Sun, C.; Mariano, A. L.; Fernandez-Bartolome, E.; Guerrero-SanVicente, E.; Vlaisavljevich, B.; Castells-Gil, J.; Marti-Gastaldo, C.; Poloni, R.; Wannemacher, R.; Cabanillas-Gonzalez, J.; Costa, J. S. Divergent Adsorption-Dependent Luminescence of Amino-Functionalized Lanthanide Metal–Organic Frameworks for Highly Sensitive NO2 Sensors. *J. Phys. Chem. Lett.* **2020**, *11* (9), 3362–3368.
118. DMello, M. E.; Sundaram, N. G.; Singh, A.; Singh, A. K.; Kalidindi, S. B. An Amine Functionalized Zirconium Metal-Organic Framework as an Effective Chemiresistive Sensor for Acidic Gases. *Chem. Commun.* **2019**, *55* (3), 349–352. https://doi.org/10.1039/C8CC06875E.
119. Bhattacharyya, S.; Chakraborty, A.; Jayaramulu, K.; Hazra, A.; Maji, T. K. A Bimodal Anionic MOF: Turn-off Sensing for CuII and Specific Sensitization of EuIII. *Chem Commun* **2014**, *50*, 13567.
120. Jiang, J.; Lu, Y.; Liu, J.; Zhou, Y.; Zhao, D.; Li, C. An Acid-Base Resistant Zn-Based Metal-Organic Framework as a Luminescent Sensor for Mercury (II). *J. Solid State Chem.* **2020**, *283*, 121153. https://doi.org/10.1016/j.jssc.2019.121153.
121. Yang, H.; Zhou, C.; Yang, Y.; Chu, Z.; Yan, W.; Nie, S.; Luo, J.; Lin, S.; Wang, Y. A New Three Sensing Channels Platform of Eu@Zn-MOF for Quantitative Detection of Cr(III). *Inorg. Chem. Commun.* **2020**, *116*, 107898. https://doi.org/10.1016/j.inoche.2020.107898.
122. Guo, H. X.; Wang, D. F.; Chen, J. H.; Weng, W.; Huang, M. Q.; Zheng, Z. S. Simple Fabrication of Flake-like NH2-MIL-53(Cr) and Its Application as an Electrochemical Sensor for the Detection of Pb2+. *Chem. Eng. J.* **2016**, *289*, 479–485.

123. Tran, H. V.; Dang, H. T. M.; Tran, L. T.; Van Tran, C.; Huynh, C. D. Metal-Organic Framework MIL-53(Fe): Synthesis, Electrochemical Characterization, and Application in Development of a Novel and Sensitive Electrochemical Sensor for Detection of Cadmium Ions in Aqueous Solutions. *Adv. Polym. Technol.* **2020**, *2020*, 1–10. https://doi.org/10.1155/2020/6279278.

124. Xu, H.; Cao, C. S.; Zhao, B. A Water-Stable Lanthanide-Organic Framework as a Recyclable Luminescent Probe for Detecting Pollutant Phosphorus Anions. *Chem Commun* **2015**, *51*, 10280–10283.

125. Lu, T.; Zhang, L. C.; Sun, M. X.; Deng, D. Y.; Su, Y. Y.; Lv, Y. Amino-Functionalized Metal–Organic Frameworks Nanoplatesbased Energy Transfer Probe for Highly Selective Fluorescence Detection of Free Chlorine. *Anal. Chem* **2016**, *88* (6), 3413–3420.

126. Karmakar, A.; Kumar, N.; Samanta, P.; Desai, A. V; Ghosh, S. K. A Post-Synthetically Modified MOF for Selective and Sensitive Aqueous-Phase Detection of Highly Toxic Cyanide Ions. **2016**, 864–868. https://doi.org/10.1002/chem.201503323.

127. Saraf, M.; Rajak, R.; Mobin, S. M. A Fascinating Multitasking Cu- MOF/RGO Hybrid for High Performance Supercapacitors and Highly Sensitive and Selective Electrochemical Nitrite Sensors. *J. Mater. Chem. A* **2016**, *4* (42), 16432–16445.

128. Xia, Y.; Wang, C.; Feng, R.; Li, K.; Chang, Z.; Bu, X. A Novel Double-Walled Cd (II) Metal – Organic Framework as Highly Selective Luminescent Sensor for Cr 2 O 2 7 Anion. *Polyhedron* **2018**, *153*, 110–114. https://doi.org/10.1016/j.poly.2018.07.001.

129. Awais, M.; Ye, J.; Wang, G.; Shi, L.; Liu, Z.; Lu, H.; Zhang, S.; Ning, G. Bifunctional Chemosensor Based on a Dye-Encapsulated Metal-Organic Framework for Highly Selective and Sensitive Detection of Cr 2 O 2 7 and Fe 3 + Ions. *Polyhedron* **2020**, *185*, 114604. https://doi.org/10.1016/j.poly.2020.114604.

130. Yang, Y.; Wang, Q.; Qiu, W.; Guo, H.; Gao, F. Covalent Immobilization of Cu3(Btc)2 at Chitosan–Electroreduced Graphene Oxide Hybrid Film and Its Application for Simultaneous Detection of Dihydroxybenzene Isomers. *J. Phys. Chem. C* **2016**, *120* (18), 9794–9803.

131. Dong, S.; Suo, G.; Li, N.; Chen, Z.; Peng, L.; Fu, Y.; Yang, Q.; Huang, T. A Simple Strategy to Fabricate High Sensitive 2,4- Dichlorophenol Electrochemical Sensor Based on Metal Organic Framework Cu3(BTC)2. *Sensors Actuators B. Chem.* **2016**, *222*, 972–979.

132. Zhou, X.; Yan, X.; Hong, Z.; Zheng, X.; Wang, F. Design of Magnetic Core–Shell Ni@graphene Composites as a Novel Electrochemical Sensing Platform. *Sensors Actuators B. Chem.* **2018**, *255*, 2959–2962.

133. Wang, J.; Si, P.; Yang, J.; Zhao, S.; Li, P.; Li, B.; Wang, S.; Lu, M.; Yu, S. La (III) -Based MOFs with 5-Aminoisophthalic Acid for Optical Detection and Degradation of Organic Molecules in Water. **2019**, *162*, 255–262. https://doi.org/10.1016/j.poly.2019.01.045.

134. Gan, Y. Le; Huang, K. R.; Li, Y. G.; Qin, D. P.; Zhang, D. M.; Zong, Z. A.; Cui, L. S. Synthesis, Structure and Fluorescent Sensing for Nitrobenzene of a Zn-Based MOF. *J. Mol. Struct.* **2021**, *1223*, 129217. https://doi.org/10.1016/j.molstruc.2020.129217.

135. Su, Y.; Qu, Y.; Fu, L.; Cui, G. An Unprecedented Binodal (4 , 6) -Connected Co (II) MOF as Dual-Responsive Luminescent Sensor for Detection of Acetylacetone and Hg 2 + Ions. *Inorg. Chem. Commun.* **2020**, *118* (April), 108013. https://doi.org/10.1016/j.inoche.2020.108013.

136. Li, D.; Wang, F.; Tian, Y.; Liu, S.; Liu, D.; Yang, B. The Selective Recognition and Luminescence Sensing Demonstrated in Yttrium - Organic Framework. *Opt. Mater. (Amst).* **2020**, *108* (September), 109757. https://doi.org/10.1016/j.optmat.2020.109757.

137. Zhou, Y.; Yang, Q.; Zhang, D.; Gan, N.; Li, Q.; Cuan, J. Detection and Removal of Antibiotic Tetracycline in Water with a Highly Stable Luminescent MOF. *Sensors Actuators B. Chem.* **2018**, *262*, 137–143.

138. Fan, L.; Wang, F. A Self-Penetrating and Chemically Stable Zinc (Ii) -Organic Framework as Multi-Responsive Chemo-Sensor to Detect Pesticide and Antibiotics in Water. **2020** (May), 1–10. https://doi.org/10.1002/aoc.5960.

139. Bhatnagar, A.; Minocha, A. K. Conventional and Non-Conventional Adsorbents for Removal of Pollutants from Water - A Review. *Indian J. Chem. Technol.* **2006**, *13* (3), 203–217.

140. Hasan, Z.; Jhung, S. H. Removal of Hazardous Organics from Water Using Metal-Organic Frameworks (MOFs): Plausible Mechanisms for Selective Adsorptions. *J. Hazard. Mater.* **2015**, *283*, 329–339. https://doi.org/10.1016/j.jhazmat.2014.09.046.

141. Adeyemo, A. A.; Adeoye, I. O.; Bello, O. S. Metalorganicframeworksasadsor- Bents Fordyeadsorption:Overview,Prospectsandfuturechallenges. *Toxicol. Environ. Chem* **2012**, *94*, 1846–1863.

142. Khan, N. A.; Jhung, S. H. Adsorptive Removal and Separation of Chemicals with Metal-Organic Frameworks: Contribution of π-Complexation. *J. Hazard. Mater.* **2017**, *325*, 198–213. https://doi.org/10.1016/j.jhazmat.2016.11.070.

143. Ahmed, I.; Jhung, S. H. Applications of Metal-Organic Frameworks in Adsorption/Separation Processes via Hydrogen Bonding Interactions. *Chem. Eng. J.* **2017**, *310*, 197–215. https://doi.org/10.1016/j.cej.2016.10.115.

144. Jun, J. W.; Tong, M.; Jung, B. K.; Hasan, Z.; Zhong, C.; Jhung, S. H. Effect of Central Metal Ions of Analogous Metal-Organic Frameworks on Adsorption of Organoarsenic Compounds from Water: Plausible Mechanism of Adsorption and Water Purification. *Chem. - A Eur. J.* **2015**, *21* (1), 347–354. https://doi.org/10.1002/chem.201404658.

145. Li, B.; Wen, H. M.; Cui, Y.; Zhou, W.; Qian, G.; Chen, B. Emerging Multifunctional Metal–Organic Framework Materials. *Adv. Mater.* **2016**, *28* (40), 8819–8860. https://doi.org/10.1002/adma.201601133.

146. Lu, M.; Li, L.; Shen, S.; Chen, D.; Han, W. Highly Efficient Removal of Pb2+ by a Sandwich Structure of Metal-Organic Framework/GO Composite with Enhanced Stability. *New J. Chem.* **2019**, *43* (2), 1032–1037. https://doi.org/10.1039/c8nj05091k.

147. Mehdinia, A.; Jahedi Vaighan, D.; Jabbari, A. Cation Exchange Superparamagnetic Al-Based Metal Organic Framework (Fe3O4/MIL-96(Al)) for High Efficient Removal of Pb(II) from Aqueous Solutions. *ACS Sustain. Chem. Eng.* **2018**, *6* (3), 3176–3186. https://doi.org/10.1021/acssuschemeng.7b03301.

148. Chaikittisilp, W.; Ariga, K.; Yamauchi, Y. A New Family of Carbon Materials: Synthesis of MOF-Derived Nanoporous Carbons and Their Promising Applications. *J. Mater. Chem. A* **2013**, *1* (1), 14–19. https://doi.org/10.1039/c2ta00278g.

149. Kumar, S.; Jain, S.; Nehra, M.; Dilbaghi, N.; Marrazza, G.; Kim, K. H. Green Synthesis of Metal–Organic Frameworks: A State-of-the-Art Review of Potential Environmental and Medical Applications. *Coord. Chem. Rev.* **2020**, *420*, 213407. https://doi.org/10.1016/j.ccr.2020.213407.

150. Bobbitt, N. S.; Mendonca, M. L.; Howarth, A. J.; Islamoglu, T.; Hupp, J. T.; Farha, O. K.; Snurr, R. Q. Metal-Organic Frameworks for the Removal of Toxic Industrial Chemicals and Chemical Warfare Agents. *Chem. Soc. Rev.* **2017**, *46* (11), 3357–3385. https://doi.org/10.1039/c7cs00108h.
151. Tsivadze, A. Y.; Aksyutin, O. E.; Ishkov, A. G.; Knyazeva, M. K.; Solovtsova, O. V; Men'shchikov, I. E.; Fomkin, A. A.; Shkolin, A. V; Khozina, E. V; Grachev, V. A. Metal-Organic Framework Structures: Adsorbents for Natural Gas Storage. *Russ. Chem. Rev.* **2019,** *88* (9), 925–978. https://doi.org/10.1070/rcr4873.
152. Mohamedali, M.; Henni, A.; Ibrahim, H. Markedly Improved CO2 Uptake Using Imidazolium-Based Ionic Liquids Confined into HKUST-1 Frameworks. *Microporous Mesoporous Mater* **2019**, *284*, 98–110. https://doi.org/10.1016/j.micromeso.2019.04.004.
153. Zhou, L.; Niu, Z.; Jin, X.; Tang, L.; Zhu, L. Effect of Lithium Doping on the Structures and CO2 Adsorption Properties of Metal-Organic Frameworks HKUST-1. *ChemistrySelect* **2018**, *3* (45), 12865–12870. https://doi.org/10.1002/slct.201803164.
154. Bao, Z.; Yu, L.; Ren, Q.; Lu, X.; Deng, S. Adsorption of CO2 and CH4 on a Magnesium-Based Metal Organic Framework. *J. Colloid Interface Sci.* **2011**, *353* (2), 549–556. https://doi.org/10.1016/j.jcis.2010.09.065.
155. Edubilli, S.; Gumma, S. A Systematic Evaluation of UiO-66 Metal Organic Framework for CO2/N2 Separation. *Sep. Purif. Technol.* **2019**, *224*, 85–94. https://doi.org/10.1016/j.seppur.2019.04.081.
156. Ghanbari, T.; Abnisa, F.; Wan Daud, W. M. A. A Review on Production of Metal Organic Frameworks (MOF) for CO2 Adsorption. *Sci. Total Environ.* **2020**, *707*, 135090. https://doi.org/10.1016/j.scitotenv.2019.135090.
157. Wen, M.; Li, G.; Liu, H.; Chen, J.; An, T.; Yamashita, H. Metal-Organic Framework-Based Nanomaterials for Adsorption and Photocatalytic Degradation of Gaseous Pollutants: Recent Progress and Challenges. *Environ. Sci. Nano* **2019**, *6* (4), 1006–1025. https://doi.org/10.1039/c8en01167b.
158. Bloch, E. D.; Hudson, M. R.; Mason, J. A.; Chavan, S.; Crocellà, V.; Howe, J. D.; Lee, K.; Dzubak, A. L.; Queen, W. L.; Zadrozny, J. M.; Geier, S. J.; Lin, L. C.; Gagliardi, L.; Smit, B.; Neaton, J. B.; Bordiga, S.; Brown, C. M.; Long, J. R. Reversible CO Binding Enables Tunable CO/H2 and CO/N2 Separations in Metal-Organic Frameworks with Exposed Divalent Metal Cations. *J. Am. Chem. Soc.* **2014**, *136* (30), 10752–10761. https://doi.org/10.1021/ja505318p.
159. Danaci, D.; Singh, R.; Xiao, P.; Webley, P. A. Assessment of ZIF Materials for CO2 Capture from High Pressure Natural Gas Streams. *Chem. Eng. J.* **2015**, *280*, 486–493. https://doi.org/10.1016/j.cej.2015.04.090.
160. Wongsakulphasatch, S.; Kiatkittipong, W.; Saupsor, J.; Chaiwiseshphol, J.; Piroonlerkgul, P.; Parasuk, V.; Assabumrungrat, S. Effect of Fe Open Metal Site in Metal-Organic Frameworks on Post-Combustion CO2 Capture Performance. *Greenh. Gases Sci. Technol.* **2017**, 7 (2), 383–394. https://doi.org/10.1002/ghg.1662.
161. Chen, Y.; Qiao, Z.; Huang, J.; Wu, H.; Xiao, J.; Xia, Q.; Xi, H.; Hu, J.; Zhou, J.; Li, Z. Unusual Moisture-Enhanced CO 2 Capture within Microporous PCN-250 Frameworks. *ACS Appl. Mater. Interfaces* **2018**, *10* (44), 38638–38647. https://doi.org/10.1021/acsami.8b14400.

162. Kim, S. N.; Lee, Y. R.; Hong, S. H.; Jang, M. S.; Ahn, W. S. Pilot-Scale Synthesis of a Zirconium-Benzenedicarboxylate UiO-66 for CO2 Adsorption and Catalysis. *Catal. Today* **2015**, *245*, 54–60. https://doi.org/10.1016/j.cattod.2014.05.041.
163. Niu, Z.; Guan, Q.; Shi, Y.; Chen, Y.; Chen, Q.; Kong, Z.; Ning, P.; Tian, S.; Miao, R. A Lithium-Modified Zirconium-Based Metal Organic Framework (UiO-66) for Efficient CO2 Adsorption. *New J. Chem.* **2018**, *42* (24), 19764–19770. https://doi.org/10.1039/c8nj04945a.
164. Ullah, S.; Bustam, M. A.; Assiri, M. A.; Al-Sehemi, A. G.; Sagir, M.; Abdul Kareem, F. A.; Elkhalifah, A. E. I.; Mukhtar, A.; Gonfa, G. Synthesis, and Characterization of Metal-Organic Frameworks -177 for Static and Dynamic Adsorption Behavior of CO2 and CH4. *Microporous Mesoporous Mater* **2019**, *288*. https://doi.org/10.1016/j.micromeso.2019.109569.
165. Kloutse, F. A.; Hourri, A.; Natarajan, S.; Benard, P.; Chahine, R. Systematic Study of the Excess and the Absolute Adsorption of N2/H2 and CO2/H2 Mixtures on Cu-BTC. *Adsorption* **2019**. https://doi.org/10.1007/s10450-019-00124-3.
166. Mohamedali, M.; Henni, A.; Ibrahim, H. Investigation of CO2 Capture Using Acetate-Based Ionic Liquids Incorporated into Exceptionally Porous Metal–Organic Frameworks. *Adsorption* **2019**, *25* (4), 675–692. https://doi.org/10.1007/s10450-019-00073-x.
167. Su, X.; Bromberg, L.; Martis, V.; Simeon, F.; Huq, A.; Hatton, T. A. Postsynthetic Functionalization of Mg-MOF-74 with Tetraethylenepentamine: Structural Characterization and Enhanced CO2 Adsorption. *ACS Appl. Mater. Interfaces* **2017**, *9* (12), 11299–11306. https://doi.org/10.1021/acsami.7b02471.
168. Brandt, P.; Nuhnen, A.; Lange, M.; Möllmer, J.; Weingart, O.; Janiak, C. Metal-Organic Frameworks with Potential Application for SO2 Separation and Flue Gas Desulfurization. *ACS Appl. Mater. Interfaces* **2019**, *11* (19), 17350–17358. https://doi.org/10.1021/acsami.9b00029.
169. Ebrahim, A. M.; Levasseur, B.; Bandosz, T. J. Interactions of NO2 with Zr-Based MOF: Effects of the Size of Organic Linkers on NO2 Adsorption at Ambient Conditions. *Langmuir* **2013**, *29* (1), 168–174. https://doi.org/10.1021/la302869m.
170. Dathe, H.; Peringer, E.; Roberts, V.; Jentys, A.; Lercher, J. A. Metal Organic Frameworks Based on Cu2+ and Benzene-1,3,5-Tricarboxylate as Host for SO2 Trapping Agents. *Comptes Rendus Chim.* **2005**, *8* (3–4), 753–763. https://doi.org/10.1016/j.crci.2004.10.018.
171. Zhou, L.; Zhang, X.; Chen, Y. Facile Synthesis of Al-Fumarate Metal–Organic Framework Nano-Flakes and Their Highly Selective Adsorption of Volatile Organic Compounds. *Mater. Lett.* **2017**, *197*, 224–227. https://doi.org/10.1016/j.matlet.2017.01.120.
172. Zhang, X.; Yang, Y.; Song, L.; Chen, J.; Yang, Y.; Wang, Y. Enhanced Adsorption Performance of Gaseous Toluene on Defective UiO-66 Metal Organic Framework: Equilibrium and Kinetic Studies. *J. Hazard. Mater.* **2019**, *365*, 597–605. https://doi.org/10.1016/j.jhazmat.2018.11.049.
173. Yang, K.; Xue, F.; Sun, Q.; Yue, R.; Lin, D. Adsorption of Volatile Organic Compounds by Metal-Organic Frameworks MOF-177. *J. Environ. Chem. Eng.* **2013**, *1* (4), 713–718. https://doi.org/10.1016/j.jece.2013.07.005.
174. Kampouraki, Z. C.; Giannakoudakis, D. A.; Nair, V.; Hosseini-Bandegharaei, A.; Colmenares, J. C.; Deliyanni, E. A. Metal Organic Frameworks as Desulfurization Adsorbents of DBT and

4,6-DMDBT from Fuels. *Molecules* **2019**, *24* (24), 1–23. https://doi.org/10.3390/molecules24244525.

175. Lin, C.; Cheng, Z.; Li, B.; Chen, T.; Zhang, W.; Chen, S.; Yang, Q.; Chang, L.; Che, G.; Ma, H. High-Efficiency Separation of Aromatic Sulfide from Liquid Hydrocarbon Fuel in Conjugated Porous Organic Framework with Polycarbazole Unit. *ACS Appl. Mater. Interfaces* **2019**, *11* (43), 40970–40979. https://doi.org/10.1021/acsami.9b15815.
176. Radwan, D. R.; Matloob, A.; Mikhail, S.; Saad, L.; Guirguis, D. Metal Organic Framework-Graphene Nano-Composites for High Adsorption Removal of DBT as Hazard Material in Liquid Fuel. *J. Hazard. Mater.* **2019**, *373*, 447–458. https://doi.org/10.1016/j.jhazmat.2019.03.098.
177. Jasuja, H.; Peterson, G. W.; Decoste, J. B.; Browe, M. A.; Walton, K. S. Evaluation of MOFs for Air Purification and Air Quality Control Applications: Ammonia Removal from Air. *Chem. Eng. Sci.* **2015**, *124*, 118–124. https://doi.org/10.1016/j.ces.2014.08.050.
178. Shah, M. S.; Tsapatsis, M.; Siepmann, J. I. Hydrogen Sulfide Capture: From Absorption in Polar Liquids to Oxide, Zeolite, and Metal-Organic Framework Adsorbents and Membranes. *Chem. Rev.* **2017**, *117* (14), 9755–9803. https://doi.org/10.1021/acs.chemrev.7b00095.
179. Rieth, A. J.; Tulchinsky, Y.; Dincă, M. High and Reversible Ammonia Uptake in Mesoporous Azolate Metal-Organic Frameworks with Open Mn, Co, and Ni Sites. *J. Am. Chem. Soc.* **2016**, *138* (30), 9401–9404. https://doi.org/10.1021/jacs.6b05723.
180. Moghadam, P. Z.; Fairen-Jimenez, D.; Snurr, R. Q. Efficient Identification of Hydrophobic MOFs: Application in the Capture of Toxic Industrial Chemicals. *J. Mater. Chem. A* **2015**, *4* (2), 529–536. https://doi.org/10.1039/c5ta06472d.
181. Kumar, P.; Kim, K. H.; Kwon, E. E.; Szulejko, J. E. Metal-Organic Frameworks for the Control and Management of Air Quality: Advances and Future Direction. *J. Mater. Chem. A* **2015**, *4* (2), 345–361.
182. Mocniak, K. A.; Kubajewska, I.; Spillane, D. E. M.; Williams, G. R.; Morris, R. E. Incorporation of Cisplatin into the Metal-Organic Frameworks UiO66-NH2 and UiO66-Encapsulation vs. Conjugation. *RSC Adv.* **2015**, *5* (102), 83648–83656. https://doi.org/10.1039/c5ra14011k.
183. He, H.; Li, R.; Yang, Z.; Chai, L.; Jin, L.; Alhassan, S. I.; Ren, L.; Wang, H.; Huang, L. Preparation of MOFs and MOFs Derived Materials and Their Catalytic Application in Air Pollution: A Review. *Catal. Today* **2020**, *139* (February), 12125–12128. https://doi.org/10.1016/j.cattod.2020.02.033.
184. Wang, S.; Wu, D.; Huang, H.; Yang, Q.; Tong, M.; Liu, D.; Zhong, C. Computational Exploration of H2S/CH4 Mixture Separation Using Acid-Functionalized UiO-66(Zr) Membrane and Composites. *Chinese J. Chem. Eng.* **2015**, *23* (8), 1291–1299. https://doi.org/10.1016/j.cjche.2015.04.017.
185. Zhang, H. Y.; Yang, C.; Geng, Q.; Fan, H. L.; Wang, B. J.; Wu, M. M.; Tian, Z. Adsorption of Hydrogen Sulfide by Amine-Functionalized Metal Organic Framework (MOF-199): An Experimental and Simulation Study. *Appl. Surf. Sci.* **2019**, *497*, 143815. https://doi.org/10.1016/j.apsusc.2019.143815.
186. Pokhrel, J.; Bhoria, N.; Wu, C.; Reddy, K. S. K.; Margetis, H.; Anastasiou, S.; George, G.; Mittal, V.; Romanos, G.; Karonis, D.; Karanikolos, G. N. Cu- and Zr-Based Metal Organic Frameworks and Their Composites with Graphene Oxide for Capture of Acid Gases at Ambient

Temperature. *J. Solid State Chem.* **2018**, *266*, 233–243. https://doi.org/10.1016/j.jssc.2018.07.022.

187. Li, Z.; Liao, F.; Jiang, F.; Liu, B.; Ban, S.; Chen, G.; Sun, C.; Xiao, P.; Sun, Y. Capture of H2S and SO2 from Trace Sulfur Containing Gas Mixture by Functionalized UiO-66(Zr) Materials: A Molecular Simulation Study. *Fluid Phase Equilib.* **2016**, *427* (2), 259–267. https://doi.org/10.1016/j.fluid.2016.07.020.
188. Georgiadis, A. G.; Charisiou, N.; Yentekakis, I. V.; Goula, M. A. Hydrogen Sulfide (H2S) Removal via MOFs. *Materials (Basel)* **2020**, *13* (16), 12125–12128. https://doi.org/10.3390/MA13163640.
189. Ethiraj, J.; Bonino, F.; Lamberti, C.; Bordiga, S. H2S Interaction with HKUST-1 and ZIF-8 MOFs: A Multitechnique Study. *Microporous Mesoporous Mater.* **2015**, *207*, 90–94. https://doi.org/10.1016/j.micromeso.2014.12.034.
190. Shekhah, O.; Belmabkhout, Y.; Adil, K.; Bhatt, P. M.; Cairns, A. J.; Eddaoudi, M. A Facile Solvent-Free Synthesis Route for the Assembly of a Highly CO2 Selective and H2S Tolerant NiSIFSIX Metal-Organic Framework. *Chem. Commun.* **2015**, *51* (71), 13595–13598. https://doi.org/10.1039/c5cc04487a.
191. Gargiulo, N.; Peluso, A.; Caputo, D. MOF-Based Adsorbents for Atmospheric Emission Control: A Review. *Processes* **2020**, *8* (5), 12125–12128. https://doi.org/10.3390/PR8050613.
192. Zheng, C.; Shen, J.; Zhang, Y.; Huang, W.; Zhu, X.; Wu, X.; Chen, L.; Gao, X.; Cen, K. Quantitative Assessment of Industrial VOC Emissions in China: Historical Trend, Spatial Distribution, Uncertainties, and Projection. *Atmos. Environ.* **2017**, *150*, 116–125. https://doi.org/10.1016/j.atmosenv.2016.11.023.
193. Vellingiri, K.; Szulejko, J. E.; Kumar, P.; Kwon, E. E.; Kim, K. H.; Deep, A.; Boukhvalov, D. W.; Brown, R. J. C. Metal Organic Frameworks as Sorption Media for Volatile and Semi-Volatile Organic Compounds at Ambient Conditions. *Sci. Rep.* **2016**, *6*. https://doi.org/10.1038/srep27813.
194. Saha, D.; Deng, S. Structural Stability of Metal Organic Framework MOF-177. *J. Phys. Chem. Lett.* **2010**, *1* (1), 73–78. https://doi.org/10.1021/jz900028u.
195. Hua, Y.; Gargiulo, N.; Peluso, A.; Aprea, P.; Eić, M.; Caputo, D. Adsorption Behavior of Halogenated Anesthetic and Water Vapor on Cr-Based MOF (MIL-101) Adsorbent. Part II. Multiple-Cycle Breakthrough Tests. *Chemie-Ingenieur-Technik* **2016**, *88* (11), 1739–1745. https://doi.org/10.1002/cite.201600052.
196. Chakraborty, A.; Bhattacharyya, S.; Hazra, A.; Ghosh, A. C.; Maji, T. K. Post-Synthetic Metalation in an Anionic MOF for Efficient Catalytic Activity and Removal of Heavy Metal Ions from Aqueous Solution. *Chem. Commun.* **2016**, *52* (13), 2831–2834. https://doi.org/10.1039/c5cc09814a.
197. Pearson, R. G. Hard and Soft Acids and Bases. *J. Am. Chem. Soc.* **1963**, *85* (22), 3533–3539. https://doi.org/10.1021/ja00905a001.
198. Huang, Z.; Zhao, M.; Wang, C.; Wang, S.; Dai, L.; Zhang, L.; Xu, L. Selective Removal Mechanism of the Novel Zr-Based Metal Organic Framework Adsorbents for Gold Ions from Aqueous Solutions. *Chem. Eng. J.* **2020**, *384*, 123343. https://doi.org/10.1016/j.cej.2019.123343.

199. Jiang, S. Y.; He, W. W.; Li, S. L.; Su, Z. M.; Lan, Y. Q. Introduction of Molecular Building Blocks to Improve the Stability of Metal-Organic Frameworks for Efficient Mercury Removal. *Inorg. Chem.* **2018**, *57* (10), 6118–6123. https://doi.org/10.1021/acs.inorgchem.8b00704.
200. Zhang, R.; Wang, Z.; Zhou, Z.; Li, D.; Wang, T.; Su, P.; Yang, Y. Highly Effective Removal of Pharmaceutical Compounds from Aqueous Solution by Magnetic Zr-Based MOFs Composites. *Ind. Eng. Chem. Res.* **2019**, *58* (9), 3876–3884. https://doi.org/10.1021/acs.iecr.8b05244.
201. Fu, L.; Wang, S.; Lin, G.; Zhang, L.; Liu, Q.; Zhou, H.; Kang, C.; Wan, S.; Li, H.; Wen, S. Post-Modification of UiO-66-NH2 by Resorcyl Aldehyde for Selective Removal of Pb(II)in Aqueous Media. *J. Clean. Prod.* **2019**, *229*, 470–479. https://doi.org/10.1016/j.jclepro.2019.05.043.
202. Wang, C.; Lin, G.; Zhao, J.; Wang, S.; Zhang, L.; Xi, Y.; Li, X.; Ying, Y. Highly Selective Recovery of Au(III) from Wastewater by Thioctic Acid Modified Zr-MOF: Experiment and DFT Calculation. *Chem. Eng. J.* **2020**, *380* (August 2019), 122511. https://doi.org/10.1016/j.cej.2019.122511.
203. Song, J. Y.; Ahmed, I.; Seo, P. W.; Jhung, S. H. UiO-66-Type Metal-Organic Framework with Free Carboxylic Acid: Versatile Adsorbents via H-Bond for Both Aqueous and Nonaqueous Phases. *ACS Appl. Mater. Interfaces* **2016**, *8* (40), 27394–27402. https://doi.org/10.1021/acsami.6b10098.
204. Luo, Z.; Fan, S.; Liu, J.; Liu, W.; Shen, X.; Wu, C.; Huang, Y.; Huang, G.; Huang, H.; Zheng, M. A 3D Stable Metal-Organic Framework for Highly Efficient Adsorption and Removal of Drug Contaminants from Water. *Polymers (Basel)* **2018**, *10* (2), 209–222. https://doi.org/10.3390/polym10020209.
205. Abdullah, N. H.; Shameli, K.; Abdullah, E. C.; Abdullah, L. C. Solid Matrices for Fabrication of Magnetic Iron Oxide Nanocomposites: Synthesis, Properties, and Application for the Adsorption of Heavy Metal Ions and Dyes. *Compos. Part B Eng.* **2019**, *162* (December 2018), 538–568. https://doi.org/10.1016/j.compositesb.2018.12.075.
206. Li, L.; Xu, Y.; Zhong, D.; Zhong, N. CTAB-Surface-Functionalized Magnetic MOF@MOF Composite Adsorbent for Cr(VI) Efficient Removal from Aqueous Solution. *Colloids Surfaces A Physicochem. Eng. Asp.* **2020**, *586*, 124255. https://doi.org/10.1016/j.colsurfa.2019.124255.
207. Wang, K.; Tao, X.; Xu, J.; Yin, N. Novel Chitosan-MOF Composite Adsorbent for the Removal of Heavy Metal Ions. *Chem. Lett.* **2016**, *45* (12), 1365–1368. https://doi.org/10.1246/cl.160718.
208. Liu, B.; Yang, F.; Zou, Y.; Peng, Y. Adsorption of Phenol and p -Nitrophenol from Aqueous Solutions on Metal-Organic Frameworks: Effect of Hydrogen Bonding. *J. Chem. Eng. Data* **2014**, *59* (5), 1476–1482. https://doi.org/10.1021/je4010239.
209. Tella, A. C.; Owalude, S. O.; Olatunji, S. J.; Adimula, V. O.; Elaigwu, S. E.; Alimi, L. O.; Ajibade, P. A.; Oluwafemi, O. S. Synthesis of Zinc-Carboxylate Metal-Organic Frameworks for the Removal of Emerging Drug Contaminant (Amodiaquine) from Aqueous Solution. *J. Environ. Sci.* **2017**, *64* (June), 1–12. https://doi.org/10.1016/j.jes.2017.06.015.
210. Xu, G. R.; An, Z. H.; Xu, K.; Liu, Q.; Das, R.; Zhao, H. L. Metal Organic Framework (MOF)-Based Micro/Nanoscaled Materials for Heavy Metal Ions Removal: The Cutting-Edge Study on Designs, Synthesis, and Applications. *Coord. Chem. Rev.* **2021**, *427*. https://doi.org/10.1016/j.ccr.2020.213554.

211. Song, J. Y.; Jhung, S. H. Adsorption of Pharmaceuticals and Personal Care Products over Metal-Organic Frameworks Functionalized with Hydroxyl Groups: Quantitative Analyses of H-Bonding in Adsorption. *Chem. Eng. J.* **2017**, *322*, 366–374. https://doi.org/10.1016/j.cej.2017.04.036.
212. Seo, P. W.; Bhadra, B. N.; Ahmed, I.; Khan, N. A.; Jhung, S. H. Adsorptive Removal of Pharmaceuticals and Personal Care Products from Water with Functionalized Metal-Organic Frameworks: Remarkable Adsorbents with Hydrogen-Bonding Abilities. *Sci. Rep.* **2016**, *6* (September), 1–11. https://doi.org/10.1038/srep34462.
213. Lv, Y.; Zhang, R.; Zeng, S.; Liu, K.; Huang, S.; Liu, Y.; Xu, P.; Lin, C.; Cheng, Y.; Liu, M. Removal of P-Arsanilic Acid by an Amino-Functionalized Indium-Based Metal–Organic Framework: Adsorption Behavior and Synergetic Mechanism. *Chem. Eng. J.* **2018**, *339* (January), 359–368. https://doi.org/10.1016/j.cej.2018.01.139.
214. Andrew Lin, K. Y.; Yang, H.; Lee, W. Der. Enhanced Removal of Diclofenac from Water Using a Zeolitic Imidazole Framework Functionalized with Cetyltrimethylammonium Bromide (CTAB). *RSC Adv.* **2015**, *5* (99), 81330–81340. https://doi.org/10.1039/c5ra08189k.
215. Li, X.; Yuan, H.; Quan, X.; Chen, S.; You, S. Effective Adsorption of Sulfamethoxazole, Bisphenol A and Methyl Orange on Nanoporous Carbon Derived from Metal-Organic Frameworks. *J. Environ. Sci. (China)* **2018**, *63*, 250–259. https://doi.org/10.1016/j.jes.2017.10.019.
216. Bhadra, B. N.; Ahmed, I.; Kim, S.; Jhung, S. H. Adsorptive Removal of Ibuprofen and Diclofenac from Water Using Metal-Organic Framework-Derived Porous Carbon. *Chem. Eng. J.* **2017**, *314*, 50–58. https://doi.org/10.1016/j.cej.2016.12.127.
217. Bhadra, B. N.; Jhung, S. H. Adsorptive Removal of Wide Range of Pharmaceuticals and Personal Care Products from Water Using Bio-MOF-1 Derived Porous Carbon. *Microporous Mesoporous Mater.* **2018**, *270*, 102–108. https://doi.org/10.1016/j.micromeso.2018.05.005.
218. Alqadami, A. A.; Naushad, M.; Alothman, Z. A.; Ahamad, T. Adsorptive Performance of MOF Nanocomposite for Methylene Blue and Malachite Green Dyes: Kinetics, Isotherm and Mechanism. *J. Environ. Manage.* **2018**, *223* (February), 29–36. https://doi.org/10.1016/j.jenvman.2018.05.090.
219. Hasanzadeh, M.; Simchi, A.; Shahriyari Far, H. Nanoporous Composites of Activated Carbon-Metal Organic Frameworks for Organic Dye Adsorption: Synthesis, Adsorption Mechanism and Kinetics Studies. *J. Ind. Eng. Chem.* **2020**, *81*, 405–414. https://doi.org/10.1016/j.jiec.2019.09.031.
220. Torad, N. L.; Hu, M.; Ishihara, S.; Sukegawa, H.; Belik, A. A.; Imura, M.; Ariga, K.; Sakka, Y.; Yamauchi, Y. Direct Synthesis of MOF-Derived Nanoporous Carbon with Magnetic Co Nanoparticles toward Efficient Water Treatment. *Small* **2014**, *10* (10), 2096–2107. https://doi.org/10.1002/smll.201302910.
221. Shi, Z.; Xu, C.; Guan, H.; Li, L.; Fan, L.; Wang, Y.; Liu, L.; Meng, Q.; Zhang, R. Magnetic Metal Organic Frameworks (MOFs) Composite for Removal of Lead and Malachite Green in Wastewater. *Colloids Surfaces A Physicochem. Eng. Asp.* **2018**, *539*, 382–390. https://doi.org/10.1016/j.colsurfa.2017.12.043.

Chapter 12

Environmental Applications of Metal–Organic Frameworks: Recent Advances and Challenges

Meena Nemiwal,[*,1] Vijayalakshmi Gosu,[2] Ankita Dhillon,[3] and Dinesh Kumar[4]

[1]Department of Chemistry, Malaviya National Institute of Technology, Jaipur, India 302017

[2]Department of Chemical Engineering, Malaviya National Institute of Technology, Jaipur, India 302017

[3]Department of Chemistry, Banasthali Vidyapith, Rajasthan, India 304022

[4]School of Chemical Sciences, Central University of Gujarat, Gandhinagar, India 382030

[*]Email author: meena.chy@mnit.ac.in

Nanomaterials have gained tremendous attention in the field of research owing to their fantastic and unique properties. Metal-Organic Frameworks (MOFs) are the class of porous nanomaterials constituting organic ligands and inorganic metal ions. MOFs are hybrids of organic-inorganic materials and have attracted researcher's attention due to their large surface area, tunable pore structure, and high porosity. These properties make MOFs applicable in a wide range of applications in various fields, including catalysis, sensing, gas adsorption, and separation etc. Currently, MOFs have been employed for several environmental applications and resulted in outstanding performances. This chapter begins with the introduction of MOFs, then it covers the different applications of MOFs for the environment, including recent advances and challenges.

Introduction

Nanomaterials have become an indispensable part of the research and have been exploited extensively in catalysis, adsorption, sensing (*1*), drug delivery etc. There are different types of nanoporous materials such as titanium oxide nanotubes (*2*), zinc oxide nanotubes (*3*), carbon nanotubes (*4*), zeolites (*5*), MOFs (*6*), covalent-organic frameworks (*7*, *8*) etc. Currently, MOFs have shown outstanding performances in catalysis (*9*), sensing (*10*), adsorption (*11*), and reducing carbon dioxide (*12*) and being used for many environmental applications (*13*). Researchers have used MOFs to harvest water from the atmosphere, solving water scarcity in a desert area (*14*). To save the environment from pollution, clean energy sources are explored, like hydrogen production.

MOFs have shown potential performance in the production of hydrogen both by electrocatalysis and photocatalysis (*15*).

MOFs possess unique characteristics such as a high surface area with a tunable porous framework, making them suitable for diverse fields (*16*). MOFs are developed by coordinating organic ligands with inorganic metal ions or metal clusters. Metal ions or clusters are also called secondary building units (SBU), and organic ligands used to link these units are called linkers, forming a periodic porous structure. Many MOFs have been synthesized by altering SBUs and linkers based on desired properties like pore size, surface area etc. Composites of MOFs are also prepared by combining other nanomaterials like graphene oxide (GO), TiO_2 etc. to improve properties like stability, porosity, and adsorption.

Their diverse structure demonstrates flexible modifications to improve properties like thermal stability, porosity, and catalytic properties by altering the metal ion centers or organic ligands (*17*). The structure of MOFs allows the feasible addition and removal of desirable functional groups within the porous structures.

The explosive growth of population and technology has severely impacted the environment. It has raised issues such as increased heavy metal ion toxicity in water (*18*), enhanced radioactive disposal in water (*19*), effluents of textiles and dye industries (*20*) etc. MOFs are useful for the remediation of the environment in many ways.

In this chapter, recent advances in MOFs application for the environment will also be discussed with the inclusion of challenges.

MOFs for Environmental Applications

MOFs have gained attention in the field of catalysis, sensing, adsorption, and sequestration due to their remarkable properties, as shown in Figure 1.

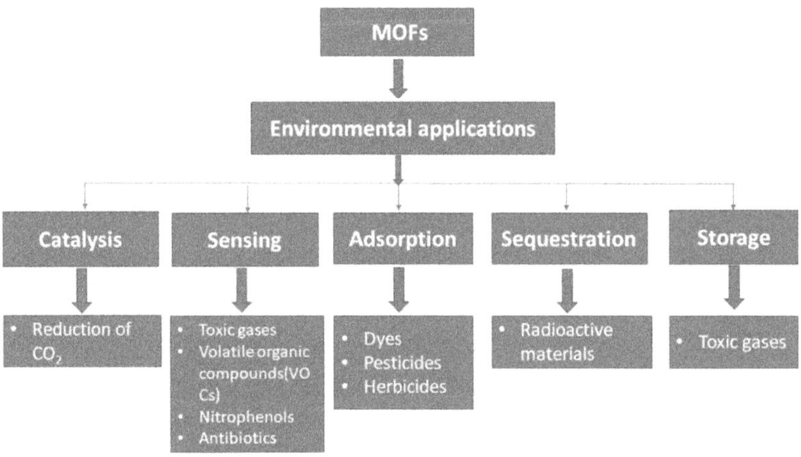

Figure 1. Schematic illustration of MOFs for their environmental applications.

Reduction of Carbon Dioxide

In the present scenario, CO_2 emission in the environment is the main cause of global warming (*21*). Various technologies are being developed for the capture and separation/sequestration of CO_2 and its utilization to produce valuable fuels that can alleviate the greenhouse effect and shortage of fuel (*22*). Since CO_2 is a very stable form of carbon, and its reduction is a challenge. Various materials such as zeolites, homogeneous molecular complexes, MOFs etc. have been employed for reducing CO_2. In particular, MOFs have been observed as a fantastic catalyst for the reduction of CO_2 and have the capacity to overcome the limitations faced with other conventional materials. MOFs have been explored with photocatalytic and electrocatalytic methods to efficiently and successfully reduce the CO_2 into various useful hydrocarbons like CH_4, CO, and CH_3OH (*23–25*) (Figure 2).

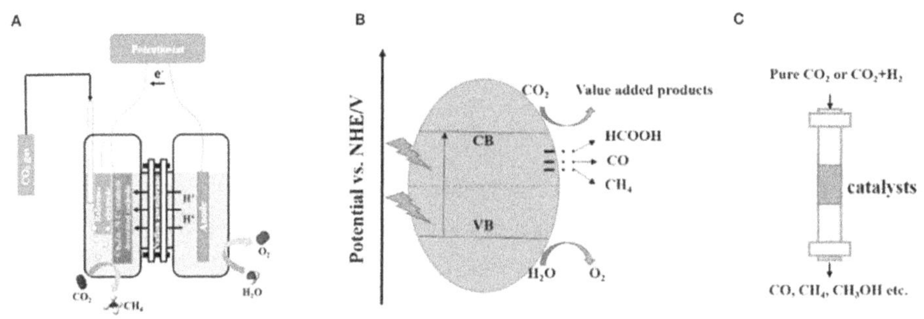

Figure 2. Schematic representation of CO_2 reduction systems: (a) electrochemical reduction, (b) photocatalytic reduction, and (c) thermal catalytic reduction. Reproduced with permission from reference (25). Copyright 2020 Frontiers in Chemistry.

Reduction of CO_2 by Photocatalytic Catalysis

MOFs-based photocatalysts can be easily developed due to their modular nature, making easy immobilization of the photoactive catalytic sites. The metal nodes and organic linkers can easily be tuned to be used as photocatalysts. Metal nodes in MOFs act as semiconductor quantum dots that are sensitized by the organic ligands. Dyes and photoactive ligands can easily be incorporated with organic ligands by grafting or building blocks. Upon light exposure, the MOFs' crystalline nature facilitates the charge transfer from linker to metal nodes. The adjustable pores also support mass transfer with various topologies. The solar light can be efficiently utilized by using the rich coordination chemistry of metal ions (transition) that offers flexibility for desirable tuning of light adsorption. TiO_2 and many other similar inorganic semiconductors have been employed to enhance the light-absorbing capacity of MOFs (*26, 27*). It is very important to note that semiconductors' particle size should be less than twice the Bohr's radius of the charge carriers, resembling the size of quantum dots (2–20 nm) and improving photocatalytic activity through multiple exciton generation and quantum size effect.

The mechanism of photocatalytic reduction of CO_2 includes 1) the irradiation of sunlight having the energy equal to or greater than the bandgap energy in between valence band (VB) and conduction band (CB) of metal ion, which leads to excitation of an electron from VB and holes are generated in CB, 2) electron-hole pair migration to the surface takes place and finally, 3) the charge carriers have interaction with the substrates adsorbed ate surface. The MOFs' electronic properties depend on the highest occupied molecular orbital of the organic linkers and the lowest unoccupied molecular orbital of the metal ions. Generally, organic linkers in MOFs act as antennas upon irradiation by solar light and absorbs photons. The electron density of the organic linkers is very high due to an excess of the aromatic polycarboxylates and can easily transfer the electrons to the metal ions. As the recombination of electron-hole pairs is being solved in MOFs, extensive research has been done in this direction. A large number of MOFs have been employed to reduce CO_2 into valuable products. Performance of photocatalysts based on MOFs was improved by various strategies such as dye sensitization, modification of organic ligands, composite formation, derivatives of MOFs etc. For instance, Wei et al. introduced electron-withdrawing and electron-attracting groups through the post-synthetic method and developed three new MOFs named UiO-68-OCH_3, UiO-68-CH_3 and UiO-68-F, and used for CO_2 reduction to CO with high selectivity (*28*). The selective and efficient photocatalytic reduction was improved by metal ion incorporation into the matrix of Zr-based MOF (Zr-DMBD), which carries a free thiol group. The thiol group was found to be bonded with Cu metal ion, and upon irradiation of visible-light efficient and selective (98%) CO_2 reduction was reported (*29*). Metal nanoparticles, semiconductors, and photosensitizers can be used to prepare MOFs' composite to enhance their efficiency. Initially, graphitic carbon nitrides (g-C_3N_4) (*30*) were used, and subsequently, various materials were employed such as TiO_2 (*31*), (rGO) (*32*), CDs (*33*), Zn_2GeO_4 (*34*), Ag (35), Pt, NH_2-rGO (*36*), N-carbon dots, $Cd_{0.2}Zn_{0.8}S$ (37) etc.

Electrocatalytic Reduction of CO_2

The reduction of CO_2 by electrocatalysis has attracted researchers due to its feasibility at ambient temperature and can be easily driven by renewable sources such as wind, solar tide (*38*). However, chemically CO_2 is very stable, high energy is required to remove the barrier of high kinetic and thermodynamic energy. The conversion of CO_2 into different valuable products includes multiple protons and electrons transfer and involves complex intermediates. The catalysts' composition, coordination geometry, surface electronic structure, atom distribution, and several active sites are the deciding factors for the catalysts' activity and selectivity. Thus, to synthesize highly efficient electrocatalysts, rational designing is very important to understand the relationship between the reaction mechanism and structure of the catalyst (39). In this context, it is very important to design catalysts with a tailorable structure (*40*). MOFs are the class of porous materials constructed by arranging metal ions and clusters with periodic arrangements through rational design through post-synthesis modification and pre-assembly (*41*). For instance, Co, Ni, and Zn transitions metals have been incorporated into MOF-74 to improve performance as an electrocatalyst to reduce CO_2 into valuable products (*42*). In this addition, Albo et al. developed Cu-based metal-organic porous materials (HKUST-1) impregnated on electrodes, having high accessibilities and high surface area. The exposure of the Cu catalytic centers was also increased with high- performance for electrocatalytic reduction of CO_2 (43).

MOFs-Based Materials for Sensing of Pollutants

Sensing of Toxic Gases

With a rapid increase in air pollution due to explosive increase in automobiles and industrialization, toxic gases (SO_3, NO_2, SO_2, NH_3 etc.) concentrations have been enhanced into the atmosphere. In view of this, several technologies have been implemented to detect and remove toxic gases from the atmosphere. MOFs have worked well in this direction and showed excellent performances for sensing and capture of toxic gases (*44*). In this view, chemiresistive MOFs was developed (*45*) for sensing purpose. Cu–Zn-MOF was used for sensing H_2S with high efficiency (*46*). Fluorescent film of Eu^{3+}/Cu^{2+} incorporated MOF (MIL-100(In)@Eu^{3+}/Cu^{2+}) was developed by Zhang et al. and used to detect H_2S gas with a meager limit of detection (0.535 ppm). It was based on the fluorescence turn-on phenomena of Eu^{3+} emission due to the formation of CuS (*47*). Similarly, the turn-on mechanism was explored with Fe^{3+}-MIL-88-NH_2 MOF for H_2S gas sensing with high efficiency (*48*). Ammonia is a toxic and poisonous gas, and its sensing by using MOF has resulted in a big environmental problem (*49*). Through chemisorption and physisorption, Cu^{2+}-based MOF (SION-10) was developed for NH_3 sensing. Cu^{2+} site acted as an active site for gas bonding, and large pores worked for physisorption. Color change was noticed from green to blue due to a shift in ultraviolet/visible (UV/Vis) absorption that could be observed even less than 300 ppm of ammonia (*50*). Other gases like SO_2, NO, CO_2, CS_2 were detected using modified MOFs.

Sensing of Volatile Organic Pollutants

Organic pollutant like volatile organic compounds (VOCs) tends to volatize easily into the air and subsequently into the ecosystem which is deleterious for the environment and human health. VOCs include methanol, ethanol, ethane, ethylene, benzene, toluene, formaldehyde, acetaldehyde, dichloromethane, chloroform etc. Their rapid detection is urgently required, and MOFs have been attracting researchers due to their structural advantages (*51*). Yao et al. developed a core-shell nanowire array of MOX@catalytic MOF to detect the acetone with high efficiency and selectivity (*52*). Another toxic VOC with mutagenic characteristics is ally alcohol which is also very hazardous and detected by the MOFs through the luminescence pathway (*53*). Indoor formaldehyde was detected by using MOF functionalized with 4d-4f Ag^{+}-Eu^{3+} via luminescence. In this MOF, Ag^{+} ions were used to tune the dual–emissive property of the material and easily interact with the folic acid (FA). The resultant nanoprobe of MOF exhibited many FA properties such as low-cost, easy preparation, testing ability at room temperature, high selectivity and sensitivity etc. The nanoprobe showed a fast response to detect FA with a limit of detection 51 ppb (*54*).

Adenine-based MOF named as bio-MOF-11 [$Co_2(ad)_2(CH_3CO_2)_2 \cdot 0.3EtOH \cdot 0.6H_2O$, ad = adeninate] was developed for the adsorption of VOCs like toluene, benzene, acetone, and methanol and the saturated adsorption capacity was 0.73-3.57 mmol/g. The adsorption observed was following the order: toluene < benzene < acetone < methanol (Figure 3). But the adsorption capacity decreased from 7.3 to 35.6% with an increase in temperature from 288 to 308 K. The abundance of nitrogen groups and nitrogen atoms with cavities facilitated the adsorption of VOCs (*55*).

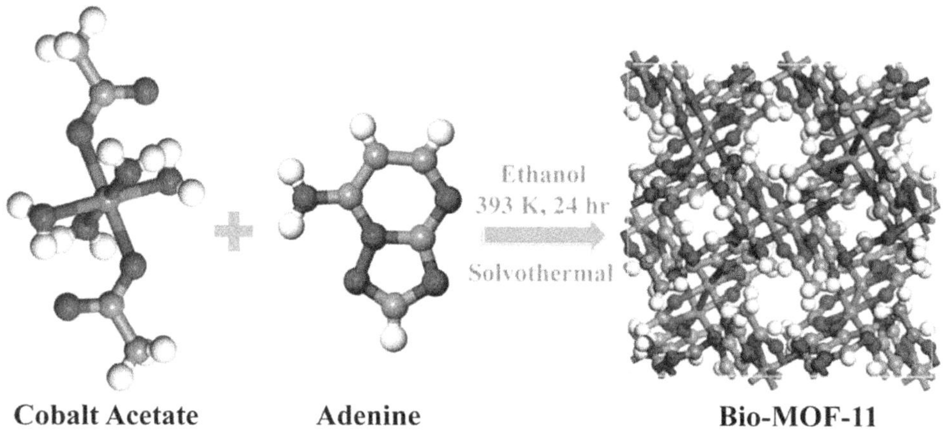

Figure 3. Schematic representation for the synthesis of bio-MOF-1. Reproduced with permission from reference (55). Copyright 2020 American Chemical Society.

Detection of Nitrophenol

Diesel exhaust releases nitrophenols, and their particles in the air impact environment adversely. There are many derivatives of the nitrophenols exhausted from the diesel like nitrophenol, 4-nitrophenol and 2-methyl-4-nitrophenol. These nitrophenols are also produced during the manufacturing process of the paints, pesticides, explosive materials, and vasodilator compounds. Methods such as capillary electrophoresis, flow-injection analysis, high-performance liquid chromatography, spectrophotometric methods have been employed to detect nitrophenols. However, MOFs low-costwas developed for efficient oxidation of the nitrophenol through electrocatalysis (56). Composite zirconium-based Zr-MOF was synthesized by incorporating dye eosin (EY) via synthetic encapsulation. The developed composite had a good overlap EY and Zr-MOF, due to which a facile energy transfer took place from Zr-MOF to EY. This composite was used to detect of nitrophenols through fluorescence which was highly stable (57).

Detection of Antibiotics

Antibiotics are being used for the treatment of bacterial infections in microorganisms. Still, their abuse has enhanced the organic pollutant content in water and became the source of many health problems like genetic defects, cancer, allergic reactions, immunity decline, etc. Many detection methods such as catalytic oxidation, thermolysis, hydrolysis, and photolysis have been used. Still, low sensitivity and low absorption limit their use. In this context, MOFs have attracted researchers due to their high surface area and tunable properties for adsorption and selectivity. However, very high adsorption capacities, simple synthesis steps, low-cost, and high stabilities are still challenged to overcome. MOFs have been modified by incorporating metals, active functional groups, and forming composites to enhance the performances. For instance, zinc-based (MOFs) $\{[Zn_2(bcob)(OH)(H_2O)]\cdot DMA\}n$ (ROd-Zn1) and $\{[Zn(Hbcob)]\cdot(solvent)\}n$ (ROd-Zn_2) (H_3bcob = 1,3-bis((4′-carboxylbenzyl)oxy)benzoic acid) were prepared, and their composites were developed with melamine foam (MF). The resultant composite was a microporous-macroporous MOF@MF system and was used to detect the antibiotics in water with high chemical and water stabilities (*58*). Zhou et al. used zirconium-based MOF(PCY-128 Y) to detect and remove the tetracycline (TC) from water. The detection was based on the luminescence quenching of the PCN-128Y for the TC, which was due to the strong metal-ligand bond between TC and Zr_6 nodes via solvent-assisted ligand corporation (*59*).

Furthermore, tetracycline was detected by a composite of MOFs (Fe-BTC, HKUST-1, ZIF-67, and ZIF-8) with chitosan, which shaped MOF into beads form and enhanced the adsorption capacity. The removal capacity for tetracycline was more than 90%, even after the ten cycles (*60*). Similarly, zirconium-based MOFs (UiO-66) were synthesized by a) doping GO sheets into the solution of membrane casting having polyacrylonitrile (PAN) to form GO@PAN), b) immersing the GO@PAN substrate into dopamine solution for the detection and removal of the antibiotics ciprofloxacin, oxytetracycline, tetracycline hydrochloride, and adsorptions were more than 94% (*61*).

MOFs-Based Materials for Adsorption of Organic Pollutants in Water

A rapid increase in urbanization and population has resulted in adverse effects on the environment of the earth. Water and soil are affected and need to be treated due to accumulation of biological pollutants. Many technologies have been used to detect and remove organic pollutants in water and soil. Many pollutants have been reported like organic dyes, pesticides, insecticides, antibiotics, plasticizers etc. These organic pollutants are very hard to get metabolized in the human body and are toxic upon accumulation in the body. Organic pollutants are being treated in the sewage plants, including adsorption, anaerobic digestion, oxidation, and disinfection by using UV rays or chlorine directly. But owing to the high energy consumption, incomplete disinfection, and disposal problem of sludge, new methods are being used by the researchers. Porous materials like zeolites, activated carbon, etc., are potential candidates for treating organic pollutants. But low selectivity is obtained due to the limited possibility of adjusting the desired pore sizes and shapes. Due to controlled functionality, tunable porosity, and designable structure, MOFs have emerged as promising porous materials to remediate organic pollutants. Organic pollutants may be adsorbed onto the MOFs through various interactions like hydrogen bonding, π-π effect, electrostatic, hydrophobic, and acid-base interaction. Interaction of organic pollutants and MOFs can be engineered by scientists due to MOFs' controllable functionalities. The MOFs favor adsorption of organic pollutants in the liquid and gaseous phase due to high accessibility for the adsorption sites and easy diffusion through the porous framework. Many organic pollutants are non-biodegradable in the water and cannot be removed by conventional methods. The traces of these organic pollutants are toxic to the human body. Most organic pollutants have groups like -COOH, -OH, NH_2, NHR, NR_2, SO_3^{2-} etc., which form electrostatic interaction with the MOFs and results in high adsorption.

Adsorption and Removal of Dyes

One of the major causes of environmental pollution is the excessive utilization of organic dyes, which is highly toxic for aquatic life and humankind. Various dyes are used in textile industries such as Malachite green (MG), methylene blue (MB) and Congo red (CR) etc. Removal of these dyes is urgently required to protect environment (*62*), Many technologies like physical, chemical, and biological have been used, but adsorption is the most efficient and economical method. MOFs played a significant role as adsorbents due to the easy accessibility and availability of adsorption sites.

For instance, Luo et al. used a sulphonic group containing MOF (MIL-101-SO_3H) to adsorb cationic dyes (MB, MG). The anionic group on the MOF played a significant role in the cationic dye adsorption (*63*). The adsorption's excellent performance may be attributed to the π-π interaction between the aromatic rings of the MB and MG with MOF, which is found to be high at lower pH conditions. As the pH increases, the electrostatic interaction increases between MOF and dye. Similarly, cationic dye (MB) was efficiently adsorbed using Cu-BTC MOF synthesized by the hydrothermal method. Cu-BTC (copper benzene-1,3,5-tricarboxylate) exhibited octahedral morphology and followed the model of pseudo-second-order (PSO) kinetics. It showed high adsorption of MB (200 mg/L), which remained 42.3 mg/g after the fourth cycle (*64*). 2D Ni-based MOF NiCu-BTC has also been used for the removal of MB from wastewater (*65*).

Although opposite charges are maintained on the MOFs for the charged (cationic/anionic) dye removal, the simultaneous removal of positively and negatively charged dyes is still challenging. In this context, Li et al. developed zirconium-metalloporphyrin meso MOF (PCN-222/MOF-545) by solvothermal method and used for adsorption and removal of cationic and anionic dyes simultaneously and individually. The synthesized MOF exhibited a large pore size, high surface area, and suitable potentials of 23.5 and -13.6 mV within the range of pH 3-10. It showed the highest loading efficacy of 906 mg/g and 589 mg/g for MB (anionic) and MO (cationic) respectively in a single dye system (*66*).

Anionic dye (CR) was adsorbed by a series of four MOFs synthesized by microwave-assisted ball milling method using copper(II) acetate tetrahydrate and cobalt(II) acetate tetrahydrate metal salts. The MOF-2 showed the highest adsorption of 85.54% after 300 min due to the strong π-π stacking and electrostatic interactions (*67*). Another MOF NH_2-UiO-66 was synthesized using the hydrothermal method for adsorption of safranin dye, and results were compared with that of pristine UiO-66, and enhancement of adsorption was observed with NH_2-UiO-66. A significant hike in adsorption at neutral pH was observed due to the hydrogen bond formation between the dye and MOF, as revealed by the Fourier transform infrared spectroscopy (FTIR) data. The regeneration study showed that the synthesized MOF could be used many times without any loss inefficiency (*68*).

Arora et al. used iron-based MOF (Fe-BDC) to remove MB from industrial waste and optimizing the results with different parameters such as concentration of the dye, the effect of contact time, quantity of adsorbent, and temperature. Maximum 94.74% of dye was obtained to be removed with 25 gm of MOF and 5 mg/L of dye concentration after 24 h (*69*). In many cases, MOFs' composites with other porous materials have been developed for successful adsorption and removal of organic pollutants. Hasanzadeh et al. used a composite of activated charcoal and chromium-based MOF(MIL-101(Cr)) for the adsorption and removal of anionic dyes (Direct Red 31 and Acid Blue 92). This composite exhibited high specific surface area and high total pore volume than bare MIL-101(Cr) (*70*).

Similarly, copper-based MOF was immobilized on the GO and used for MB adsorption. Through the molecular dynamics simulations, kinetics data for the removal of dye was investigated. The synthesized composite exhibited a high adsorption capacity of 173, 251, and 262 mg/g at 25 °C, 45 °C, and 65 °C compared to 106, 117, and 142 mg/g adsorption capacity of Cu-MOF at the same temperatures (*71*). In another example hybrid of MOF and covalent-organic framework (COF) was explored in which zinc-based MOF (MOF-5) was integrated with melamine-terephthaldehyde-based COF was developed as MOF-5/COF for the fast and high-efficient removal of auramine O (AO) and rhodamine B (RB) cationic dyes using acid-base interaction, H-bonding, π-π stacking and electrostatic interaction. Amide bonds between the COF and MOF lead to the high stability of hybrid and high adsorption for the dyes (Figure 4) (*72*).

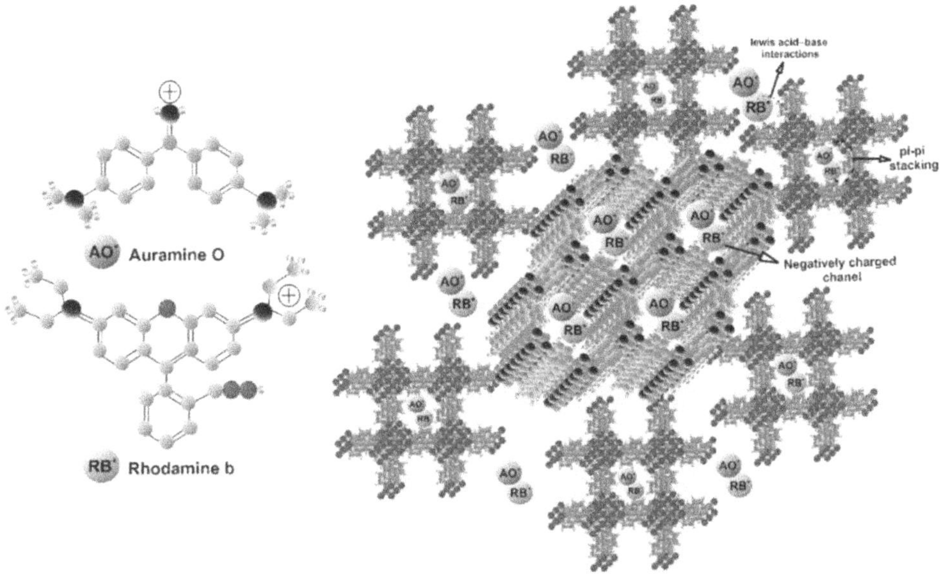

Figure 4. Interactions of the Auramine O and Rhodamine B with the MOF-COF hybrid and proposed mechanism. Reproduced with permission from reference (72). Copyright 2020 American Chemical Society.

Absorption and Removal of Pesticides and Herbicides

Rapid utilization of agrochemicals in the modern era has increased water pollution, adversely affecting the environment and the human community (73). To ameliorate this issue, adsorbents like zeolite, activated etc. have been used. Advanced organic-inorganic hybrid porous material MOFs have shown favorable performance in the absorption and removal of many pesticides and herbicides (74). He et al. developed a stable and low-cost Zr-MOF constituted of Zr^{4+} and 1,2,4,5-tetrakis(4-carboxyphenyl)benzene for the trace detection of the organophosphorus (OPPs) with high selectivity (75).

MOFs-Based Materials for Sequestration of Radioactive Materials

MOFs have shown outstanding performance in the sequestration of radioactive materials. Radioactive materials such as uranium, thorium, strontium etc. have been very disastrous, and their sequestration is urgently required. MOFs in various forms are being used, like pristine, composite, and functionalized. However, the adsorption capacity of MOFs is good. Still, researchers are working to improve the adsorption capacity(76). Coumarin was functionalized on Zn-MOF-74 by the post-synthetic method. The adsorption capacity was found to enhance as 360 mg/g at pH value of 4. (77). In another example, Bai et al. improved the absorption capacity of MOF by incorporating amine, ethylenediamine, and diethylenetriamine into MIL-101 to develop MIL-101-NH_2, MIL-101-ED, and MIL-101-DETA, respectively. These amine grafted MOFs showed high sorption in comparison to the bare MIL-101, and the order obtained was MIL-101-DETA > MIL-101-ED > MIL-101-NH_2 > MIL-101. The desorption of U^{6+} was also very easy at pH ≤ 3.0 (78). Similarly, amidoxime groups were introduced into MIL-53(Al)-AO, and sorption capacity for U^{6+} was improved 2.36 times more than the MIL-53(Al)-NH_2 at pH 6.0 owing to the strong chelation of the aldoxime (79).

Amine derivative (UiO-66-NH_2) with high chemical stability was found to have a high U^{6+}capture capacity of more than 100 mg/g at pH 5.5 (*80*).

Like uranium, thorium has also been employed for nuclear fuel, and its separation from an aqueous medium is urgently required. Th^{4+} capture was first explored in weak acidic solution by using carboxyl derivative of UiO-66 MOF to form UiO-66-COOH and UiO-66-$(COOH)_2$. The maximum absorption of 350 mg/g was observed at pH 3.0. carboxylate anions tend to coordinate with the thorium cations in the pores, facilitating the Th^{4+} uptake (*81*). Strontium from nuclear wastewater has been successfully removed by using oxalate or sulfate-modified MOFs. The developed MOFs MOF-808, giving MOF-808-SO_4 and MOF-808-C_2O_4 were employed for Sr^{2+} adsorption, and high adsorption capacities were obtained.

Remediation of Heavy Metal Ions (HMIs) by MOF-Based Materials

HMIs are one the most serious cause of water pollution. The main toxicants are lead (Pb), arsenic (As), cobalt (Co), mercury (Hg), etc., which are deleterious to human health if present in excess in the human body. Many remedial methods such as reverse osmosis, biological treatment, membrane filtration, photocatalytic degradation, and adsorption have been employed yet. Adsorption using MOFs has attracted researchers for the adsorption of HMIs due to high surface areas, high design capabilities, high porosity, and optimum physical and chemical properties. Table 1 is showing few examples of MOFs used for remediation of HMIs in recent years.

Table 1. Summary of MOFs-Based Materials For Heavy Metal Ion Remediation

MOFs	*Metal Ions*	*Adsorption kinetics*	*Adsorption mechanism*	*Maximum Adsorption (Mg/g)*	*Time (min)*	*Ref*
MIL-88 A	As^{5+}	PSO	Chemical adsorption	98.72	20	(*82*)
Zn-MOF-74	As^{5+}	PSO	Chemical adsorption	325	150	(*83*)
UiO-66	As^{5+}	PSO	Chemical adsorption	9.2	20	(*84*)
UiO-66-NH_2	Cd^{2+}	PSO	Chemical adsorption	415.6	60	(*85*)
MIL-100(Fe)	Cr^{6+}	PSO	Physical adsorption	46.02	120	(*86*)
UiO-66-NH_2	Hg^{2+}	PSO	Chemical adsorption	145.1	360	(*87*)
Bi-MOF (CAU-7-TATB)	Pb^{2+}	PSO	Chemical adsorption	63	45	(*88*)

MOFs for Toxic Gas Storage

Industrialization and excessive use of automobiles are the main causes of toxic gases in the environment. Gas storage is an efficient method to control toxic gases present in the environment. In this context, MOFs with desired chemical properties have been exploited for storage of gases like CO, NOx, SO2, H2S, and NH3 (*89*). Although MOFs are not stable with these corrosive gases and limited MOFs have been explored for toxic gases adsorption. Table 2 is summarizing the applicability of MOFs for the storage of toxic gases.

Table 2. Applicability of MOFs for the Storage of Toxic Gases

MOFs	*Toxic gas*	*Gas uptake (mmol/g)*	*Ref*
UiO-66-NH_3Cl	NH_3NH_3	12[a] 4[a]	*(90)*
UiO-66-NH_2	NH_3 NH_3	10.5[a] 2[a]	*(90)*
Fe-MIL-101-SO_3 H	NH_3 NH_3	18[a] 5[a]	*(90)*
Ni_2Cl_2(BTDD)-$(H_2O)_2$	NH_3	15.47	*(91)*
Co_2Cl_2(BTDD)-$(H_2O)_2$	NH_3	12.00	*(91)*
Mn_2Cl_2(BTDD)-$(H_2O)_2$	NH_3	12.02	*(91)*
Zn-MOF-74	CO	1.95	*(92)*
Ni-MOF-74	CO	5.79	*(92)*
Co-MOF-74	CO	5.95	*(92)*
Fe-MOF-74	CO	6.04	*(92)*
Mn-MOF-74	CO	3.24	*(92)*
Mg-MOF-74	CO	4.58	*(92)*

[a] Specific values are not given, so estimations are made from the adsorption isotherm

Challenges in Using MOF-Based Materials for Environmental Applications

MOFs are emerging novel hybrid compounds explored extensively in the research area, but understanding their toxicity is in the infant stage. It is very important to know the impact of MOFs on human health as well as the environment. The metal ions and organic linkers present in MOFs crystals may be the key source of toxicity. In addition, the solvent such as acetone, diethylene formamide (DMF), dimethyl sulfoxide (DMSO), dimethylformamide (DMF), water etc. which are used during fabrication of MOFs may remain filled in the pores of the MOFs and can cause toxicity *(93)*. Mainly, the metal ions present in MOFs are $Fe^{(2+/3+)}$, Ni^{2+}, Cd^{2+}, Zn^{2+}, Co^{2+}, Cu^{2+}, in which Cd^{2+} is very toxic, and zinc and iron are less toxic *(94)*. Human health can get harmed on exposure to these materials either by inhalation or by ingestion. This material is in nanoparticle size and non-biodegradable, which is deleterious to the environment and human health. The occupational exposure may be reduced by following good lab practices, which could control the working environment and adopt specific safety equipment for personal safety.

MOFs have shown remarkable performances in the biomedical field for therapeutic and diagnostic purposes and are a potential candidate for targeted drug delivery due to their tunable structure and pore size *(95)*. For instance, iron-based MOFs like MIL-100, MIL-88A etc. have shown their toxicity in the human body because it degrades in the body, and iron causes oxidative stress in the liver and spleen *(96)*. However, organic linkers like imidazolates, sulphonates, phosphonates etc. do not cause toxicity. However, some linkers like terephthalic acid and trisomic acid may irritate the nose, skin, eyes, and skin on direct exposure.

Generally, synthesis methods of MOFs such as solvent exchange, vacuum, and thermal treatment require solvents like ethanol, DEF, DMSO, and DMF for various activation processes and liquid state reactions (*97*). For instance, exposure of chloroform as solvent trapped into the MOFs network's pores may cause drowsiness, renal and hepatic toxicity, acute hypoventilation, transient hypotension, and in some cases instantaneous unconsciousness (*98*). In addition, the exposure of chloroform-like solvents may take place through various pathways such as inhalation, dermal contact, and ingestion. Similarly, DMF on exposure also causes liver damage, vomiting, nausea, abdominal pain, and alcohol intolerance (*99*). Acetone is the most common solvent used to fabricate MOFs due to its low-cost, causes dermal and ocular irritation, and severe exposure causes respiratory tract irritation or neurotoxicity. Therefore, considering all the factors, solvent-based MOFs are deleterious to the environment.

MOFs have nanoscale dimensions which is again a threat to the human body because nanoparticles (NPs) less than 100 nm can easily pass through the cell membrane, and NPs less than 40 nm can enter into the cell nucleus. Researchers presume that the blood-brain barrier can be penetrated by NPs less than 35 nm, which is also a grand global challenge (*100*).

Conclusion

MOFs are the potential candidates for application in wide areas of the environment. Its pore size, tunable structure, and nano dimensions make it suitable for several applications like gas sensing, catalysis, adsorption etc. To prevent pollution, CO_2 reduction by MOFs has shown remarkable performance. Many water pollutants like toxic dyes, pesticides, and herbicides are also explored for their degradation by using MOFs. However, many applications have been explored for MOFs. However, the toxic effects of solvents used in the synthesis of MOFs and metal ions should also be considered and is still a challenge presently. The biocompatible and biodegradable organic linkers can substitute bulky organic linkers to produce sustainable and eco-friendly MOFs. Therefore, to reduce the harmful effects of MOFs, various approaches can be opted such as recycling, reuse, regeneration, and degradation of MOFs. The economic aspect of MOF synthesis should also be explored for large-scale production by developing new synthetic techniques and substituting conventional solvents with novel solvents.

Acknowledgments

Meena Nemiwal is thankful to MNIT Jaipur for providing support to carry out this work. Dinesh Kumar is thankful to DST, New Delhi, for this work's financial support (sanctioned vide project Sanction Order F. No. DST/TM/WTI/WIC/2K17/124).

References

1. Nemiwal, M.; Kumar, D. Recent Progress on Electrochemical Sensing Strategies as Comprehensive Point-Care Method. *Monatsh. Chem.* **2021**, *152*, 1–18. https://doi.org/10.1007/s00706-020-02732-0.
2. Ge, M.; Li, Q.; Cao, C.; Huang, J.; Li, S.; Zhang, S.; Chen, Z.; Zhang, K.; Al-Deyab, S. S.; Lai, Y. One-Dimensional TiO_2 Nanotube Photocatalysts for Solar Water Splitting. *Adv. Sci. (Weinh)* **2017**, *4*, 1600152. https://doi.org/10.1002/advs.201600152.
3. Hatamie, A.; Khan, A.; Golabi, M.; Turner, A. P. F.; Beni, V.; Mak, W. C.; Sadollahkhani, A.; Alnoor, H.; Zargar, B.; Bano, S.; Nur, O.; Willander, M. Zinc Oxide Nanostructure-Modified

Textile and Its Application to Biosensing, Photocatalysis, and as Antibacterial Material. *Langmuir* **2015**, *31*, 10913–10921. https://doi.org/10.1021/acs.langmuir.5b02341.

4. Hasnain, M. S.; Nayak, A. K. Applications of Carbon Nanotubes. *In SpringerBriefs Appl. Sci. Technol.*. **2019**, 33–36ISBN 978-981-15-0910-0. https://doi.org/10.1007/978-981-15-0910-0_6.
5. Güntner, A. T.; Abegg, S.; Wegner, K.; Pratsinis, S. E. Zeolite Membranes for Highly Selective Formaldehyde Sensors. *Sens. Actuators B* **2018**, *257*, 916–923. https://doi.org/10.1016/j.snb.2017.11.035.
6. Dhakshinamoorthy, A.; Li, Z.; Garcia, H. Catalysis and Photocatalysis by Metal–Organic Frameworks. *Chem. Soc. Rev.* **2018**, *47*, 8134–8172. https://doi.org/10.1039/c8cs00256h.
7. Zeng, Y.; Zou, R.; Zhao, Y. Covalent Organic Frameworks for CO_2 Capture. *Adv. Mater.* **2016**, *28*, 2855–2873. https://doi.org/10.1002/adma.201505004.
8. Nemiwal, M.; Sharma, V.; Kumar, D. Improved Designs of Multifunctional Covalent-Organic Frameworks: Hydrogen Storage, Methane Storage and Water Harvesting. *Mini Rev. Org. Chem.* **2020**, *17*. https://doi.org/10.2174/1570193X17999201127105752.
9. Chen, Y. Z.; Zhang, R.; Jiao, L.; Jiang, H. L. Metal–Organic Framework-Derived Porous Materials for Catalysis. *Coord. Chem. Rev.* **2018**, *362*, 1–23. https://doi.org/10.1016/j.ccr.2018.02.008.
10. Lustig, W. P.; Mukherjee, S.; Rudd, N. D.; Desai, A. V.; Li, J.; Ghosh, S. K. Metal–Organic Frameworks: Functional Luminescent and Photonic Materials for Sensing Applications. *Chem. Soc. Rev.* **2017**, *46*, 3242–3285. https://doi.org/10.1039/c6cs00930a.
11. Alqadami, A. A.; Naushad, M.; Alothman, Z. A.; Ahamad, T. Adsorptive Performance of MOF Nanocomposite for Methylene Blue and Malachite Green Dyes: Kinetics, Isotherm and Mechanism. *J. Environ. Manage.* **2018**, *223*, 29–36. https://doi.org/10.1016/j.jenvman.2018.05.090.
12. Nemiwal, M.; Subbaramaiah, V.; Zhang, T. C.; Kumar, D. Recent Advances in Visible-Light-Driven Carbon Dioxide Reduction by Metal–Organic Frameworks. *Sci. Total Environ.* **2021**, *762*, 144101. https://doi.org/10.1016/j.scitotenv.2020.144101.
13. Maina, J. W.; Pozo-Gonzalo, C.; Kong, L.; Schütz, J.; Hill, M.; Dumée, L. F. Metal–Organic Framework Based Catalysts for CO_2 Conversion. *Mater. Horiz.* **2017**, *4*, 345–361. https://doi.org/10.1039/C6MH00484A.
14. Nemiwal, M.; Kumar, D. Metal–Organic Frameworks as Water Harvester from Air: Hydrolytic Stability and Adsorption Isotherms. *Inorg. Chem. Commun.* **2020**, *122*. https://doi.org/10.1016/j.inoche.2020.108279. 108279.
15. Nemiwal, M.; Gosu, V.; Zhang, T. C.; Kumar, D. Metal–Organic Frameworks as Electrocatalysts: Hydrogen Evolution Reactions and Overall Water Splitting. *Int. J. Hydr. Energy* **2021**, *46*, 10216–10238. https://doi.org/10.1016/j.ijhydene.2020.12.146.
16. Zhao, M.; Huang, Y.; Peng, Y.; Huang, Z.; Ma, Q.; Zhang, H. Two-Dimensional Metal-Organic Framework Nanosheets: Synthesis and Applications. *Chem. Soc. Rev.* **2018**, *47*, 6267–6295. https://doi.org/10.1039/c8cs00268a.
17. Yap, M. H.; Fow, K. L.; Chen, G. Z. Synthesis and Applications of MOF-Derived Porous Nanostructures. *Green Energy Environ.* **2017**, 2, 218–245. https://doi.org/10.1016/j.gee.2017.05.003.

18. Tchounwou, P. B.; Yedjou, C. G.; Patlolla, A. K.; Sutton, D. J. Heavy Metal Toxicity and the Environment. *Exp. Suppl.* **2012**, *101*, 133–164. https://doi.org/10.1007/978-3-7643-8340-4_6.
19. Ahmad, T.; Ahmad, K.; Alam, M. Sustainable Management of Water Treatment Sludge Through 3'R' Concept. *J. Cleaner Prod.* **2016**, *124*, 1–13. https://doi.org/10.1016/j.jclepro.2016.02.073.
20. Katheresan, V.; Kansedo, J.; Lau, S. Y. Efficiency of Various Recent Wastewater Dye Removal Methods: A Review. *J. Environ. Chem. Eng.* **2018**, *6*, 4676–4697. https://doi.org/10.1016/j.jece.2018.06.060.
21. Zhu, D. D.; Liu, J. L.; Qiao, S. Z. Recent Advances in Inorganic Heterogeneous Electrocatalysts for Reduction of Carbon Dioxide. *Adv. Mater.* **2016**, *28*, 3423–3452. https://doi.org/10.1002/adma.201504766.
22. Feng, D. M.; Zhu, Y. P.; Chen, P.; Ma, T. Y. Recent Advances in Transition-Metal-Mediated Electrocatalytic CO_2 Reduction: From Homogeneous to Heterogeneous Systems. *Catalysts* **2017**, 7, 373. https://doi.org/10.3390/catal7120373.
23. Shao, P.; Yi, L.; Chen, S.; Zhou, T.; Zhang, J. Metal–Organic Frameworks for Electrochemical Reduction of Carbon Dioxide: The Role of Metal Centers. *J. Energy Chem.* **2020**, *40*, 156–170. https://doi.org/10.1016/j.jechem.2019.04.013.
24. Li, D.; Kassymova, M.; Cai, X.; Zang, S. Q.; Jiang, H. L. Photocatalytic CO_2 Reduction over Metal–Organic Framework-Based Materials. *Coord. Chem. Rev.* **2020**, *412*. https://doi.org/10.1016/j.ccr.2020.213262. 213262.
25. Xuan, X.; Chen, S.; Zhao, S.; Yoon, J. Y.; Boczkaj, G.; Sun, X. Carbon Nanomaterials from Metal–Organic Frameworks: A New Material Horizon for CO_2 Reduction. *Front. Chem.* **2020**, *8*, 573797. https://doi.org/10.3389/fchem.2020.573797.
26. Crake, A.; Christoforidis, K. C.; Kafizas, A.; Zafeiratos, S.; Petit, C. CO_2 Capture and Photocatalytic Reduction Using Bifunctional TiO_2/MOF Nanocomposites Under UV–Vis Irradiation. *Appl. Catal. B* **2017**, *210*, 131–140. https://doi.org/10.1016/j.apcatb.2017.03.039.
27. Abdullah, H.; Khan, M. M. R.; Ong, H. R.; Yaakob, Z. Modified TiO_2 Photocatalyst for CO_2 Photocatalytic Reduction: An Overview. *J. Co* **2017**, *22*, 15–32. https://doi.org/10.1016/j.jcou.2017.08.004.
28. Wei, Y. P.; Liu, Y.; Guo, F.; Dao, X. Y.; Sun, W. Y. Different Functional Group Modified Zirconium Frameworks for the Photocatalytic Reduction of Carbon Dioxide. *Dalton Trans.* **2019**, *48*, 8221–8226. https://doi.org/10.1039/c9dt01767d.
29. Liu, D. C.; Ouyang, T.; Xiao, R.; Liu, W. J.; Zhong, D. C.; Xu, Z.; Lu, T. B. Anchoring CoII Ions into a Thiol-Laced Metal–Organic Framework for Efficient Visible-Light-Driven Conversion of CO_2 into CO. *ChemSusChem* **2019**, *12*, 2166–2170. https://doi.org/10.1002/cssc.201900338.
30. Wang, Y.; Guo, L.; Zeng, Y.; Guo, H.; Wan, S.; Ou, M.; Zhang, S.; Zhong, Q. Amino-Assisted NH_2-UiO-66 Anchored on Porous g-C_3N_4 for Enhanced Visible-Light-Driven CO2 Reduction. *ACS Appl. Mater. Interfaces* **2019**, *11*, 30673–30681. https://doi.org/10.1021/acsami.9b04302.

31. Shang, Q. C.; Fang, X.; Jiang, H.; Zhang, Q. Location Effect in a Photocatalytic Hybrid System of Metal–Organic Framework Interfaced with Semiconductor Nanoparticles. *Chin. J. Chem. Phys.* **2018**, *31*, 613–618. https://doi.org/10.1063/1674-0068/31/cjcp1803044.
32. Zhao, Y.; Cai, W.; Chen, J.; Miao, Y.; Bu, Y. A Highly Efficient Composite Catalyst Constructed from NH_2-MIL-125(Ti) and Reduced Graphene Oxide for CO_2 Photoreduction. *Front. Chem.* **2019**, 7, 789. https://doi.org/10.3389/fchem.2019.00789.
33. Liu, Y.; Deng, L.; Sheng, J.; Tang, F.; Zeng, K.; Wang, L.; Liang, K.; Hu, H.; Liu, Y. N. Photostable Core-Shell CdS/ZIF-8 Composite for Enhanced Photocatalytic Reduction of CO_2. *Appl. Surf. Sci.* **2019**, *498*. https://doi.org/10.1016/j.apsusc.2019.143899. 143899.
34. Zhao, H.; Wang, X.; Feng, J.; Chen, Y.; Yang, X.; Gao, S.; Cao, R. Synthesis and Characterization of Zn_2GeO_4/Mg-MOF-74 Composites with Enhanced Photocatalytic Activity for CO_2 Reduction. *Catal. Sci. Technol.* **2018**, *8*, 1288–1295. https://doi.org/10.1039/C7CY02286G.
35. Al-Rowaili, F. N.; Jamal, A.; Ba Shammakh, M. S.; Rana, A. A Review on Recent Advances for Electrochemical Reduction of Carbon Dioxide to Methanol Using Metal–Organic Framework (MOF) and Non-MOF Catalysts: Challenges and Future Prospects. *ACS Sustainable Chem. Eng.* **2018**, *6*, 15895–15914. https://doi.org/10.1021/acssuschemeng.8b03843.
36. Han, Y.; Xu, H.; Su, Y.; Xu, Z.; Wang, K.; Wang, W. Noble Metal (Pt, Au@Pd) Nanoparticles Supported on Metal–Organic Framework (MOF-74) Nanoshuttles as High-Selectivity CO_2 Conversion Catalysts. *J. Catal.* **2019**, *370*, 70–78. https://doi.org/10.1016/j.jcat.2018.12.005.
37. Su, Y.; Zhang, Z.; Liu, H.; Wang, Y. Cd. $Cd_{0.2}Zn_{0.8}S$@UiO-66-NH_2 Nanocomposites as Efficient and Stable Visible-Light-Driven Photocatalyst for H2 Evolution and CO_2 Reduction. *Appl. Catal. B* **2017**, *200*, 448–457. https://doi.org/10.1016/j.apcatb.2016.07.032.
38. Varela, A. S.; Ju, W.; Strasser, P. Molecular Nitrogen–Carbon Catalysts, Solid Metal–Organic Framework Catalysts, and Solid Metal/Nitrogen-Doped Carbon (MNC) Catalysts for the Electrochemical CO_2 Reduction. *Adv. Energy Mater.* **2018**, *8*. https://doi.org/10.1002/aenm.201703614. 1703614.
39. Zhang, W.; Hu, Y.; Ma, L.; Zhu, G.; Wang, Y.; Xue, X.; Chen, R.; Yang, S.; Jin, Z. Progress and Perspective of Electrocatalytic CO_2 Reduction for Renewable Carbonaceous Fuels and Chemicals. *Adv. Sci. (Weinh)* **2018**, *5*, 1700275. https://doi.org/10.1002/advs.201700275.
40. Gao, D.; Arán-Ais, R. M.; Jeon, H. S.; Roldan Cuenya, B. Rational Catalyst and Electrolyte Design for CO_2 Electroreduction Towards Multicarbon Products. *Nat. Cat.* **2019**, *2*, 198–210. https://doi.org/10.1038/s41929-019-0235-5.
41. Younis, S. A.; Kwon, E. E.; Qasim, M.; Kim, K. H.; Kim, T.; Kukkar, D.; Dou, X.; Ali, I. Metal–Organic Framework as a Photocatalyst: Progress in Modulation Strategies and Environmental/Energy Applications. *Prog. Energy Combust. Sci.* **2020**, *81*. https://doi.org/10.1016/j.pecs.2020.100870. 100870.
42. Choi, I.; Jung, Y. E.; Yoo, S. J.; Kim, J. Y.; Kim, H. J.; Lee, C. Y.; Jang, J. H. Facile Synthesis of M-MOF-74 (M=Co, Ni, Zn) and Its Application as an Electrocatalyst for Electrochemical CO_2 Conversion and H_2 Production. *J. Electrochem. Sci. Technol.* **2017**, *8*, 61–68. https://doi.org/10.5229/JECST.2017.8.1.61.

43. Albo, J.; Vallejo, D.; Beobide, G.; Castillo, O.; Castaño, P.; Irabien, A. Copper-Based Metal–Organic Porous Materials for CO_2 Electrocatalytic Reduction to Alcohols. *ChemSusChem* **2017**, *10*, 1100–1109. https://doi.org/10.1002/cssc.201600693.
44. Wang, H.; Lustig, W. P.; Li, J. Sensing and Capture of Toxic and Hazardous Gases and Vapors by Metal–Organic Frameworks. *Chem. Soc. Rev.* **2018**, *47*, 4729–4756. https://doi.org/10.1039/c7cs00885f.
45. Smith, M. K.; Jensen, K. E.; Pivak, P. A.; Mirica, K. A. Direct Self-Assembly of Conductive Nanorods of Metal–Organic Frameworks into Chemiresistive Devices on Shrinkable Polymer Films. *Chem. Mater.* **2016**, *28*, 5264–5268. https://doi.org/10.1021/acs.chemmater.6b02528.
46. Ling, P.; Qian, C.; Yu, J.; Gao, F. Metal–Organic Framework Nanosheets with Flower-Like Structure as Probes for H_2S Detection and In Situ Singlet-Oxygen Production. *Chem. Commun. (Camb)* **2019**, *55*, 6385–6388. https://doi.org/10.1039/c9cc02285f.
47. Zhang, J.; Liu, F.; Gan, J.; Cui, Y.; Li, B.; Yang, Y.; Qian, G. Metal–Organic Framework Film for Fluorescence Turn-On H_2S Gas Sensing and Anti-Counterfeiting Patterns. *Sci. China Mater.* **2019**, *62*, 1445–1453. https://doi.org/10.1007/s40843-019-9457-5.
48. Cao, Y. Y.; Guo, X. F.; Wang, H. High Sensitive Luminescence Metal–Organic Framework Sensor for Hydrogen Sulfide in Aqueous Solution: A Trial of Novel Turn-On Mechanism. *Sens. Actuators B* **2017**, *243*, 8–13. https://doi.org/10.1016/j.snb.2016.11.085.
49. Yao, M. S.; Zheng, J. J.; Wu, A. Q.; Xu, G.; Nagarkar, S. S.; Zhang, G.; Tsujimoto, M.; Sakaki, S.; Horike, S.; Otake, K.; Kitagawa, S. A Dual-Ligand Porous Coordination Polymer Chemiresistor with Modulated Conductivity and Porosity. *Angew. Chem. Int. Ed. Engl.* **2020**, *59*, 172–176. https://doi.org/10.1002/anie.201909096.
50. Gładysiak, A.; Nguyen, T. N.; Navarro, J. A. R.; Rosseinsky, M. J.; Stylianou, K. C. A Recyclable Metal–Organic Framework as a Dual Detector and Adsorbent for Ammonia. *Chemistry* **2017**, *23*, 13602–13606. https://doi.org/10.1002/chem.201703510.
51. Jackson, S. L.; Rananaware, A.; Rix, C.; Bhosale, S. V.; Latham, K. Highly Fluorescent Metal–Organic Framework for the Sensing of Volatile Organic Compounds. *Cryst. Growth Des.* **2016**, *16*, 3067–3071. https://doi.org/10.1021/acs.cgd.6b00428.
52. Yao, M. S.; Tang, W. X.; Wang, G. E.; Nath, B.; Xu, G.; Thin, M. O. F. MOF Thin Film-Coated Metal Oxide Nanowire Array: Significantly Improved Chemiresistor Sensor Performance. *Adv. Mater.* **2016**, *28*, 5229–5234. https://doi.org/10.1002/adma.201506457.
53. Desai, A. V.; Sharma, S.; Roy, A.; Shirolkar, M. M.; Ghosh, S. K. Specific Recognition of Toxic Allyl Alcohol by Pore-Functionalized Metal–Organic Frameworks. *Mol. Syst. Des. Eng.* **2020**, *5*, 469–476. https://doi.org/10.1039/C9ME00133F.
54. Hao, J. N.; Yan, B. A Dual-Emitting 4d-4f Nanocrystalline Metal–Organic Framework as a Self-Calibrating Luminescent Sensor for Indoor Formaldehyde Pollution. *Nanoscale* **2016**, *8*, 12047–12053. https://doi.org/10.1039/c6nr02446g.
55. Chen, R.; Yao, Z.; Han, N.; Ma, X.; Li, L.; Liu, S.; Sun, H.; Wang, S.; Li, L.; Wang, S. Insights into the Adsorption of VOCs on a Cobalt-Adeninate Metal–Organic Framework (Bio-MOF-11). *ACS Omega* **2020**, *5*, 15402–15408. https://doi.org/10.1021/acsomega.0c01504.
56. Yadav, D. K.; Ganesan, V.; Marken, F.; Gupta, R.; Sonkar, P. K. Metal@MOF Materials in Electroanalysis: Silver-Enhanced Oxidation Reactivity Towards Nitrophenols Adsorbed into a

Zinc Metal–Organic Framework—Ag@MOF-5(Zn). *Electrochim. Acta* **2016**, *219*, 482–491. https://doi.org/10.1016/j.electacta.2016.10.009.

57. Li, Y.; Wei, Z.; Zhang, Y.; Guo, Z.; Chen, D.; Jia, P.; Chen, P.; Xing, H. Dual-Emitting EY@Zr-MOF Composite as Self-Calibrating Luminescent Sensor for Selective Detection of Inorganic Ions and Nitroaromatics. *ACS Sustainable Chem. Eng.* **2019**, *7*, 6196–6203. https://doi.org/10.1021/acssuschemeng.8b06500.
58. Gai, S.; Zhang, J.; Fan, R.; Xing, K.; Chen, W.; Zhu, K.; Zheng, X.; Wang, P.; Fang, X.; Yang, Y. Highly Stable Zinc-Based Metal–Organic Frameworks and Corresponding Flexible Composites for Removal and Detection of Antibiotics in Water. *ACS Appl. Mater. Interfaces* **2020**, *12*, 8650–8662. https://doi.org/10.1021/acsami.9b19583.
59. Zhou, Y.; Yang, Q.; Zhang, D.; Gan, N.; Li, Q.; Cuan, J. Detection and Removal of Antibiotic Tetracycline in Water with a Highly Stable Luminescent MOF. *Sens. Actuators B* **2018**, *262*, 137–143. https://doi.org/10.1016/j.snb.2018.01.218.
60. Zhao, R.; Ma, T.; Zhao, S.; Rong, H.; Tian, Y.; Zhu, G. Uniform and Stable Immobilization of Metal–Organic Frameworks into Chitosan Matrix for Enhanced Tetracycline Removal from Water. *Chem. Eng. J.* **2020**, *382*. https://doi.org/10.1016/j.cej.2019.122893. 122892.
61. Fang, S. Y.; Zhang, P.; Gong, J. L.; Tang, L.; Zeng, G. M.; Song, B.; Cao, W. C.; Li, J.; Ye, J. Construction of Highly Water-Stable Metal–Organic Framework UiO-66 Thin-Film Composite Membrane for Dyes and Antibiotics Separation. *Chem. Eng. J.* **2020**, *385*. https://doi.org/10.1016/j.cej.2019.123400. 123400.
62. Nemiwal, M.; Zhang, T. C.; Kumar, D. Recent Progress in g-C_3N_4, TiO_2 and ZnO Based Photocatalysts for Dye Degradation: Strategies to Improve Photocatalytic Activity. *Sci. Total Environ.* **2021**, *767*, 144896. https://doi.org/10.1016/j.scitotenv.2020.144896.
63. Luo, X. P.; Fu, S. Y.; Du, Y. M.; Guo, J. Z.; Li, B. Adsorption of Methylene Blue and Malachite Green from Aqueous Solution by Sulfonic Acid Group Modified MIL-101. *Micropor. Mesopor. Mater.* **2017**, *237*, 268–274. https://doi.org/10.1016/j.micromeso.2016.09.032.
64. Kaur, R.; Kaur, A.; Umar, A.; Anderson, W. A.; Kansal, S. K. Metal–Organic Framework (MOF) Porous Octahedral Nanocrystals of Cu-BTC: Synthesis, Properties and Enhanced Absorption Properties. *Mater. Res. Bull.* **2019**, *109*, 124–133. https://doi.org/10.1016/j.materresbull.2018.07.025.
65. Abd El Salam, H. M.; Zaki, T. Removal of Hazardous Cationic Organic Dyes from Water Using Nickel-Based Metal-Organic Frameworks. *Inorg. Chim. Acta* **2018**, *471*, 203–210. https://doi.org/10.1016/j.ica.2017.10.040.
66. Li, H.; Cao, X.; Zhang, C.; Yu, Q.; Zhao, Z.; Niu, X.; Sun, X.; Liu, Y.; Ma, L.; Li, Z. Enhanced Adsorptive Removal of Anionic and Cationic Dyes from Single or Mixed Dye Solutions Using MOF PCN-222. *RSC Adv.* **2017**, *7*, 16273–16281. https://doi.org/10.1039/C7RA01647F.
67. Wei, F. H.; Chen, D.; Liang, Z.; Zhao, S. Q.; Luo, Y. Synthesis and Characterization of Metal-Organic Frameworks Fabricated by Microwave-Assisted Ball Milling for Adsorptive Removal of Congo Red from Aqueous Solutions. *RSC Adv.* **2017**, *7*, 46520–46528. https://doi.org/10.1039/C7RA09243A.
68. Tambat, S. N.; Sane, P. K.; Suresh, S.; Varadan O., N.; Pandit, A. B.; Sontakke, S. M. Hydrothermal Synthesis of NH_2-UiO-66 and Its Application for Adsorptive Removal of Dye. *Adv. Powder Technol.* **2018**, *29*, 2626–2632. https://doi.org/10.1016/j.apt.2018.07.010.

69. Arora, C.; Soni, S.; Sahu, S.; Mittal, J.; Kumar, P.; Bajpai, P. K. Iron-Based Metal-Organic Framework for Efficient Removal of Methylene Blue Dye from Industrial Waste. *J. Mol. Liq.* **2019**, *284*, 343–352. https://doi.org/10.1016/j.molliq.2019.04.012.
70. Hasanzadeh, M.; Simchi, A.; Shahriyari Far, H. Nanoporous Composites of Activated Carbon-Metal–Organic Frameworks for Organic Dye Adsorption: Synthesis, Adsorption Mechanism and Kinetics Studies. *J. Ind. Eng. Chem.* **2020**, *81*, 405–414. https://doi.org/10.1016/j.jiec.2019.09.031.
71. Dadashi Firouzjaei, M.; Akbari Afkhami, F.; Rabbani Esfahani, M.; Turner, C. H.; Nejati, S. Experimental and Molecular Dynamics Study on Dye Removal from Water by a Graphene Oxide–Copper-Metal–Organic Framework Nanocomposite. *J. Water Process Eng.* **2020**, *34*. https://doi.org/10.1016/j.jwpe.2020.101180. 101180.
72. Firoozi, M.; Rafiee, Z.; Dashtian, K. New MOF/COF Hybrid as a Robust Adsorbent for Simultaneous Removal of Auramine O and Rhodamine B Dyes. *ACS Omega* **2020**, *5*, 9420–9428. https://doi.org/10.1021/acsomega.0c00539.
73. Vikrant, K.; Tsang, D. C. W.; Raza, N.; Giri, B. S.; Kukkar, D.; Kim, K. H. Potential Utility of Metal–Organic Framework-Based Platform for Sensing Pesticides. *ACS Appl. Mater. Interfaces* **2018**, *10*, 8797–8817. https://doi.org/10.1021/acsami.8b00664.
74. Hasan, Z.; Jhung, S. H. Removal of Hazardous Organics from Water Using Metal–Organic Frameworks (MOFs): Plausible Mechanisms for Selective Adsorptions. *J. Hazard. Mater.* **2015**, *283*, 329–339. https://doi.org/10.1016/j.jhazmat.2014.09.046.
75. He, K.; Li, Z.; Wang, L.; Fu, Y.; Quan, H.; Li, Y.; Wang, X.; Gunasekaran, S.; Xu, X.; Water-Stable, A. A Water-Stable Luminescent Metal–Organic Framework for Rapid and Visible Sensing of Organophosphorus Pesticides. *ACS Appl. Mater. Interfaces* **2019**, *11*, 26250–26260. https://doi.org/10.1021/acsami.9b06151.
76. Feng, Y.; Jiang, H.; Li, S.; Wang, J.; Jing, X.; Wang, Y.; Chen, M. Metal–Organic Frameworks HKUST-1 for Liquid-Phase Adsorption of Uranium. *Colloids Surf. A Physicochem. Eng. Asp.* **2013**, *431*, 87–92. https://doi.org/10.1016/j.colsurfa.2013.04.032.
77. Zhang, L.; Wang, L. L.; Le Gong, L.; Feng, X. F.; Luo, M. B.; Luo, F. Coumarin-Modified Microporous-Mesoporous Zn-MOF-74 Showing Ultra-High Uptake Capacity and Photo-Switched Storage/Release of UVI Ions. *J. Hazard. Mater.* **2016**, *311*, 30–36. https://doi.org/10.1016/j.jhazmat.2016.01.082.
78. Bai, Z. Q.; Yuan, L. Y.; Zhu, L.; Liu, Z. R.; Chu, S. Q.; Zheng, L. R.; Zhang, J.; Chai, Z. F.; Shi, W. Q. Introduction of Amino Groups into Acid-Resistant MOFs for Enhanced U(VI) Sorption. *J. Mater. Chem. A* **2015**, 3, 525–534. https://doi.org/10.1039/C4TA04878D.
79. Liu, J.; Liu, T.; Wang, C.; Yin, X.; Xiong, Z. Introduction of Amidoxime Groups into Metal-Organic Frameworks to Synthesize MIL-53(Al)-AO for Enhanced U(VI) Sorption. *J. Mol. Liq.* **2017**, *242*, 531–536. https://doi.org/10.1016/j.molliq.2017.07.024.
80. Luo, B. C.; Yuan, L. Y.; Chai, Z. F.; Shi, W. Q.; Tang, Q. U. U(VI) Capture from Aqueous Solution by Highly Porous and Stable MOFs: UiO-66 and Its Amine Derivative. *J. Radioanal. Nucl. Chem.* **2016**, *307*, 269–276. https://doi.org/10.1007/s10967-015-4108-3.
81. Zhang, N.; Yuan, L. Y.; Guo, W. L.; Luo, S. Z.; Chai, Z. F.; Shi, W. Q. Extending the Use of Highly Porous and Functionalized MOFs to Th(IV) Capture. *ACS Appl. Mater. Interfaces* **2017**, 9, 25216–25224. https://doi.org/10.1021/acsami.7b04192.

82. Wu, H.; Ma, M. D.; Gai, W. Z.; Yang, H.; Zhou, J. G.; Cheng, Z.; Xu, P.; Deng, Z. Y. Arsenic Removal from Water by Metal–Organic Framework MIL-88A Microrods. *Environ. Sci. Pollut. Res. Int.* **2018**, *25*, 27196–27202. https://doi.org/10.1007/s11356-018-2751-2.
83. Yu, W.; Luo, M.; Yang, Y.; Wu, H.; Huang, W.; Zeng, K.; Luo, F. Metal–Organic Framework (MOF) Showing Both Ultrahigh As(V) and As($^{3+}$) Removal from Aqueous Solution. *J. Solid State Chem.* **2019**, *269*, 264–270. https://doi.org/10.1016/j.jssc.2018.09.042.
84. He, X.; Deng, F.; Shen, T.; Yang, L.; Chen, D.; Luo, J.; Luo, X.; Min, X.; Wang, F. Exceptional Adsorption of Arsenic by Zirconium Metal–Organic Frameworks: Engineering Exploration and Mechanism Insight. *J. Colloid Interface Sci.* **2019**, *539*, 223–234. https://doi.org/10.1016/j.jcis.2018.12.065.
85. Jamshidifard, S.; Koushkbaghi, S.; Hosseini, S.; Rezaei, S.; Karamipour, A.; Jafari Rad, A.; Irani, M. Incorporation of UiO-66-NH_2 MOF into the PAN/Chitosan Nanofibers for Adsorption and Membrane Filtration of Pb(II), Cd(II) and Cr(VI) Ions from Aqueous Solutions. *J. Hazard. Mater.* **2019**, *368*, 10–20. https://doi.org/10.1016/j.jhazmat.2019.01.024.
86. Fang, Y.; Wen, J.; Zeng, G.; Jia, F.; Zhang, S.; Peng, Z.; Zhang, H. Effect of Mineralizing Agents on the Adsorption Performance of Metal-Organic Framework MIL-100(Fe) Towards Chromium(VI). *Chem. Eng. J.* **2018**, *337*, 532–540. https://doi.org/10.1016/j.cej.2017.12.136.
87. Fu, L.; Wang, S.; Lin, G.; Zhang, L.; Liu, Q.; Fang, J.; Wei, C.; Liu, G. Post-Functionalization of UiO-66-NH_2 by 2,5-Dimercapto-1,3,4-Thiadiazole for the High Efficient Removal of Hg(II) in Water. *J. Hazard. Mater.* **2019**, *368*, 42–51. https://doi.org/10.1016/j.jhazmat.2019.01.025.
88. Zhang, R.; Liu, Y.; An, Y.; Wang, Z.; Wang, P.; Zheng, Z.; Qin, X.; Zhang, X.; Dai, Y.; Huang, B.; Water-Stable, A. Triazine-Based Metal–Organic Framework as an Efficient Adsorbent of Pb(II) Ions. *Colloids Surf. A Physicochem. Eng. Asp.* **2019**, *560*, 315–322. https://doi.org/10.1016/j.colsurfa.2018.10.011.
89. Georgiadis, A. G.; Charisiou, N.; Yentekakis, I. V.; Goula, M. A. Hydrogen Sulfide (H_2S) Removal via MOFs. *Materials (Basel)* **2020**, 13. https://doi.org/10.3390/MA13163640.
90. Van Humbeck, J. F.; McDonald, T. M.; Jing, X.; Wiers, B. M.; Zhu, G.; Long, J. R. Ammonia Capture in Porous Organic Polymers Densely Functionalized with Brønsted Acid Groups. *J. Am. Chem. Soc.* **2014**, *136*, 2432–2440. https://doi.org/10.1021/ja4105478.
91. Rieth, A. J.; Tulchinsky, Y.; Dincă, M. High and Reversible Ammonia Uptake in Mesoporous Azolate Metal–Organic Frameworks with Open Mn, Co, and Ni Sites. *J. Am. Chem. Soc.* **2016**, *138*, 9401–9404. https://doi.org/10.1021/jacs.6b05723.
92. Bloch, E. D.; Hudson, M. R.; Mason, J. A.; Chavan, S.; Crocellà, V.; Howe, J. D.; Lee, K.; Dzubak, A. L.; Queen, W. L.; Zadrozny, J. M.; Geier, S. J.; Lin, L. C.; Gagliardi, L.; Smit, B.; Neaton, J. B.; Bordiga, S.; Brown, C. M.; Long, J. R. Reversible CO Binding Enables Tunable CO/H_2 and CO/N_2 Separations in Metal–Organic Frameworks with Exposed Divalent Metal Cations. *J. Am. Chem. Soc.* **2014**, *136*, 10752–10761. https://doi.org/10.1021/ja505318p.
93. Ouyang, H.; Chen, N.; Chang, G.; Zhao, X.; Sun, Y.; Chen, S.; Zhang, H.; Yang, D. Selective Capture of Toxic Selenite Anions by Bismuth-Based Metal–Organic Frameworks. *Angew. Chem. Int. Ed. Engl.* **2018**, *57*, 13197–13201. https://doi.org/10.1002/anie.201807891.

94. Zhang, Y.; Yuan, S.; Day, G.; Wang, X.; Yang, X.; Zhou, H. C. Luminescent Sensors Based on Metal–Organic Frameworks. *Coord. Chem. Rev.* **2018**, *354*, 28–45. https://doi.org/10.1016/j.ccr.2017.06.007.
95. Malonzo, C. D.; Wang, Z.; Duan, J.; Zhao, W.; Webber, T. E.; Li, Z.; Kim, I. S.; Kumar, A.; Bhan, A.; Platero-Prats, A. E.; Chapman, K. W.; Farha, O. K.; Hupp, J. T.; Martinson, A. B. F.; Penn, R. L.; Stein, A. Application and Limitations of Nanocasting in Metal–Organic Frameworks. *Inorg. Chem.* **2018**, *57*, 2782–2790. https://doi.org/10.1021/acs.inorgchem.7b03181.
96. Sajid, M. Toxicity of Nanoscale Metal–Organic Frameworks: A Perspective. *Environ. Sci. Pollut. Res. Int.* **2016**, *23*, 14805–14807. https://doi.org/10.1007/s11356-016-7053-y.
97. Kumar, P.; Pournara, A.; Kim, K. H.; Bansal, V.; Rapti, S.; Manos, M. J. Metal–Organic Frameworks: Challenges and Opportunities for Ion-Exchange/Sorption Applications. *Prog. Mater. Sci.* **2017**, *86*, 25–74. https://doi.org/10.1016/j.pmatsci.2017.01.002.
98. Jayaweera, D.; Islam, S.; Gunja, N.; Cowie, C.; Broska, J.; Poojara, L.; Roberts, M. S.; Isbister, G. K. Chloroform Ingestion Causing Severe Gastrointestinal Injury, Hepatotoxicity and Dermatitis Confirmed with Plasma Chloroform Concentrations. *Clin. Toxicol. (Phila)* **2017**, 55, 147–150. https://doi.org/10.1080/15563650.2016.1249795.
99. Ma, L.; Dong, J. X.; Fu, W. R.; Li, X. Y.; Chen, J.; Liu, Y. Mitochondrial Morphology and Function Impaired by Dimethyl Sulfoxide and Dimethyl Formamide. *J. Bioenerg. Biomembr.* **2018**, *50* (4), 297–305. https://doi.org/10.1007/s10863-018-9759-7.
100. Furtado, D.; Björnmalm, M.; Ayton, S.; Bush, A. I.; Kempe, K.; Caruso, F. Overcoming the Blood–Brain Barrier: The Role of Nanomaterials in Treating Neurological Diseases. *Adv. Mater.* **2018**, *30*, e1801362. https://doi.org/10.1002/adma.201801362. 30066406.

Editors' Biographies

Smita S. Kumar

Dr. Smita is an assistant professor in the department of environmental sciences at J.C. Bose University of Science & Technology, YMCA, Faridabad. She is a strong research professional with a doctor of philosophy (Ph.D.) in environmental science and engineering focused on "Bioelectrochemical Stabilization of Landfill Leachate" via microbial fuel cell Technology (MFC). Prior to her current position, she pursued her postdoctoral research at the Centre for Rural Development and Technology, Indian Institute of Technology Delhi. Here, her work was based on the conversion of emerging water pollutants from agrochemical/pesticide industries into green energy by MFC technology. Her major MTech (environmental science and engineering) project work at CFEES, DRDO focussed on the bioremediation of effluent generated from defence manufacturing facilities. Her research interests include environmental pollution assessment and abatement; microbial fuel cells; bioelectrochemical waste and wastewater treatment; microalgae; sustainable bioenergy production; landfill leachate treatment; and bioremediation. She has to her credit more than 60 publications in high-impact peer-reviewed journals and book chapters (more than 1,600 citations), with an h-index of 21 and i10-index of 40 (per Google Scholar).

Pooja Ghosh

Dr. Pooja Ghosh is an assistant professor in the Centre for Rural Development and Technology at the Indian Institute of Technology Delhi (IITD), India. She obtained her Ph.D. in environmental sciences from Jawaharlal Nehru University, India in 2015. Prior to joining IIT Delhi as an assistant professor in September 2020, she was the recipient of the prestigious DST-Inspire Faculty Award and conducted her post-doctoral research in IIT Delhi between 2016 to 2020. Her area of domain is environmental risk assessment and management with a keen interest on waste management, bioenergy, and remediation of persistent and emerging contaminants. She has published more than 30 peer-reviewed articles in highimpact international journals and has more than 1,000 citations with an h-index of 18 and i10-index of 22 (per Google Scholar).

Lakhveer Singh

Dr. Lakhveer Singh is a faculty member in the department of environment science, SRM-University Amaravati, AP. Earlier, he worked as an associate professor at the University Malaysia Pahang, Malaysia and as a post-doctoral researcher at Oregon State University, Corvallis, USA. His major areas of research expertise are energy production, bioelectrochemical systems, wastewater treatment, and nanomaterial synthesis for sustainable applications. He has three patents to his credit, has edited 10 books (ACS, Elsevier, CRC, etc.), and published over 80 scientific papers (more than 2,500 citations; h-index 30; i10-index 55) and 15 book chapters. Dr. Singh has supervised five Ph.D. theses and has completed more than 10 R&D projects. He also serves as an editorial board member for the *Journal of Biomass Conversion and Biorefinery* (Springer I.F 4.987).

Indexes

Author Index

Subject Index

H

M

S

Printed in the USA/Agawam, MA
September 6, 2022

798162.002